Revised and Expanded...

HVAC

Design Criteria, Options, Selection

William H. Rowe III, AIA, PE

Revised and Expanded...

HVAC

Design Criteria, Options, Selection

New in the Second Edition:

- *Indoor Air Quality*
- *CFC Removal*
- *Energy Conservation*
- *Special Systems by Building Type*
- *Metric Units Added*

William H. Rowe III, AIA, PE

Copyright 1994

R.S. MEANS COMPANY, INC.
CONSTRUCTION PUBLISHERS & CONSULTANTS

100 Construction Plaza
P.O. Box 800
Kingston, MA 02364-0800
(617) 585-7880

R.S. Means Company, Inc. ("R.S. Means"), its authors, editors and engineers, apply diligence and judgment in locating and using reliable sources for the information published. **However, R.S. Means makes no express or implied warranty or guarantee in connection with the content of the information contained herein, including the accuracy, correctness, value, sufficiency, or completeness of the data, methods and other information contained herein. R.S. Means makes no express or implied warranty of merchantability or fitness for a particular purpose.** R.S. Means shall have no liability to any customer or third party for any loss, expense, or damage including consequential, incidental, special or punitive damages, including lost profits or lost revenue, caused directly or indirectly by any error or omission, or arising out of, or in connection with, the information contained herein.

No part of this publication may be reproduced, stored in a retrieval system, or transmitted in any form or by any means without prior written permission of R.S. Means Company, Inc.

The editors for this book were Suzanne Morris and Reinhold A. Carlson, PE; the managing editor was Mary Greene; the production coordinator was Wayne Anderson. Composition was supervised by Joan C. Marshman and Karen O'Brien. The book and jacket were designed by Norman R. Forgit. Some illustrations by Carl Linde.

Printed in the United States of America

10   9   8   7   6   5   4   3   2

Library of Congress Catalog Number Pending
ISBN 0-87629-347-X

For Marie, Boa, Gil, and Sil.

TABLE OF CONTENTS

Foreword	xvii
PART ONE – INITIAL DESIGN CONSIDERATIONS	1
CHAPTER ONE: BASIC HVAC SYSTEMS	3
Heating	6
Cooling	14
Ductwork	28
Ventilation	28
Air Conditioning	29
Problems	29
CHAPTER TWO: CODES, REGULATIONS AND STANDARDS	33
Code Structure	33
The Fire Safety Evaluation System (FSES)	35
Building Permits	37
Licensed Designers and Contractors	37
General Code Requirements	38
Typical Energy Requirements	43
Professional Organizations	44
Problems	46
CHAPTER THREE: HEATING AND COOLING LOADS	53
Heating Loads	53
Cooling Loads	85
Latent Loads (ΔW)	107
The BIN Method for Estimating Energy Usage	119
Psychrometrics	131
Problems	144
CHAPTER FOUR: ENERGY CONSERVATION	149
Conservation and Sacrifice	149
Sustainable Design	150
Indigenous Architecture: Using Natural Energy for Heating, Ventilating, and Air Conditioning	150
Energy Conservation Techniques	155
Energy Audits	172

Rebate Programs	190
Problems	191
CHAPTER FIVE: HOW TO SELECT A SYSTEM	**193**
Selection Criteria	193
Selecting the Type of System	195
Selecting the Type of Fuel	203
Selecting the Distribution System	206
Selecting the Generation Equipment	207
Equipment Selection Examples	207
Summary	212
Problems	216
PART TWO: EQUIPMENT SELECTION	**217**
CHAPTER SIX: GENERATION EQUIPMENT	**219**
Hydronic Systems — Heating Only	219
Warm Air Systems — Heating Only	227
Hydronic Systems — Cooling Only	229
Combined Heating and Cooling Systems	233
Generation Equipment Data Sheets	237
Problems	239
CHAPTER SEVEN: GENERATING EQUIPMENT ASSEMBLIES	**247**
Primary Equipment	247
Typical Assemblies	248
Problems	254
CHAPTER EIGHT: DRIVING SYSTEMS	**259**
Pumps	259
Fans	266
Equipment Data Sheets	272
Problems	272
CHAPTER NINE: HEAT EXCHANGERS	**277**
Shell and Tube Heat Exchangers	277
Fin and Tube Heat Exchangers	279
Steam Traps	281
Equipment Data Sheets	284
Problems	284
CHAPTER TEN: DISTRIBUTION AND DRIVING SYSTEMS	**287**
Steam Distribution	288
Water Distribution	299
Air Distribution Systems	313
Miscellaneous Fluids	339
Distribution Equipment Data Sheets	339
Problems	339
CHAPTER ELEVEN: TERMINAL UNITS	**351**
Steam Systems	351
Water Systems	352
Air Systems	359
Electric Systems	363
Floor Heating System	366
Equipment Data Sheets	367
Problems	372

CHAPTER TWELVE: CONTROLS — 373
- Types of Operation — 373
- Sensors, Controllers, and Actuators — 376
- Internal and External Controls — 376
- Zoning — 377
- Control Diagrams and Sequences — 379
- Energy Management Systems — 382
- Control of Building System — 389
- Equipment Data Sheets — 389
- Summary — 390
- Problems — 396

CHAPTER THIRTEEN: ACCESSORIES — 397
- Flues and Chimneys — 397
- Draft Inducers — 399
- Motor Starters — 401
- Noise and Vibration — 402
- Air Cleaning and Filtration — 404
- Hangers and Supports — 409
- Insulation — 414
- Equipment Data Sheets — 415
- Problems — 416

CHAPTER FOURTEEN: SPECIAL SYSTEMS — 423
- Fire and Smoke Control — 423
- High-Rise Buildings — 431
- Atrium and Covered Malls — 435
- Design of HVAC Systems for Health Care Facilities — 436
- Commercial Kitchens — 454
- Specifications — 458
- Building Management Costs — 467
- Summary — 474
- Problems — 474

CHAPTER FIFTEEN: ENVIRONMENTAL CONSIDERATIONS — 477
- CFC's and HCFC's: An Introduction — 477
- Ozone Depletion — 478
- Alternatives to CFC's and HCFC's — 484
- Indoor Air Quality: An Introduction — 487
- IAQ and Energy Conservation — 494
- IAQ and HVAC Controls — 494
- Oil Storage and Containment Piping — 494
- Radon Gas — 495
- Summary — 499
- Problems — 499

PART THREE — SAMPLE ESTIMATES — 501

CHAPTER SIXTEEN: MULTI-FAMILY HOUSING MODEL — 503
- Initial Selection — 503
- Design Criteria — 508
- Computation of Loads — 508
- Equipment Selection — 510
- Summary — 525

CHAPTER SEVENTEEN: COMMERCIAL BUILDING MODEL	527
Initial Selection	527
Computation of Loads	531
Equipment Selection	542
Summary	559
APPENDICES	563
Appendix A: HVAC Symbols	566
Appendix B: Abbreviations	569
Appendix C: SI Conversion Tables	572
Appendix D: Solutions to Problems	581
GLOSSARY	583
INDEX	589

TABLE OF FIGURES

Figure

1.1	Basic Hydronic Heating System	4
1.2	Cooling System Examples	5
1.3	Examples of Ventilation	6
1.4	Basic Rooftop HVAC Unit	7
1.5	Schematic Diagram: Generation – Distribution – Termination	8
1.6	Heating Ladder	9
1.7	Hot Water Heating System Without Heat Exchanger	10
1.8	Steam to Air System with Two Heat Exchangers	11
1.9	Heat Exchangers	12
1.10	Comparison of Steam, Water, and Air Systems for Space Heating	13
1.11	Cooling Ladder	15
1.12	Basic Cooling Layout	16
1.13	Heating and Cooling – Methods of Transferring Heat	17
1.14	Selection Criteria for Refrigeration Equipment	19
1.15	Guidelines for Cooling System Selection	20
1.16	The Compression Refrigeration Cycle	21
1.17	Types of Positive Displacement Refrigeration Compressors	22
1.18	Centrifugal, Water-Cooled, Hermetic Water Chiller	23
1.19	The Absorption Cycle	25
1.20	Selection Criteria for Condensers	27
1.21	Ideal Locations for Heating and Cooling Supply Ducts	28
1.22	Methods of Ventilation	31
1.23	Air Conditioning Ladder	32
2.1	FSES Categories and Criteria	36
2.2	Mandatory Safety Scores for Various Occupancy Groups	37
2.3	Health and Life Safety Standards	41
2.4	Design Condition Standards for Energy Conservation	43
2.5	Reference Standards	47
3.1	Basic Heating Diagram	54
3.2	Conduction, Convection, and Radiation	56
3.3	Common Building Heat Losses	57
3.4	Computation of the U Value and Temperature Gradient	58
3.5	Conductivity, Conductance, and the U Value	59
3.6	U Values for Common Building Materials	61

Figure

3.7	C, K, and R Values for Common Building Materials	66
3.7a	C, K, and R Values for Common Building Materials—In SI Metric Units	70
3.8	Table of Indoor Design Temperatures	74
3.9	Table of Outdoor Design Temperatures	75
3.10	ΔT for Adjoining Unheated Spaces	76
3.11	Perimeter Edge Slab Conditions	77
3.11a	Perimeter Edge Slab Conditions—In SI Metric Units	78
3.12	Infiltration Through Windows and Doors	79
3.13	Overall Effect of Infiltration in Buildings	84
3.14	Heat Loss Formulas	85
3.15	Sensible and Latent Heat Gains	87
3.16	Common Building Heat Gains	88
3.17	Cooling Load Temperature Differences	90
3.17a	Cooling Load Temperature Differences—In SI Metric Units	99
3.18	ASHRAE Psychrometric Chart	108
3.18a	ASHRAE Psychrometric Chart—In SI Metric Units	109
3.19	Solar Cooling Load Factors	111
3.19a	Solar Cooling Load Factors—In SI Metric Units	113
3.20	Shading Coefficients	115
3.21	Cooling Load Factors	118
3.21a	Cooling Load Factors—In SI Metric Units	121
3.22	Sensible and Latent Heat Loads for Internal Heat Gains	124
3.22a	Sensible and Latent Heat Loads for Internal Heat Gains—In SI Metric Units	127
3.23	Sensible and Latent Heat Gains for People	130
3.23a	Sensible and Latent Heat Gains for People—In SI Metric Units	131
3.24	Heat Gain Formulas	132
3.25	Air Conditioning Requirements	133
3.26	Bin Method Heat Pump Table	134
3.27	Bin Method Computation	134
3.28a	ASHRAE Psychrometric Chart (Showing Seven Properties of Air)	135
3.28b	ASHRAE Psychrometric Chart (Showing Four Different HVAC Processes)	136
3.28c	ASHRAE Psychrometric Chart (Showing Sensible and Latent Energy, Change in Moisture, and Dry Bulb Temperature Change)	137
3.28d	ASHRAE Psychrometric Chart (Showing Sensible and Latent Energy, Change in Moisture, and Dry Bulb Temperature Change)—In SI Metric Units	138
3.28e	ASHRAE Psychrometric Chart (Showing a Simple HVAC Cycle)	139
3.28f	ASHRAE Psychrometric Chart (Showing a Simple HVAC Cycle)—In SI Metric Units	140
3.29	Problem No. 3.2	146
3.30	Problem No. 3.4	147
3.31	Problem No. 3.8	147
4.1	Indigenous Building Types	152
4.2	How Animals Control Their Environment	154
4.3	Cogeneration Systems	158
4.4	Passive Solar Energy Systems	161
4.5	Active Solar Energy Systems	163
4.6	Typical Ice Storage System	164

Figure		
4.7	Modular Boiler Setup	167
4.8	Common Heat Recovery Equipment	171
4.9	Heating Plant Checklist	175
4.10	Lighting Fixture Survey Form	176
4.11	Door and Window Survey Form	177
4.12	Building Occupancy Pattern Chart	177
4.13	Annuel Fuel Consumption Chart	178
4.14	Current Energy Price Form	180
4.15	Annual Energy Consumption Chart	180
4.16	Estimated Energy Consumption Chart	181
4.17	O&M Checklist	182
4.18	Typical O&M and ECM Paybacks	185
4.19	O&M/ECM Detail Sheet	187
4.20	Combustion Efficiency Report	189
5.1	Decision Path for Selecting an HVAC Generating System	196
5.2	Types of Heating and Cooling Systems for Buildings	197
5.3	System Selection Chart	199
5.4	Computer Room System Choices	204
5.5	Basic Properties and Costs of Fuels	205
5.6	Preliminary Selection of Distribution Systems	207
5.7	Generating Equipment Selection Chart	208
5.8	Separate Heating System Example	210
5.9	Separate Cooling System Example	211
5.10	Centralized Systems—Combined Heating and Cooling	213
5.11	Combined Heating and Cooling—Cooling Side	214
5.12	Combined Heating and Cooling—Heating Side	215
5.13	Problem No. 5.1	216
6.1	Boiler Selection Chart	219
6.2	Typical Boiler Installation	221
6.3	Comparative C Values for Fuel Consumption	224
6.4	Typical Warm Air Heating System	228
6.5	Typical Chiller, Condenser, and Cooling Tower Installation	230
6.6	Typical Condenser Water System	232
6.7	Typical Rooftop Air Handling System	234
6.8	Free Cooling Using an Economizer Cycle	235
6.9	Riser Diagram—Split System	236
6.10	Water Source Heat Pump System	238
6.11a	Generation Equipment—Heating	239
6.11b	Generation Equipment—Cooling	242
7.1	Supplied Steam to Hot Water Distribution	249
7.2	Steam Boiler/Two-Pipe Vapor System	250
7.3	Gas-Fired Hot Water Boiler	251
7.4	Warm Air Furnace	252
7.5	Liquid Chiller—Cooling Tower	253
7.6	Packaged Rooftop Multizone Air Conditioning Unit	255
7.7	Built Up Air Handler Unit with Evaporative Condenser	256
7.8	Problem No. 7.1	257
8.1	Pressure and Head	260
8.2	Selection Guide for Pump Types	261
8.3	Characteristic Pump Curves	263
8.4	Pump Selection Graph	265
8.5	Pump Laws	266
8.6	Pump Arrangements	267
8.7	Selection Guide for Fan Types	269
8.8	Typical Fan Performance Curves	271

Figure

8.9	Fan Laws	271
8.10a	Driving Equipment Data Sheets for Pumps	273
8.10b	Driving Equipment Data Sheets for Fans	274
9.1	Types of Heat Exchangers	278
9.2	Heat Exchanger Performance	280
9.3	Capacities of Thermostatic Radiator Traps	283
9.4	Heat Exchangers Equipment Data Sheets	285
10.1	Open and Closed Systems	288
10.2	Selection Guide for Steam, Water, and Air Distribution Systems	289
10.3	Basic Steam Piping Systems	292
10.4	Equivalent Pipe Lengths for Steam Fittings	294
10.4a	Equivalent Pipe Lengths for Steam Fittings—In SI Metric Units	295
10.5	Steam Supply Pipe Sizes When Condensate Flows in Same Direction as Steam	296
10.5a	Steam Supply Pipe Sizes When Condensate Flows in Same Direction as Steam—In SI Metric Units	297
10.6	Steam Supply Pipe Sizes When Condensate Flow is Opposite to Steam	298
10.6a	Steam Supply Pipe Sizes When Condensate Flow is Opposite to Steam—In SI Metric Units	299
10.7	Steam Pipe Sizes for Condensate Returns	300
10.7a	Steam Pipe Sizes for Condensate Returns—In SI Metric Units	301
10.8	Eccentric and Concentric Reducers	302
10.9	Basic Water Piping Systems	303
10.10	Pipe Layout and Sizing Example	305
10.11	Friction Loss for Steel, Copper, and Plastic Pipe	307
10.11a	Friction Loss for Steel, Copper, and Plastic Pipe—In SI Metric Units	308
10.12	Correction Factors for Glycol Mixtures	309
10.13	Equivalent Lengths of Pipe Fittings	310
10.13a	Equivalent Lengths of Pipe Fittings—In SI Metric Units	312
10.14	Layouts of Distribution Piping	314
10.15	Single and Dual Air Duct Systems	316
10.16	Diffuser Spacing and Throw	318
10.17	Basic Duct Sizing Methods	319
10.18	Recommended Maximum Duct Velocities for Low Velocity Systems	320
10.19	The Velocity Method	321
10.20	Duct Sizing Chart for Circular Ducts	324
10.20a	Duct Sizing Chart for Circular Ducts—In SI Metric Units	325
10.21	Equivalent Rectangular Duct Dimension	326
10.21a	Equivalent Rectangular Duct Dimension—In SI Metric Units	329
10.22	Constant Pressure Drop Method	331
10.23	Equivalent Length Factors for Duct Fittings	333
10.24	Balanced Pressure Drop Duct Sizing Example	337
10.25	Duct System for Static Pressure Regain Method	339
10.25a	Static Pressure Regain Method of Duct Sizing	340
10.26	Static Regain Charts	343
10.27	Characteristics of Typical Refrigerants	344
10.28	Distribution Equipment Data Sheets	345
10.29	Problem No. 10.1	348

Figure		
10.30	Problem No. 10.8	349
10.31	Problem No. 10.10	350
11.1	Sizes for Typical Hydronic Terminal Heating Units	353
11.2	Steam Terminal Units and Accessories	354
11.3	Performance Characteristics of Hydronic Unit Heating and Cooling Coils	355
11.4	Hydronic Terminal Units—Valves and Fittings	356
11.5	Two-Way and Three-Way Valves—Two-, Three-, and Four-Pipe Systems	360
11.6	Typical Air Terminal Systems	361
11.7	Air Terminal Equipment	364
11.8	Electric Terminal Units	365
11.9	Portable Spot Cooler	366
11.10	Terminal Units Equipment Data Sheets	368
12.1	Electrical, Electronic, and Pneumatic Control Circuits	374
12.2	Typical DDC Controls for Make-up Air	375
12.3	Simple Control Loops	377
12.4	Basic Control Diagrams	378
12.5	Oneline Schematic (DDC) for Air Handling Unit	380
12.6	Oneline Schematic (DDC) for Heat Exchanger and Pumps	381
12.7	Oneline Schematic (DDC) for Controllers	382
12.8	Fan Coil Unit Piping and Controls	383
12.9	Ladder Diagram for Simple Economizer Cycle	384
12.10	Typical Algorithm	385
12.11	Temperature and Control Settings Chart	386
12.12	Typical Building Load Profile	388
12.13	Schematic of Typical Building Automation System	390
12.14	Typical Controls in "Smart" Buildings	391
12.15	Control Equipment Data Sheets	393
13.1	Flues and Chimneys	400
13.2	Motor Starters	403
13.3	Vibration Isolation and Transmission	405
13.4	Noise Control Details	406
13.5	Methods of Odor Control	408
13.6	Air Filters	410
13.7	Hangers and Supports	411
13.8	Pipe and Alignment Guides	413
13.9	Duct and Pipe Insulation	417
13.10	Accessory Equipment Data Sheets	418
14.1	Stair Pressurization and Smoke Control	425
14.2	Single Injection Stair Pressurization System	427
14.3	Sandwich Pressure-Zoned Smoke Control System	428
14.4	Typical Fire, Smoke, and Fire and Smoke Dampers	432
14.5	Typical High-Rise Requirements	434
14.6	High-Rise Code Comparison	436
14.7	Atrium Code Comparison	437
14.8	Covered Malls Code Comparison	438
14.9	Ventilation Requirements for Areas Affecting Patient Care in Hospitals and Outpatient Facilities	442
14.10	Filter Efficiencies for Central Ventilation and Air Conditioning Systems in General Hospitals and Psychiatric Facilities	446
14.11	Typical Isolation Room	449
14.12a	Acute Patient Care Unit Without Fire Protection per 1988 NFPA/LSC, 101	451

Figure		
14.12b	Acute Patient Care Unit With Fire Protection per 1988 NFPA/LSC 101	452
14.13	Exhaust-Only Kitchen Hood	456
14.14	Exhaust/Makeup Air Kitchen Hood	456
14.15	Coordination of Electrical Disconnects	463
14.16	Coordination of Trades on a Smoke Exhaust System	465
14.17	Coordination Between the Trades	468
14.18	Typical Annual O&M Expenses	471
14.19	Typical Capital Improvement Expenses	472
14.20	Annual Capital Improvement Costs as a Percentage of Original Building Cost	473
15.1	Required Levels of Evacuation for Refrigeration Appliances	481
15.2	CFC and HCFC Production Phaseout Schedule	483
15.3	Class I Substance Production Phaseout Schedule	484
15.4	CFC and HCFC Alternative Refrigerants	486
15.5	Building Layout for Good IAQ	492
15.6	Section of an Above-Grade Fuel Oil Tank	496
15.7	Section of Buried Fuel Oil Tank	497
15.8	Methods of Venting Radon Gas	498
16.1	Low-Rise Multi-Family Building Example – Plans	504
16.1a	First Floor Plan	505
16.2	Multi-Family Building Elevations	506
16.3	Multi-Family Building – Sections and Details	507
16.4	Basic Oil Fired Hot Water Heating System	509
16.5	Design Criteria	510
16.6	U Values and Constants	511
16.7	Areas and Crack Calculations	512
16.8	Heat Loss Calculations	513
16.9	Layout and Detail of Baseboard Radiation	514
16.10	Types of Systems to Consider	515
16.11	Riser Diagram	517
16.12	Mechanical Room Details	521
16.13	Heating Load Breakdown by Floor	522
16.14	Loads and Pipe Sizes for Risers	523
16.15	System Summary Sheet	526
17.1	Commercial Building Example – Plans	528
17.2	Commercial Building Elevations	529
17.3	Commercial Building – Sections and Details	530
17.4	Air Conditioning Ladder and Riser Diagram	532
17.5	Design Criteria	534
17.6	U Values and Coefficients	535
17.7	Areas and Crack Calculations	536
17.8	Heat Loads	537
17.9	Cooling Loads	538
17.10	Mechanical Layout	543
17.11	Mechanical Room Piping	545
17.12	Load Breakdown by Floor	552
17.13	Loads and Pipe Sizes for Distribution Systems	553
17.14	Ductwork Layout and Sizes	554
17.15	System Summary Sheet	560

FOREWORD

This book is designed to be used as a guide in the analysis and selection of HVAC systems. A step-by-step process is used to determine the most appropriate type of system for a project, and to estimate the loads that HVAC system components must handle and the procedures used to size equipment. The selection process includes choosing the appropriate generation equipment, distribution piping and/or ductwork, accessories, controls, and terminal units.

In the years since the first edition of *HVAC* was published, there has been a revolution in certain areas of design. Environmental concerns, particularly energy conservation, have become more important than ever. New standards and technologies have continued to develop at a rapid pace. Accordingly, the new and updated topics for this new edition are as follows:

- Latest ASHRAE methods for simplified cooling calculations and updated ventilation standards for air quality, including metric units.

- Environmental considerations such as CFC's, indoor air quality, and energy conservation.

- Sustainable design; indigenous building types as clues to modern, earth-conscious design.

- Specialty topics such as smoke evacuation and issues related to structures such as high-rise buildings, atriums, covered malls, hospitals, and kitchens (including hoods).

- Greater coverage of national codes, regulations, and standards; latest federal, CABO, MEC, UBC, SBCCI, EPACT, and ASHRAE requirements.

- Greater coverage of VAV systems, controls, and related areas such as dual fuel boilers and energy and operations costs.

- Hospital and facility design and related environmental issues: lighting conversions, variable speed drives, cogeneration, insulation, chiller upgrades, time and occupancy controls, and modular boiler and compressor systems.

- Issues particular to hospitals such as infectious disease rooms, DDC controls, upgrades, and money-saving suggestions for cooling tower and chiller upgrades.

Issues particular to facility managers such as when and why engineers are required, how capital improvements tie into code upgrades, and cost guidelines on predicting capital and operating expenditures.

Integrative systems such as energy management that control all building systems.

A list of problems at the conclusion of each chapter.

Throughout the book, metric equivalents are provided for most equations and measures. These metric units appear in the text, tables, and figures in square brackets.

The book is divided into three sections. Part One is a review of basic HVAC systems and principles, such as heating and cooling loads, codes, and section criteria. Part Two provides descriptions of the system components that supply the heating and cooling for a building. Equipment data sheets at the end of each chapter in this section summarize the important characteristics of each piece of equipment in an HVAC system. In Part Three, the selection of appropriate HVAC systems for two complete projects – multi-family housing and a commercial office building – are analyzed in detail.

Author's Note

As both a practitioner and an educator in architecture and engineering, I have tried to combine the two fields. In design, the architect is more than a general synthesizer of concepts; and, for many building types, the engineer's experience extends beyond basic details of design. In the end, a design must function well on both a large and a small scale. A building's environment is comprised of both its image and its function – the challenge of designing is to meet and match both qualities. When the owner also understands and encourages a dedicated design team, the formula for success is initiated. This book tries to approach HVAC design in this way, clarifying what the choices are and how to make them, so that the final design is as successful as possible.

Acknowledgments

For use of valuable copyright information:

American Society of Heating, Refrigerating, and Air Conditioning Engineers

American Institute of Architects

Building Officials and Code Administrators International, Inc.

Carrier Corporation, McGraw-Hill Book Company

ITT Fluid Handling Sales

R.S. Means Company, Inc.

For defining the goals and championing the development of the new edition:

    Mary Greene, R.S. Means Company.

For interpreting the manuscript as submitted and editing the entirety:

    Suzanne Morris, R.S. Means Company.

For technical reviewing:

    Reinhold A. Carlson, PE.

For research into old and new topics:

    The staff at William H. Rowe and Associates–
    Sam Bird
    Scott Hall
    Conleth O'Flynn
    Alphonso Sira
    Peter Brooks
    Terri Mansur
    Gordon Rowe
    Bonnie Wilson
    Lucy Dong

For being role models and tyrants of quality, for providing great war stories, and for serving as good company:

    Robert O. Smith, Consulting Engineer

    John and Paul Kennedy, Contractors, P.J. Kennedy and Son

    Kevin Cotter and Joe Walsh, Officers, Plumbers and Gasfitters Union Local No. 12

    George Harlow, New England Medical Center

    Michael O'Connell, Project Manager

    Ernest Weber, Consulting Engineer

    George Wingblade, Amherst College

    Gene Sullivan, R.W. Sullivan, Inc.

    Anand Seth, Massachusetts General Hospital

    Chad Maurer, Town Manager, Town of Winchester, MA

    Len Jones, Assistant Director, Boston Housing Authority

    Walter Adams and Chet Penza, Building Commissioners

    Jay Gentile, Project Architect, Bank of Boston

    Joe McCoy, Project Manager, Digital Equipment Corporation

    Alan Beveridge, Akira Yamashika and Associates

    Vic Walker and Cheri Klensing, Walker Klensing Design Group

For 25-plus years:

    June Rowe.

Part One
INITIAL DESIGN CONSIDERATIONS

In order to design, price, and lay out an HVAC system, it is necessary to understand heating, ventilating, and air conditioning systems, the calculation of heating and cooling loads, and the codes and regulations to be observed. Part One addresses these initial design considerations. In this section of the book, an overview of heating and cooling systems and how they operate, and the basic components that comprise a system, from generation through termination, are explained for heating, cooling, and ventilating systems. The procedures for selection and quantitative analyses provide the basis for the rest of the book.

Chapter One

BASIC HVAC SYSTEMS

Nature does not produce an environment that is always ideally comfortable to man. Excessive wind, rain, heat, humidity, and cold are undesirable; mechanical and natural energy systems are used to create more acceptable climates by heating, ventilating, and/or air conditioning (HVAC). An understanding of the fundamentals of these functions is necessary in order to select appropriate HVAC systems.

Heating is required in a building when the ambient temperatures are low enough to demand additional warmth for comfort. Boilers or furnaces *generate* the heat for a building; solar devices *capture*, *store*, and *release* heat; pipes or ducts *distribute* the heat; and convectors/radiators or diffusers are the terminal units that *deliver* the heat. A typical hydronic (hot water or steam) heating system is shown in Figure 1.1.

Cooling systems utilize cooler outdoor air when available, a refrigeration cycle, or other heat rejection method to supply cool air to occupied spaces. Chilled water or cool air is distributed by pipes and ducts throughout the building to terminal units (diffusers or fan coils). These end units deliver the cooling to the desired spaces. While cooling is rarely required by code, it is almost universally expected in commercial environments. Cooling systems may be independent of heating systems (such as a simple window air conditioner) or integrated with the heating system (such as a rooftop unit). Examples of cooling systems are shown in Figure 1.2.

Ventilating systems operate to provide fresh outdoor air to minimize odors and to reduce unhealthy dust or fumes. In many spaces, simple operable windows satisfy ventilation requirements. On the other hand, ventilation may be provided to a building by exhaust fans or fresh air intakes. Some of these methods are illustrated in Figure 1.3.

Air conditioning usually combines all of the features of heating, cooling, and ventilating systems, and may also provide additional "conditioning" of the overall environment such as noise control, air cleaning (filtration), humidity control, and energy-efficient controls (free-cooling options). A basic rooftop HVAC unit, illustrated in Figure 1.4, is a common example of air conditioning.

The process of producing heating or cooling from generation to distribution to terminal units is shown in Figure 1.5, and is common to all systems.

Generation equipment produces heat (heating) or removes heat (cooling) to or from the building. Boilers, furnaces, or supplied steam add heat; cooling towers, chillers, or heat pumps reject heat. The equipment for generation systems is the most expensive component of the HVAC system and is generally located in the mechanical equipment room. (See Chapters 6 and 7 for more information on generation and generation systems.)

The warm air or cold water that the various pieces of generation equipment produce is *distributed* throughout the building. A distribution system basically consists of pipes (water or steam) or ducts (air) that take the heated or cooled medium from the equipment that generated it through the building to the terminal unit. In addition, a distribution system may have valves, dampers, and fittings. (Distribution is further explained in Chapter 10.)

The **terminal units**, located in the conditioned spaces, include convectors (radiators), air diffusers, and fan coil units. These units receive the air or water from the distribution system and utilize it to warm or cool the air in the space. (Terminal units are discussed in detail in Chapter 11.)

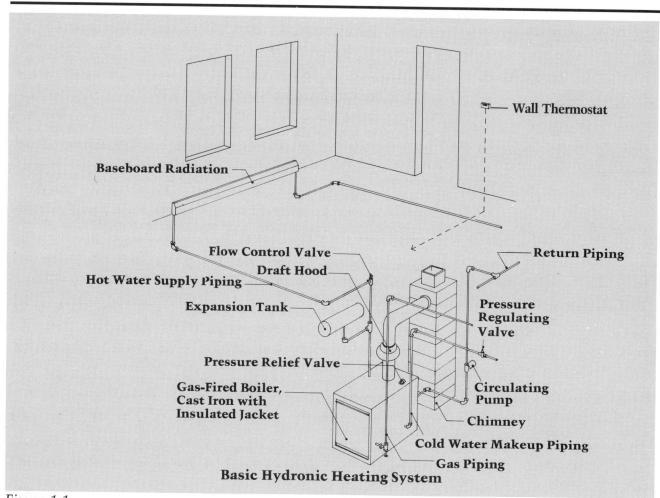

Figure 1.1

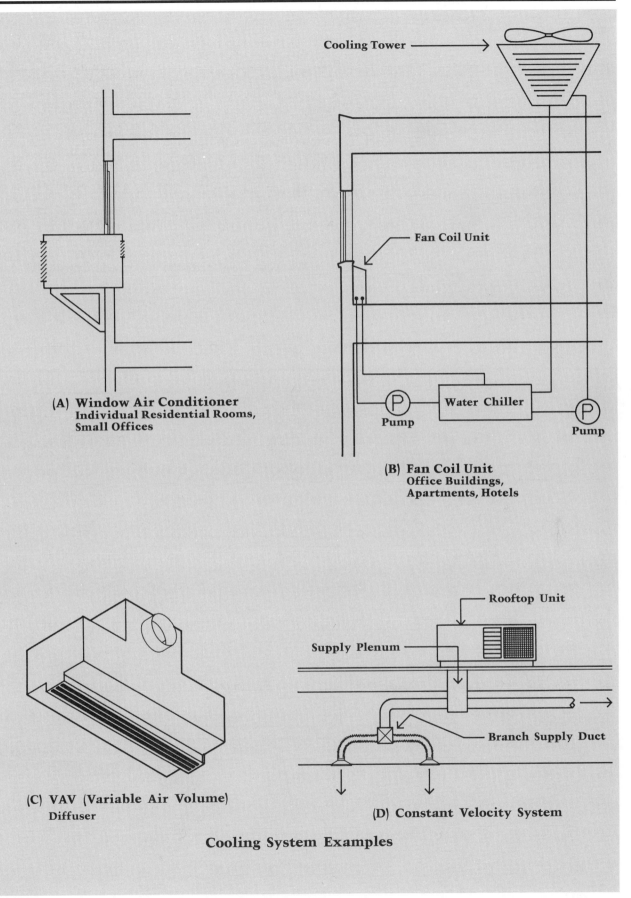

Figure 1.2

Heating

The most common heating systems warm a space by obtaining heat from a source and moving it through the building until it eventually becomes warm air in the space. Purchased steam from a utility company and oil or gas boilers located in a building's boiler room are typical sources of heat. The flow of heat from source to termination is shown in Figure 1.5.

The English Imperial unit of measure for heat is the British thermal unit, or **Btu**. One Btu is the amount of heat necessary to heat one pound of water 1°F. For simplicity, one thousand Btu's per hour is written as 1 MBH. The metric or SI unit of measure for heat (energy) is the Joule, or J. One thousand J's per second is written 1 kJ/s or 1 kW (1 kilowatt). A typical heat loss for a single-family house is often an average of 60,000 Btu's [17 500 W] each hour, or 60 MBH [17.5 kW].*

All systems that move heat downgrade the temperature of the heat as it moves through the system. In the process of moving from a generation source to a terminal unit (in a room), the temperature of heat is reduced

*Throughout the book, SI metric equivalents are provided for most imperial units. The SI units appear in brackets [ ].

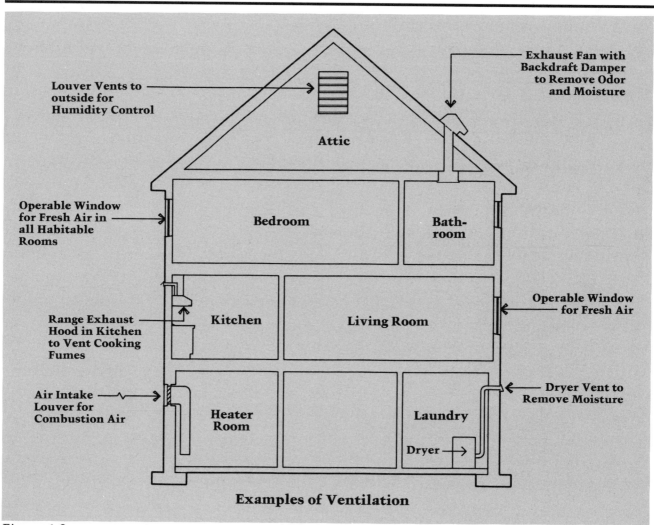

Figure 1.3

from high energy levels (high temperatures) to lower energy levels (lower temperatures). For example, the 1,800°F [970 °C] temperature in the combustion chamber of a boiler becomes 400°F [200 °C]. 150 psi [1136 kPa] high-pressure steam loses temperature and pressure along the route from source to termination. The heat reaches the terminal unit at about 130°F [150 °C] and finally downgrades to the room air design temperature of 68°F [54 °C].

The heating ladder shown in Figure 1.6 illustrates the common steps in moving heat from generation to termination. A heating system may use the same medium, such as hot water, throughout, or may "step down" from hot water to warm air. On the heating ladder, lateral movement is represented by a straight line, such as from a hot water boiler to a hot water pipe. The line continues to a hot water convector (radiator), which produces warm air for the room. This hot water system does not involve any intermediate heat exchangers (see Figure 1.7).

On the heating ladder it is only possible to move either across or across and down. A system that moves across and down the heating ladder is shown in Figure 1.8. A steam system utilizes steam from a utility company and converts it to hot water in the mechanical room. The hot water is distributed to heating coils that transfer the heat from hot water to warm air, and the warm air is then blown directly into the room. Each time the medium changes, such as from steam to hot water and then from hot water to warm air, a heat exchanger and a driving device (pump or

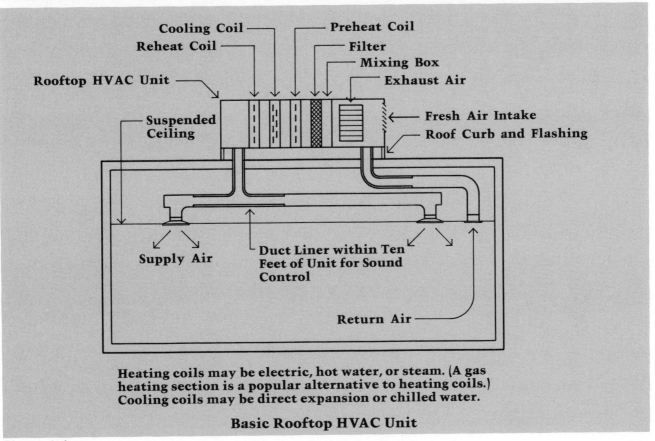

Basic Rooftop HVAC Unit

Figure 1.4

fan) are required. Heat exchangers are illustrated in Figure 1.9. (See Chapter 9 for more information on heat exchangers.)

Steam Systems

Steam distribution carries the greatest amount of heat per unit of volume for conventional systems, because each pound of steam delivers approximately 1,000 Btu's [1 kg steam delivers 2,200 Btu's] as it condenses from steam to water (the latent heat of vaporization). Steam systems do not need pumps or fans to drive them. Instead, the condensing steam creates a vacuum, which draws new steam into the system; gravity then draws the condensed steam (hot condensate in water form) back to the boiler or to waste. Condensate pumps and/or vacuum pumps can be used to increase the flow around a system and to improve overall performance. The American Society of Mechanical Engineers (ASME) *Code for Low-Pressure Steam Heating Boilers* limits the working pressure

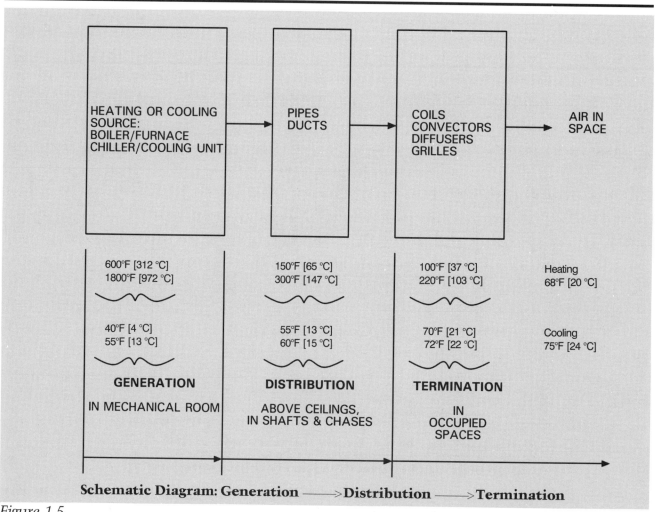

Figure 1.5

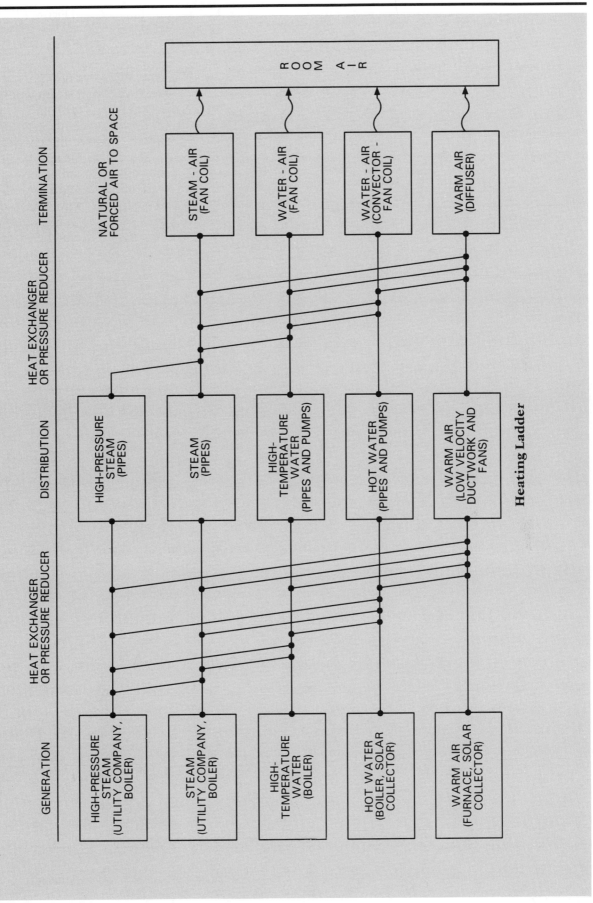

Figure 1.6

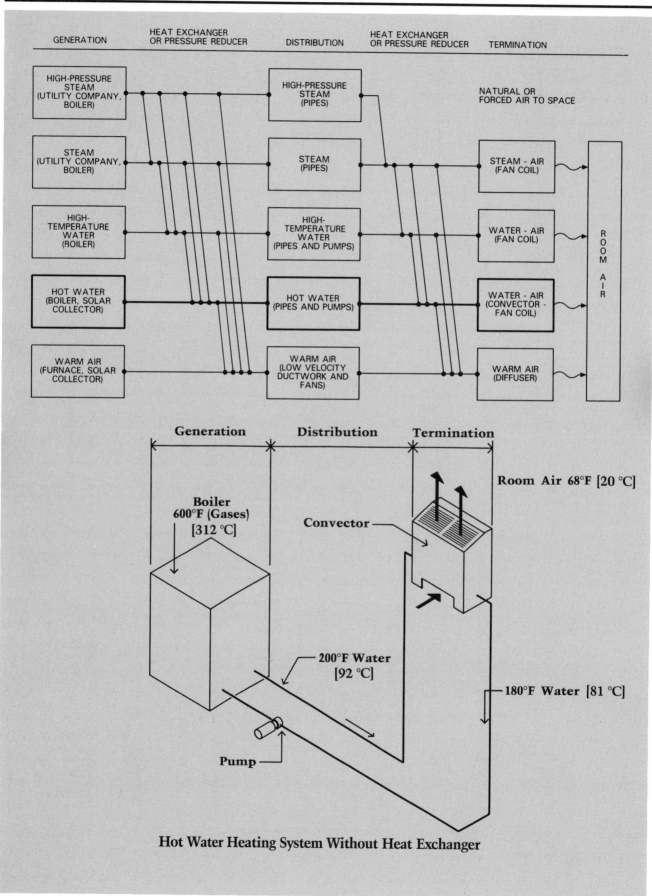

Figure 1.7

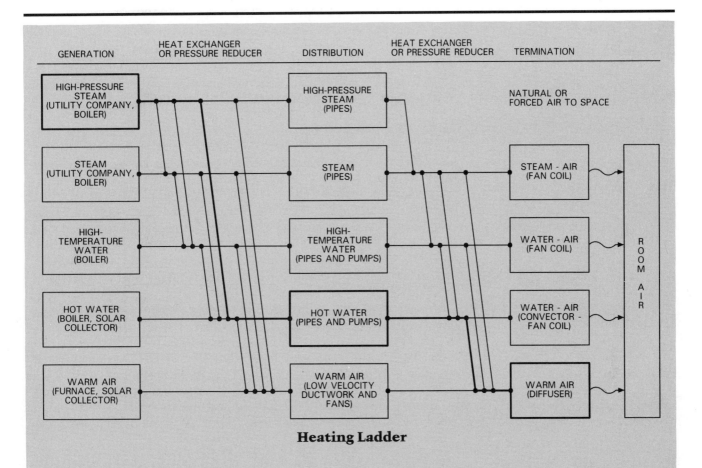

Heating Ladder

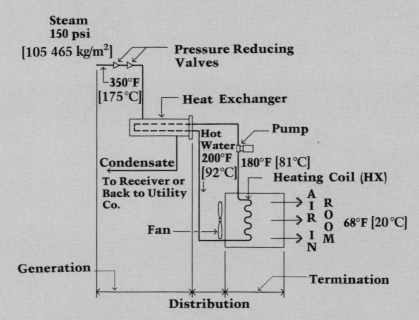

For the steam system shown, note how the energy is "stepped down" from 350°F [175 °C] to 68°F [20 °C].

Steam to Air System with Two Heat Exchangers

Figure 1.8

of steam heating boilers to 15 psi [205 kPa] and 250°F [120 °C]. Figure 1.10 compares some of the basic characteristics of steam, water, and air systems.

Steam systems have the advantage of having few mechanical parts. They are well suited to large buildings because of overall competitive installation costs and an ability to supply large quantities of heat.

Hot Water Systems

Hot water systems typically supply water that is between 180°F [81 °C] and 220°F [103 °C] at pressures of 15 to 30 pounds per square inch gauge pressure, or psig [205 – 410 kPa]. Atmospheric pressure, which is not usually read by the gauge, is 14.7 psi [101 kPa] at sea level. When the gauge pressure is combined with the atmospheric pressure, it equals 29.7 psia [205 kPa]. This is the *absolute pressure* of the 15 psig [205 kPa] reading. Water above the normal boiling temperature of 212°F [100 °C] can be supplied because the boiling point rises with more pressure (water at 15 psig [205 kPa] boils at about 250°F [120 °C]). A supply temperature of 200°F [93 °C] is common.

Hot water systems are normally designed for a 20°F [11 °C] temperature drop between the supply water temperature and the return water temperature. (Fan coils can be sized for 30 – 40°F [16 – 22 °C] temperature drop.) The quantities of water required are considerable when compared to steam. One pound of water with a 20° temperature drop delivers 20 Btu's [1 kg water delivers 45.9 kJ], or 1/50 of the heat transmitted in a pound of steam (which equals 1,000 Btu's and occupies only 1/13 of the volume). For this reason, pumps are usually required in hot water systems, which

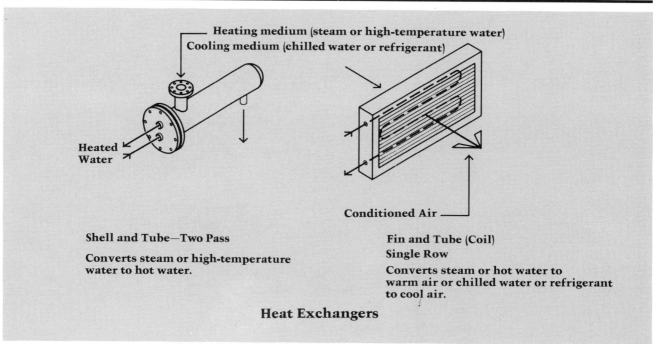

Figure 1.9

Comparison of Steam, Water, and Air Systems for Space Heating

Medium	Heat Delivered Per Pound [kg]	Volume Per Pound (Ft.³/lb.) [m³/kg]	Volume for 10,000 BTU [10 548 kJ]	Advantages	Disadvantages	Common Uses
Steam 230°F [109° C] Steam to Condensate	1,000 Btu/lb. [2.3 kJ/kg]	19.0 ft.³/lb. [1.2 m³/kg] Steam	$\frac{10,000}{1,000} = 10$ lbs. $[\frac{10\,548}{2300} = 4.6$ kg$]$ 10 lbs. × $\frac{19\,ft.^3}{lbs.}$ = 190 ft.³ $[4.6$ kg × $\frac{1.2\,m^3}{kg} = 5.52\,m^3]$	No Pumps	Scalding Risk	Institutional and Commercial
Water 20° [11 °C] ΔT 200° [92° C] Boiler 180° [81° C] Return	1 Btu/lb. × 20°F = 20 Btu/lb. [45.9 kJ/kg]	0.0166 ft.³/lb. [.001 m³/kg] Water	$\frac{10,000}{20} = 500$ lbs. $[\frac{10\,548}{45.9} = 23]$ 500 lbs. × 0.0166 ft.³/lb. = 83 ft.³ $[23$ kg × $\frac{.001\,m^3}{kg} = .23\,m^3]$	Small Pipes	Pump Cost	Residential
Air 60° [33° C] ΔT 130° [54° C] Supply 70° [21° C] Return	15 Btu/lb. [34.4 kJ/kg]	14.5 ft.³/lb. [.9 m³/kg] Air	$\frac{10,000}{15} = 667$ lbs. $[\frac{10\,548}{34.4} = 307]$ 667 lbs. × 14.5 = 9,760 ft.³ $[307$ kg × $\frac{.9\,m^3}{kg} = 276\,m^3]$	No Freezing Humidification Filtration and Ventilation Possible	Large Duct Space Required	Combined with Cooling Temporary Heat

All numbers shown in brackets are metric equivalents.

Note: Water pipes have the smallest, and air ducts have the largest, space requirements for transporting equivalent amounts of heat.

Figure 1.10

are commonly known as *forced hot water systems*. The ASME *Code for Low-Pressure Steam Heating Boilers* limits the working pressure of hot-water heating boilers to 30 psi [308 kPa] and 250°F [120 °C].

High-pressure boilers are used for a combination of heating and process loads and are usually designed for 150 psig [1136 kPa], found in systems with temperatures in the range of 350°F [175 °C]. The ASME *Code for Power Boilers* is used for boilers operating over 15 psig [205 kPa].

Warm Air Systems

Warm air systems that utilize electric heating elements or gasfoil warm air furnaces do not have the freeze-up problems that occasionally occur with piping systems. Warm air systems typically supply air from a furnace to the terminal units at 130°F [54 °C]. Because air has the lowest capacity to hold heat, it requires the greatest volume of distribution ductwork. Heating coils in the ductwork continue to warm the air as it passes through the system. The somewhat unwieldy size of the ductwork (when compared to pipe) is often compensated for by the benefits of allowing humidification or dehumidification and the ability to provide fresh outdoor air for ventilation.

Warm air systems with water or steam coils may need some protection from freezing.

Cooling

Cooling is almost universally accomplished by blowing cool air into a room or building. For example, by supplying the proper quantity of air at 60°F [15 °C] to the room temperature of a space, the air will mix and the temperature will be maintained at the design temperature (plus or minus 78°F [25 °C]). The cool air absorbs the heat gained from transmission through walls, windows, and doors; infiltration of warm air from around doors and windows; solar radiant energy; and internal heat gains from lights, people, and equipment.

The common unit of measure for cooling is the ton, which is derived from an earlier period when ice was used for cooling. One ton of cooling is equal to 12,000 Btu's [3.5 kW] per hour. It takes 144 Btu's [152 kJ] to melt one pound of ice at 32°F [0 °C]; when one ton of ice melts in one 24-hour period, it has absorbed 12,000 Btu's [3.5 kW] per hour.

Just as boilers and furnaces produce heating, a refrigeration system produces cooling. Heating equipment adds heat to a building; cooling equipment, in contrast, subtracts heat, or more simply, pushes heat away from the building. Basic cooling systems are shown on the cooling ladder in Figure 1.11.

Cool air is delivered by the air conditioning unit, neutralizes the warm-air buildup (heat gain), and maintains the desired room temperature. Some of the return air from the space is exhausted to the outdoors. The balance of the return air is mixed with the outdoor air, is cooled, and is redistributed to the space. Figure 1.12 illustrates a basic cooling layout.

There is a significant difference between *making* heat (heating) and *transferring* heat (cooling). The principle of heating is easy to understand. It is just like boiling a cup of water. In heating a building, fuel is burned in the combustion chamber of a boiler or furnace, and the heat energy from the combustion warms the water, steam, or air for the system. Each unit of energy from combustion is added to the system, starting at a high temperature and stepping down along the distribution path.

Cooling is more complicated. It is obviously not a problem to cool a space in the winter – simply opening the windows will cool off a space,

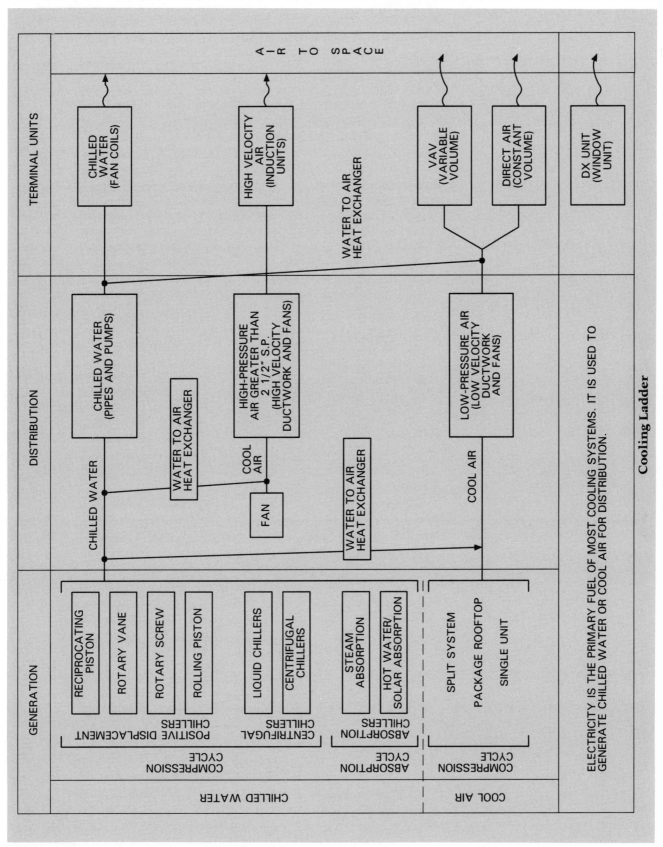

Figure 1.11

Cooling Ladder

as warm air flows "downhill" (outside to colder temperatures) to be replaced by the cooler outside air. Opening windows in the summer, however, will not cool a space. The heat inside must be pushed "uphill" to the outdoors. In the summer, the supply air temperature of 60°F [15 °C] is made by taking the heat from the return air mixture and rejecting it to the warmer outdoor air.

This is very similar to what happens with a household refrigerator. The inside of a refrigerator is kept cooler than the surrounding kitchen. To maintain cooling, the refrigeration cycle takes the heat to be removed from the inside of the refrigerator and pumps it outside. The process of taking the heat from inside the space and putting it into the refrigerant warms the refrigerant gas. The added energy of the compressor raises the temperature and pressure of the refrigerant gas further, to about 110°F [43 °C]. The exposed condenser coils on the back of the refrigerator are cooled by the room air temperature, which turns the refrigerant gas back into a liquid because of the heat reduction. This process keeps the inside of the refrigerator at about 40°F [4 °C]. (The compressor is the only major device that consumes energy in the refrigerator.) The temperatures and layouts for basic heating and cooling systems are shown in Figure 1.13.

In heating, each Btu [Joule] of heat burned produces nearly one Btu [Joule] of building heat. In cooling, each Btu [Joule] to be rejected requires only about one-third to one-fifth of a Btu [Joule] compressor energy. The effectiveness of cooling is measured by the coefficient of performance. The **coefficient of performance** (C.O.P.) equals the amount of cooling

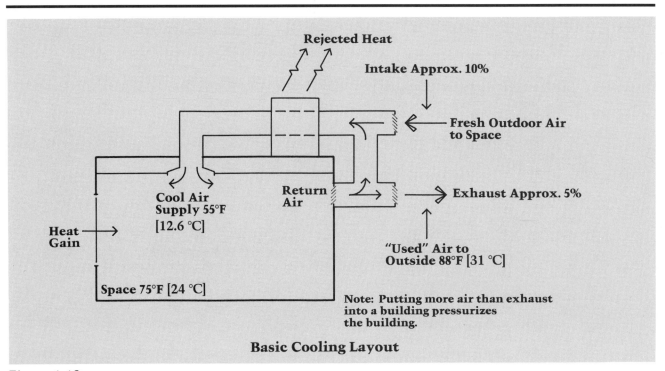

Figure 1.12

energy produced, or heat gain removed (in Btu/hour), divided by the energy consumed to produce the work of the compressor cooling (in Btu/hour).

$$\text{C.O.P.} = \frac{\text{amount of cooling energy produced}}{\text{energy consumed to produce the cooling}}$$

The coefficient of performance generally ranges from 3.5 to 5 for most applications.

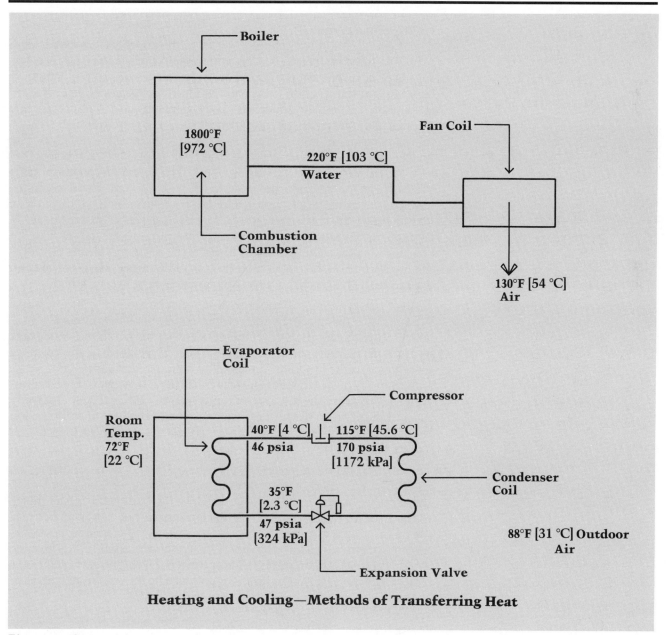

Figure 1.13

Refrigeration Cycles

All cooling systems use refrigeration equipment to produce cooling. There are six refrigeration systems:

- Compression cycle
- Absorption cycle
- Steam jet cycle
- Air cycle
- Thermoelectric cycle
- Solar cycle

Since compression and absorption refrigeration systems account for over 98 percent of all installations, this discussion is limited to these two areas. The steam jet, air, thermoelectric, and solar cycles are rarely installed.

The type of refrigeration cycle selected depends on cost and capacity considerations. Figures 1.14 and 1.15 are selection guides for refrigeration equipment and all types of cooling systems.

Compression Cycle

The most common refrigeration system is the **compression cycle**, which is used in most installations. The compression cycle is the most effective means of heat removal per pound of refrigerant. Figure 1.16 illustrates the compression refrigeration cycle with its four basic pieces of equipment: evaporator, compressor, condenser, and expansion valve. A refrigerant flows through the four pieces of equipment, forming a loop in which the refrigerant changes from a liquid to a gas and then back to a liquid. This refrigerant has a low boiling point. Nearly all refrigerants boil below 0°F [−17.8 °C]. Beyond the expansion valve, the boiling (evaporation) takes place. In expanding from the liquid to the gaseous state, the refrigerant absorbs heat from the evaporator and the space surrounding the evaporator. This changes the refrigerant from a liquid to a gas as it absorbs heat from the air surrounding the evaporator (room temperature), causing it to "boil." This phase-change heat transfer is the reverse of steam heating; the medium in cooling changes from a liquid to a gas, whereas in heating, the medium changes from a gas to a liquid as heat is transferred.

The gas temperature and pressure must then be raised significantly above the outdoor temperature by the compressor to allow the outdoor temperature to cool the refrigerant, condense it back to liquid form, and return it for recycling around the loop. The refrigerant has absorbed heat twice: once in the evaporator, where the building heat gain is removed; and again from the compressor, which has added energy to raise the gas temperature above the outdoor air temperature. The total heat absorbed by the refrigerant, from the heat of evaporation and the heat of compression, is finally rejected to the outside by the condenser, and the refrigerant repeats the loop.

The building heat gain is absorbed by the evaporator at no operating cost other than that used by the fan evaporator, which, while it may be considerable, is still less than that used by the compressor. It is the compressor that uses the most significant amount of energy and costs the most money to run. As previously explained, the effectiveness of a cooling system is measured by the coefficient of performance. Another term used in cooling is the **energy efficiency ratio** (E.E.R.). The energy efficiency ratio equals the heat gain in Btu/hour removed by the equipment, divided by the watts of energy consumed to cool the space.

$$\text{E.E.R.} = \frac{\text{heat gain removed by equipment (Btu/hr.)}}{\text{watts of energy consumed to cool the space}}$$

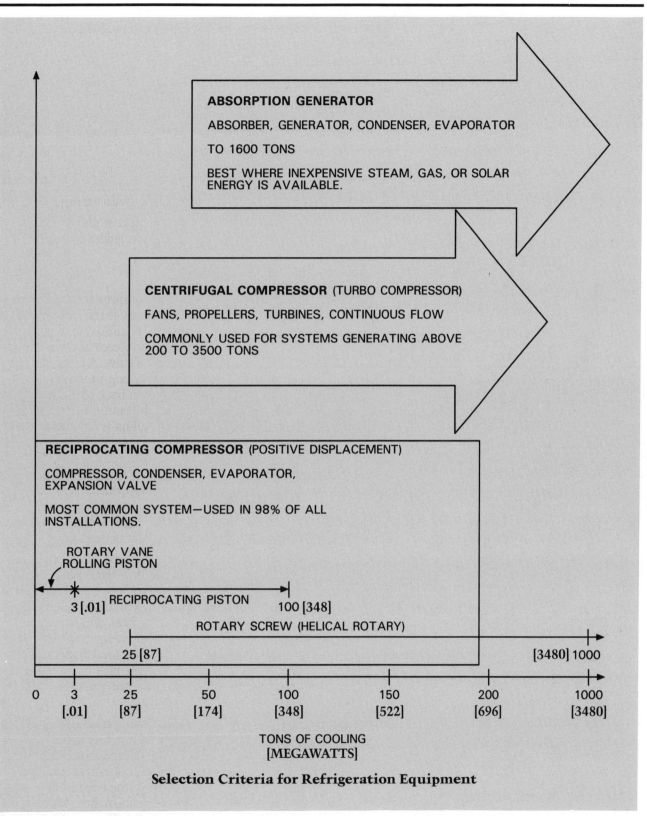

Figure 1.14

Since 1 watt equals 3.41 Btu/hour, the energy efficient ratio equals 3.41 times the value of the coefficient of performance.

1 watt = 3.41 Btu/hr., therefore,
E.E.R. = 3.41 × C.O.P.

Energy efficient ratio values of 9 to 12 are typical, and a value of over 10.5 is considered good.

The type of compression system selected depends on the overall cooling load required, the type of facility, the cost of the system, and the way in which the requirements of each system can be accommodated. Compression cycle systems are either reciprocating compressors (positive displacement) or centrifugal compressors (turbo).

Because there are many criteria that establish system selection, more than one type of compressor may be adequate for a given application. When

Guidelines for Cooling System Selection

System		Subsystem	Range of Tonnage [megawatts]	Common Applications	Remarks
Compression Cycle	Positive Displacement	Reciprocating piston	3 – 100 [.01 – 348]	All buildings	Noise isolation recommended.
		Rotary vane	0 – 3 [0 – .01]	Refrigerators Window AC units	
		Rotary screw	25 – 1,000 [87 – 3480]	Commercial Institutional	Variable speed possible.
		Rolling piston	0 – 3 [0 – .01]	Refrigerators Window AC units	
	Centrifugal compressors (turbocompressors)		50 – 1,000 [174 – 3480]	All buildings	Produce chilled water or brine for circulation to coils and terminal units.
Absorption Cycle			3 – 3,500 [.01 – 12 180]	All buildings	Inexpensive steam, gas, solar most common requirement.
Steam Jet Cycle			10 – 1,000 [34.8 – 3480]	Foods and chemicals Freeze drying	Special applications.
Air Cycle			10 – 150 [34.8 – 522]	Aircraft	Special applications.
Thermoelectric Cycle			5 – 50 [17.4 – 174]	Low-rise buildings	Thermoelectric solar panels generate electricity. Good in remote sunny locations.
Solar Cycle			50 – 100 [174 – 348]	Low-rise buildings	Hot water solar panels supply absorption units.

All numbers shown in brackets are metric equivalents.

Figure 1.15

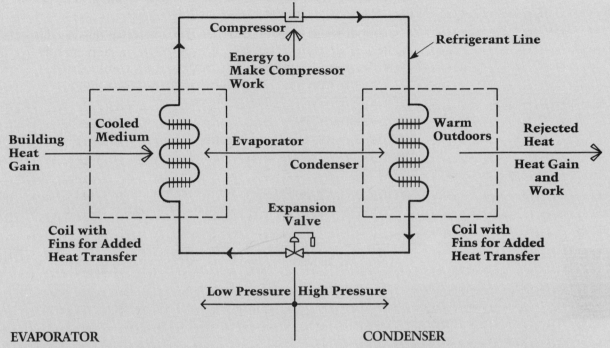

COMPRESSOR
With each stroke of the piston, the pressure of the refrigerant gas is increased as it passes to the condenser. This causes the temperature of the refrigerant to be simultaneously raised and set significantly above the outdoor temperature where it can reject heat.

EVAPORATOR

In the evaporator the refrigerant, which has a low boiling point, changes from a liquid to a gas (evaporates, boils). This change of phase from liquid to gas causes it to absorb heat from the medium to be cooled. The gas then carries the absorbed heat away from the medium and to the compressor.

CONDENSER

The condenser receives the hot gases from the compressor and cools them with the cooler outdoor air. This causes the gases to liquify or condense, some heat to be absorbed from the medium, and the compressor to be rejected to the outdoors.

EXPANSION VALVE
The hot liquid refrigerant from the condenser, which is at a high pressure level, flows through the orifice of the expansion valve. The orifice causes some of the hot liquid refrigerant to turn to gas at the lower pressure of the evaporator side. This "misting", or boiling, of some of the refrigerant lowers the temperature and pressure of the remaining liquid.

The Compression Refrigeration Cycle

Figure 1.16

this occurs, the final selection is based on either the initial cost or the life cycle cost of the system.

Reciprocating compressors, or positive displacement compressors, are the most common type of cooling equipment, and are used almost exclusively for systems with up to 50 tons [180 kW] capacity. Positive displacement compressors "squeeze" the refrigerant by using a piston, vane, or screw to compress the gas.

The four different types of reciprocating compressors are listed below. Figure 1.17 illustrates examples of these types.
- Reciprocating piston
- Rotary vane
- Rotary screw
- Rolling piston

Because they are small, reciprocating compressor units are usually single packaged units, which are air-cooled and located on the roof or through a wall.

Chillers produce chilled water for distribution to terminal cooling units. They often contain multiple compressors and are able to provide multi-stage cooling.

Some reciprocating compressors are used in split systems, which means that they are divided into two parts. An evaporator, expansion valve, and

Types of Positive Displacement Refrigeration Compressors

Type	Description	Comments
Reciprocating Piston Compressor	A cylinder, piston-type compressor with crankshaft and inlet and outlet valves utilizing the two-cycle method of intake and compression.	May be electric drive or powered by an internal combustion engine or a steam turbine. This is the most common compressor used in the 3 to 50 ton [.01 – 174 megawatt] range.
Rotating Vane Rotary Compressor	An off-center roller with two oscillating vanes attached rotates within the housing, compressing the gas and discharging through exhaust vane-type valves.	High volumetric efficiency. Electric drive. Most common use is 3 tons [.01 megawatts] and smaller.
Rotary Screw (Helical Rotary) Compressor	The single main helical rotor acts against and in conjunction with the two-gate rotors to compress the gas and discharges through vane-type valves.	Wide load range possible. Common in 25 to 1,000 ton [87 – 3480 megawatt] capacities. May be electric drive or powered by an internal combustion engine.
Rolling Piston-Type Rotary Compressor	The eccentric shaft and the rotating roller combine with a single oscillating vane (attached to the housing) in compressing the gas and discharging through vane-type valves.	High volumetric efficiency. Most common usage in 3 tons [.01 megawatts] and smaller. Electric drive.

All numbers shown in brackets are metric equivalents.

Compressors may be open-type or hermetic. In an open-type, the shaft connects the motor and compressor. The integrity is maintained through a shaft seal. In hermetic design, the motor and compressor are self-contained and are sealed within an air-tight, gas-cooled housing.

Rotary compressors maintain a continuous suction pressure, negating the need for inlet or suction valves.

Halocarbons and ammonia are the two common types of refrigerant used in these compressors.

Figure 1.17

supply air fan make up one half of the unit, which is located in or near the space to be cooled. The compressor and air-cooled condenser (condensing unit) make up the other half of the unit, which is usually located outdoors away from the space, typically on the roof. Reciprocating compressors are best suited to low-rise office and residential buildings with many tenants, since each tenant can be held responsible for the individual split system serving the space. The practical distance between the evaporator/valve/fan and the compressor/condenser unit is approximately 50 feet [15.25 m], limiting pressure and temperature changes in the refrigerant lines that connect the two halves of the system.

Centrifugal compressors are used in larger installations, typically above 200 tons [720 kW], although some manufacturers have models for loads as low as 50 tons [180 kW]. The centrifugal compressors continually "squeeze" the refrigerant gases by forcing a continual stream of the gas through a fan-like device. This forces the gas into a high-pressure line, which simultaneously raises its temperature. Figure 1.18 illustrates a typical centrifugal chiller.

Absorption Cycle

The absorption cycle is typically used in buildings where there is both a high cooling load (over 200 tons [720 kW]) and a heat source that is inexpensive in the summer, such as solar or steam. Absorption cycle is the second most popular cooling method after compression. Because absorption machines use water and salt solutions and do not use refrigerants (specifically CFC's), they are environmentally more attractive than compressors.

The absorption cycle involves the principle that a salt or salt solution has an affinity for water or water vapor. A salt shaker that clogs in the summer

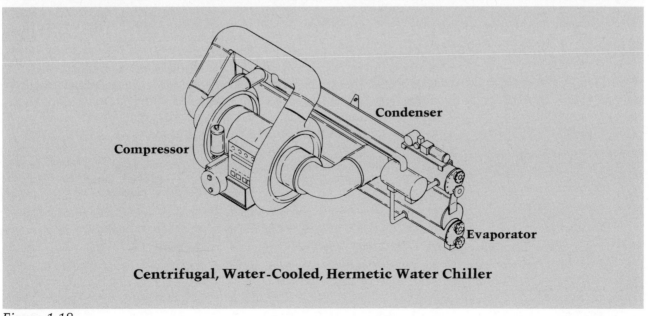

Centrifugal, Water-Cooled, Hermetic Water Chiller

Figure 1.18

when the moisture content of the air is high is a simple example of this. Thus, as a highly concentrated salt solution, such as anhydrous lithium bromide, is sprayed over a pool of water, some of the water is induced to vaporize and mix with the salt spray. This happens because chemically, the salt attracts water, and because the concentrated salt spray displaces some of the normally occurring water vapor, creating a partial vacuum of water vapor above the pool of water. These factors combine to draw the water from a liquid to a vapor, which mixes with the salt spray. Whenever water changes from a liquid to a gas, approximately 1,000 Btu's [300 W] of heat are required to vaporize each pound of water. As the water vaporizes, it takes the 1,000 Btu's/pound [650 W/kg] from the liquid pool and energizes the water into vapor. The heat leaves the water pool, chilling it considerably below 50°F [9 °C].

Thus, absorption systems produce chilled water by a method different than compression refrigeration machines, but in the same overall tonnage as centrifugal chillers. Depending on relative installation costs, efficiencies, structural loads, and operating costs, compression or absorption systems can be interchanged while other piping and HVAC systems are left intact.

Refer to the schematic of an absorption chiller in Figure 1.19. The salt solution or absorbent (lithium bromide) is sprayed in one chamber – the absorber – and the refrigerant (water) is sprayed in another chamber – the evaporator. Both elements are sprayed to increase their surface area, thereby increasing the potential for attraction. As the refrigerant is attracted to the absorbent chamber (via a connecting tube), the refrigerant changes from a liquid to a vapor and mixes with the absorbent, diluting it. This vaporization achieves two things. First, by drawing some refrigerant away from the refrigerant chamber, a partial vacuum is created, which draws more refrigerant into the vapor stage by lowering its boiling point. Second, because energy is required to convert a liquid to a vapor, and the vapor carries this potential energy, heat is drawn from the refrigerant that remains, and it is chilled considerably.

Water, for example, requires approximately 1,000 Btu's [650 W/kg] per pound to convert from a liquid to a vapor. The chilled water return from the building passes through the evaporator chamber to be cooled, and supplies the building with chilled water. In the absorber, the water vapor (refrigerant) releases the heat of vaporization as it is converted back into a liquid and absorbed by the weak salt solution. This heat tends to warm up the absorber chamber but is carried away by the cooling tower water, which dumps this heat at the rooftop cooling tower. The weak salt solution in the absorber is pumped up by the generator, which will "regenerate" the strong salt solution by separating the water and salt for recirculation. In the generator, the weak salt solution is heated in order to separate the water (volatile refrigerant) from the salt (absorbent). The concentrated absorbent returns to the absorber, where it may be reused and may capture more water vapor from the evaporator. The heat exchanger between the absorber and the generator recaptures some of the heat for the weak absorbent solution, and removes some heat from the strong absorbent on its way back to the absorber. The warm water vapor travels from the generator to the condenser, where it is condensed. The heat is removed via the cooling tower water. The condensed refrigerant then returns to the evaporator to start the cycle again.

Absorption equipment is large, heavy, and requires a heat source to provide the energy for the system. It is most advantageous when the summer cost of the heat source is low. Absorption systems have a natural potential for use with solar energy, because the heat from the sun increases as the need for cooling and the availability of heat to drive the absorption system increase.

There are several elements that must be carefully considered in the design and construction of an absorption chiller.

- The absorbent should have a strong affinity for the refrigerant.
- The refrigerant should be sufficiently more volatile than the absorbent so that a minimal amount of heat can separate them in the generator.
- Corrosive substances should be matched by materials that are corrosion-resistant.
- Because CFC's are not used in absorbent chillers, the negative environmental impact of these machines is limited. The designer should note, however, that some absorbent-refrigerant pairs are not permitted in residential applications because of the toxic or volatile nature of the chemicals used.
- Since there is no solid phase in the absorption cycle, some absorbents have the potential to crystallize, clogging the machine or upsetting

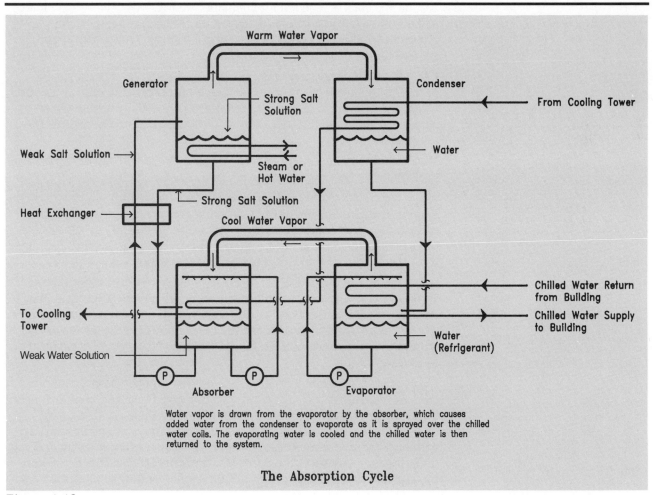

Figure 1.19

the delicate balance between the chemicals involved. Many machines on the market have integral safety features that can both detect and inhibit the formation of crystals in the cycle; the designer should be careful to include such features to prevent blockages in the cycle.
- Absorption equipment is large and heavy, requiring special space and structural considerations.
- The primary power requirement for absorption equipment, aside from electricity to run the pumps, is a heat source. The heat sources that are most commonly used are steam, gas, and solar; each potential source should be evaluated.

The selection of the refrigeration cycle usually establishes the principal component parts. The parts are normally furnished by the equipment manufacturer, either as a package unit or in separate, but compatible, component parts.

Condensers

After the type of refrigeration cycle and compressor (where applicable) are chosen, the condenser must be selected. There are three principal types of condensers: air-cooled condensers, water-cooled condensers, and evaporative condensers (see Figure 1.20).

Air-cooled condensers are typically used for systems where refrigerant is the medium in the equipment. The refrigerant coils are placed directly in the outdoor air. Mounted outdoors on a roof or through a window, air-cooled condensers are suitable for loads up to 50 tons [180 kW]. Air around the coils is required for circulation to reject the heat. Fans are usually provided to increase performance. Household refrigerators, common window air conditioners, simple package rooftop units, and split systems all use air-cooled condensers.

Closed-circuit condenser water coolers are similar in appearance and operation to air-cooled condensers. These units cool condenser water or a water glycol solution rather than a refrigerant gas. This type of water cooler is used for winter operation in climates subject to below-freezing temperatures.

Water-cooled condensers cool the refrigerant coils with water. By using a shell and tube heat exchanger, water from sources such as cooling towers (described below), or river, lake, spray pond, or ground water can be used.

Evaporative condensers utilize water that is sprayed across refrigerant coils to boost the cooling effect by evaporative cooling. They are similar to air-cooled condensers in that a fan is used to remove the heat and vapor from the surface of the coils, thus cooling the refrigerant.

A special type of condenser arrangement that is very common in large applications involves a **cooling tower** for condenser heat removal. The condenser water at the water-cooled condenser is sent to the cooling tower, where it is cooled and returned to the condenser. Cooling towers operate similarly to evaporative condensers, in that sprayed water and a fan are used to remove heat.

Cooling towers operate with either a closed or an open loop. In an open loop, the water that cools the condenser is sprayed in a large chamber (open to the atmosphere), where air is used to remove the heat. In a closed loop, the water from the condenser is enclosed in a coil and is sprayed with water from a separate loop. Cooling towers are very efficient, and can bring the temperature of the water to within 5 – 10°F [2.75 – 5.5 °C] of the ambient dry bulb temperature. Common operating temperatures for

Selection Criteria for Condensers

Type	Tonnage [megawatt] Range	Common Uses	Location	Special Requirements
Air-Cooled Condenser	0 – 50 [0 – 174]	Package units using refrigerant only Split system	Rooftop Through wall	Limit run and pipe for split systems.
Water-Cooled Condenser	3 – 150 [.01 – 522]	Small water cooled units using domestic water (where permitted) or water from ponds, lakes, rivers, or groundwater	Rooftop Ground mount	Cooling tower or other water source is required in urban setting.
Evaporative Condenser	10-250 [34.8 – 870]	Intermediate facilities	Rooftop Ground mount	Freeze protection.
Cooling Tower	50 – 1,000 [174 – 3430]	High-rise buildings	Rooftop Ground mount	Cools water for water-cooled condenser.

Figure 1.20

condenser water are 95°F [35 °C] to the cooling tower and 85°F [29 °C] to the condenser.

Ductwork

The ideal size and location of the ductwork for air-supply systems is different for heating and cooling. Even when the heating and cooling loads are approximately equal, as they are in temperate climates, the cooling ductwork is always larger. Therefore, it is difficult for one system to perfectly meet heating and cooling requirements.

Because warm air rises above cooler air, heating supply ducts should be placed low and under windows. The rising heated air will stop down-drafts and cold air spill from the windows, will help to prevent condensation on the glass, and will inject the heating supply at the source of greatest heat loss. This helps to maintain overall even room temperature.

In contrast, cooling supply ducts should be located at the ceiling and slightly away from the perimeter of the room so the cool air can mix with the warm air. The cool air will drift downward over windows and over the lights and occupants, which are other sources of heat gain.

The ideal locations of air ducts are illustrated in Figure 1.21.

Using a system that combines heating and cooling in the same air supply ducts involves a compromise because of the inherent conflict between heating and cooling duct sizes and locations. This can be accomplished by larger duct sizes and increased fan capacity. (For more information on ductwork, see Chapter 10.)

Ventilation

Building spaces must be vented for a variety of reasons. The main purpose of ventilation is to provide fresh outdoor air for the occupants. Fresh outdoor air replenishes indoor air for breathing and most noticeably reduces odors, smoke, and fumes caused by people, cooking, and manufacturing processes. All occupied rooms (apartments, offices, stores, schools,

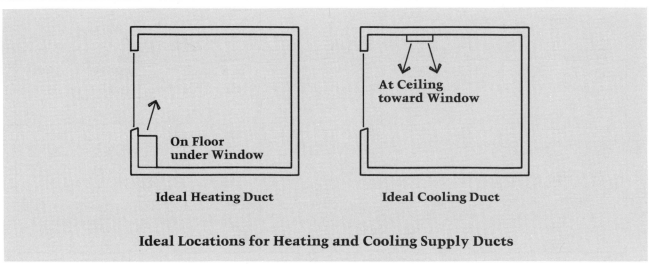

Ideal Locations for Heating and Cooling Supply Ducts

Figure 1.21

hospitals) must be properly ventilated. There are several acceptable methods of providing ventilation: operable windows, fresh outdoor air supply, exhaust air, supply air with exhaust air, and purging. These methods are illustrated in Figure 1.22.

The effect of ventilation on the heating and cooling system is to increase the loads that the heating and cooling system must carry, since the outside air must be treated before it is introduced into the space. The computation of these loads is described in Chapter 3.

Air Conditioning

When a building combines heating and cooling in one system, a rudimentary air conditioning system is established. In addition to heating and cooling, air conditioning includes humidification and dehumidification, cleaning of air, and providing fresh outdoor air for ventilation. When all of these functions are included in one system, the building can be considered to have a full heating, ventilating, and air conditioning system. An air conditioning ladder is shown in Figure 1.23. This ladder incorporates portions of the heating and cooling ladders previously noted.

Problems

1.1 In a given building, where are the following pieces of equipment typically located?
   a. Boiler
   b. Hot water pumps
   c. Cooling tower
   d. Heat pumps
   e. Packaged HVAC unit

1.2 Gas is contained in a vertical cylinder with a frictionless piston. What is the pressure of the gas if a 2,000-pound [920 kg] weight is placed on its 3-inch-diameter [7.6 cm] piston?

1.3 Define each of the following:
   a. Dew point
   b. Btu
   c. Specific heat of air
   d. Shell and tube heat exchanger

1.4 Describe the basic heating of your residence. Be sure to identify the type of fuel; basic system (water, steam, forced hot water); how the heat gets from the central area to your living space (pipes, ducts, or not at all); and the terminal units in your spaces (radiators, baseboard, or other). Sketch it.

1.5 A building is to use steam from the utility company to generate hot water baseboard heat. Using Figures 1.5 and 1.6 in the text as a model, draw the heating ladder and sketch the generation, distribution, and terminal equipment.

1.6 Assume you have 100,000 Btu's [105 480 kJ] of fuel energy available for either heating or cooling. Explain why this fuel can produce only 80,000 Btu's [84 384 kJ] of heating but could produce 350,000 Btu's [369 180 kJ] of cooling. Be sure to include a description of coefficient of performance.

1.7 Identify each of the following as part of the generation system, distribution system, terminal unit, controls, or miscellaneous equipment.
   a. Steam to hot water heat exchanger in boiler room

b. Hot water coil in air duct
c. Condensate piping
d. Rooftop HVAC package unit
e. Fancoil unit

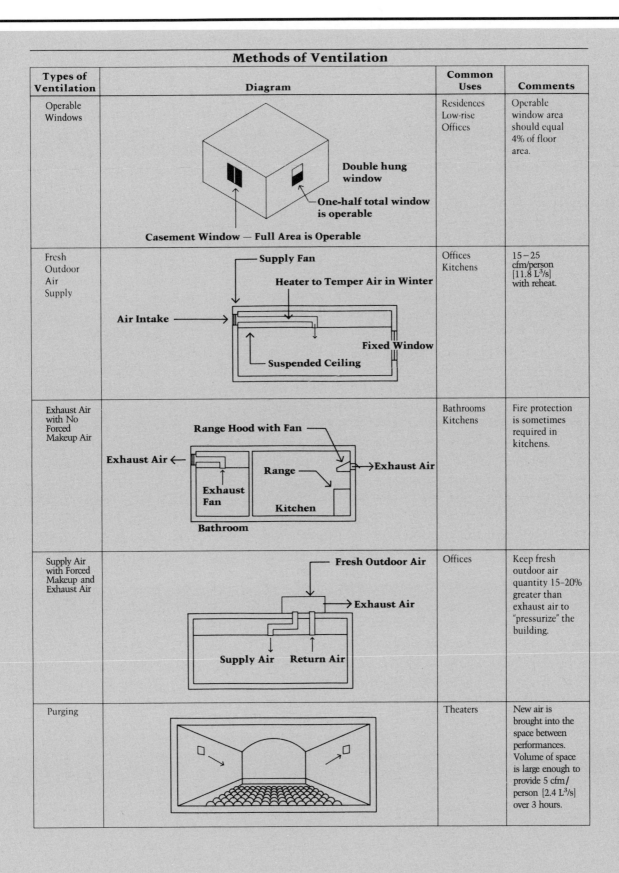

Figure 1.22

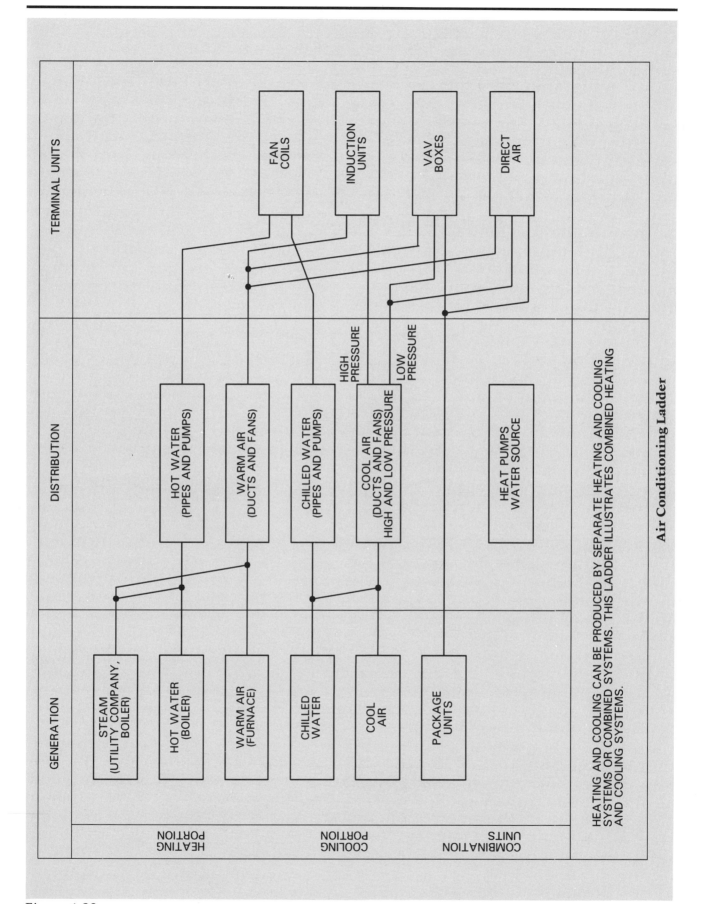

Figure 1.23

Chapter Two
CODES, REGULATIONS, AND STANDARDS

Virtually every building in a city, town, or municipality is governed by a building code. Building codes are minimum standards that are written to protect public health, safety, and welfare. Building codes typically establish minimum requirements for the design of selected building systems, including HVAC systems and components. The owner is legally obligated to establish and install heating, ventilating, and air conditioning systems that conform to the governing code.

Building codes regulate design, methods of construction, quality of materials, and building use for the structures within a political jurisdiction. The local code may be state- or city-drafted, or it may be adapted from one of several recognized national standards. The Building Officials and Code Administrators International (BOCA), the Uniform Building Code (UBC), and the Southern Building Code Congress International (SBCCI) are the principal national organizations that have established model building codes.

Issues such as fire protection, material quality, and structural integrity are addressed in detail in each building code. The aspects of HVAC design that relate to these issues must be considered in order to ensure public safety. For example, since HVAC involves combustion, proper construction of flues and ductwork is essential. Procedures for ongoing inspection and maintenance are also important. The weight of HVAC equipment must be checked against structural limits. Allowable pressures and flows for piping, proper storage of fuel containers, and adequate electrical wiring for HVAC equipment are a few of the common issues that are regulated by code. Almost every chapter of any building code has some relevance to HVAC design.

Code Structure

In addition to the codes, which are actually enabling acts passed by the legislature, there are many organizations that produce reference standards that are often incorporated into the model codes. These organizations have committees that update the standards regularly. A state code may stipulate, for example, that the provisions of the latest ASHRAE standards on ventilation apply to the requirements for fresh air in the state. Thus many standards, industry guidelines, and reference regulations are made a part of most building codes. Some of the professional organizations whose standards are incorporated by reference in many codes include:

National Standards

NFPA	National Fire Protection Association
JCAHO	Joint Commission on Accreditation of Healthcare Organizations
AIA	American Institute of Architects
UL	Underwriters' Laboratories
ASTM	American Society for Testing Materials

Trade Organizations

ASHRAE	American Society of Heating, Refrigeration and Air Conditioning Engineers
AISC	American Institute of Steel Construction
ACI	American Concrete Institute
NPA	National Plywood Association
SMACNA	Sheet Metal and Air Conditioning Contractors' National Association

Since all of these standards (approximately 44,000 in the United States) are revised periodically, it involves some work to simply determine which editions of which standard apply to a particular community.

In order for a code to apply, it must be enacted by the related legislature or governing body. There are federal codes for federal property, state codes for public and private property within a state, local city and town codes for particular localities, and separate codes for certain groups such as bridge and tunnel, subway, and airport authorities.

In addition to codes, there are laws that affect building design whether or not they are incorporated into the code itself or by reference. For example, at the federal level, congress has passed several pieces of legislation that affect buildings. Some of these laws include the following:
- Americans with Disabilities Act (ADA)
- Comprehensive National Policy Act
- Clean Air Act (CAA)
- Federal Water Pollution Control Act (FWPCA)
- Occupational Safety and Health Act (OSHA)

Some of these Acts affect only federal property; others are standards for the nation that all buildings are mandated to follow. There are also local laws that govern issues such as elevator upgrading, structural certifications of exterior facade and stairs, retroactive legislation for fire alarms, sprinklerization, and stair pressurization. These are part of the rules governing buildings even if they are not incorporated into the codes.

States can independently adopt or write building codes. Some states have written entire codes; others have adopted the model codes that are written by private organizations such as BOCA, SBCCI, NFPA, or the UBC. States that adopt model codes may also amend certain sections if they wish to change particular provisions.

Not every state requires that its code be used in all communities in the state. In some areas, each community, or selected city or town, has the authority to pass legislation that includes the right to have its own building code.

Depending on how a state is organized, one or more departments may be involved in the regulation of building construction. Departments of Public Safety, Health, Administration and Finance, Economic Affairs, Corrections, Education, Environmental Protection, and Mental Health, the Architectural Barriers Commission, and the State Historical Commission are all likely to have codes, rules, regulations, or standards to follow for their areas of interest.

The building code is usually enforced by a state's Department of Public Safety or its equivalent, which regulates certain codes, including:
- Building regulations (building code)
- Fire prevention (fire prevention regulations)
- Elevators (elevator code)
- Sprinklerfitters and Pipefitters
- Plumbing code
- Electric code

The Department of Public Safety may also be responsible for building inspections statewide, or monitoring of local building inspectors.

Code Complications

The procedures for determining what the code regulations are in a particular community, and which referenced standards apply, are sometimes difficult. Even determining the governing authority may be difficult. It may be the federal government for a federal installation, it may be the state building code, or it might be a local building code. The local fire official may also be a designated "official having jurisdiction" for certain matters. The referenced standards themselves may conflict with the code that incorporates them. Finally, there may be situations that are not covered by the code, or situations where a strict interpretation of the code would be difficult or even impossible. In situations where the resolution of a conflict is necessary, the designer (usually a registered architect or professional engineer) recommends a solution and then reviews it with the local officials for acceptance. The solution may require a variance to formalize its acceptance by the community.

Because of a growing awareness that the regulatory process is complicated, efforts toward code harmonization are being made. As currently written, most codes can prohibit occupancy or use in the most minor circumstances. One missing plug, one grille that is not yet installed, or one poorly-set thermostat can cause a delay in the acceptance and occupancy of an entire building. There is currently an effort to establish an overall rating system for building performance wherein a building would be allowed to be utilized when it meets the overall standards for safety.

The Fire Safety Evaluation System (FSES)

Several model codes have incorporated the Fire Safety Evaluation System (FSES) as part of their provisions. The FSES is an alternative system for determining a building's overall fire safety. In renovation projects, it is not always possible or practical to meet all the requirements in the building code. The FSES offers an alternative rating system to determine whether a building incorporates certain design features that make the structure functionally equivalent for life safety purposes.

Figure 2.1 provides a list of the categories that are normally evaluated in the FSES, as well as the criteria for that evaluation. The building receives scores for egress safety, life safety, and/or general safety, based on the occupancy of the building. The scores for each category are totaled and compared to a table of mandatory safety scores (see Figure 2.2). If the building's total egress, life safety, or general safety scores are greater than the mandatory scores for each safety item, the building is considered to be in compliance for that item. For example, if the total egress score for a business building (Occupancy Group B) is greater than the mandatory score of 40, then the egress for the building is considered to be in compliance.

FSES Categories and Criteria

Category	Criteria for Evaluation
Fire Area Values Evaluates a fire area based on occupancy and floor area.	• >15,000 s.f. • 10,001 to 15,000 s.f. • 7,501 to 10,000 s.f. • <7,500 s.f.
Space Division Evaluates the subdivision of a fire area by fire separation assemblies.	• No partitions • Fixed partitions to ceilings (with door closers) • Floor to floor partitions (without door closers) • Floor to floor partitions (with door closers)
Corridor Wall Values Evaluates the completeness and separation of fire partitions separating corridors from other spaces.	• No fire partitions; no self-closing doors • Less than 1-hour partitions • 1 hour to less than 2 hours with conforming doors • Greater than 2-hour partitions with conforming doors
Vertical Opening Protection Evaluates the fire resistance rating of shaft enclosures within the building and all openings between 2 or more floors.	• Based on a vertical opening value, which is the product of a protection value factor (PV) and a construction-type factor (CV) • $VO = PV \cdot CV$
HVAC System Evaluates the number of floors served by an individual HVAC system.	• Greater than 5 floors • 3 to 5 floors • 2 floors • 1 floor or central boiler/chiller system without ductwork
Automatic Fire Detection Evaluates the smoke detection capability of the building based on location and operation of a fire detection system.	• No fire detection system • HVAC returns only • Elevator lobby only • Elevator lobbies, in duct returns and single station units in residential occupancies • All corridors and elevator lobbies • Total space
Fire Protective Signaling Evaluates the capability of a fire protective signaling system if one exists.	• No fire protective system • Manual fire alarm boxes • Manual fire alarm boxes and a voice/alarm system • Fire command station
Smoke Control Evaluates the ability of a natural or mechanical smoke evacuation system.	• No smoke control system • Operable windows • 1 smokeproof enclosure • Stairway with operable windows • Complete smoke control system
Exit Capacity Evaluates the number and capacity of exit routes located within the building.	• Minimum number of required exits provided • Excess capacity provided • Horizontal exits provided • Excess exits provided
Dead End Evaluates the length of dead-end corridors.	• Dead ends are more than 20 ft. [6.1 m] but less than 50 ft. [15.3 m] • Dead ends are less than 20 ft. [6.1 m]
Elevator Controls Evaluates the elevator equipment and controls that may be used by fire departments for rescue purposes.	• Controls not provided or elevators not present for buildings >4 stories • Control provided for buildings <4 stories • Automatic recall provided • Control and recall provided for 1-story buildings
Means of Egress Emergency Lighting Evaluates the presence and reliability of emergency lighting located in means of egress.	• No emergency lighting exists • Lighting provided but no emergency power provided • Emergency lighting provided is in full compliance
Mixed Use Groups Evaluates the building's fire separation compliance with the code when two or more use groups occupy the same building.	• In compliance with the code • Not in compliance with the code
Sprinkler System Evaluates the fire suppression system within the building.	• No or partial protection • Equipped throughout with an automatic sprinkler system

All numbers shown in brackets are metric equivalents.

Figure 2.1

During a renovation or alteration project, it may be desirable to perform an FSES evaluation to determine whether the building meets acceptable life safety requirements. When a building score is determined to be acceptable, renovations may occur without requiring full compliance with the building code.

Building Permits

For work other than ordinary repairs, most jurisdictions require a building permit. It is generally necessary for a community at large to determine that adequate sewer and water are available; that the fire department can respond adequately in the event of a fire; that the uses of the property meet zoning and historical restrictions; and that the provisions of the governing building code are met before allowing construction to proceed.

The Building Commissioner or local Inspector of Buildings usually enforces the building code. There are other agencies that may also play a role in the review process. For example, the Department of Public Health may review plans for hospitals and further require a separate approval from the Department of Public Safety. Other codes may have little or no review. For example, there is currently no review agency for ADA.

Licensed Designers and Contractors

Building codes generally require that buildings other than one- or two-family residences be designed by registered architects or licensed professional engineers who stamp their drawings. Architects and engineers are licensed by each state's Division of Professional Registration. In addition to architects and engineers, professionals such as landscape architects and land surveyors may be licensed. Certain building trades are typically licensed; for example, electricians, plumbers, gasfitters, and sanitarians.

While many trades are licensed, many are not. Sheet metal workers, painters, tilesetters, ceiling installers, caulkers, insulators, control wiring specialists, and communication wiring specialists are not regulated by most building codes.

Mandatory Safety Scores for Various Occupancy Groups			
Occupancy	MFS	MME	MGS
A-1	10	24	24
A-2	5	19	19
A-3	9	21	21
A-4, E	17	29	29
B	28	40	40
F, S-1	26	53	53
M, R	23	35	35
S-2	38	65	65

MFS—Mandatory Fire Safety
MME—Mandatory Means of Egress
MGS—Mandatory General Safety

Figure 2.2

Codes have a significant benefit for architects and engineers. The codes act as a shield against frivolous lawsuits. Since codes generally are regarded as a level of quality and a standard of acceptance, designs that conform to codes are generally considered to be designed properly. Hence an architect or engineer whose design meets code requirements is generally assumed to be responsible.

As codes become more complicated, and, in some instances, contradictory, there will need to be systems to determine levels of compliance. Some areas that codes currently mandate may be replaced with more overall performance-driven criteria much like the FSES system. There needs to be a distinction between life safety, public policy, and workmanship of construction issues in the design of buildings.

The Role of Architects and Engineers: Design Requirements

The designer is responsible for designing the building to meet code requirements. Each code specifies the requirements for the design drawings. The design drawings are reviewed by officials having jurisdiction. Typically, drawings submitted to the building official for review are expected to contain the following:

- Accurate location and dimensions of all means of egress from fire
- An occupancy schedule of persons in all occupiable spaces
- Method and amount of ventilation and sanitation
- The methods of fire-stopping
- Schedule and details indicating trim and finishes
- Seal of architect or engineer if building is >35,000 cubic feet [980 m^3]*
- Site plan showing size and location of new construction, existing structures, distances from lot lines, street grades, and finished grades
- Engineering details and calculations stamped by a PE of structural, mechanical, electrical work as determined by building official
- Sizes, sections, and relative locations of all structural members with design loads
- Type of construction; fire-resistive rating of all structural elements
- Building exterior envelope, U values, R values, and lighting levels

Once the drawings are approved, the design professional is expected to review the shop drawings, the procedures for code-required quality control materials, and the critical design components for general compliance with the design requirements.

General Code Requirements

As we have mentioned, for HVAC systems there are two broad categories of code concern: health and safety considerations and energy provisions. Of the two, health and safety considerations have priority, although in fact they rarely conflict with energy concerns. There are two basic principles of health and safety: containment of a fire and provision of a means of escape (egress). Containment of fire is primarily obtained by providing floors and walls of sufficient fire rating to contain a fire. A brief discussion of fire ratings and fire separation requirements follows.

*Throughout the book, SI metric equivalents are provided for most imperial units. The SI units appear in brackets [ ].

General Life Safety Considerations: Fire Ratings and Fire Separation

Fire Separation

In general, different types of spaces must be fire-separated from one another. Occupied spaces typically have a one-hour rating separating them from the corridor, and corridors have a two-hour separation from the building egress stair. Similarly, in mixed-use buildings, a retail floor would be separated from the upper office levels with a two-hour fire-rated floor. In an apartment or other building, each tenant would be separated from the next by a fire separation of one hour (fire-rated walls and floors).

Buildings that are sprinklered are often permitted to reduce the fire ratings of fire separations, typically by one hour. Thus for many jurisdictions, if a building is sprinklered, exitway access corridors and tenant spaces that normally have a one-hour fire separation requirement are not required to be fire-separated. There are other fire-rating reductions commonly permitted for sprinklered buildings, such as for shafts, elevators, and doors.

Exterior Wall Protection

Not only must interior walls and floors have a fire rating, but some exterior walls also have restrictions. It is generally accepted that buildings separated by more than 30 feet [9.1 m] are sufficiently separated to prevent the spread of fire in a cityscape. Thus a building whose exterior walls are at least 30 feet [9.1 m] away from another structure or property line may have an exterior wall protection of 0 hours. This means the wall can have windows (windows generally have no fire rating). Buildings closer than 30 feet [9.1 m] or close to property lines must have some degree of fire protection: fire-rated shutters, deluge sprinkler systems, and limited or no window openings.

Vertical Openings

Generally, any opening in a floor of more than one square foot must be protected to prevent smoke and fire migration. Stairwells, elevator shafts, plumbing chases, dumbwaiter shafts, and duct shafts are all examples of common vertical openings. These openings generally are required to be vented to the roof and capped with a skylight or roof hatch, which when opened will vent out smoke and gases. All must be surrounded with a fire-rated separation of two hours, or one hour for one- and two-family residences.

Means of Egress

Most codes require that buildings have at least two means of egress on each story for occupants to exit a building in an emergency. In addition, dead end corridors are generally restricted to 20' [6.1 m] maximum and corridor width is determined from the occupant load on a floor with 44" [1.1 m] as a general minimum. Corridors and stairwells that are the primary egress path in an emergency have requirements in some buildings to prevent air from being excessively contaminated; these smoke control procedures are discussed in more detail in Chapter 14. Doors from corridors to stairwells have minimum dimension requirements and must be self-closing.

Boiler Rooms and Electric Switchgear Rooms

Special rooms such as boiler rooms and electric switchgear rooms may require two exits as well as special fire alarm requirements.

General HVAC System Life Safety Code Requirements

Fire and Smoke Dampers

Any breech in a fire separation must be protected. A door from a tenant space through a one-hour rated corridor must use a fire-rated door and an automatic door closer that will maintain the fire rating integrity of the wall when not in use. Fire dampers are similarly required at all penetrations of ducts through fire walls (NFPA requires fire dampers in all walls with a rating of two hours or more). Certain buildings, such as hospitals, have requirements for smoke partitions. Duct penetrations through smoke partitions require smoke dampers. Smoke and fire dampers may need to be tied to activate doors or other automatic devices. Chapter 14 discusses these requirements in more detail.

Ventilation: Natural versus Mechanical

Figure 1.22 in Chapter 1 illustrates methods for ventilation. All habitable spaces must have a minimum amount of either natural or mechanical ventilation. Natural ventilation is generally achieved if 4% of the floor area is available as operable windows. Ancillary rooms without a direct window to a prime space, such as a kitchen off a dining area, can "borrow" ventilation from the prime space if there is a sufficiently large opening (8% of ancillary space or 25 square feet) [2.3 m^3] between them.

Spaces that do not meet the criteria for natural ventilation must be vented mechanically. Toilet rooms should be mechanically vented in all cases. Other spaces are generally required to provide a minimum amount of fresh outdoor air per person. Figures 2.3 and 2.4 are tables of ventilation standards for specific occupancies and list the ventilation standards for ASHRAE and BOCA. Generally, 15 cfm [7.1 L/s] of fresh outdoor air per person is required for spaces such as classrooms, lobbies, auditoriums, and dining rooms. Offices require 20 cfm [9.4 L/s] fresh outdoor air per person. If the outdoor air quality is poor, filtration is also required.

Recirculation

Air in excess of the ventilation requirement can be completely recirculated (untreated), provided it is not from areas such as kitchens and toilet rooms. The air in these and similar contaminant-producing spaces is not permitted to recirculate.

Plenums

Generally, exits and access corridors cannot be used for supply or return air plenums. Ceilings above corridors may be used if they are fire-separated from the corridor. Plenums are generally limited to one fire area. Special insulation is required on exposed wiring systems.

Smoke Detectors

Smoke detectors are an integral part of a building's fire protection system. Smoke detectors are required in a variety of spaces, including mechanical rooms, electrical rooms, corridors, storage areas, and shafts. They are connected to the building fire alarm system. There are requirements for smoke detectors in HVAC systems, particularly ductwork and air handlers. The general requirements are as follows: If an air handler delivers >2,000 cfm [944 L/s], it must have a smoke detector in the supply air duct tied to shut down the air handler and to the fire alarm systems. If an air handler delivers >15,000 cfm [7080 L/s], it must have smoke detectors in both supply and return systems tied to shutdown and fire alarm systems as well as smoke dampers that isolate air handlers from the rest of the system.

Depending on the type of building and system, the air handlers may also need to be part of a smoke control system.

High-rise Requirements

High-rise buildings have special life safety requirements. Chapter 14 discusses the particular requirements of high-rise buildings, including smokeproof enclosures, fire protection, smoke detection, elevator recall,

Health and Life Safety Standards

Code Topic	ASHRAE 62-1989	BOCA Mechanical Code 1993	NFPA 90A 1985	CABO-MEC 1992	Remarks
Natural Ventilation (Outdoor Air)		An area equal to 4% of the floor areas must be operable. [1]			Toilets and kitchens are required to have mechanical ventilation in some codes.
Mechanical Ventilation (Outdoor Air)	10 – 35 cfm [4.7 – 16.5 L^2/s] per occupant. Refer to Figure 2.5. [2]	10 – 35 cfm [4.7 – 16.5 L^2/s] per occupant. Refer to Figure 2.4. [1,2]		References ASHRAE 62	Required if natural ventilation is not satisfied.
Natural Light		An area equal to 8% of the floor area must be glazed. [1]			
Artificial Light		[1]		General Light. 20 FC [3]	
Smoke Detectors in Air Handling Units >2,000 cfm [944 L^2/s]		Supply duct	Supply duct		
Smoke Detectors in Air Handling Units >15,000 cfm [7080 L^2/s]		Supply and return ducts. Also supply and return ducts when >50% of system air is exhausted.	Supply and return ducts.		
Sprinklers		Both the 1990 BOCA Fire Prevention Code and NFPA 96 require sprinklers in commercial kitchen hoods and ducts. The intent is to prevent grease fires.			

All numbers shown in brackets are metric equivalents.

[1] Natural Light, Artificial Light, and Natural Ventilation data is from the 1993 BOCA Building Code. The equivalent in artificial light is an acceptable substitute for natural light. The 1993 BOCA Mechanical Code is referenced for mechanical ventilation.

[2] Higher values are required for special occupancies.

[3] IES Lighting Handbook is referenced for specific lighting.

Additional reference standards:
  CABO-MEC (Council of American Building Officials – Model Energy Code) 1992 (BOCA reference CABO-93)
  ASHRAE 62 – Ventilation for Acceptable Indoor Air Quality 1989
  NFPA 96 – Vapor Removal from Cooking Equipment 1991

Figure 2.3

fire safety plans, fire alarm command center, smoke evacuation and emergency power systems.

HVAC System Energy Code Requirements

Energy Conservation

Since the oil crisis of the 1970s, most states have incorporated into their codes a chapter on energy conservation. New energy efficiency standards for buildings have also been established. These include The Comprehensive National Policy Act (EPACT), and the Council of American Building Officials (CABO) Model Energy Code (MEC), which establishes performance standards for building envelopes. It is generally stringent and far-reaching.

While there are other factors that contribute to energy use, such as lighting and electrical loads, HVAC is clearly a major focus for energy conservation. As a result, there are extensive code requirements for energy consumption in HVAC systems. Heating and cooling systems must meet criteria for health and comfort, including space temperatures, which are prescribed by code: for example, no warmer than 70°F [21 °C] for heating and no cooler than 78°F [25 °C] for cooling. Factors that affect the design space temperature include the room type (living room or gymnasium); building type (hospital or office building); and building location (e.g., New England or the Southwest). Many other aspects of building design contribute to the achievement of the required temperatures. Some of these factors include:

- the insulation of the building envelope, which, if adequate, makes the work of heating and cooling the building easier;
- lighting, which carries its own heat; and
- the use pattern of the building.

Although the code establishes specific design temperatures that the heating system must be able to maintain, heating is a dynamic system. The HVAC system design must allow for some flexibility and response mechanisms to adjust to variable loads. This flexibility is characteristic of a good design.

Energy conservation codes also cover the energy consumed by HVAC equipment. HVAC systems and related equipment are governed by certain established criteria. The required coefficients of performance (C.O.P.) and energy efficiency ratio (E.E.R.) are listed in most energy codes. These should be checked with the manufacturer – not only for code conformance, but also to compare cost savings and energy consumption.

Cooling

Cooling is generally not mandated by code except for certain health care environments, although the C.O.P. and E.E.R. are regulated by some codes. There are, however, certain accepted design standards for cooling systems. For example, a typical summer outdoor maximum design temperature in the northeast is 88°F [31 °C]. The range of humidity allowed is also established, since humidity levels play a major part in the comfort of people with respect to a given room temperature. Indoor design relative humidity for heating is approximately 40%, and for cooling, approximately 60%.

Controls

A diversity in temperatures within a single building is achieved by controls that monitor heating and cooling temperatures and direct equipment to operate to achieve the set point conditions. Controls are used to save energy by setting back the temperature at the end of the day or by shutting off equipment when it is not needed. Criteria for these controls are established by the energy conservation articles included in most codes as well as by building managers and tenant leases.

HVAC controls are typically regulated by energy codes. HVAC control systems must be capable of meeting the individual requirements for each building area and may be complex. Chapter 12 discusses controls in more detail.

Typical Energy Requirements

The material in this chapter is representative of the standards and information sources that govern and assist the HVAC designer. The chart in Figure 2.4 contains typical code requirements for HVAC systems. Each of these should be reviewed and verified for a particular building and locality.

The following items contain typical code requirements for HVAC systems. Each of these items should be reviewed and verified for a particular building locality.

Design Condition Standards for Energy Conservation				
Design Standard		ASHRAE 90A 1989	BOCA Mechanical Code 1993	CABO – MEC 1992
"U" value Btu/(°F/s.f./hr.) [W/m²·K]	Wall	1 – .045 [5.68 – .26] [4]	[2]	.55 – .20 [3.2 – 1.15] [3]
	Roof	.084 – .031 [.48 – .18] [4]		.1 – .06 [.56 – .35] [3]
	Window	1.15 – 0 [6.5] [4]		
	Floor Slab	0 – .18 [.1] [4]		.17 – .1 [.98 – .56] [3]
OTTV Valve Btu/(hr./s.f.) [W/m²·hr.]	Wall			27.8 – 32.2 [287.2 – 332.6] Based on °N Lat.
Design Temp. Indoor			70°F [21 °C] Heating 78°F [25 °C] Cooling	72°F [22 °C] Heating 78°F [25 °C] Cooling
Design R.H. Indoor		30 – 60% R.H. [1]	30 – 60% R.H. [1]	30-60% R.H. [1]
E.E.R. of Equipment			8 – 12.9 depending on type	
Maximum Lighting Energy Usage		1 – 3 watts per s.f. [5]		

All numbers shown in brackets are metric equivalents.

[1] 30% relative humidity maximum for heating. Energy may be used to prevent relative humidity from rising above 60%.
[2] 1993 BOCA Building Code references ASHRAE 90A and CABO-MEC.
[3] U Values are determined in a series of charts based on Heating Degree Days. Values shown are for buildings other than residential, in areas with HDD between 2,000 and 10,000.
[4] Values from ASHRAE 90.1 are based on data from a series of tables that are collectively called the Alternate Component Packages.
[5] 3 watts/s.f. is a good design guide.
Note: See Figure 3.9 for design exterior temperatures in summer and winter for various areas in the U.S.
The designer should note that most Codes require the stamp of a Registered Architect or a Professional Engineer on all drawings for buildings other than one- and two-family dwellings.

Figure 2.4

Drawings:
These are typically prepared by a registered architect or licensed professional engineer in all buildings other than one- or two-family residences.

Design Conditions:

	Summer		Winter	
Temperature:	*Indoor*	*Outdoor*	*Indoor*	*Outdoor*
	75–78°F	88°F (varies)	68°F	10–40°F
Humidity:	50% (varies)			

U Values:
Building Enclosure "U" Values Maximum Acceptable (Btu/hr./s.f./°F)
    Walls   0.08
    Roof   0.06
    Windows 0.53 (double insulated)

Lighting Level:
3 watts per square foot maximum

Ventilation:
2 air changes allowed per hour – 10 percent fresh outdoor air or 20 cfm [9.5 L/s] per person. Natural light and air in residential spaces except bathrooms and kitchens.

Equipment Oversizing:
25 percent maximum

Many codes are based on the component method, which prescribes specific U or R values for each building component. There is increasing acceptance of the system design method, which matches the overall energy consumption as computed from the component design method or as determined on an annual energy consumption in Btu/s.f./yr. [kWh/m^2/yr.]. This method thus allows for increases of R values in some areas (say roofs) to allow for decreases in R values in other areas (say larger windows).

In addition to the standards shown in Figure 2.4, most energy conservation codes restrict the amount of equipment oversizing to 25% maximum. ASHRAE 90.1 recommends a 10% maximum oversizing. Equipment that is too large will not run efficiently, will cycle more often, and will waste energy.

The overall objective of energy codes is to limit energy use per square foot. Currently the goal is approximately 55,000 Btu/s.f./yr. [17.8 kWh/m^2/yr.]. It is expected that overall energy consumption performance will replace much of the code as consensus develops on the actual energy consumption criteria. As with FSES, and performance-based codes in general, it is expected that overall goals will be simplified as creative solutions are implemented.

Professional Organizations

The background research for energy conservation standards is rarely performed by each individual municipality. Instead, it is done by several organizations, one of which is the American Society of Heating, Refrigeration, and Air Conditioning Engineers (ASHRAE); the information gathered is incorporated into the building code. What the local code does not cover is referenced by the standards of various professional organizations in the construction field. These references are usually listed in the back of each code and should be consulted for complete information. For HVAC, the standards are based on information supplied by ASHRAE and SMACNA (Sheet Metal and Air Conditioning Contractors National Association), among many others. These organizations form a vast information network that covers accepted engineering and industry practice and goes beyond the local code's mandate of protecting the public's health, safety, and welfare. To this extent, the criteria established are not required by law, but represent practitioners' wisdom.

The material controlled by such standards covers several different categories. It is produced by professional societies (such as ASHRAE), and is written at a certain level with appropriate detail, as represents the organization. ASME, the American Society of Mechanical Engineers, is similar in the way that they present their material. The "Boiler and Pressure Vessel Code," researched and established by ASME, is an important reference tool for HVAC designers.

Contractors also have organizations. SMACNA is a good example. Contractors' associations have a different emphasis from engineering associations. The key issues tend to be accepted industry practice for equipment and installation, rather than concern with research and engineering formulas and derivations for calculations. Each type of material has its place in the information network and is useful in its own way.

Materials Testing

All construction materials, including those for HVAC (particularly pipe and metal for ductwork), are covered in the publications of several national organizations. These include the American Society of Testing and Materials (ASTM), Underwriters Laboratories (UL), and the American National Standards Institute (ANSI). Each of these organizations also has a unique place in the information network. ASTM represents an engineering society whose main concern is to establish the characteristics of engineering materials. ASTM also conducts the testing required to establish the performance of material under different temperatures and stresses.

Underwriters Laboratories (UL) is a private testing company that tests equipment under different circumstances for various criteria. The UL label has become nationally known as an indication that a piece of equipment or material meets the high standards UL requires. The testing criteria that UL uses varies, but it is usually from accepted industry sources such as ASHRAE or ASTM. UL performs the tests on a given model of the equipment as submitted by the manufacturer. It then produces manuals listing accepted equipment. Each piece of equipment that has been built to the performance criteria accepted by UL is allowed to carry a UL label.

The American National Standards Institute (ANSI) is a national organization that brings together the variety of organizations evaluating materials and equipment in the United States. ANSI does not conduct its own tests or establish material quality. Instead, it incorporates those standards produced by other organizations. By coordinating this information, ANSI attempts to establish one system for evaluating materials in equipment.

Fire Safety Standards

An important element in both the governing codes and good engineering practice is the code and research produced by the National Fire Protection Association (NFPA). Since HVAC involves combustion, NFPA requirements are extensive. NFPA research has established standards for flammable liquids and the installation of oil-burning, gas, and air conditioning and ventilation equipment. NFPA has also established the National Electric Code and National Fuel Gas Code.

Figure 2.5 lists the regulatory agency codes, regulations, and standards generally utilized.

Problems

2.1 For your current residence, determine the governing building code. Is the governing authority local, state, or federal? Determine which national codes apply and which edition of each governs.

2.2 What are the ventilation requirements for the following (use both ASHRAE and BOCA and show differences, if any):
   a. offices
   b. theaters
   c. school laboratories

2.3 What type of work can a contractor do without a building permit? What type of work can a contractor do without having a set of plans stamped by a registered architect or licensed professional engineer?

2.4 An office building has a space of 10,000 square feet [3048 m] with a 9 foot [2.7 m] ceiling. There are strip windows on the facade 3'-0" [.9 m] high, and there is 250 linear feet [76.2 m] of window adjoining the space. Analyze the space for light and ventilation utilizing the applicable codes in your community. What must be done to meet these requirements?

2.5 An eight-story high-rise office building with two stair towers, two elevators, and one mechanical shaft is to be protected with a smoke evacuation system. In this case the main air handler on the roof is not to be used for smoke evacuation. Briefly review Chapter 14 and sketch an independent smoke evacuation system for the building, showing how it will operate in an alarm. Be sure to show how a typical floor as well as all the shafts in the building will be protected.

2.6 For your community, determine the code requirements for U values for the roof, walls and windows of a typical commercial building.

Reference Standards

American National Standards Institute, Inc. (ANSI)
1430 Broadway
New York, New York 10018

Title	Standard Reference Number
Ductile-Iron Pipe, Centrifugally Cast, in Metal Molds or Sand-Lined Molds for Gas	A21.52–82
Cast Iron Pipe Flanges and Flanged Fittings, Class 25, 125, 250 and 800	B16.1–75
Pipe Flanges and Flanged Fittings, Steel Nickel Alloy and other Special Alloys	B16.5–81
Factory-Made Wrought Steel Buttwelding Fittings 1981 Supplement B16.9a	B16.9–78
Forged Steel Fittings, Socket-Welding and Threaded	B16.11–80
Cast Copper Alloy Solder Joint Pressure Fittings	B16.18–84
Wrought Copper and Copper Alloy Solder Joint Pressure Fittings	B16.22–80
Cast Copper Alloy Solder Joint Drainage Fittings – DWV	B16.23–84
Bronze Pipe Flanges and Flanged Fittings, Class 150 and 300	B16.24–79
Cast Copper Alloy Fittings for Flared Copper Tubes	B16.26–83
Wrought Steel Buttwelding Short Radius Elbows and Returns	B16.28–78
Wrought Copper and Wrought Copper Alloy Solder Joint Drainage Fittings – DWV	B16.29–80

Air Conditioning and Refrigeration Institute (ARI)
1501 Wilson Blvd.
Suite 600
Arlington, Virginia 22209

Title	Standard Reference Number
Unitary Air-Conditioning Equipment	210–81

American Society of Heating, Refrigerating and Air Conditioning Engineers, Inc. (ASHRAE)
1791 Tullie Circle, N.E.
Atlanta, Georgia 30329

Title	Standard Reference Number
Safety Code for Mechanical Refrigeration	15–78
Number Designation of Refrigerants	34–78
Thermal Environmental Conditions for Human Occupancy	55–81
Energy Conservation in New Building Design	90A–80
Handbook, Fundamentals Volume	ASHRAE–85

American Society of Mechanical Engineers (ASME)
United Engineering Center
345 East 47th Street
New York, New York 10017

Title	Standard Reference Number
Boiler and Unfired Pressure Vessel Code, Section VIII, Division 1 & 2 Summer 83, Winter 83, Summer 84, Winter 84, Summer 85, Winter 85 and Summer 86 Addenda	ASME–83
Pipe Threads, General Purpose (inch)	B1.20.1–83
Malleable-Iron Threaded Fittings, Classes 150 and 300	B16.3–85
Cast Bronze Threaded Fittings, Class 125 and 250	B16.15–85

Figure 2.5

Reference Standards (continued)

American Society for Testing and Materials (ASTM)
1916 Race Street
Philadelphia, Pennsylvania 19103

Title	Standard Reference Number
Pipe, Steel, Black and Hot Dipped Zinc Coated, Welded and Seamless	A53–86
Seamless Carbon Steel Pipe for High Temperature Service	A106–86
Pipe, Steel, Black and Hot-Dipped Zinc-Coated (Galvanized) Welded and Seamless, for Ordinary Uses	A120–84
Gray Iron Castings for Valves, Flanges and Pipe Fittings	A126–84
Copper Brazed Steel Tubing	A254–84
Gray Iron and Ductile Iron Pressure Pipe	A377–84
Piping Fittings of Wrought Carbon Steel and Alloy Steel for Low Temperature Service	A420–85A
Steel Sheet, Zinc-Coated (Galvanized) by the Hot-Dip Process — General Requirements	A525–86
Electric-Resistance-Welded Coiled Steel Tubing for Gas and Fuel Oil Lines	A539–85
Solder Metal	B32–83
Standard Sizes of Seamless Copper Pipe	B42–85
Standard Sizes of Seamless Red Brass Pipe	B43–86
Seamless Copper Tube	B75–86
Seamless Copper Water Tube	B88–86
Seamless Brass Tube	B135–86
Aluminum-Alloy Drain Seamless Tubes	B210–82A
Aluminum-Alloy Seamless Pipe and Seamless Extruded Tube	B241–83A
Wrought Seamless Copper and Copper-Alloy Tube	B251–86
Seamless Copper Tube for Air Conditioning and Refrigeration Field Service	B280–83
Threadless Copper Pipe	B302–85
Refractories for Incinerators and Boilers	C64–77
Ground Fire Clay as a Refractory Mortar for Laying Up Fireclay Brick	C105–81
Classification of Insulating Fire Brick	C155–84
Clay Flue Linings	C315–83
Hot Surface Performance of High Temperature Thermal Insulation — Test for	C411–82
Mineral Fiber Block and Board Thermal Insulation	C612–83
Flash Point by Tag Closed Tester — Test for	D56–82
Flash Point by Pensky — Martens Closed Tester — Method of Test	D93–85
Acrylonitrile-Butadiene-Styrene (ABS) Plastic Pipe Schedules 40 and 80	D1527–82
Poly (Vinyl Chloride) (PVC) Plastic Pipe, Schedules 40, 80 and 120	D1785–83
Solvent Cement for Acrylonitrile-Butadiene-Styrene (ABS) Plastic Pipe and Fittings	D2235–81
Poly (Vinyl Chloride) (PVC) Plastic Pipe (SDR-PR)	D2241–86
Acrylonitrile-Butadiene-Styrene (ABS) Plastic Pipe, (SDR-PR)	D2282–82
Poly (Vinyl Chloride) (PVC) Plastic Pipe Fittings, Schedule 40	D2466–78
Socket-Type Poly (Vinyl Chloride) (PVC) Plastic Pipe Fittings, Schedule 80	D2467–76A
Acrylonitrile-Butadiene-Styrene (ABS) Plastic Pipe Fittings, Schedule 40	D2468–80
Socket-Type Acrylonitrile-Butadiene-Styrene (ABS) Plastic Pipe Fittings, Schedule 80	D2469–76
Thermoplastic Gas Pressure Pipe, Tubing and Fittings	D2513–85A
Reinforced Epoxy Resin Gas Pressure Pipe and Fittings	D2517–81
Solvent Cements for Poly (Vinyl Chloride) (PVC) Plastic Pipe and Fittings	D2564–84
Chlorinated Poly (Vinyl Chloride) (CPVC) Plastic Hot- and Cold-Water Distribution Systems	D2846–82

Figure 2.5 (continued)

Reference Standards (continued)

Title	Standard Reference Number
Polybutylene (PB) Plastic Hot-Water Distribution Systems	D3309-85B
Surface Burning Characteristics of Building Materials	E84-84
Behavior of Materials in a Vertical Tube Furnace @ 750° C. — Standard Test Method	E136-82
Safe Handling of Solvent Cements used for Joining Thermoplastic Pipe and Fittings — Recommended Practice	F402-80
Socket-Type Chlorinated Poly (Vinyl Chloride) (CPVC) Plastic Pipe Fittings, Schedule 40	F438-82
Socket-Type Chlorinated Poly (Vinyl Chloride) (CPVC) Plastic Pipe Fittings, Schedule 80	F439-82
Chlorinated Poly (Vinyl Chloride) (CPVC) Plastic Pipe, Schedules 40 and 80	F441-86
Chlorinated Poly (Vinyl Chloride) (CPVC) Plastic Pipe (SDR-PR)	F442-85
Solvent Cements for Chlorinated Poly (Vinyl Chloride) (CPVC) Plastic Pipe and Fittings	F493-85
Field Measurement of Soil Resistivity Using the Wenner Four-Electrode Method — Field Measurement	G57-84

American Welding Society (AWS)
P.O. Box 351040
Miami, Florida 33135

Title	Standard Reference Number
Brazing Filler Metal	A5.8-81

Building Officials and Code Administrators International (BOCA)
4051 West Flossmoor Road
Country Club Hills, Illinois 60477-5795

Title	Standard Reference Number
National Mechanical Code	NMC-87
National Building Code	NBC-87
National Fire Prevention Code	NFPC-87
National Plumbing Code	NPC-87

United States Department of Energy (DOE)
Washington, D.C. 20545

Title	Standard Reference Number
Furnaces and Boilers — U.S. Department of Energy Test Procedures	Federal Register Volume 49, FR12148, No. 61-84

Federal Specifications General Service Administration (FS)
7th and D Streets
Specifications Section
Room 6039
Washington, D.C. 20407

Title	Standard Reference Number
Pipe, Bends, Traps, Caps and Plugs (for Industrial Pressure and Soil and Waste Applications)	WW-P-325B-76

Figure 2.5 (continued)

Reference Standards (continued)		
International Institute of Ammonia Refrigeration (IIAR) 111 E. Wacker Drive Chicago, IL 60601		
Title		**Standard Reference Number**
Equipment, Design and Installation of Ammonia Mechanical Refrigeration Systems		2-84
Manufacturers Standardization Society of the Valve and Fittings Industry, Inc. (MSS) 127 Park Street, N.E. Vienna, Virginia 22180		
Title		**Standard Reference Number**
Pipe Hangers and Supports — Selection and Application		SP-69-83
National Association of Corrosion Engineers (NACE) P.O. Box 218340 Houston, TX 77218		
Title		**Standard Reference Number**
Control of External Corrosion of Underground or Submerged Metallic Piping Systems		RP-01-69-83
National Fire Protection Association (NFPA) Batterymarch Park Quincy, Massachusetts 02269		
Title		**Standard Reference Number**
Flammable and Combustible Liquids Code		30-84
Automotive and Marine Service Station Code		30A-84
Storage and Handling of Cellulose Nitrate Motion Picture Film		40-82
Storage and Handling of Liquefied Petroleum Gases		58-83
Explosion Venting Guide		68-78
National Electrical Code		70-84
Repair Garages — Standard		88B-85
Sheet Metal and Air Conditioning Contractors National Association, Inc. (SMACNA) 8224 Old Courthouse Road Vienna, Virginia 22180		
Title		**Standard Reference Number**
Fibrous Glass Ducts — Construction Standards		SMACNA-79
Metallic Ducts — HVAC Duct Construction Standards		SMACNA-85

Figure 2.5 (continued)

Reference Standards (continued)

Underwriters Laboratories, Inc. (UL)
333 Pfingsten Road
Northbrook, Illinois 60062

Title	Standard Reference Number
Flammability of Plastic Materials for Parts in Devices and Appliances	94-85
Chimneys, Factory-Built, Residential Type and Building Heating Appliance	103-86
Factory-Built Fireplaces	127-86
Pressure Regulating Valves for LP Gas	144-85
Factory Made Air Ducts and Connectors	181-81
Gas Vents	441-86
Dampers, Fire and Ceiling Dampers — 1986 Supplement	555-79
Type L Low-Temperature Venting Systems	641-86
Grease Extractors for Exhaust Ducts	710-83
Fireplace Stoves	737-86
Fire and Smoke Characteristics of Electrical and Optical-Fiber Cables Used in Air-Handling Spaces — Test Method	910-85
Medium Heat Appliance, Factory Built Chimneys	959-86
Room Heaters, Solid Fuel Type	1482-86

Office of Technical Information Bureau of Mines (USBM)
Department of the Interior
2401 E. Street, N.W.
Washington, D.C. 20241

Title	Standard Reference Number
Ringelmann Smoke Chart	USBM Circular No. 8333 May, 1967

Adapted from the Building Officials and Code Administrators International (BOCA) *National Mechanical Code 1987.*

Figure 2.5 (continued)

Chapter Three

HEATING AND COOLING LOADS

The size of heating and cooling equipment is determined by the "load" it must carry. For a boiler, the load is the amount of heat that must be pushed into the building per hour to keep it warm. For cooling, the load is the amount of heat per hour that must be taken away from the building to keep it cool. Methods for determining both heating and cooling loads are discussed in this chapter.

Heating Loads

Heat is measured in British thermal units (Btu), calories (c), or joules (J). One Btu equals the amount of energy needed to heat one pound of water 1°F (see Figure 3.1). One calorie equals the amount of energy needed to heat one gram of water 1 °C. One calorie equals 4.2 joules. The joule is the more common measure of heat content.

Heat rates are expressed in the following units:

> British thermal unit per hour – Btu/hr.
> Joules per second – J/s
> (One J/s equals 1 watt – W)

Because smaller units are easier to work with, 1,000 Btu/hr. is often written as 1 MBH. 1000 J/s is written as 1 kW.

Water

The properties of water are basically constant in the 32 – 212°F [0 – 100 °C]* range. Water is easy to work with because it is a liquid with nearly constant physical characteristics over the range of temperatures normally used in heating, and is readily available and chemically stable. For these reasons, water is the most commonly used heating fluid.

Heating and Sensible Heat

Heating systems are often designed without humidification; all of the heat energy is used to warm the air in a room. When heating systems are designed without humidification, all of the added heat is termed *sensible heat*, because every Btu added to the air is devoted only to raising the air temperature. This heat can be felt or "sensed." When humidification is incorporated into heating systems, the heat load (number of Btu's necessary) increases by approximately 30 percent. (Humidification and latent heat are discussed in more detail in the cooling section of this chapter.)

*Throughout the book, SI metric equivalents are provided for most imperial units. The SI units appear in brackets [ ].

Heat Loss

The heating load for a building is determined by calculating the amount of heat lost from the building to the outside during the winter design condition. (The common winter design condition is the cold winter night in a particular climate that is surpassed 97-1/2% of the time.)

There are three ways in which heat is lost from a building: *conduction, convection,* and *radiation*. The total MBH of these three types of heat loss equals the heat load, or the amount of heat that must be put into the building by the heating system to offset these losses and maintain the proposed comfort zone.

Conduction transfers heat in a chain-like manner, from higher-temperature to lower-temperature molecules. As one molecule heats up, it transfers heat to the molecule next to it. With this type of heat transference, the molecules are stationary and the heat moves, or is conducted, through them. For example, if you hold one end of a steel poker and use the other end to stir a fire, your hand gets warmer as the heat from the fire moves, or is conducted, along the shaft to your hand. Similarly, a building's heat is conducted to the outdoors through the solid surfaces of walls, roofs, floor slabs, glass doors, and windows. The rate of heat depends on the type of material and the temperature difference between adjacent materials. Good conductors, such as metals, transfer more heat than good insulators.

Convection transfers heat as warm molecules actually move from one place to another. If a current of air was passed over the heated poker noted above, some of the heat of the rod would be transferred to the passing air, or convected. In buildings, heat is convected from the interior to the

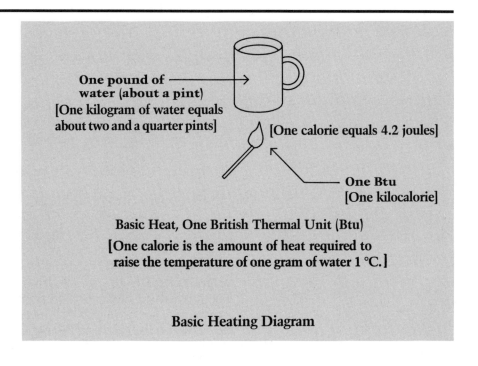

Figure 3.1

outdoors by air that leaks, or infiltrates, through cracks around windows and doors, and by the exhaust and ventilation air that moves between the interior and the exterior of the building. Most HVAC distribution systems work by convection. The hot water that moves heat from the boiler through the pipes to the fan coil units and the warm air furnace that distributes heat to the room via ductwork are moving heat around the building by convection, either by force or by gravity.

Radiation transfers heat by electromagnetic waves. Sunlight is the most common, and most spectacular, example. Heat from the sun reaches the earth by radiating through 93 million miles of space. In the steel poker example, heat is radiated when it gets "red hot," at which time the heat can be felt several feet away. Heat is radiated onto buildings by the sun during the day. Solar radiation is primarily considered as a factor in calculating the cooling load, where it must be overcome. It is not a significant factor in calculating the heating load.

Figure 3.2 illustrates the basic processes of conduction, convection, and radiation. The heat losses that commonly occur in buildings are illustrated in Figure 3.3. Each type of heat loss is described in the following sections.

There are several methods used to calculate heating and cooling loads. They are comparable. This book uses the ASHRAE method.

Conduction Heat Losses

Conduction heat losses (symbolized by "H_c" in Figure 3.3) are calculated by adding together the conduction heat losses from each separate building material. The formula for computing conduction losses for each material is shown below. Each element is then discussed individually.

$H_c = UA(\Delta T)$

H_c = Conduction heat losses from building materials (Btu/hr.)

U = Overall heat transfer coefficient

Btu/(hr. × s.f. × °F [watts/m^2 . K] (see Figures 3.5-3.7)

A = Surface area (s.f.) [m^2]

ΔT = Temperature difference (°F) [K]

Overall Heat Transfer Coefficient

The value of the overall heat transfer coefficient ("U" in the formula for computing conduction heat loss shown above) is based on the type of materials comprising the building enclosure. For example, an uninsulated wood-frame wall has a U value of 0.23 Btu/s.f. hr. °F [1.29 W/m^2.K]. This means that 0.23 Btu is conducted through each square foot of wall surface each hour for each 1°F temperature difference between the inside and outside wall surface temperatures. If four inches [100 mm] of insulation are added to this wall, the U value becomes 0.07 Btu/ s.f. hr. °F [0.4 W/m^2.K]. The value is lowered because less heat is conducted through a better insulated wall. Insulation, which is relatively inexpensive, makes the wall surface more than three times more effective. This is illustrated in Figure 3.4.

The U value is equal to the reciprocal of the sum of the resistances of the components of the building envelope. Computation of the U value is illustrated in Figure 3.5. This formula is shown below.

$$U = \frac{1}{\text{Total resistance}} = \frac{1}{R_T}$$

The **total resistance**, or R_t, equals the sum of the resistances of each individual material making up the building envelope (i.e., $R_t = R_1 + R_2 + R_3 + ...$). The resistance of a single material normally varies with its thickness; therefore, in certain cases, the resistance is computed from the material conductivity, or k.

k equals the number of Btu's [joules] conducted per square foot [square meter] per hour [second] for each one degree Fahrenheit [Kelvin]

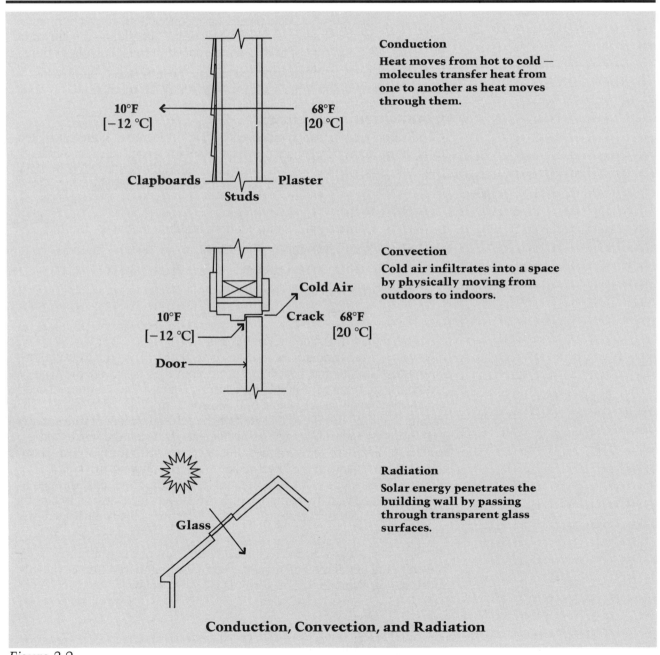

Figure 3.2

temperature difference for a one-inch-thick [1 cm] material (i.e., Btu/hr. °F or W/m.s per cm). For a material for which the thickness and conductivity is known, the conductance equals the conductivity divided by the thickness, and the resistance equals the reciprocal of the conductance. These formulas are shown below.

$$C = k/t$$
$$R = 1/C$$

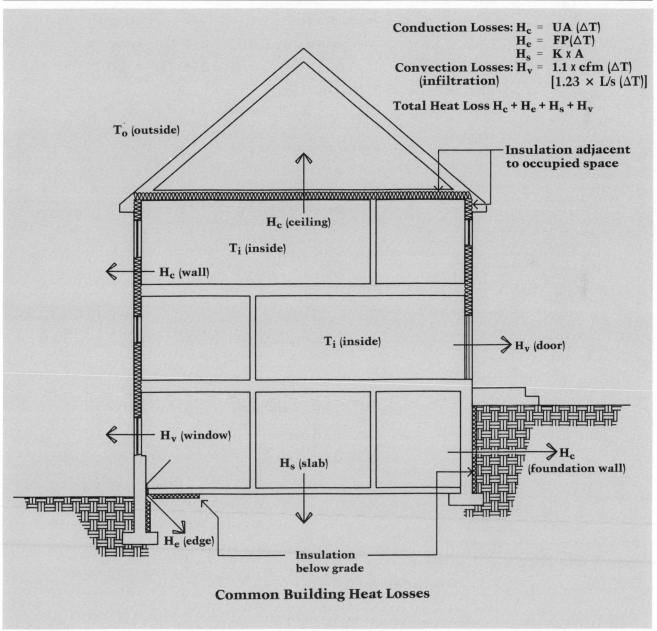

Figure 3.3

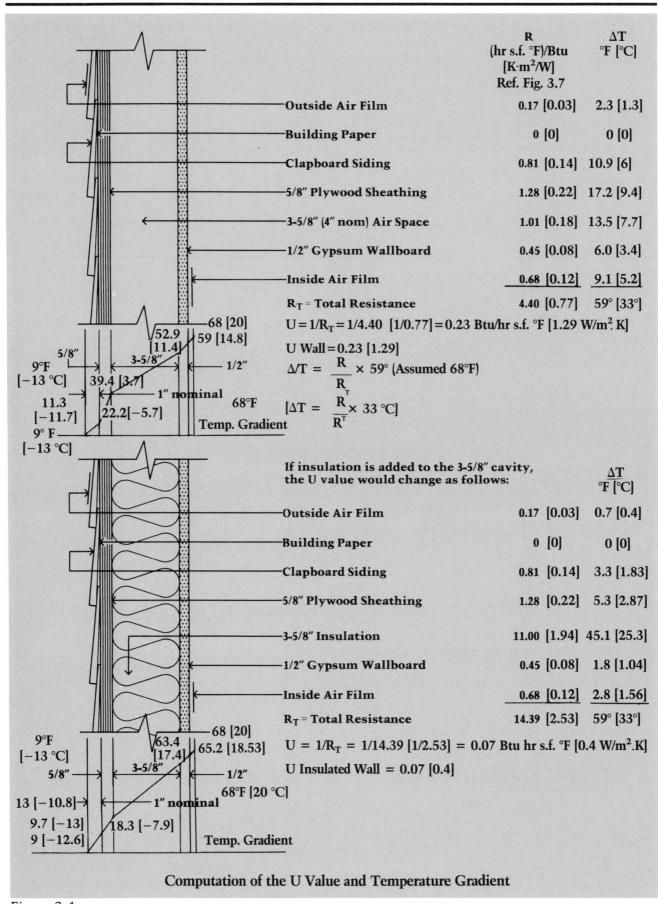

Figure 3.4

Conductivity, Conductance, and the U Value

k = Conductivity   Btu/hr sf °F/inch [W/m.K]

t = 1" [1 m] thick

k = Btu per square foot per hour for each 1°F temperature difference for a 1" thick material
[Watts per square meter for each 1 K temperature difference for a 1 m thick material]

k is listed for most materials (see Figure 3.7)

C = Conductance   Btu/hr sf °F [W/m².K]

C = Btu per square foot per hour for each 1°F temperature difference for the material thickness indicated.
[Watts per square meter for each 1 K temperature difference for the material thickness indicated.]

C = k/t

C is listed for common materials (see Figure 3.7)

t = material thickness

R = Resistance   hr sf °F/Btu [m².K/W]
R = 1/C or t/k for each material

Air resistances hr sf °F/Btu [m².K/W]	Winter	Summer
f_i = Inside air film resistance =	0.68 [0.12]	0.68 [0.12]
f_o = Outside air film resistance =	0.17 [0.03]	0.25 [0.04]
a = Air space (1" to 4" nominal) resistance =	0.97 [0.17]	0.86 [0.15]

R_T = Total Resistance

$$R_T = f_i + f_o + R_1 + R_2 + R_3 + \ldots + a \text{ (if applicable)}$$

U = Overall heat transfer coefficient Btu hr sf °F [W/m².K] = $1/R_T$

Figure 3.5

Thus, for each portion of the building envelope, the U value must be determined. Figure 3.6 lists U values for common building systems and Figure 3.7 lists conductances, conductivities, and resistances (C, k, and R) for common building materials. (Figure 3.7a is in SI metric units.)

Once the overall heat transfer coefficient, U, has been determined for each building envelope system, the total conduction loss is determined by computing the surface area of each material and the change in temperature across each material. Each of these processes is described below.

Surface Area
The surface area, A, for each building material is measured from within the occupied space. Wall areas are measured from the inside finish floor to the underside of the finish ceiling, excluding window and door areas, which are computed separately. Slanted ceilings, such as cathedral ceilings, are measured along the slant. Window frames are generally considered as window or glass area.

Temperature Difference (ΔT)
The temperature difference, ΔT, is the difference in temperature between the inside and the outside of the wall surface. Figure 3.8 lists indoor design temperatures (T_i) for some typical indoor conditions and Figure 3.9 lists outdoor design temperatures (T_o) for several locations. In winter, this is the difference between the indoor design temperature, 68 – 72°F [20 – 22 °C], and the design outdoor temperature, which varies with location. For example, 9°F [– 13 °C] is the outdoor design temperature for Boston; thus the temperature difference is calculated as shown below.

$\Delta T = T_i - T_o$, in winter; therefore,

$\Delta T = 68 - 9$, or 59°F [20 – (–13), or 33 °C]

In summer,

$\Delta T = T_o - T_i$

The temperature drop through any portion of a wall is proportional to the resistance of that portion to the entire wall. Figure 3.4 illustrates the computation of temperature gradient for two walls.

In some cases, ΔT is measured from another enclosed or unheated space that has a temperature that differs from the outdoor design temperature (T_o). Examples include party walls, attics, basements, attached garages, slabs and walls below grade, and rooms deliberately kept at a special temperature, such as gymnasiums, operating rooms, freezers, or computer rooms. The temperature of the attached space is used instead of the T_o, and in some cases must be estimated. Where the enclosing wall is not an exterior wall, ΔT is equal to the indoor temperature minus the temperature of the adjoining surfaces, or

$T_i - T_{adj}$

Figure 3.10 is a guide to establishing ΔT for adjoining surfaces.

Perimeter Losses
When an exterior wall meets the grade, there are additional conductive losses at the edge of the slab. (These are noted as H_e in Figure 3.3.) Figure 3.11 illustrates perimeter edge slab conditions. The formula for calculating perimeter edge losses is shown below.

$H_e = FP(\Delta T)$
H_e = Heat loss through edge (Btu/hr.) [watts]
F = Edge factor (Btu/hr. ft. °F) [watts/m².k/m]

U Values for Common Building Materials

TRANSMISSION COEFFICIENT U—LIGHT CONSTRUCTION, INDUSTRIAL WALLS*†
FOR SUMMER AND WINTER
Btu/(hr) (sq ft) (deg F temp diff)

All numbers in parentheses indicate weight per sq ft. Total weight per sq ft is sum of wall and finishes.

					INTERIOR FINISH			
		WEIGHT (lb per sq ft)	None	Flat Iron (1)	Insulating Board		Wood	
EXTERIOR FINISH	SHEATHING				½″ (2)	25⁄32″ (3)	¾″ (2)	
⅜″ Corrugated Transite	None	(1)	1.16	.55	.32	.26	.36	
	½″ Ins. Board	(2)	.34	.26	.19	.17	.21	
	25⁄32″ Ins. Board	(2)	.27	.21	.17	.15	.18	
24 Gauge Corrugated Iron	None	(1)	1.40	.60	.33	.27	.38	
	½″ Ins. Board	(2)	.36	.27	.20	.17	.21	
	25⁄32″ Ins. Board	(2)	.28	.22	.17	.15	.18	
	¾″ Wood	(3)	.46	.33	.22	.19	.24	
¾″ Wood Siding	None	(2)	.58	.37	.25	.21	.27	

1958 ASHRAE Guide

Equations: Heat Gain, Btu/hr = (Area, sq ft) × (U value) × (equivalent temp diff, Table 19).
Heat Loss, Btu/hr = (Area, sq ft) × (U value) × (outdoor temp − inside temp).

*For addition of air spaces and insulation to walls, refer to Table 31, page 75.

†Values apply when sealed with calking compound between sheets, and at ground and roof lines. When sheets are not sealed, increase U factors by 10%. These values may be used for roofs, heat flow up-winter; for heat flow down-summer, multiply above factors by 0.8.

TRANSMISSION COEFFICIENT U—LIGHTWEIGHT, PREFABRICATED CURTAIN TYPE WALLS*
FOR SUMMER AND WINTER
Btu/(hr) (sq ft) (deg F temp diff)

All numbers in parentheses indicate weight per sq ft. Total weight per sq ft is sum of wall and finishes.

INSULATING CORE MATERIAL	DENSITY† (lb/cu ft)	METAL FACING (3)				METAL FACING WITH ¼″ AIR SPACE (3)			
		Core Thickness (in.)				Core Thickness (in.)			
		1	2	3	4	1	2	3	4
Glass, Wood, Cotton Fibers	3	.21	.12	.08	.06	.19	.11	.08	.06
Paper Honeycomb	5	.39	.23	.17	.13	.32	.20	.15	.12
Paper Honeycomb with Perlite Fill, Foamglas	9	.29	.17	.12	.09	.25	.15	.11	.09
Fiberboard	15	.36	.21	.15	.12	.29	.19	.14	.11
Wood Shredded (Cemented in Preformed Slabs)	22	.31	.18	.13	.10	.25	.16	.12	.09
Expanded Vermiculite	7	.34	.20	.14	.11	.28	.18	.13	.10
Vermiculite or Perlite Concrete	20	.44	.27	.19	.15	.35	.23	.18	.14
	30	.51	.32	.24	.19	.39	.27	.21	.17
	40	.58	.38	.29	.23	.43	.31	.25	.20
	60	.69	.49	.38	.31	.49	.38	.31	.26

1958 ASHRAE Guide

Equations: Heat Gain, Btu/hr = (Area, sq ft) × (U value) × (equivalent temp diff, Table 19).
Heat Loss, Btu/hr = (Area, sq ft) × (U value) × (outdoor temp − inside temp).

*For addition of insulation and air spaces to walls, refer to Table 31, page 75.

†Total weight per sq ft = $\dfrac{\text{core density} \times \text{core thickness}}{12}$ + 3 lb/sq ft

Figure 3.6

U Values for Common Building Materials (continued)

TRANSMISSION COEFFICIENT U—FRAME WALLS AND PARTITIONS*

FOR SUMMER AND WINTER

Btu/(hr) (sq ft) (deg F temp diff)

All numbers in parentheses indicate weight per sq ft. Total weight per sq ft is sum of component materials.

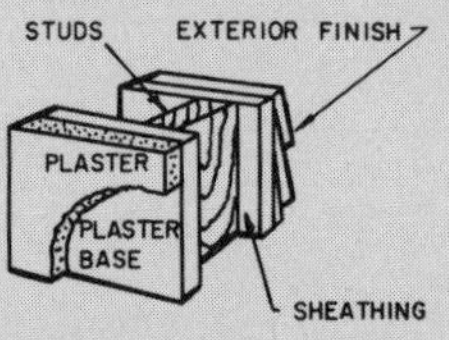

		INTERIOR FINISH							
		None	¾" Wood Panel (2)	⅜" Gypsum Board (Plaster Board) (2)	Metal Lath Plastered		⅜" Gypsum or Wood Lath Plastered		Insulating Board Plain or Plastered	
					¾" Sand Plaster (7)	¾" Lt Wt Plaster (3)	½" Sand Plaster (7)	½" Lt Wt Plaster (2)	½" Board (2)	1" Board (4)
EXTERIOR FINISH	**SHEATHING**									
1" Stucco (10) OR Asbestos Cement Siding (1) OR Asphalt Roll Siding (2)	None, Building Paper	.91	.33	.42	.45	.39	.40	.37	.29	.20
	5/16" Plywood (1) or ½" Gyp (2)	.68	.30	.37	.40	.35	.36	.33	.26	.19
	25/32" Wood & Bldg Paper (2)	.48	.25	.30	.31	.28	.29	.27	.22	.17
	½" Insulating Board (2)	.42	.23	.27	.29	.26	.27	.25	.21	.16
	25/32" Insulating Board (3)	.32	.20	.23	.24	.22	.22	.21	.18	.14
4" Face Brick Veneer (43) OR ⅜" Plywood (1) OR Asphalt Siding (2)	None, Building Paper	.73	.30	.37	.40	.35	.36	.33	.26	.19
	5/16" Plywood (1) or ½" Gyp (2)	.57	.28	.33	.36	.32	.32	.30	.24	.18
	25/32" Wood & Bldg Paper (2)	.42	.23	.27	.29	.26	.27	.25	.21	.16
	½" Insulating Board (2)	.38	.22	.25	.27	.25	.25	.24	.20	.15
	25/32" Insulating Board (3)	.30	.19	.21	.22	.21	.21	.20	.17	.14
Wood Siding (3) OR Wood Shingles (2) OR ¾" Wood Panels (3)	None, Building Paper	.57	.27	.33	.35	.31	.32	.30	.24	.18
	5/16" Plywood (1) or ½" Gyp (2)	.48	.25	.30	.31	.28	.29	.27	.22	.17
	25/32" Wood & Bldg Paper	.36	.22	.25	.26	.24	.24	.23	.19	.15
	½" Insulating Board (2)	.33	.20	.23	.24	.22	.23	.22	.18	.14
	25/32" Insulating Board (3)	.27	.18	.20	.21	.19	.19	.19	.16	.13
Wood Shingles Over 5/16" Insul Backer Board (3) OR Asphalt Insulated Siding (4)	None, Building Paper	.43	.24	.28	.29	.27	.27	.25	.21	.16
	5/16" Plywood (1) or ½" Gyp (2)	.38	.22	.25	.27	.24	.25	.23	.19	.15
	25/32" Wood & Bldg Paper	.30	.19	.22	.23	.21	.21	.20	.17	.14
	½" Insulating Board (2)	.28	.18	.20	.21	.20	.20	.19	.16	.13
	25/32" Insulating Board (3)	.23	.16	.18	.18	.17	.18	.17	.15	.12
Single Partition (Finish on one side only)			.43	.60	.67	.55	.57	.50	.36	.23
Double Partition (Finish on both sides)			.24	.34	.39	.31	.32	.28	.19	.12

1958 ASHRAE Guide

Equations: Walls—Heat Gain, Btu/hr = (Area, sq ft) × (U value) × (equivalent temp diff, *Table 19*).
—Heat Loss, Btu/hr = (Area, sq ft) × (U value) × (outdoor temp—inside temp).
Partitions, unconditioned space adjacent—Heat Gain or Loss, Btu/hr = (Area sq ft) × (U value) × (outdoor temp—inside temp—5 F).
Partitions, kitchen or boiler room adjacent—Heat Gain, Btu/hr = (Area sq ft) × (U value)
× (actual temp diff or outdoor temp—inside temp + 15 F to 25 F).

*For addition of insulation and air spaces to partitions, refer to *Table 31*, page 75.

Figure 3.6 (continued)

U Values for Common Building Materials (continued)

TRANSMISSION COEFFICIENT U—FLAT ROOFS COVERED WITH BUILT-UP ROOFING*
FOR HEAT FLOW DOWN—SUMMER. FOR HEAT FLOW UP—WINTER (See Equation at Bottom of Page).

Btu/(hr) (sq ft) (deg F temp diff)

All numbers in parentheses indicate weight per sq ft. Total weight per sq ft is sum of roof, finish and insulation.

TYPE OF DECK	THICKNESS OF DECK (inches) and WEIGHT (lb per sq ft)	CEILING†	INSULATION ON TOP OF DECK, INCHES						
			No Insulation	½ (1)	1 (1)	1½ (2)	2 (3)	2½ (3)	3 (4)
Flat Metal	1 (5)	None or Plaster (6)	.67	.35	.23	.18	.15	.12	.10
		Suspended Plaster (5)	.32	.22	.17	.14	.12	.10	.09
		Suspended Acou Tile (2)	.23	.18	.14	.12	.11	.09	.08
Preformed Slabs—Wood Fiber and Cement Binder	2 (4)	None or Plaster (6)	.20	.16	.13	.11	.10	.09	.08
		Suspended Plaster (5)	.15	.12	.11	.09	.08	.08	.07
		Suspended Acou Tile (2)	.13	.10	.09	.08	.08	.07	.06
	3 (7)	None or Plaster (6)	.14	.11	.10	.09	.08	.08	.07
		Suspended Plaster (5)	.12	.10	.09	.07	.07	.06	.05
		Suspended Acou Tile (2)	.10	.09	.08	.07	.07	.06	.05
Concrete (Sand & Gravel Agg)	4, 6, 8 (47),(70),(93)	None or Plaster (6)	.51	.30	.21	.16	.14	.12	.10
		Suspended Plaster (5)	.28	.20	.16	.13	.12	.10	.09
		Suspended Acou Tile (2)	.21	.16	.13	.11	.10	.09	.08
(Lt Wt Agg on Gypsum Board)	2 (9)	None or Plaster (6)	.27	.20	.15	.13	.11	.10	.08
		Suspended Plaster (5)	.18	.14	.12	.10	.09	.09	.08
		Suspended Acou Tile (2)	.15	.12	.11	.09	.08	.08	.07
	3 (13)	None or Plaster (6)	.21	.16	.13	.11	.10	.09	.08
		Suspended Plaster (5)	.15	.12	.11	.09	.08	.08	.07
		Suspended Acou Tile (2)	.13	.11	.10	.08	.08	.07	.06
	4 (16)	None or Plaster (6)	.17	.14	.11	.10	.09	.08	.07
		Suspended Plaster (5)	.13	.11	.10	.08	.08	.07	.06
		Suspended Acou Tile (2)	.12	.10	.09	.07	.07	.06	.05
Gypsum Slab on ½" Gypsum Board	2 (11)	None or Plaster (6)	.32	.22	.17	.14	.12	.10	.09
		Suspended Plaster (5)	.21	.17	.13	.11	.10	.09	.08
		Suspended Acou Tile (2)	.17	.13	.12	.10	.09	.08	.07
	3 (15)	None or Plaster (6)	.27	.19	.15	.13	.11	.10	.08
		Suspended Plaster (5)	.19	.15	.13	.11	.10	.09	.08
		Suspended Acou Tile (2)	.15	.12	.11	.09	.08	.08	.07
	4 (19)	None or Plaster (6)	.23	.17	.14	.12	.10	.09	.08
		Suspended Plaster (5)	.17	.13	.12	.10	.09	.08	.07
		Suspended Acou Tile (2)	.14	.12	.11	.09	.08	.08	.07
Wood	1 (3)	None or Plaster (6)	.40	.26	.19	.15	.13	.11	.09
		Suspended Plaster (5)	.24	.18	.14	.12	.11	.09	.08
		Suspended Acou Tile (2)	.19	.15	.13	.11	.10	.08	.07
	2 (5)	None or Plaster (6)	.28	.20	.16	.13	.11	.10	.08
		Suspended Plaster (5)	.19	.15	.13	.11	.10	.09	.07
		Suspended Acou Tile (2)	.16	.13	.11	.10	.09	.08	.07
	3 (8)	None or Plaster (6)	.21	.16	.13	.11	.10	.09	.08
		Suspended Plaster (5)	.16	.13	.11	.09	.09	.08	.07
		Suspended Acou Tile (2)	.13	.11	.10	.09	.08	.07	.06

1958 ASHRAE Guide

Equations: Summer—(Heat Flow Down) Heat Gain, Btu/hr = (Area, sq ft) × (U value) × (equivalent temp diff, Table 20).
Winter—(Heat Flow Up) Heat Loss, Btu/hr = (Area, sq ft) × (U value × 1.1) × (outdoor temp—inside temp).

*For addition of air spaces or insulation to roofs, refer to Table 31, page 75.
†For suspended ½" insulation board, plain (.6) or with ½" sand aggregate plaster (5). use values of suspended acou tile.

Figure 3.6 (continued)

U Values for Common Building Materials (continued)

TRANSMISSION COEFFICIENT U—WITH INSULATION & AIR SPACES
SUMMER AND WINTER
Btu/(hr) (sq ft) (deg F temp diff)

U Value Before Adding Insul. Wall, Ceiling, Roof Floor	Addition of Fibrous Insulation Thickness (Inches)			Add'n of Air Space ¾" or more *	Addition of Reflective Sheets to Air Space (Aluminum Foil Average Emissivity = .05) Direction of Heat Flow								
					Winter and Summer Horizontal			Summer Down			Winter Up		
	1	2	3		Added to one or both sides	One sheet in air space	Two sheets in air space	Added to one or both sides	One sheet in air space	Two sheets in air space	Added to one or both sides	One sheet in air space	Two sheets in air space
.60	.19	.11	.08	.38	.34	.18	.11	.12	.06	.05	.36	.20	.14
.58	.19	.11	.08	.37	.33	.18	.11	.12	.06	.05	.36	.20	.14
.56	.18	.11	.08	.36	.32	.18	.11	.11	.06	.05	.35	.20	.14
.54	.18	.11	.08	.36	.31	.17	.11	.11	.06	.05	.34	.19	.14
.52	.18	.11	.08	.35	.30	.17	.10	.11	.06	.05	.33	.19	.14
.50	.18	.11	.08	.34	.29	.17	.10	.11	.06	.05	.32	.19	.13
.48	.17	.11	.08	.33	.28	.16	.10	.11	.06	.04	.31	.18	.13
.46	.17	.10	.08	.32	.28	.16	.10	.11	.06	.04	.30	.18	.13
.44	.17	.10	.07	.31	.27	.16	.10	.11	.06	.04	.29	.18	.13
.42	.16	.10	.07	.30	.26	.15	.10	.11	.06	.04	.28	.17	.13
.40	.16	.10	.07	.29	.26	.15	.10	.10	.06	.04	.27	.17	.12
.38	.16	.10	.07	.28	.25	.15	.09	.10	.06	.04	.26	.17	.12
.36	.15	.10	.07	.27	.24	.14	.09	.10	.06	.04	.25	.16	.12
.34	.15	.10	.07	.26	.23	.14	.09	.10	.06	.04	.24	.16	.12
.32	.15	.10	.07	.25	.22	.13	.09	.10	.05	.04	.23	.15	.11
.30	.14	.09	.07	.23	.21	.13	.09	.10	.05	.04	.22	.15	.11
.28	.14	.09	.07	.22	.20	.13	.08	.09	.05	.04	.20	.14	.10
.26	.13	.09	.07	.21	.19	.12	.08	.09	.05	.04	.19	.13	.10
.24	.13	.09	.07	.20	.17	.12	.08	.09	.05	.04	.18	.13	.10
.22	.12	.08	.06	.18	.16	.11	.08	.08	.05	.04	.16	.12	.09
.20	.12	.08	.06	.17	.15	.10	.07	.08	.05	.04	.15	.11	.09
.18	.11	.08	.06	.15	.14	.10	.07	.08	.05	.04	.14	.11	.08
.16	.10	.07	.06	.14	.12	.09	.07	.07	.05	.04	.13	.10	.08
.14	.09	.07	.05	.12	.11	.08	.06	.07	.04	.04	.12	.09	.07
.12	.08	.06	.05	.11	.10	.08	.06	.06	.04	.03	.10	.08	.07
.10	.07	.06	.05	.09	.08	.07	.05	.06	.04	.03	.09	.07	.06

1958 ASHRAE Guide

Insulation Added	Air Space Added	Reflective Sheets Added to One or Both Sides	Reflective Sheet in Air Space	Reflective Sheets in Air Space

*Checked for summer conditions for up, down and horizontal heat flow. Error from above values is less than 1%.

Figure 3.6 (continued)

U Values for Common Building Materials (continued)

TRANSMISSION COEFFICIENT U—FLAT ROOFS WITH ROOF-DECK INSULATION
SUMMER AND WINTER

Btu/(hr) (sq ft) (deg F temp diff)

U VALUE OF ROOF BEFORE ADDING ROOF DECK INSULATION	Addition of Roof-Deck Insulation Thickness (in.)					
	½	1	1½	2	2½	3
.60	.33	.22	.17	.14	.12	.10
.50	.29	.21	.16	.14	.12	.10
.40	.26	.19	.15	.13	.11	.09
.35	.24	.18	.14	.12	.10	.09
.30	.21	.16	.13	.12	.10	.09
.25	.19	.15	.12	.11	.09	.08
.20	.16	.13	.11	.10	.09	.08
.15	.12	.11	.09	.08	.08	.07
.10	.09	.08	.07	.07	.06	.05

1958 ASHRAE Guide

TRANSMISSION COEFFICIENT U—WINDOWS, SKYLIGHTS, DOORS & GLASS BLOCK WALLS

Btu/(hr) (sq ft) (deg F temp diff)

GLASS

	Vertical Glass						Horizontal Glass			
	Single	Double			Triple			Single		Double (¼")
Air Space Thickness (in.)		¼	½	¾-4	¼	½	¾-4	Summer	Winter	Summer	Winter
Without Storm Windows	1.13	0.61	0.55	0.53	0.41	0.36	0.34	0.86	1.40	0.50	0.70
With Storm Windows	0.54							0.43	0.64		

DOORS

Nominal Thickness of Wood (inches)	U Exposed Door	U With Storm Door
1	0.69	0.35
1¼	0.59	0.32
1½	0.52	0.30
1¾	0.51	0.30
2	0.46	0.28
2½	0.38	0.25
3	0.33	0.23
Glass (¾" Herculite)	1.05	0.43

HOLLOW GLASS BLOCK WALLS

Description*	U
5¾x5¾x3⅞" Thick—Nominal Size 6x6x4 (14)	0.60
7¾x7¾x3⅞" Thick—Nominal Size 8x8x4 (14)	0.56
11¾x11¾x3⅞" Thick—Nominal Size 12x12x4 (16)	0.52
7¾x7¾x3⅞" Thick with glass fiber screen dividing the cavity (14)	0.48
11¾x11¾x3⅞" Thick with glass fiber screen dividing the cavity (16)	0.44

1958 ASHRAE Guide

Equation: Heat Gain or Loss, Btu/hr = (Area, sq ft) × (U value) × (outdoor temp − inside temp)

*Italicized numbers in parentheses indicate weight in lb per sq ft.

(Copyright 1993 by ASHRAE, from 1993 Fundamentals. Used by permission.)

Footnote: Cross-references in table footnotes pertain to the source from which tables were taken; they do not refer to chapters or pages in this text.

Figure 3.6 (continued)

Typical Thermal Properties of Common Building and Insulating Materials—Design Values[a]

Description	Density, lb/ft³	Conductivity[b] (k), Btu·in / h·ft²·°F	Conductance (C), Btu / h·ft²·°F	Resistance[c] (R) Per Inch Thickness (1/k), °F·ft²·h / Btu·in	Resistance[c] (R) For Thickness Listed (1/C), °F·ft²·h / Btu	Specific Heat, Btu / lb·°F
BUILDING BOARD						
Asbestos-cement board	120	4.0	—	0.25	—	0.24
Asbestos-cement board ..... 0.125 in.	120	—	33.00	—	0.03	
Asbestos-cement board ..... 0.25 in.	120	—	16.50	—	0.06	
Gypsum or plaster board ..... 0.375 in.	50	—	3.10	—	0.32	0.26
Gypsum or plaster board ..... 0.5 in.	50	—	2.22	—	0.45	
Gypsum or plaster board ..... 0.625 in.	50	—	1.78	—	0.56	
Plywood (Douglas Fir)[d]	34	0.80	—	1.25	—	0.29
Plywood (Douglas Fir) ..... 0.25 in.	34	—	3.20	—	0.31	
Plywood (Douglas Fir) ..... 0.375 in.	34	—	2.13	—	0.47	
Plywood (Douglas Fir) ..... 0.5 in.	34	—	1.60	—	0.62	
Plywood (Douglas Fir) ..... 0.625 in.	34	—	1.29	—	0.77	
Plywood or wood panels ..... 0.75 in.	34	—	1.07	—	0.93	0.29
Vegetable fiber board						
Sheathing, regular density[e] ..... 0.5 in.	18	—	0.76	—	1.32	0.31
..... 0.78125 in.	18	—	0.49	—	2.06	
Sheathing intermediate density[e] ..... 0.5 in.	22	—	0.92	—	1.09	0.31
Nail-base sheathing[e] ..... 0.5 in.	25	—	0.94	—	1.06	0.31
Shingle backer ..... 0.375 in.	18	—	1.06	—	0.94	0.31
Shingle backer ..... 0.3125 in.	18	—	1.28	—	0.78	
Sound deadening board ..... 0.5 in.	15	—	0.74	—	1.35	0.30
Tile and lay-in panels, plain or acoustic	18	0.40	—	2.50	—	0.14
..... 0.5 in.	18	—	0.80	—	1.25	
..... 0.75 in.	18	—	0.53	—	1.89	
Laminated paperboard	30	0.50	—	2.00	—	0.33
Homogeneous board from repulped paper	30	0.50	—	2.00	—	0.28
Hardboard[e]						
Medium density	50	0.73	—	1.37	—	0.31
High density, service-tempered grade and service grade	55	0.82	—	1.22	—	0.32
High density, standard-tempered grade	63	1.00	—	1.00	—	0.32
Particleboard[e]						
Low density	37	0.71	—	1.41	—	0.31
Medium density	50	0.94	—	1.06	—	0.31
High density	62.5	1.18	—	0.85	—	0.31
Underlayment ..... 0.625 in.	40	—	1.22	—	0.82	0.29
Waferboard	37	0.63	—	1.59	—	—
Wood subfloor ..... 0.75 in.	—	—	1.06	—	0.94	0.33
BUILDING MEMBRANE						
Vapor—permeable felt	—	—	16.70	—	0.06	
Vapor—seal, 2 layers of mopped 15-lb felt	—	—	8.35	—	0.12	
Vapor—seal, plastic film	—	—	—	—	Negl.	
FINISH FLOORING MATERIALS						
Carpet and fibrous pad	—	—	0.48	—	2.08	0.34
Carpet and rubber pad	—	—	0.81	—	1.23	0.33
Cork tile ..... 0.125 in.	—	—	3.60	—	0.28	0.48
Terrazzo ..... 1 in.	—	—	12.50	—	0.08	0.19
Tile—asphalt, linoleum, vinyl, rubber	—	—	20.00	—	0.05	0.30
vinyl asbestos						0.24
ceramic						0.19
Wood, hardwood finish ..... 0.75 in.	—	—	1.47	—	0.68	
INSULATING MATERIALS						
Blanket and Batt[f,g]						
Mineral fiber, fibrous form processed from rock, slag, or glass						
approx. 3-4 in.	0.4-2.0	—	0.091	—	11	
approx. 3.5 in.	0.4-2.0	—	0.077	—	13	
approx. 3.5 in.	1.2-1.6	—	0.067	—	15	
approx. 5.5-6.5 in.	0.4-2.0	—	0.053	—	19	
approx. 5.5 in.	0.6-1.0	—	0.048	—	21	
approx. 6-7.5 in.	0.4-2.0	—	0.045	—	22	
approx. 8.25-10 in.	0.4-2.0	—	0.033	—	30	
approx. 10-13 in.	0.4-2.0	—	0.026	—	38	
Board and Slabs						
Cellular glass	8.0	0.33	—	3.03	—	0.18
Glass fiber, organic bonded	4.0-9.0	0.25	—	4.00	—	0.23
Expanded perlite, organic bonded	1.0	0.36	—	2.78	—	0.30
Expanded rubber (rigid)	4.5	0.22	—	4.55	—	0.40
Expanded polystyrene, extruded (smooth skin surface) (CFC-12 exp.)	1.8-3.5	0.20	—	5.00	—	0.29
Expanded polystyrene, extruded (smooth skin surface) (HCFC-142b exp.)[h]	1.8-3.5	0.20	—	5.00	—	0.29

C, K, and R Values for Common Building Materials

Figure 3.7

Typical Thermal Properties of Common Building and Insulating Materials—Design Values[a] (Continued)

Description	Density, lb/ft³	Conductivity[b] (k), Btu·in / h·ft²·°F	Conductance (C), Btu / h·ft²·°F	Resistance[c] (R) Per Inch Thickness (1/k), °F·ft²·h / Btu·in	Resistance[c] (R) For Thickness Listed (1/C), °F·ft²·h / Btu	Specific Heat, Btu / lb·°F
Expanded polystyrene, molded beads	1.0	0.26	—	3.85	—	—
	1.25	0.25	—	4.00	—	—
	1.5	0.24	—	4.17	—	—
	1.75	0.24	—	4.17	—	—
	2.0	0.23	—	4.35	—	—
Cellular polyurethane/polyisocyanurate[i] (CFC-11 exp.) (unfaced)	1.5	0.16–0.18	—	6.25–5.56	—	0.38
Cellular polyisocyanurate[j] (CFC-11 exp.)(gas-permeable facers)	1.5–2.5	0.16–0.18	—	6.25–5.56	—	0.22
Cellular polyisocyanurate[j] (CFC-11 exp.) (gas-impermeable facers)	2.0	0.14	—	7.04	—	0.22
Cellular phenolic (closed cell)(CFC-11, CFC-113 exp.)	3.0	0.12	—	8.20	—	—
Cellular phenolic (open cell)	1.8–2.2	0.23	—	4.40	—	—
Mineral fiber with resin binder	15.0	0.29	—	3.45	—	0.17
Mineral fiberboard, wet felted						
Core or roof insulation	16–17	0.34	—	2.94	—	—
Acoustical tile	18.0	0.35	—	2.86	—	0.19
Acoustical tile	21.0	0.37	—	2.70	—	—
Mineral fiberboard, wet molded						
Acoustical tile[k]	23.0	0.42	—	2.38	—	0.14
Wood or cane fiberboard						
Acoustical tile[k] .... 0.5 in.	—	—	0.80	—	1.25	0.31
Acoustical tile[k] .... 0.75 in.	—	—	0.53	—	1.89	—
Interior finish (plank, tile)	15.0	0.35	—	2.86	—	0.32
Cement fiber slabs (shredded wood with Portland cement binder)	25–27.0	0.50–0.53	—	2.0–1.89	—	—
Cement fiber slabs (shredded wood with magnesia oxysulfide binder)	22.0	0.57	—	1.75	—	0.31
Loose Fill						
Cellulosic insulation (milled paper or wood pulp)	2.3–3.2	0.27–0.32	—	3.70–3.13	—	0.33
Perlite, expanded	2.0–4.1	0.27–0.31	—	3.7–3.3	—	0.26
	4.1–7.4	0.31–0.36	—	3.3–2.8	—	—
	7.4–11.0	0.36–0.42	—	2.8–2.4	—	—
Mineral fiber (rock, slag, or glass)[g]						
approx. 3.75–5 in.	0.6–2.0	—	—	—	11.0	0.17
approx. 6.5–8.75 in.	0.6–2.0	—	—	—	19.0	—
approx. 7.5–10 in.	0.6–2.0	—	—	—	22.0	—
approx. 10.25–13.75 in.	0.6–2.0	—	—	—	30.0	—
Mineral fiber (rock, slag, or glass)[g]						
approx. 3.5 in. (closed sidewall application)	2.0–3.5	—	—	—	12.0–14.0	—
Vermiculite, exfoliated	7.0–8.2	0.47	—	2.13	—	0.32
	4.0–6.0	0.44	—	2.27	—	—
Spray Applied						
Polyurethane foam	1.5–2.5	0.16–0.18	—	6.25–5.56	—	—
Ureaformaldehyde foam	0.7–1.6	0.22–0.28	—	4.55–3.57	—	—
Cellulosic fiber	3.5–6.0	0.29–0.34	—	3.45–2.94	—	—
Glass fiber	3.5–4.5	0.26–0.27	—	3.85–3.70	—	—
METALS (See Chapter 36, Table 3)						
ROOFING						
Asbestos-cement shingles	120	—	4.76	—	0.21	0.24
Asphalt roll roofing	70	—	6.50	—	0.15	0.36
Asphalt shingles	70	—	2.27	—	0.44	0.30
Built-up roofing .... 0.375 in.	70	—	3.00	—	0.33	0.35
Slate .... 0.5 in.	—	—	20.00	—	0.05	0.30
Wood shingles, plain and plastic film faced	—	—	1.06	—	0.94	0.31
PLASTERING MATERIALS						
Cement plaster, sand aggregate	116	5.0	—	0.20	—	0.20
Sand aggregate .... 0.375 in.	—	—	13.3	—	0.08	0.20
Sand aggregate .... 0.75 in.	—	—	6.66	—	0.15	0.20
Gypsum plaster:						
Lightweight aggregate .... 0.5 in.	45	—	3.12	—	0.32	—
Lightweight aggregate .... 0.625 in.	45	—	2.67	—	0.39	—
Lightweight aggregate on metal lath .... 0.75 in.	—	—	2.13	—	0.47	—
Perlite aggregate	45	1.5	—	0.67	—	0.32
Sand aggregate	105	5.6	—	0.18	—	0.20
Sand aggregate .... 0.5 in.	105	—	11.10	—	0.09	—
Sand aggregate .... 0.625 in.	105	—	9.10	—	0.11	—
Sand aggregate on metal lath .... 0.75 in.	—	—	7.70	—	0.13	—
Vermiculite aggregate	45	1.7	—	0.59	—	—
MASONRY MATERIALS						
Masonry Units						
Brick, fired clay	150	8.4–10.2	—	0.12–0.10	—	—
	140	7.4–9.0	—	0.14–0.11	—	—
	130	6.4–7.8	—	0.16–0.12	—	—
	120	5.6–6.8	—	0.18–0.15	—	0.19
	110	4.9–5.9	—	0.20–0.17	—	—

C, K, and R Values for Common Building Materials (continued)

Figure 3.7 (continued)

Typical Thermal Properties of Common Building and Insulating Materials—Design Values[a] (*Continued*)

Description	Density, lb/ft³	Conductivity[b] (k), Btu·in / h·ft²·°F	Conductance (C), Btu / h·ft²·°F	Resistance[c] (R) Per Inch Thickness (1/k), °F·ft²·h / Btu·in	Resistance[c] (R) For Thickness Listed (1/C), °F·ft²·h / Btu	Specific Heat, Btu / lb·°F
Brick, fired clay *continued*	100	4.2-5.1	—	0.24-0.20	—	—
	90	3.6-4.3	—	0.28-0.24	—	—
	80	3.0-3.7	—	0.33-0.27	—	—
	70	2.5-3.1	—	0.40-0.33	—	—
Clay tile, hollow						
1 cell deep .... 3 in.	—	—	1.25	—	0.80	0.21
1 cell deep .... 4 in.	—	—	0.90	—	1.11	—
2 cells deep .... 6 in.	—	—	0.66	—	1.52	—
2 cells deep .... 8 in.	—	—	0.54	—	1.85	—
2 cells deep .... 10 in.	—	—	0.45	—	2.22	—
3 cells deep .... 12 in.	—	—	0.40	—	2.50	—
Concrete blocks[i]						
Limestone aggregate						
8 in., 36 lb, 138 lb/ft³ concrete, 2 cores	—	—	—	—	—	—
Same with perlite filled cores	—	—	0.48	—	2.1	—
12 in., 55 lb, 138 lb/ft³ concrete, 2 cores	—	—	—	—	—	—
Same with perlite filled cores	—	—	0.27	—	3.7	—
Normal weight aggregate (sand and gravel)						
8 in., 33-36 lb, 126-136 lb/ft³ concrete, 2 or 3 cores	—	—	0.90-1.03	—	1.11-0.97	0.22
Same with perlite filled cores	—	—	0.50	—	2.0	—
Same with verm. filled cores	—	—	0.52-0.73	—	1.92-1.37	—
12 in., 50 lb, 125 lb/ft³ concrete, 2 cores	—	—	0.81	—	1.23	0.22
Medium weight aggregate (combinations of normal weight and lightweight aggregate)						
8 in., 26-29 lb, 97-112 lb/ft³ concrete, 2 or 3 cores	—	—	0.58-0.78	—	1.71-1.28	—
Same with perlite filled cores	—	—	0.27-0.44	—	3.7-2.3	—
Same with verm. filled cores	—	—	0.30	—	3.3	—
Same with molded EPS (beads) filled cores	—	—	0.32	—	3.2	—
Same with molded EPS inserts in cores	—	—	0.37	—	2.7	—
Lightweight aggregate (expanded shale, clay, slate or slag, pumice)						
6 in., 16-17 lb 85-87 lb/ft³ concrete, 2 or 3 cores	—	—	0.52-0.61	—	1.93-1.65	—
Same with perlite filled cores	—	—	0.24	—	4.2	—
Same with verm. filled cores	—	—	0.33	—	3.0	—
8 in., 19-22 lb, 72-86 lb/ft³ concrete,	—	—	0.32-0.54	—	3.2-1.90	0.21
Same with perlite filled cores	—	—	0.15-0.23	—	6.8-4.4	—
Same with verm. filled cores	—	—	0.19-0.26	—	5.3-3.9	—
Same with molded EPS (beads) filled cores	—	—	0.21	—	4.8	—
Same with UF foam filled cores	—	—	0.22	—	4.5	—
Same with molded EPS inserts in cores	—	—	0.29	—	3.5	—
12 in., 32-36 lb, 80-90 lb/ft³ concrete, 2 or 3 cores	—	—	0.38-0.44	—	2.6-2.3	—
Same with perlite filled cores	—	—	0.11-0.16	—	9.2-6.3	—
Same with verm. filled cores	—	—	0.17	—	5.8	—
Stone, lime, or sand						
Quartzitic and sandstone	180	72	—	0.01	—	—
	160	43	—	0.02	—	—
	140	24	—	0.04	—	—
	120	13	—	0.08	—	0.19
Calcitic, dolomitic, limestone, marble, and granite	180	30	—	0.03	—	—
	160	22	—	0.05	—	—
	140	16	—	0.06	—	—
	120	11	—	0.09	—	0.19
	100	8	—	0.13	—	—
Gypsum partition tile						
3 by 12 by 30 in., solid	—	—	0.79	—	1.26	0.19
3 by 12 by 30 in., 4 cells	—	—	0.74	—	1.35	—
4 by 12 by 30 in., 3 cells	—	—	0.60	—	1.67	—
Concretes						
Sand and gravel or stone aggregate concretes (concretes with more than 50% quartz or quartzite sand have conductivities in the higher end of the range)	150	10.0-20.0	—	0.10-0.05	—	—
	140	9.0-18.0	—	0.11-0.06	—	0.19-0.24
	130	7.0-13.0	—	0.14-0.08	—	—
Limestone concretes	140	11.1	—	0.09	—	—
	120	7.9	—	0.13	—	—
	100	5.5	—	0.18	—	—
Gypsum-fiber concrete (87.5% gypsum, 12.5% wood chips)	51	1.66	—	0.60	—	0.21
Cement/lime, mortar, and stucco	120	9.7	—	0.10	—	—
	100	6.7	—	0.15	—	—
	80	4.5	—	0.22	—	—
Lightweight aggregate concretes						
Expanded shale, clay, or slate; expanded slags; cinders; pumice (with density up to 100 lb/ft³); and scoria (sanded concretes have conductivities in the higher end of the range)	120	6.4-9.1	—	0.16-0.11	—	—
	100	4.7-6.2	—	0.21-0.16	—	0.20
	80	3.3-4.1	—	0.30-0.24	—	0.20
	60	2.1-2.5	—	0.48-0.40	—	—
	40	1.3	—	0.78	—	—

C, K, and R Values for Common Building Materials (continued)

Figure 3.7 (continued)

Typical Thermal Properties of Common Building and Insulating Materials—Design Values[a] (*Concluded*)

Description	Density, lb/ft³	Conductivity[b] (k), Btu·in / h·ft²·°F	Conductance (C), Btu / h·ft²·°F	Resistance[c] (R) Per Inch Thickness (1/k), °F·ft²·h / Btu·in	Resistance[c] (R) For Thickness Listed (1/C), °F·ft²·h / Btu	Specific Heat, Btu / lb·°F
Perlite, vermiculite, and polystyrene beads	50	1.8-1.9	—	0.55-0.53	—	—
	40	1.4-1.5	—	0.71-0.67	—	0.15-0.23
	30	1.1	—	0.91	—	—
	20	0.8	—	1.25	—	—
Foam concretes	120	5.4	—	0.19	—	—
	100	4.1	—	0.24	—	—
	80	3.0	—	0.33	—	—
	70	2.5	—	0.40	—	—
Foam concretes and cellular concretes	60	2.1	—	0.48	—	—
	40	1.4	—	0.71	—	—
	20	0.8	—	1.25	—	—
SIDING MATERIALS (on flat surface)						
Shingles						
Asbestos-cement	120	—	4.75	—	0.21	—
Wood, 16 in., 7.5 exposure	—	—	1.15	—	0.87	0.31
Wood, double, 16-in., 12-in. exposure	—	—	0.84	—	1.19	0.28
Wood, plus insul. backer board, 0.3125 in.	—	—	0.71	—	1.40	0.31
Siding						
Asbestos-cement, 0.25 in., lapped	—	—	4.76	—	0.21	0.24
Asphalt roll siding	—	—	6.50	—	0.15	0.35
Asphalt insulating siding (0.5 in. bed.)	—	—	0.69	—	1.46	0.35
Hardboard siding, 0.4375 in.	—	—	1.49	—	0.67	0.28
Wood, drop, 1 by 8 in.	—	—	1.27	—	0.79	0.28
Wood, bevel, 0.5 by 8 in., lapped	—	—	1.23	—	0.81	0.28
Wood, bevel, 0.75 by 10 in., lapped	—	—	0.95	—	1.05	0.28
Wood, plywood, 0.375 in., lapped	—	—	1.59	—	0.59	0.29
Aluminum or Steel[m], over sheathing						
Hollow-backed	—	—	1.61	—	0.61	0.29
Insulating-board backed nominal 0.375 in.	—	—	0.55	—	1.82	0.32
Insulating-board backed nominal 0.375 in., foil backed	—	—	0.34	—	2.96	—
Architectural (soda-lime float) glass	158	6.9	—	—	—	0.21
WOODS (12% moisture content)[e,n]						
Hardwoods						
Oak	41.2-46.8	1.12-1.25	—	0.89-0.80	—	0.39[o]
Birch	42.6-45.4	1.16-1.22	—	0.87-0.82	—	—
Maple	39.8-44.0	1.09-1.19	—	0.92-0.84	—	—
Ash	38.4-41.9	1.06-1.14	—	0.94-0.88	—	—
Softwoods						0.39[o]
Southern Pine	35.6-41.2	1.00-1.12	—	1.00-0.89	—	—
Douglas Fir-Larch	33.5-36.3	0.95-1.01	—	1.06-0.99	—	—
Southern Cypress	31.4-32.1	0.90-0.92	—	1.11-1.09	—	—
Hem-Fir, Spruce-Pine-Fir	24.5-31.4	0.74-0.90	—	1.35-1.11	—	—
West Coast Woods, Cedars	21.7-31.4	0.68-0.90	—	1.48-1.11	—	—
California Redwood	24.5-28.0	0.74-0.82	—	1.35-1.22	—	—

[a] Values are for a mean temperature of 75°F. Representative values for dry materials are intended as design (not specification) values for materials in normal use. Thermal values of insulating materials may differ from design values depending on their in-situ properties (*e.g.*, density and moisture content, orientation, etc.) and variability experienced during manufacture. For properties of a particular product, use the value supplied by the manufacturer or by unbiased tests.

[b] To obtain thermal conductivities in Btu/h·ft·°F, divide the k-factor by 12 in./ft.

[c] Resistance values are the reciprocals of C before rounding off C to two decimal places.

[d] Lewis (1967).

[e] U.S. Department of Agriculture (1974).

[f] Does not include paper backing and facing, if any. Where insulation forms a boundary (reflective or otherwise) of an airspace, see Tables 2 and 3 for the insulating value of an airspace with the appropriate effective emittance and temperature conditions of the space.

[g] Conductivity varies with fiber diameter. (See Chapter 20, Factors Affecting Thermal Performance.) Batt, blanket, and loose-fill mineral fiber insulations are manufactured to achieve specified R-values, the most common of which are listed in the table. Due to differences in manufacturing processes and materials, the product thicknesses, densities, and thermal conductivities vary over considerable ranges for a specified R-value.

[h] This material is relatively new and data are based on limited testing.

[i] For additional information, see Society of Plastics Engineers (SPI) *Bulletin* U108. Values are for aged, unfaced board stock. For change in conductivity with age of expanded polyurethane/polyisocyanurate, see Chapter 20, Factors Affecting Thermal Performance.

[j] Values are for aged products with gas-impermeable facers on the two major surfaces. An aluminum foil facer of 0.001 in. thickness or greater is generally considered impermeable to gases. For change in conductivity with age of expanded polyisocyanurate, see Chapter 20, Factors Affecting Thermal Performance, and SPI *Bulletin* U108.

[k] Insulating values of acoustical tile vary, depending on density of the board and on type, size, and depth of perforations.

[l] Values for fully grouted block may be approximated using values for concrete with a similar unit weight.

[m] Values for metal siding applied over flat surfaces vary widely, depending on amount of ventilation of airspace beneath the siding; whether airspace is reflective of non-reflective; and on thickness, type, and application of insulating backing-board used. Values given are averages for use as design guides, and were obained from several guarded hot box tests (ASTM C236) or calibrated hot box (ASTM C976) on hollow-backed types and types made using backing-boards of wood fiber, foamed plastic, and glass fiber. Departures of ±50% or more from the values given may occur.

[n] See Adams (1971), MacLean (1941), and Wilkes (1979). The conductivity values listed are for heat transfer across the grain. The thermal conductivity of wood varies linearly with the density, and the density ranges listed are those normally found for the wood species given. If the density of the wood species is not known, use the mean conductivity value. For extrapolation to other moisture contents, the following empirical equation developed by Wilkes (1979) may be used:

$$k = 0.1791 + \frac{(1.874 \times 10^{-2} + 5.753 \times 10^{-4}M)\rho}{1 + 0.01M}$$

where ρ is density of the moist wood in lb/ft³, and M is the moisture content in percent.

[o] From Wilkes (1979), an empirical equation for the specific heat of moist wood at 75°F is as follows:

$$c_p = \frac{(0.299 + 0.01M)}{(1 + 0.01M)} + \Delta c_p$$

where Δc_p accounts for the heat of sorption and is denoted by

$$\Delta c_p = M(1.921 \times 10^{-3} - 3.168 \times 10^{-5}M)$$

where M is the moisture content in percent by mass.

(Copyright 1993 by ASHRAE, from *1993 Fundamentals*. Used by permission.)

Footnote: Cross-references in table footnotes pertain to the source from which tables were taken; they do not refer to chapters or pages in this text.

C, K, and R Values for Common Building Materials (continued)

Figure 3.7 (continued)

Typical Thermal Properties of Common Building and Insulating Materials—Design Values[a]

Description	Density, kg/m³	Conductivity[b] (k), W/(m·K)	Conductance (C), W/(m²·K)	Resistance[c](R) (1/k), K·m/W	Resistance[c](R) For Thickness Listed (1/C), K·m²/W	Specific Heat, kJ/(kg·K)
BUILDING BOARD						
Asbestos-cement board	1900	0.58	—	1.73	—	1.00
Asbestos-cement board ....3.2 mm	1900	—	187.4	—	0.005	—
Asbestos-cement board ....6.4 mm	1900	—	93.7	—	0.011	—
Gypsum or plaster board ....9.5 mm	800	—	17.6	—	0.056	1.09
Gypsum or plaster board ....12.7 mm	800	—	12.6	—	0.079	—
Gypsum or plaster board ....15.9 mm	800	—	10.1	—	0.099	—
Plywood (Douglas Fir)[d]	540	0.12	—	8.66	—	1.21
Plywood (Douglas Fir) ....6.4 mm	540	—	18.2	—	0.055	—
Plywood (Douglas Fir) ....9.5 mm	540	—	12.1	—	0.083	—
Plywood (Douglas Fir) ....12.7 mm	540	—	9.1	—	0.11	—
Plywood (Douglas Fir) ....15.9 mm	540	—	7.3	—	0.14	—
Plywood or wood panels ....19.0 mm	540	—	6.1	—	0.16	1.21
Vegetable fiber board						
Sheathing, regular density........12.7 mm	290	—	4.3	—	0.23	1.30
..............................19.8 mm	290	—	2.8	—	0.36	—
Sheathing intermediate density...12.7 mm	350	—	5.2	—	0.19	1.30
Nail-base sheathing ....12.7 mm	400	—	5.3	—	0.19	1.30
Shingle backer ....9.5 mm	290	—	6.0	—	0.17	1.30
Shingle backer ....7.9 mm	290	—	7.3	—	0.14	—
Sound deadening board..........12.7 mm	240	—	4.2	—	0.24	1.26
Tile and lay-in panels, plain or acoustic	290	0.058	—	17.3	—	0.59
............................12.7 mm	290	—	4.5	—	0.22	—
............................19.0 mm	290	—	3.0	—	0.33	—
Laminated paperboard	480	0.072	—	13.9	—	1.38
Homogeneous board from repulped paper	480	0.072	—	13.9	—	1.17
Hardboard						
Medium density	800	0.105	—	9.50	—	1.30
High density, service-tempered grade and service grade	880	0.82	—	8.46	—	1.34
High density, standard-tempered grade	1010	0.144	—	6.93	—	1.34
Particleboard						
Low density	590	0.102	—	9.77	—	1.30
Medium density	800	0.135	—	7.35	—	1.30
High density	1000	0.170	—	5.90	—	1.30
Underlayment ....15.9 mm	640	—	6.9	—	0.14	1.21
Waferboard	590	0.01	—	11.0	—	—
Wood subfloor ....19.0 mm	—	—	6.0	—	0.17	1.38
BUILDING MEMBRANE						
Vapor—permeable felt	—	—	94.9	—	0.011	
Vapor—seal, 2 layers of mopped 0.73 kg/m² felt	—	—	47.4	—	0.21	
Vapor—seal, plastic film	—	—	—	—	Negl.	
FINISH FLOORING MATERIALS						
Carpet and fibrous pad	—	—	2.73	—	0.37	1.42
Carpet and rubber pad	—	—	4.60	—	0.22	1.38
Cork tile ....3.2 mm	—	—	20.4	—	0.049	2.01
Terrazzo ....25 mm	—	—	71.0	—	0.014	0.80
Tile—asphalt, linoleum, vinyl, rubber	—	—	113.6	—	0.009	1.26
vinyl asbestos						1.01
ceramic						0.80
Wood, hardwood finish ....19 mm			8.35	—	0.12	
INSULATING MATERIALS						
Blanket and Batt[f,g]						
Mineral fiber, fibrous form processed from rock, slag, or glass						
approx. 75–100 mm	6.4–32	—	0.52	—	1.94	
approx. 90 mm	6.4–32	—	0.44	—	2.29	
approx. 90 mm	19–26	—	0.38	—	2.63	
approx. 140–165 mm	6.4–32	—	0.30	—	3.32	
approx. 140 mm	10–16	—	0.27	—	3.67	
approx. 150–190 mm	6.4–32	—	0.26	—	3.91	
approx. 210–250 mm	6.4–32	—	0.19	—	5.34	
approx. 250–330 mm	6.4–32	—	0.15	—	6.77	
Board and Slabs						
Cellular glass	136	0.050	—	19.8	—	0.75
Glass fiber, organic bonded	64–140	0.036	—	27.7	—	0.96
Expanded perlite, organic bonded	16	0.052	—	19.3	—	1.26
Expanded rubber (rigid)	72	0.032	—	31.6	—	1.68
Expanded polystyrene extruded (smooth skin surface) (CFC-12 exp.)	29–56	0.029	—	34.7	—	1.22
Expanded polystyrene, extruded (smooth skin surface) (HCFC-142b exp.)[h]	29–56	0.029	—	34.7	—	1.21

C, K, and R Values for Common Building Materials—In SI Metric Units

Figure 3.7a

Typical Thermal Properties of Common Building and Insulating Materials—Design Values[a] (*Continued*)

Description	Density, kg/m³	Conductivity[b] (k), W/(m·K)	Conductance (C), W/(m²·K)	Resistance[c] (R) (1/k), K·m/W	Resistance[c] (R) For Thickness Listed (1/C), K·m²/W	Specific Heat, kJ/(kg·K)
Expanded polystyrene, molded beads	16	0.037	—	26.7	—	—
	20	0.036	—	27.7	—	—
	24	0.035	—	28.9	—	—
	28	0.035	—	28.9	—	—
	32	0.033	—	30.2	—	—
Cellular polyurethane/polyisocyanurate[j] (CFC-11 exp.) (unfaced)	24	0.023–0.026	—	43.3–38.5	—	1.59
Cellular polyisocyanurate[i] (CFC-11 exp.) (gas-permeable facers)	24–40	0.023–0.026	—	43.3–38.5	—	0.92
Cellular polyisocyanurate[j] (CFC-11 exp.) (gas-impermeable facers)	32	0.020	—	48.8	—	0.92
Cellular phenolic (closed cell) (CFC-11, CFC-113 exp.)	32	0.017	—	56.8	—	—
Cellular phenolic (open cell)	29–35	0.033	—	30.5	—	—
Mineral fiber with resin binder	240	0.042	—	23.9	—	0.71
Mineral fiberboard, wet felted						
Core or roof insulation	260–270	0.049	—	20.4	—	—
Acoustical tile	290	0.050	—	19.8	—	0.80
Acoustical tile	340	0.053	—	18.7	—	
Mineral fiberboard, wet molded						
Acoustical tile[k]	370	0.060	—	16.5	—	0.59
Wood or cane fiberboard						
Acoustical tile[k] ....12.7 mm	—	—	0.80	—	1.25	1.30
Acoustical tile[k] ....19.0 mm	—	—	0.53	—	1.89	
Interior finish (plank, tile)	240	0.050	—	19.8	—	1.34
Cement fiber slabs (shredded wood with Portland cement binder)	400–430	0.072–0.076	—	13.9–13.1	—	—
Cement fiber slabs (shredded wood with magnesia oxysulfide binder)	350	0.082	—	12.1	—	1.30
Loose Fill						
Cellulosic insulation (milled paper or wood pulp)	37–51	0.039–0.046	—	25.6–21.7	—	1.38
Perlite, expanded	32–66	0.039–0.045	—	25.6–22.9	—	1.09
	66–120	0.045–0.052	—	22.9–19.4	—	—
	120–180	0.052–0.060	—	19.4–16.6	—	—
Mineral fiber (rock, slag, or glass)[g]						
approx. 95–130 mm	9.6–32	—	—	—	1.94	0.71
approx. 170–220 mm	9.6–32	—	—	—	3.35	—
approx. 190–250 mm	9.6–32	—	—	—	3.87	—
approx. 260–350 mm	9.6–32	—	—	—	5.28	—
Mineral fiber (rock, slag, or glass)[g]						
approx. 90 mm (closed sidewall application)	32–56	—	—	—	2.1–2.5	—
Vermiculite, exfoliated	110–130	0.068	—	14.8	—	1.34
	64–96	0.063	—	15.7	—	—
Spray Applied						
Polyurethane foam	24–40	0.023–0.026	—	43.3–38.5	—	—
Ureaformaldehyde foam	11–26	0.032–0.040	—	31.5–24.7	—	—
Cellulosic fiber	56–96	0.042–0.049	—	23.9–20.4	—	—
Glass fiber	56–72	0.038–0.039	—	26.7–25.6	—	—
METALS (See Chapter 36, Table 3)						
ROOFING						
Asbestos-cement shingles	1900	—	27.0	—	0.037	1.00
Asphalt roll roofing	1100	—	36.9	—	0.026	1.51
Asphalt shingles	1100	—	12.9	—	0.077	1.26
Built-up roofing ....10 mm	1100	—	17.0	—	0.058	1.46
Slate ....13 mm	—	—	114	—	0.009	1.26
Wood shingles, plain and plastic film faced	—	—	6.0	—	0.166	1.30
PLASTERING MATERIALS						
Cement plaster, sand aggregate	1860	0.72	—	1.39	—	0.84
Sand aggregate ....9.5 mm	—	—	75.5	—	0.08	0.84
Sand aggregate ....19 mm	—	—	37.8	—	0.15	0.84
Gypsum plaster:						
Low density aggregate ....12.7 mm	720	—	17.7	—	0.32	—
Low density aggregate ....16 mm	720	—	15.2	—	0.39	—
Low density agg. on metal lath ....19 mm	—	—	12.1	—	0.47	—
Perlite aggregate	720	0.22	—	4.64	—	1.34
Sand aggregate	1680	0.81	—	1.25	—	0.84
Sand aggregate ....12.7 mm	1680	—	63.0	—	0.09	—
Sand aggregate ....16 mm	1680	—	51.7	—	0.11	—
Sand aggregate on metal lath ....19 mm	—	—	43.7	—	0.13	—
Vermiculite aggregate	720	0.24	—	4.09	—	—
MASONRY MATERIALS						
Masonry Units						
Brick, fired clay	2400	1.21–1.47	—	0.83–0.68	—	—
	2240	1.07–1.30	—	0.94–0.77	—	—
	2080	0.92–1.12	—	1.08–0.89	—	—
	1920	0.81–0.98	—	1.24–1.02	—	0.79
	1760	0.71–0.85	—	1.42–1.18	—	—

C, K, and R Values for Common Building Materials – In SI Metric Units (continued)

Figure 3.7a (continued)

Typical Thermal Properties of Common Building and Insulating Materials—Design Values[a] (*Continued*)

Description	Density, kg/m³	Conductivity[b] (k), W/(m·K)	Conductance (C), W/(m²·K)	Resistance[c] (R) (1/k), K·m/W	Resistance[c] (R) For Thickness Listed (1/C), K·m²/W	Specific Heat, kJ/(kg·K)
Brick, fired clay *continued*	1600	0.61–0.74	—	1.65–1.36	—	—
	1440	0.52–0.62	—	1.93–1.61	—	—
	1280	0.43–0.53	—	2.31–1.87	—	—
	1120	0.36–0.45	—	2.77–2.23	—	—
Clay tile, hollow						
1 cell deep ....76 mm	—	—	7.10	—	0.14	0.88
1 cell deep ....102 mm	—	—	5.11	—	0.20	—
2 cells deep ....152 mm	—	—	3.75	—	0.27	—
2 cells deep ....203 mm	—	—	3.07	—	0.33	—
2 cells deep ....254 mm	—	—	2.56	—	0.39	—
3 cells deep ....305 mm	—	—	2.27	—	0.44	—
Concrete blocks[l]						
Limestone aggregate						
200 mm, 16.3 kg, 2210 kg/m³ concrete, 2 cores ...	—	—	—	—	—	—
Same with perlite filled cores	—	—	2.73	—	0.37	—
300 mm, 25 kg, 2210 kg/m³ concrete, 2 cores	—	—	—	—	—	—
Same with perlite filled cores	—	—	1.53	—	0.65	—
Normal mass aggregate (sand and gravel) 200 mm,						
15–16 kg, 2020–2180 kg/m³ concrete, 2 or 3 cores..	—	—	5.1–5.8	—	0.20–0.17	0.92
Same with perlite filled cores	—	—	2.84	—	0.35	—
Same with vermiculite filled cores	—	—	3.0–4.1	—	0.34–0.24	—
300 mm, 22.7 kg, 2000 kg/m³ concrete, 2 cores ...	—	—	4.60	—	0.217	0.92
Medium mass aggregate (combinations of normal and low mass aggregate) 200 mm, 12–13 kg,						
1550–1790 kg/m³ concrete, 2 or 3 cores	—	—	3.3–4.4	—	0.30–0.22	—
Same with perlite filled cores	—	—	1.5–2.5	—	0.65–0.41	—
Same with vermiculite filled cores	—	—	1.70	—	0.58	—
Same with molded EPS (beads) filled cores	—	—	1.82	—	0.56	—
Same with molded EPS inserts in cores	—	—	2.10	—	0.47	—
Low mass aggregate (expanded shale, clay, slate or slag, pumice) 150 mm,						
7.3–7.7 kg, 1360–1390 kg/m³ concrete, 2 or 3 cores	—	—	3.0–3.5	—	0.34–0.29	—
Same with perlite filled cores	—	—	1.36	—	0.74	—
Same with vermiculite filled cores	—	—	1.87	—	0.53	—
200 mm, 8.6–10.0 mm, 1150–1380 kg/m³ concrete,	—	—	1.8–3.1	—	0.56–0.33	0.88
Same with perlite filled cores	—	—	0.9–1.3	—	1.20–0.77	—
Same with vermiculite filled cores	—	—	1.1–1.5	—	0.93–0.69	—
Same with molded EPS (beads) filled cores	—	—	1.19	—	0.85	—
Same with ureaformaldehyde foam filled cores	—	—	1.25	—	0.79	—
Same with molded EPS inserts in cores	—	—	1.65	—	0.62	—
300 mm, 14.5–16.3 kg, 1280–1440 kg/m³ concrete,						
2 or 3 cores	—	—	2.2–2.5	—	0.46–0.40	—
Same with perlite filled cores	—	—	0.6–0.9	—	1.6–1.1	—
Same with vermiculite filled cores	—	—	0.97	—	1.0	—
Stone, lime, or sand						
Quartzitic and sandstone	2880	10.4	—	0.10	—	—
	2560	6.2	—	0.16	—	—
	2240	3.5	—	0.29	—	—
	1920	1.9	—	0.53	—	0.79
Calcitic, dolomitic, limestone, marble, and granite..	2880	4.3	—	0.23	—	—
	2560	3.2	—	0.32	—	—
	2240	2.3	—	0.43	—	—
	1920	1.6	—	0.63	—	0.79
	1600	1.1	—	0.90	—	—
Gypsum partition tile						
76 by 305 by 760, solid	—	—	4.50	—	0.222	0.79
76 by 305 by 760, 4 cells	—	—	4.20	—	0.238	—
102 by 305 by 760, 3 cells	—	—	3.40	—	0.294	—
Concretes						
Sand and gravel or stone aggregate concretes (concretes	2400	1.4–2.9	—	0.69–0.35	—	—
with more than 50% quartz or quartzite sand have	2240	1.3–2.6	—	0.77–0.39	—	0.8–1.0
conductivities in the higher end of the range)	2080	1.0–1.9	—	0.99–0.53	—	—
Limestone concretes	2240	1.60	—	0.62	—	—
	1920	1.14	—	0.88	—	—
	1600	0.79	—	1.26	—	—
Gypsum-fiber concrete (87.5% gypsum, 12.5% wood chips)	816	0.24	—	4.18	—	0.88
Cement/lime, mortar, and stucco	1920	1.40	—	0.71	—	—
	1600	0.97	—	1.04	—	—
	1280	0.65	—	1.54	—	—
Low density aggregate concretes						
Expanded shale, clay, or slate; expanded slags;	1920	0.9–1.3	—	1.08–0.76	—	—
cinders; pumice (with density up to 1600 kg/m³);	1600	0.68–0.89	—	1.48–1.12	—	0.84
and scoria (sanded concretes have conductivities	1280	0.48–0.59	—	2.10–1.69	—	0.84
in the higher end of the range)	960	0.30–0.36	—	3.30–2.77	—	—
	640	0.18	—	5.40	—	—

C, K, and R Values for Common Building Materials—In SI Metric Units (continued)

Figure 3.7a (continued)

Typical Thermal Properties of Common Building and Insulating Materials—Design Values[a] (Concluded)

Description	Density, kg/m³	Conductivity[b] (k), W/(m·K)	Conductance (C), W/(m²·K)	Resistance[c] (R) (1/k), K·m/W	Resistance[c] (R) For Thickness Listed (1/C), K·m²/W	Specific Heat, kJ/(kg·K)
Perlite, vermiculite, and polystyrene beads	800	0.26–0.27	—	3.81–3.68	—	—
	640	0.20–0.22	—	4.92–4.65	—	0.63–0.96
	480	0.16	—	6.31	—	—
	320	0.12	—	8.67	—	—
Foam concretes	1920	0.75	—	1.32	—	—
	1600	0.60	—	1.66	—	—
	1280	0.44	—	2.29	—	—
	1120	0.36	—	2.77	—	—
Foam concretes and cellular concretes	960	0.30	—	3.33	—	—
	640	0.20	—	4.92	—	—
	320	0.12	—	8.67	—	—
SIDING MATERIALS (on flat surface)						
Shingles						
Asbestos-cement	1900	—	27.0	—	0.037	—
Wood, 400 mm, 190-mm exposure	—	—	6.53	—	0.15	1.30
Wood, double, 400 mm, 300-mm exposure	—	—	4.77	—	0.21	1.17
Wood, plus insul. backer board, 8 mm	—	—	4.03	—	0.25	1.30
Siding						
Asbestos-cement, 6.4 mm, lapped	—	—	27.0	—	0.037	1.01
Asphalt roll siding	—	—	36.9	—	0.026	1.47
Asphalt insulating siding (12.7 mm bed.)	—	—	3.92	—	0.26	1.47
Hardboard siding, 11 mm	—	—	8.46	—	0.12	1.17
Wood, drop, 20 by 200 mm	—	—	7.21	—	0.14	1.17
Wood, bevel, 13 by 200 mm, lapped	—	—	6.98	—	0.14	1.17
Wood, bevel, 19 by 250 mm, lapped	—	—	5.40	—	0.18	1.17
Wood, plywood, 9.5 mm, lapped	—	—	9.03	—	0.10	1.22
Aluminum or Steel[m], over sheathing						
Hollow-backed	—	—	9.14	—	0.11	1.22
Insulating-board backed						
9.5 mm nominal	—	—	3.12	—	0.32	1.34
9.5 mm nominal, foil backed	—	—	1.93	—	0.52	—
Architectural (soda lime float) glass	—	—	56.8	—	0.018	0.84
WOODS (12% moisture content)[e,n]						
Hardwoods						1.63[o]
Oak	659–749	0.16–0.18	—	6.2–5.5	—	
Birch	682–726	0.167–0.176	—	6.0–5.7	—	
Maple	637–704	0.157–0.171	—	6.4–5.8	—	
Ash	614–670	0.153–0.164	—	6.5–6.1	—	
Softwoods						1.63[o]
Southern Pine	570–659	0.144–0.161	—	6.9–6.2	—	
Douglas Fir-Larch	536–581	0.137–0.145	—	7.3–6.9	—	
Southern Cypress	502–514	0.130–0.132	—	7.7–7.6	—	
Hem-Fir, Spruce-Pine-Fir	392–502	0.107–0.130	—	9.3–7.7	—	
West Coast Woods, Cedars	347–502	0.098–0.130	—	10.3–7.7	—	
California Redwood	392–448	0.107–0.118	—	9.4–8.5	—	

[a] Values are for a mean temperature of 24 °C. Representative values for dry materials are intended as design (not specification) values for materials in normal use. Thermal values of insulating materials may differ from design values depending on their in-situ properties (e.g., density and moisture content, orientation, etc.) and variability experienced during manufacture. For properties of a particular product, use the value supplied by the manufacturer or by unbiased tests.

[b] The symbol λ is also used to represent thermal conductivity.

[c] Resistance values are the reciprocals of C before rounding off C.

[d] Lewis (1967).

[e] U.S. Department of Agriculture (1974).

[f] Does not include paper backing and facing, if any. Where insulation forms a boundary (reflective or otherwise) of an air space, see Tables 2 and 3 for the insulating value of an air space with the appropriate effective emittance and temperature conditions of the space.

[g] Conductivity varies with fiber diameter. (See Chapter 20, Factors Affecting Thermal Performance.) Batt, blanket, and loose-fill mineral fiber insulations are manufactured to achieve specified R-values, the most common of which are listed in the table. Due to differences in manufacturing processes and materials, the product thicknesses, densities, and thermal conductivities vary over considerable ranges for a specified R-value.

[h] This material is relatively new and data are based on limited testing.

[i] For additional information, see Society of Plastics Engineers (SPI) *Bulletin* U108. Values are for aged, unfaced board stock. For change in conductivity with age of expanded polyurethane/polyisocyanurate, see Chapter 20, Factors Affecting Thermal Performance.

[j] Values are for aged products with gas-impermeable facers on the two major surfaces. An aluminum foil facer of 25 μm thickness or greater is generally considered impermeable to gases. For change in conductivity with age of expanded polyisocyanurate, see Chapter 20, Factors Affecting Thermal Performance, and SPI *Bulletin* U108.

[k] Insulating values of acoustical tile vary, depending on density of the board and on type, size, and depth of perforations.

[l] Values for fully grouted block may be approximated using values for concrete with a similar density.

[m] Values for metal siding applied over flat surfaces vary widely, depending on amount of ventilation of air space beneath the siding; whether air space is reflective of nonreflective; and on thickness, type, and application of insulating backing-board used. Values given are averages for use as design guides, and were obained from several guarded hot box tests (ASTM C236) or calibrated hot box (ASTM C976) on hollow-backed types and types made using backing-boards of wood fiber, foamed plastic, and glass fiber. Departures of ±50% or more from the values given may occur.

[n] See Adams (1971), MacLean (1941), and Wilkes (1979). The conductivity values listed are for heat transfer across the grain. The thermal conductivity of wood varies linearly with the density, and the density ranges listed are those normally found for the wood species given. If the density of the wood species is not known, use the mean conductivity value. For extrapolation to other moisture contents, the following empirical equation developed by Wilkes (1979) may be used:

$$k = 0.7494 + \frac{(4.895 \times 10^{-3} + 1.503 \times 10^{-4}M)\rho}{1 + 0.01M}$$

where ρ is density of the moist wood in kg/m³, and M is the moisture content in percent.

[o] From Wilkes (1979), an empirical equation for the specific heat of moist wood at 24 °C is as follows:

$$c_p = 0.1442 \left[\frac{(0.299 + 0.01M)}{(1 + 0.01M)} \right] + \Delta c_p$$

where Δc_p accounts for the heat of sorption and is denoted by

$$\Delta c_p = M(0.008037 - 1.325 \times 10^{-4}M)$$

where M is the moisture content in percent by mass.

(Copyright 1993 by ASHRAE, from *1993 Fundamentals, SI Edition*. Used by permission.)

Footnote: Cross-references in table footnotes pertain to the source from which tables were taken; they do not refer to chapters or pages in this text.

C, K, and R Values for Common Building Materials—In SI Metric Units (continued)

Figure 3.7a (continued)

P = Perimeter of slab (ft.) [m]
ΔT = Indoor temperature minus outdoor temperature

The edge factor, F, values are shown in Figure 3.11 (and Figure 3.11a in SI metric units) and vary from 0.47 [0.81] for edge slabs with perimeter insulation to 2.73 [4.72] for uninsulated edge slabs.

Slab Losses: Slabs on grade lose heat to the ground below. The loss is calculated as follows:

$H_s = K \times A$
H_s = Heat loss through slab
K = Constant (Btu/hr. s.f.) [W/m^2]
    2 [6.32 W/m^2] for floors and basement walls 4 feet [1.2 m] below grade *or*
    4 [12.6 W/m^2] for floors and walls from 0 to 4 feet [1.2 m] below grade
A = Area (s.f.) [m^2]

Convection Losses

Convection losses occur when outside air passes into the building. The two primary sources of convection losses are *infiltration* and *ventilation*. Infiltration results from the leakage of air into the building from cracks around windows, doors, and openings. Ventilation occurs when fresh

Table of Indoor Design Temperatures			
	Design Indoor Temperature °F		Design Relative Humidity %
Space	Winter	Summer	Summer
Assembly			
Museums	68–72	68–72	40–55
Restaurants	70–74	74–78	50–60
Gymnasiums	55–65	55–65	40–50
Auditoriums	72–76	76–80	50–60
Business	70–74	74–78	40–50
Commercial			
Retail	70–74	72–80*	30–55
Factories and			
Computer Rooms	68–74	77–85	45–55
High Hazard	special conditions		
Institutional (special applications — hospitals)	74–77	74–79	45–50
Mercantile	70–74	74–78	40–50
Residential			
Apt., House, Hotel	74–77	74–79	45–50
Storage (actual conditions depend upon materials stored)	above 32°	below 95°	N/A

Note: Typical energy code — Winter temp. = 68°F, Summer temp. = 78°F
*Short-term occupancy

Figure 3.8

Weather Data and Design Conditions (winter design @ 97.5% - summer design @ 2.5%)

City	Latitude (1) 0	Latitude (1) 1'	Winter Temperatures (1) Med. of Annual Extremes	Winter Temperatures (1) 99%	Winter Temperatures (1) 97½%	Winter Degree Days (2)	Summer (Design Dry Bulb) Temperatures and Relative Humidity 1%	Summer (Design Dry Bulb) Temperatures and Relative Humidity 2½%	Summer (Design Dry Bulb) Temperatures and Relative Humidity 5%
UNITED STATES									
Albuquerque, NM	35	0	6	12	16	4,400	96/61	94/61	92/61
Atlanta, GA	33	4	14	17	22	3,000	95/74	92/74	90/73
Baltimore, MD	39	2	12	14	17	4,600	94/75	92/75	89/74
Birmingham, AL	33	3	17	17	21	2,600	97/74	94/75	93/74
Bismarck, ND	46	5	-31	-23	-19	8,800	95/68	91/68	88/67
Boise, ID	43	3	0	3	10	5,800	96/65	93/64	91/64
Boston, MA	42	2	-1	6	9	5,600	91/73	88/71	85/70
Burlington, VT	44	3	-18	-12	-7	8,200	88/72	85/70	83/69
Charleston, WV	38	2	1	7	11	4,400	92/74	90/73	88/72
Charlotte, NC	35	1	13	18	22	3,200	96/74	94/74	92/74
Casper, WY	42	5	-20	-11	-5	7,400	92/58	90/57	87/57
Chicago, IL	41	5	-5	-3	2	6,600	94/75	91/74	88/73
Cincinnati, OH	39	1	2	1	6	4,400	94/73	92/72	90/72
Cleveland, OH	41	2	-2	1	5	6,400	91/73	89/72	86/71
Columbia, SC	34	0	16	20	24	2,400	98/76	96/75	94/75
Dallas, TX	32	5	14	18	22	2,400	101/75	99/75	97/75
Denver, CO	39	5	-9	-5	1	6,200	92/59	90/59	89/59
Des Moines, IA	41	3	-13	-10	-5	6,600	95/75	92/74	89/73
Detroit, MI	42	2	0	3	6	6,200	92/73	88/72	85/71
Great Falls, MT	47	3	-29	-21	-15	7,800	91/60	88/60	85/59
Hartford, CT	41	5	-4	3	7	6,200	90/74	88/73	85/72
Houston, TX	29	5	24	28	33	1,400	96/77	94/77	92/77
Indianapolis, IN	39	4	-2	-2	2	5,600	93/74	91/74	88/73
Jackson, MS	32	2	17	21	25	2,200	98/76	96/76	94/76
Kansas City, MO	39	1	-2	2	6	4,800	100/75	97/74	94/74
Las Vegas, NV	36	1	18	25	28	2,800	108/66	106/65	104/65
Lexington, KY	38	0	0	3	8	4,600	94/73	92/72	90/72
Little Rock, AR	34	4	13	15	20	3,200	99/76	96/77	94/77
Los Angeles, CA	34	0	38	41	43	2,000	94/70	90/70	87/69
Memphis, TN	35	0	11	13	18	3,200	98/77	96/76	94/76
Miami, FL	25	5	39	44	47	200	92/77	90/77	89/77
Milwaukee, WI	43	0	-11	-8	-4	7,600	90/74	87/73	84/71
Minneapolis, MN	44	5	-19	-16	-12	8,400	92/75	89/73	86/71
New Orleans, LA	30	0	29	29	33	1,400	93/78	91/78	90/77
New York, NY	40	5	6	11	15	5,000	94/74	91/73	88/72
Norfolk, VA	36	5	18	20	22	3,400	94/77	91/76	89/76
Oklahoma City, OK	35	2	4	9	13	3,200	100/74	97/74	95/73
Omaha, NE	41	2	-12	-8	-3	6,600	97/76	94/75	91/74
Philadelphia, PA	39	5	7	10	14	4,400	93/75	90/74	87/72
Phoenix, AZ	33	3	25	31	34	1,800	108/71	106/71	104/71
Pittsburgh, PA	40	3	1	3	7	6,000	90/72	88/71	85/70
Portland, ME	43	4	-14	-6	-1	7,600	88/72	85/71	81/69
Portland, OR	45	4	17	17	23	4,600	89/68	85/67	81/65
Portsmouth, NH	43	1	-8	-2	2	7,200	88/73	86/71	83/70
Providence, RI	41	4	0	5	9	6,000	89/73	86/72	83/70
Rochester, NY	43	1	-5	1	5	6,800	91/73	88/71	85/70
Salt Lake City, UT	40	5	-2	3	8	6,000	97/62	94/62	92/61
San Francisco, CA	37	5	38	38	40	3,000	80/63	77/62	83/61
Seattle, WA	47	4	22	22	27	5,200	81/68	79/66	76/65
Sioux Falls, SD	43	4	-21	-15	-11	7,800	95/73	92/72	89/71
St. Louis, MO	38	4	1	3	8	5,000	96/75	94/75	92/74
Tampa, FL	28	0	32	36	40	680	92/77	91/77	90/76
Trenton, NJ	40	1	7	11	14	5,000	92/75	90/74	87/73
Washington, DC	38	5	12	14	17	4,200	94/75	92/74	90/74
Wichita, KS	37	4	-1	3	7	4,600	102/72	99/73	96/73
Wilmington, DE	39	4	6	10	14	5,000	93/74	93/74	20/73
ALASKA									
Anchorage	61	1	-29	-23	-18	10,800	73/59	70/58	67/56
Fairbanks	64	5	-59	-51	-47	14,280	82/62	78/60	75/59
CANADA									
Edmonton, Alta.	53	3	-30	-29	-25	11,000	86/66	83/65	80/63
Halifax, N.S.	44	4	-4	1	5	8,000	83/66	80/65	77/64
Montreal, Que.	45	3	-20	-16	-10	9,000	88/73	86/72	84/71
Saskatoon, Sask.	52	1	-35	-35	-31	11,000	90/68	86/66	83/65
St. Johns, Nwf.	47	4	1	3	7	8,600	79/66	77/65	75/64
Saint John, N.B.	45	2	-15	-12	-8	8,200	81/67	79/65	77/64
Toronto, Ont.	43	4	-10	-5	-1	7,000	90/73	87/72	85/71
Vancouver, B.C.	49	1	13	15	19	6,000	80/67	78/66	76/65
Winnipeg, Man.	49	5	-31	-30	-27	10,800	90/73	87/71	84/70

(1) Handbook of Fundamentals, ASHRAE, Inc., NY 1972/1985
(2) Local Climatological Annual Survey, USDC Env. Science Services Administration, Ashville, NC

Table of Outdoor Design Temperatures

Figure 3.9

outdoor air is brought into the building to meet fresh air requirements. In both cases, the cold outdoor air must be heated to bring it to room temperature. It takes 0.24 Btu to heat 1 pound of air 1°F [4180 J/kg · K]; 1 pound of air at normal room temperature occupies 13.5 cubic feet [1 kg of air equals 0.83 m³]. The amount of heat necessary to heat the convected air is shown below.

$$H = \left(0.24 \; \frac{Btu}{lb.} \times \frac{1}{13.5} \; \frac{\#}{F^3} \times 60 \; min/hr.\right) \times \frac{F^3}{min.} \times °F$$

H_v = 1.1 cfm (ΔT) [1.23 L/s (ΔT)]
H_v = Convection heat loss (Btu/hr.) [W]
1.1 = Constant [1.23]
cfm = Air entering building (cubic feet per minute) [L/s] (see Figure 3.12)
ΔT = Temperature difference (°F) [°C]

The factor 1.1 [1.23] adjusts for the units and for the air density in the winter.

The cubic feet per minute (cfm) [L/s] of convected air into a building is calculated separately for infiltration and ventilation. For infiltration in the winter, a fifteen mile per hour [24 km/hr.] wind is assumed to be directed at the building. For each crack, a crack coefficient is used to determine how many cfm will infiltrate the building (per foot of crack).

Figure 3.12 lists crack coefficients for windows and doors. For each opening the air infiltrates, use the equation shown below.

cfm = crack coefficient × length of crack

ΔT for Adjoining Unheated Spaces		
	Winter	Summer
Adjoining Space	$T_i - T_{adj}$	$T_{adj} - T_i$
Party Wall	$T_i - T_{adj}$	$T_{adj} - T_i$
Attic (no insulation)*	$(T_i - T_o)/2$	$95 - T_i$ [35 °C $- T_i$]
Basement (no insulation)*	$(T_i - T_o)/2$	0
Attached Garage	$T_i - T_i$	$T_o - T_i$
Sun Porch	$T_i - T_o$	—
Slab on Grade and Walls 0 – 4' below Grade	4 Btu/hr/s.f. [1.2 W/m²]	0
Slab and Wall below Grade more than 4'	2 Btu/hr/s.f. [0.6 W/m²]	0
Crawl Spaces	$T_i - T_o$	0

All numbers shown in brackets are metric equivalents.

T_i = Indoor temperature °F [°C]
T_o = Outdoor temperature °F [°C]
T_{adj} = Temperature of adjoining space
* With insulation, the attic and basement temperature will be closer to outdoor temperature.

Figure 3.10

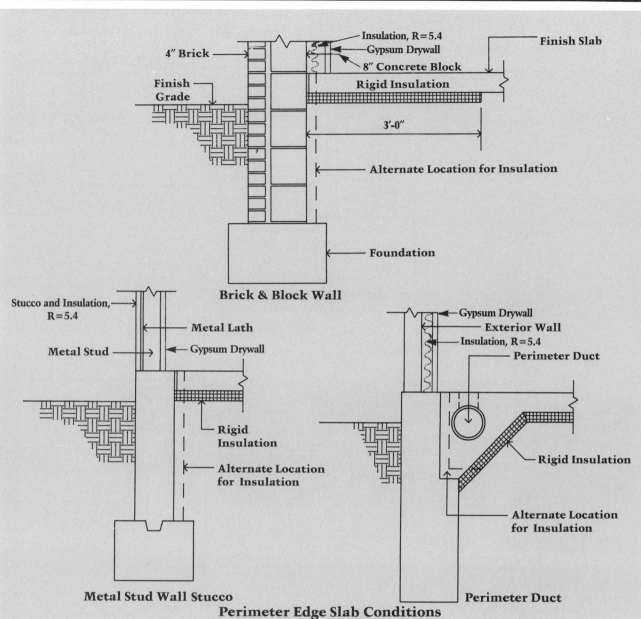

Perimeter Edge Slab Conditions

Heat Loss Coefficient of Slab Floor Construction, F_2
(Btu/h·F per ft of perimeter)

Construction	Insulated	Degree Days (65 °F Base)		
		2950	5350	7433
8-in. block wall, brick facing	Uninsulated	0.62	0.68	0.72
	R = 5.4 from edge to footer	0.48	0.50	0.56
4-in block wall, brick facing	Uninsulated	0.80	0.84	0.93
	R = 5.4 from edge to footer	0.47	0.49	0.54
Metal stud wall, stucco	Uninsulated	1.15	1.20	1.34
	R = 5.4 from edge to footer	0.51	0.53	0.58
Poured concrete wall, with duct near perimeter[a]	Uninsulated	1.84	2.12	2.73
	R = 5.4 from edge to footer, 3 ft under floor	0.64	0.72	0.90

[a] Weighted average temperature of the heating duct was assumed at 110 °F during the heating season (outdoor air temperature less than 65 °F)

(copyright 1993 by ASHRAE; from *1993 Fundamentals*. Used by permission.)

Figure 3.11

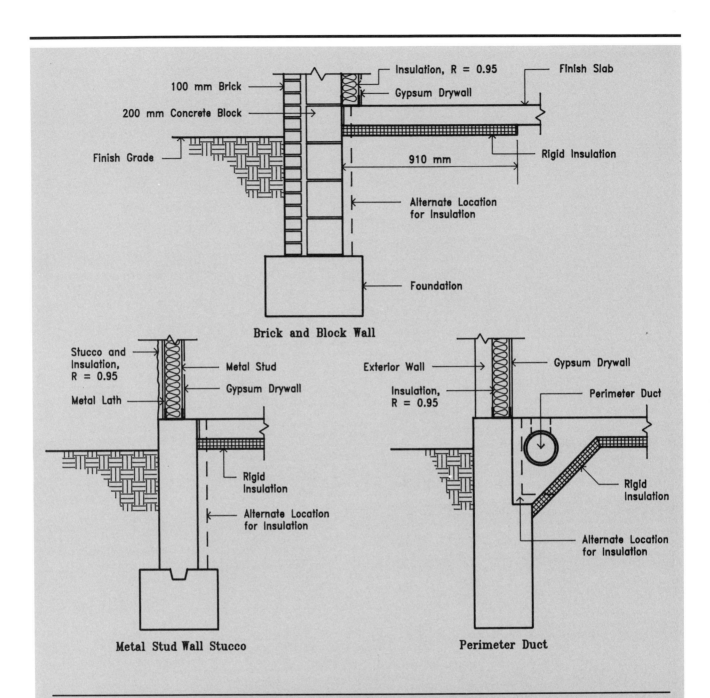

Perimeter Edge Slab Conditions—In SI Metric Units
Heat Loss Coefficient of Slab Floor Construction, F_2 [W/(m²·K) per Meter of Perimeter]

Construction	Insulation	Kelvin Days (18 °C Base)		
		4130 K·d/yr.	2970 K·d/yr.	1640 K·d/yr.
200 mm block wall, brick facing	Uninsulated	1.07	1.17	1.24
	$R = 0.95$ K·m²/W from edge to footer	0.83	0.86	0.97
100 mm block wall, brick facing	Uninsulated	1.38	1.45	1.61
	$R = 0.85$ from edge to footer	0.81	0.85	0.93
Metal stud wall, stucco	Uninsulated	1.99	2.07	2.32
	$R = 0.95$ from edge to footer	0.88	0.92	1.00
Poured concrete wall with duct near perimeter [a]	Uninsulated	3.18	3.67	4.72
	$R = 0.95$ from edge to footer, 910 mm under floor	1.11	1.24	1.56

[a] Weighted average temperature of the heating duct was assumed at 43 °C during the heating season (outdoor air temperature less than 18 °C).

Figure 3.11a

Infiltration

The wind blows from only one direction at a time. The cold air that infiltrates one side of a building usually forces warm air to exfiltrate the other side. Although it is impossible for air to infiltrate a building from all four sides at once, infiltration calculations are based on the cracks on all four sides of the building. Most designers correct for this inconsistency by sizing the distribution and terminal units for the full load and reducing the load from infiltration on the generating system by one-third to one-half. In this way, the boiler heats only the actual imposed load, while the piping and radiators are properly sized for their maximum possible load. The infiltration diversity is shown in Figure 3.13.

Ventilation

The other source of convection heat losses in buildings is ventilation air. For buildings with heating only, operable windows often satisfy the ventilation requirements. Ventilation exhaust fans used for kitchens and bathrooms operate intermittently and do not exceed the infiltrated air convection losses. Ventilated air that does exceed infiltrated air should be added to the convected cfm [L/s] quantity to be heated. In large buildings,

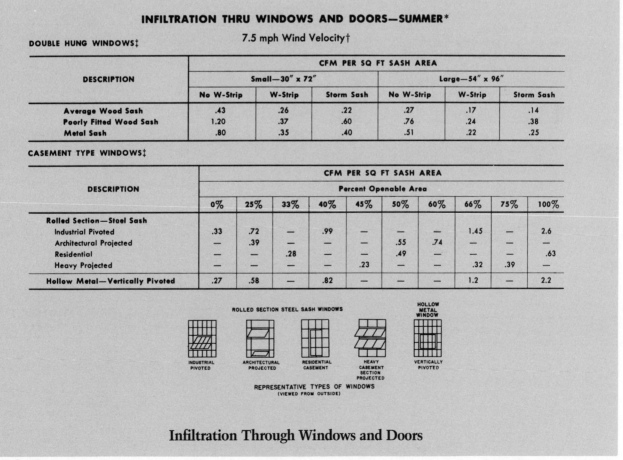

Infiltration Through Windows and Doors

Figure 3.12

INFILTRATION THRU WINDOWS AND DOORS—SUMMER* (Contd)

7.5 mph Wind Velocity†

DOORS ON ONE OR ADJACENT WALLS, FOR CORNER ENTRANCES

DESCRIPTION	CFM PER SQ FT AREA**		CFM Standing Open	
	No Use	Average Use	No Vestibule	Vestibule
Revolving Doors—Normal Operation	.8	5.2	—	—
Panels Open	—	—	1,200	900
Glass Door—3/16" Crack	4.5	10.0	700	500
Wood Door (3' x 7')	1.0	6.5	700	500
Small Factory Door	.75	6.5	—	—
Garage & Shipping Room Door	2.0	4.5	—	—
Ramp Garage Door	2.0	6.75	—	—

SWINGING DOORS ON OPPOSITE WALLS

% Time 2nd Door is Open	CFM PER PAIR OF DOORS % Time 1st Door is Open				
	10	25	50	75	100
10	100	250	500	750	1,000
25	250	625	1250	1875	2,500
50	500	1250	2500	3750	5,000
75	750	1875	3750	5625	7,500
100	1000	2500	5000	7500	10,000

DOORS

APPLICATION	CFM PER PERSON IN ROOM PER DOOR		
	72" Revolving Door	36" Swinging Door	
		No Vestibule	Vestibule
Bank	6.5	8.0	6.0
Barber Shop	4.0	5.0	3.8
Candy and Soda	5.5	7.0	5.3
Cigar Store	20.0	30.0	22.5
Department Store (Small)	6.5	8.0	6.0
Dress Shop	2.0	2.5	1.9
Drug Store	5.5	7.0	5.3
Hospital Room	—	3.5	2.6
Lunch Room	4.0	5.0	3.8
Men's Shop	2.7	3.7	2.8
Restaurant	2.0	2.5	1.9
Shoe Store	2.7	3.5	2.6

*All values in Table 41 are based on the wind blowing directly at the window or door. When the wind direction is oblique to the window or door, multiply the above values by 0.60 and use the total window and door area on the windward side(s).

†Based on a wind velocity of 7.5 mph. For design wind velocities different from the base, multiply the above values by the ratio of velocities.

‡Includes frame leakage where applicable.

**Vestibules may decrease the infiltration as much as 30% when the door usage is light. When door usage is heavy, the vestibule is of little value for reducing infiltration.

Example 1 — Infiltration in Tall Buildings, Summer

Given:
  A 20-story building in New York City oriented true north. Building is 100 ft long and 100 ft wide with a floor-to-floor height of 12 ft. Wall area is 50% residential casement windows having 50% fixed sash. There are ten 7 ft x 3 ft swinging glass doors on the street level facing south.

Find:
  Infiltration into the building thru doors and windows, disregarding outside air thru the equipment and the exhaust air quantity.

Solution:
  The prevailing wind in New York City during the summer is south, 13 mph (Table 1, page 10).

Infiltration Through Windows and Doors (continued)

Figure 3.12 (continued)

INFILTRATION THRU WINDOWS AND DOORS—WINTER*

15 mph Wind Velocity†

DOUBLE HUNG WINDOWS ON WINDWARD SIDE‡

	CFM PER SQ FT AREA					
	Small—30" x 72"			Large—54" x 96"		
DESCRIPTION	No W-Strip	W-Strip	Storm Sash	No W-Strip	W-Strip	Storm Sash
Average Wood Sash	.85	.52	.42	.53	.33	.26
Poorly Fitted Wood Sash	2.4	.74	1.2	1.52	.47	.74
Metal Sash	1.60	.69	.80	1.01	.44	.50

NOTE: W-Strip denotes weatherstrip.

CASEMENT TYPE WINDOWS ON WINDWARD SIDE‡

	CFM PER SQ FT AREA									
DESCRIPTION	Percent Ventilated Area									
	0%	25%	33%	40%	45%	50%	60%	66%	75%	100%
Rolled Section—Steel Sash										
Industrial Pivoted	.65	1.44	—	1.98	—	—	—	2.9	—	5.2
Architectural Projected	—	.78	—	—	—	1.1	1.48	—	—	—
Residential	—	—	.56	—	—	.98	—	—	—	1.26
Heavy Projected	—	—	—	—	.45	—	—	.63	.78	—
Hollow Metal—Vertically Pivoted	.54	1.19	—	1.64	—	—	—	2.4	—	4.3

DOORS ON ONE OR ADJACENT WINDWARD SIDES‡

	CFM PER SQ FT AREA**				
DESCRIPTION	Infrequent Use	Average Use			
		1 & 2 Story Bldg.	Tall Building (ft)		
			50	100	200
Revolving Door	1.6	10.5	12.6	14.2	17.3
Glass Door—(³⁄₁₆" Crack)	9.0	30.0	36.0	40.5	49.5
Wood Door 3' x 7'	2.0	13.0	15.5	17.5	21.5
Small Factory Door	1.5	13.0			
Garage & Shipping Room Door	4.0	9.0			
Ramp Garage Door	4.0	13.5			

Infiltration Through Windows and Doors (continued)

Figure 3.12 (continued)

INFILTRATION THRU WINDOWS AND DOORS—CRACK METHOD—SUMMER—WINTER*

DOUBLE HUNG WINDOWS—UNLOCKED ON WINDWARD SIDE

TYPE OF DOUBLE HUNG WINDOW	CFM PER LINEAR FOOT OF CRACK											
	Wind Velocity—Mph											
	5		10		15		20		25		30	
	No W-Strip	W-Strip	No W-Strip	W-Strip	No W-Strip	W-Strip	No W-Strip	W-Strip	No W-Strip	W-Strip	No W-Strip	W-Strip
Wood Sash												
Average Window	.12	.07	.35	.22	.65	.40	.98	.60	1.33	.82	1.73	1.05
Poorly Fitted Window	.45	.10	1.15	.32	1.85	.57	2.60	.85	3.30	1.18	4.20	1.53
Poorly Fitted—with Storm Sash	.23	.05	.57	.16	.93	.29	1.30	.43	1.60	.59	2.10	.76
Metal Sash	.33	.10	.78	.32	1.23	.53	1.73	.77	2.3	1.00	2.8	1.27

CASEMENT TYPE WINDOWS ON WINDWARD SIDE

TYPE OF CASEMENT WINDOW AND TYPICAL CRACK SIZE		CFM PER LINEAR FOOT OF CRACK					
		Wind Velocity—Mph					
		5	10	15	20	25	30
Rolled Section—Steel Sash							
Industrial Pivoted	1/16" crack	.87	1.80	2.9	4.1	5.1	6.2
Architectural Projected	1/32" crack	.25	.60	1.03	1.43	1.86	2.3
Architectural Projected	3/64" crack	.33	.87	1.47	1.93	2.5	3.0
Residential Casement	1/64" crack	.10	.30	.55	.78	1.00	1.23
Residential Casement	1/32" crack	.23	.53	.87	1.27	1.67	2.10
Heavy Casement Section Projected	1/64" crack	.05	.17	.30	.43	.58	.80
Heavy Casement Section Projected	1/32" crack	.13	.40	.63	.90	1.20	1.53
Hollow Metal—Vertically Pivoted		.50	1.46	2.40	3.10	3.70	4.00

*Infiltration caused by stack effect must be calculated separately during the winter.
†No allowance has been made for usage. See Table 43 for infiltration due to usage.

Infiltration Through Windows and Doors (continued)

Figure 3.12 (continued)

INFILTRATION THRU WINDOWS AND DOORS—CRACK METHOD—SUMMER—WINTER*
(Contd)

DOORS† ON WINDWARD SIDE

TYPE OF DOOR		CFM PER LINEAR FOOT OF CRACK					
		Wind Velocity— mph					
		5	10	15	20	25	30
Glass Door—Herculite							
Good Installation	⅛″ crack	3.2	6.4	9.6	13.0	16.0	19.0
Average Installation	3⁄16″ crack	4.8	10.0	14.0	20.0	24.0	29.0
Poor Installation	¼″ crack	6.4	13.0	19.0	26.0	26.0	38.0
Ordinary Wood or Metal							
Well Fitted—W-Strip		.45	.60	.90	1.3	1.7	2.1
Well Fitted—No W-Strip		.90	1.2	1.8	2.6	3.3	4.2
Poorly Fitted—No W-Strip		.90	2.3	3.7	5.2	6.6	8.4
Factory Door	⅛″ crack	3.2	6.4	9.6	13.0	16.0	19.0

(Courtesy Carrier Corporation, McGraw-Hill Book Company)

Footnote: Cross-references in table footnotes pertain to the source from which tables were taken; they do not refer to chapters or pages in this text.

Infiltration Through Windows and Doors (continued)

Figure 3.12 (continued)

the amount of ventilated air typically exceeds the infiltrated air. In small buildings the reverse is common.

Radiation Losses

Most heat losses are calculated based on the coldest winter night for a given climate. In the daytime during the winter, there is a solar heat gain, which some designers account for by partially offsetting the heat loss. Factors such as electric lights and body heat are also incorporated by some designers in comprehensive energy designs.

Heating Load Summary

The maximum design heating load is usually based on the demand of a cold winter night in a given climate. This is true in a residence, a hospital, or an auditorium where the facility is typically in use at night. For other building types, such as office buildings, two different design conditions are used. One condition is daytime-occupied use and the other is unoccupied (setback) night use.

The outdoor design temperature depends on the geographic location (see Figure 3.9). The inside design temperature is usually 68°F [20 °C]. The

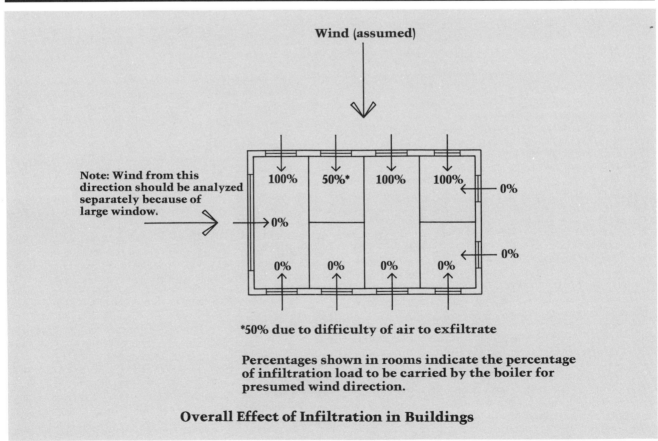

Figure 3.13

heating load consists of conduction and convection heat losses (see Figures 3.3 and 3.14).

Cooling Loads

The size of a cooling system is based on the load it must carry. In cooling, the load is expressed in Btu/hour or tons [watts or kilowatts]. One ton of cooling equals 12,000 Btu/hour, or 12 MBH [one kilowatt of cooling equals 1000 watts]. The cooling load represents the amount of heat that must be removed from a building, usually referred to as heat gain. In addition to the heat gain from conduction and convection (computed similarly to heat losses), cooling loads must take into account radiation and internal heat gain from lights, appliances, power, and people. All of these factors represent additional loads to a cooling system.

Heat Loss Formulas

Conduction Heat Losses
Losses Through Structure (Walls, Roofs, Doors, Windows)
  $H_C = UA (\Delta T)$
  H_C = Conduction loss (Btu/hr.) [W]
  U = Overall heat transfer coeffecient
      (Btu/hr. × ft.^{2} × °F) [W/m^2.K] (see Figures 3.5 – 3.7)
  A = Surface area (s.f.) [m^2]
  ΔT = Temperature difference (°F) [K]
Losses Through Edges of Slab
  $H_E = FP (\Delta T)$
  H_E = Heat loss through edge (Btu/hr.) [W]
  F = Edge factor (Btu/hr. ft. °F) [W/m^2.K/m] (see Figures 3.11 and 3.11a)
  P = Perimeter of slab (ft.) [m]
  ΔT = Indoor temperature minus outdoor temperature
Losses Through Slabs and Basement Walls
  $H_S = K \times A$ *or*
      = 2 [0.6] × Area for floors and basement walls 4 feet below grade *or*
      = 4 [1.2] × Area for floors and walls from 0 to 4 feet below grade
  H_S = Heat loss through slab
  K = Constant (Btu/hr. s.f.) [W/m^2]
  A = Area (s.f.) [m^2]

Convection Heat Losses
  $H_V = 1.1$ cfm (ΔT) [1.23 × L/s × ΔT]
  H_V = Convection loss (Btu/hr.) [W]
  1.1 = Constant
  cfm = Air entering building (cubic feet per minute) [m^3/s] (see Figure 3.12)
  ΔT = Temperature difference (°F) [°C or K]

Total Heat Loss (Btu/hr.)
  $H_T = H_C + H_E + H_V + H_S$
  H_C = Conduction losses (walls, windows, floor, roof, attics, crawl spaces, garages, basements)
  H_E = Edge losses, when applicable
  H_S = Slab losses (grades, foundations, walls)
  H_V = Infiltration and ventilation

All numbers shown in brackets are metric equivalents.

Figure 3.14

Cooling loads are based on the statistics for a hot summer day that is surpassed only 2-1/2% of the time in a particular climate. The critical time and date (which varies with each building) is determined by designers, because cooling load calculations are more complicated than heating load calculations. The primary reason for this complexity is the solar energy heat gain factor, computations for which are elaborate. The solar orientation of the building and the amount of glass receiving radiant energy varies not only from building to building, but also day by day, and even hour by hour. Furthermore, the heat "stored" in a building depends on the overall mass of the building, the hours of operation, and the length of time that lights are on in the building each day. The color of the building and type of sunscreen are also important factors. Light colors reflect more sun and lower the building heat load. Awnings, shades, and venetian blinds also cut out solar radiation and lower solar heat gain.

The actual cooling load calculation can involve many iterations or tests during the cooling season. The resulting design conditions may include many variations. For example, one room facing east may have its heaviest heat gain at 10:00 a.m. on June 19, while another room facing west may have its design load (heaviest heat gain) at 5:00 p.m. on August 21; the maximum overall load for the building as a whole may occur at 3:00 p.m. on July 20. The interior temperature may swing with the varied loads as the cooling equipment responds.

In addition to sensible heat (a change in the air temperature that is felt or "sensed"), cooling systems must also account for *latent heat gains*. Latent heat is the energy that is required to change a solid to a liquid or a liquid to a gas; no temperature change occurs during this process. In cooling, latent heat is the energy resulting from condensation of moisture in the air. Condensing water vapor is the reverse of boiling. It takes approximately 1,000 Btu's, or 1 MBH, to boil or condense a pound of water. [It takes approximately 2300 joules to boil or condense 1 kg of water.] Boiling normally occurs at 212°F [100 °C]. However, it is possible to vaporize 50°F [10 °C] water. It is also possible to pass humid air at 80°F [27 °C] over a cooling coil and produce 50°F [10 °C] air with a much lower humidity.

The energy required to vaporize the 50°F [10 °C] water and dehumidify the 80°F [27 °C] air is the *latent heat* of vaporization. Latent heat for cooling occurs when outdoor air (which contains moisture) is introduced into a building. People and cooking equipment also add moisture (latent heat) to a space. Sensible and latent heat gains are illustrated in Figure 3.15. Figure 3.16 illustrates the five categories of common building heat gains, which are listed below.

- Conduction heat gain
- Convection heat gain
- Solar heat gain
- Internal heat gain (lights, motors, cooking, etc.)
- People

Conduction Heat Gain

Conduction heat gain ("H_c" in Figure 3.16) is calculated in a manner similar to heat loss. The conductive heat gain through each envelope system (roof, walls, windows, and doors) is calculated separately, as shown in the following formula.

H_c = UA (CLTD)
H_c = Conduction heat gain (Btu/hr.) [W]

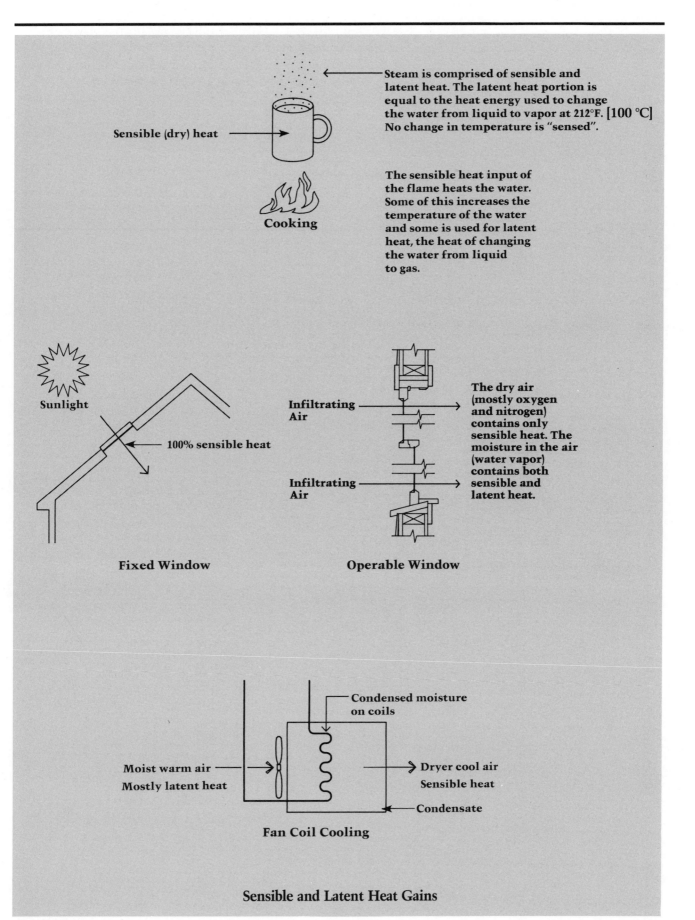

Figure 3.15

U = Overall heat transfer coefficient (Btu/s.f./°F) [W/m².K]
(see Figures 3.5 – 3.7)
A = Surface area (s.f.) [m²]
CLTD = Cooling load temperature difference (°F) [K]
(see Figures 3.17 and 3.17a)

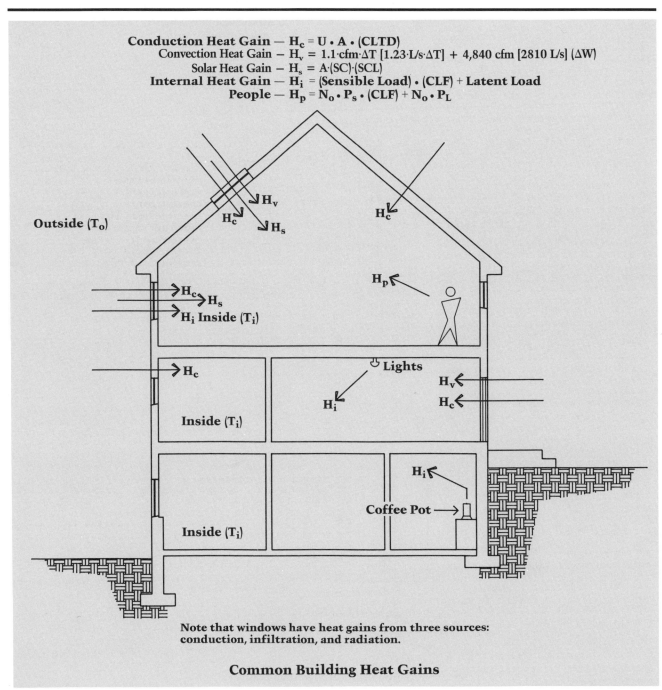

Figure 3.16

U The overall heat transfer coefficient for cooling is computed in the same way as it is for heating. The U values are used for both heating and cooling. Figures 3.5, 3.6, and 3.7 can be used to compute the U value for cooling. Because the wind velocity and air density are both lower in the summer, some designers choose to recompute the cooling U values by adjusting for the changes in f_i and f_o, shown in Figure 3.5.

A The surface area is computed in the same manner as for heating.

CLTD Calculating the temperature difference (between indoor design and actual outdoor conditions) for cooling is more complicated than the procedure for heating. Interior design temperatures can be read from the recommended indoor design temperatures in Figure 3.8. The temperature on the outside of the building envelope, however, is not necessarily the outdoor temperature. For example, a tar and gravel roof may reach a temperature of 250°F [120 °C] on a summer day, substantially warmer than the ambient 88°F [31 °C] outside design temperature. Similarly, wall color, the daily temperature range, the inside design temperature, the weight of construction, the latitude, orientation, time of day, and month all affect the actual outside surface temperature. (See Figures 3.8 and 3.9 for T_i and T_o.) For glass, CLTD equals $T_i - T_o$. The American Society of Heating, Refrigeration, and Air Conditioning Engineers (ASHRAE) has a more complete method for calculating the CLTD of glass, but many designers prefer the slightly conservative value stated here.

July Cooling Load Temperature Differences for Calculating Cooling Load from Flat Roofs at 40° North Latitude

Roof No.	1	2	3	4	5	6	7	8	9	10	11	Hour 12	13	14	15	16	17	18	19	20	21	22	23	24
1	0	−2	−4	−5	−6	−6	0	13	29	45	60	73	83	88	88	83	73	60	43	26	15	9	5	2
2	2	0	−2	−4	−5	−6	−4	4	17	32	48	62	74	82	86	85	80	70	56	39	25	15	9	5
3	12	8	5	2	0	−2	0	5	13	24	35	47	57	66	72	74	73	67	59	48	38	30	23	17
4	17	11	7	3	1	−1	−3	−3	0	7	17	29	42	54	65	73	77	78	74	67	56	45	34	24
5	21	16	12	8	5	3	1	2	6	12	21	31	41	51	60	66	69	69	65	59	51	42	34	27
8	28	24	21	17	14	12	10	10	12	16	21	28	35	42	48	53	56	57	56	52	48	43	38	33
9	32	26	21	16	13	9	6	4	4	7	12	19	27	36	45	53	59	63	64	63	58	52	45	38
10	37	32	27	23	19	15	12	10	9	10	12	17	23	30	37	44	50	55	57	58	56	52	47	42
13	34	31	28	25	22	20	18	16	16	17	20	24	28	33	38	42	46	48	49	48	46	44	40	37
14	35	32	30	27	25	23	21	20	19	20	22	24	28	32	36	39	42	44	45	45	44	42	40	37

Note 1. Direct application of data
- Dark surface
- Indoor temperature of 78°F
- Outdoor maximum temperature of 95°F with mean temperature of 85°F and daily range of 21°F
- Solar radiation typical of clear day on 21st day of month
- Outside surface film resistance of 0.333 (h·ft^2·°F)/Btu
- With or without suspended ceiling but no ceiling plenum air return systems
- Inside surface resistance of 0.685 (h·ft^2·°F)/Btu

Note 2. Adjustments to table data
- Design temperatures: Corr. CLTD = CLTD + (78 − t_r) + (t_m − 85)

where

t_r = inside temperature and t_m = mean outdoor temperature

t_m = maximum outdoor temperature − (daily range)/2

- No adjustment recommended for color
- No adjustment recommended for ventilation of air space above a ceiling

Roof Numbers Used in Table

Mass Location**	Suspended Ceiling	R-Factor, ft^2·h·°F/Btu	B7, Wood 1 in.	C12, HW Concrete 2 in.	A3, Steel Deck	Attic-Ceiling Combination
Mass inside the insulation	Without	0 to 5	*	2	*	*
		5 to 10	*	2	*	*
		10 to 15	*	4	*	*
		15 to 20	*	4	*	*
		20 to 25	*	5	*	*
		25 to 30	*	*	*	*
	With	0 to 5	*	5	*	*
		5 to 10	*	8	*	*
		10 to 15	*	13	*	*
		15 to 20	*	13	*	*
		20 to 25	*	14	*	*
		25 to 30	*	*	*	*
Mass evenly placed	Without	0 to 5	1	2	1	1
		5 to 10	2	*	1	2
		10 to 15	2	*	1	2
		15 to 20	4	*	2	2
		20 to 25	4	*	2	4
		25 to 30	*	*	*	*
	With	0 to 5	*	3	1	*
		5 to 10	4	*	1	*
		10 to 15	5	*	2	*
		15 to 20	9	*	2	*
		20 to 25	10	*	4	*
		25 to 30	10	*	*	*
Mass outside the insulation	Without	0 to 5	*	2	*	*
		5 to 10	*	3	*	*
		10 to 15	*	4	*	*
		15 to 20	*	5	*	*
		20 to 25	*	5	*	*
		25 to 30	*	*	*	*
	With	0 to 5	*	3	*	*
		5 to 10	*	3	*	*
		10 to 15	*	4	*	*
		15 to 20	*	5	*	*
		20 to 25	*	*	*	*
		25 to 30	*	*	*	*

*Denotes a roof that is not possible with the chosen parameters.
**The 2 in. concrete is considered massive and the others nonmassive.

Cooling Load Temperature Differences

Figure 3.17

Thermal Properties and Code Numbers of Layers Used in Wall and Roof Descriptions

Code Number	Description	Thickness and Thermal Properties					
		L	k	ρ	c_p	R	Mass
A0	Outside surface resistance	0.0	0.0	0.0	0.0	0.33	0.0
A1	1 in. Stucco	0.0833	0.4	116.0	0.20	0.21	9.7
A2	4 in. Face brick	0.333	0.77	125.0	0.22	0.43	41.7
A3	Steel siding	0.005	26.0	480.0	0.10	0.00	2.4
A4	1/2 in. Slag	0.0417	0.11	70.0	0.40	0.38	2.2
A5	Outside surface resistance	0.0	0.0	0.0	0.0	0.33	0.0
A6	Finish	0.0417	0.24	78.0	0.26	0.17	3.3
A7	4 in. Face brick	0.333	0.77	125.0	0.22	0.43	41.7
B1	Air space resistance	0.0	0.0	0.0	0.0	0.91	0.0
B2	1 in. Insulation	0.083	0.025	2.0	0.2	3.33	0.2
B3	2 in. Insulation	0.167	0.025	2.0	0.2	6.67	0.3
B4	3 in. Insulation	0.25	0.025	2.0	0.2	10.00	0.5
B5	1 in. Insulation	0.0833	0.025	5.7	0.2	3.33	0.5
B6	2 in. Insulation	0.167	0.025	5.7	0.2	6.67	1.0
B7	1 in. Wood	0.0833	0.07	37.0	0.6	10.00	3.1
B8	2.5 in. Wood	0.2083	0.07	37.0	0.6	2.98	7.7
B9	4 in. Wood	0.333	0.07	37.0	0.6	4.76	12.3
B10	2 in. Wood	0.167	0.07	37.0	0.6	2.39	6.2
B11	3 in. Wood	0.25	0.07	37.0	0.6	3.57	9.3
B12	3 in. Insulation	0.25	0.025	5.7	0.2	10.00	1.4
B13	4 in. Insulation	0.333	0.025	5.7	0.2	13.33	1.9
B14	5 in. Insulation	0.417	0.025	5.7	0.2	16.67	2.4
B15	6 in. Insulation	0.500	0.025	5.7	0.2	20.00	2.9
B16	0.15 in. Insulation	0.0126	0.025	5.7	0.2	0.50	0.1
B17	0.3 in. Insulation	0.0252	0.025	5.7	0.2	1.00	0.1
B18	0.45 in. Insulation	0.0379	0.025	5.7	0.2	1.50	0.2
B19	0.61 in. Insulation	0.0505	0.025	5.7	0.2	2.00	0.3
B20	0.76 in. Insulation	0.0631	0.025	5.7	0.2	2.50	0.4
B21	1.36 in. Insulation	0.1136	0.025	5.7	0.2	4.50	0.6
B22	1.67 in. Insulation	0.1388	0.025	5.7	0.2	5.50	0.8
B23	2.42 in. Insulation	0.2019	0.025	5.7	0.2	8.00	1.2
B24	2.73 in. Insulation	0.2272	0.025	5.7	0.2	9.00	1.3
B25	3.33 in. Insulation	0.2777	0.025	5.7	0.2	11.00	1.6
B26	3.64 in. Insulation	0.3029	0.025	5.7	0.2	12.00	1.7
B27	4.54 in. Insulation	0.3786	0.025	5.7	0.2	15.00	2.2
C1	4 in. Clay tile	0.333	0.33	70.0	0.2	1.01	23.3
C2	4 in. Lightweight concrete block	0.333	0.22	38.0	0.2	1.51	12.7
C3	4 in. Heavyweight concrete block	0.333	0.47	61.0	0.2	0.71	20.3
C4	4 in. Common brick	0.333	0.42	120.0	0.2	0.79	40.0
C5	4 in. Heavyweight concrete	0.333	1.0	140.0	0.2	0.33	46.7
C6	8 in. Clay tile	0.667	0.33	70.0	0.2	2.00	46.7
C7	8 in. Lightweight concrete block	0.667	0.33	38.0	0.2	2.00	25.3
C8	8 in. Heavyweight concrete block	0.667	0.6	61.0	0.2	1.11	40.7
C9	8 in. Common brick	0.667	0.42	120.0	0.2	1.59	80.0
C10	8 in. Heavyweight concrete	0.667	1.0	140.0	0.2	0.67	93.4
C11	12 in. Heavyweight concrete	1.0	1.0	140.0	0.2	1.00	140.0
C12	2 in. Heavyweight concrete	0.167	1.0	140.0	0.2	0.17	23.3
C13	6 in. Heavyweight concrete	0.5	1.0	140.0	0.2	0.50	70.0
C14	4 in. Lightweight concrete	0.333	0.1	40.0	0.20	3.33	13.3
C15	6 in. Lightweight concrete	0.5	0.1	40.0	0.2	5.00	20.0
C16	8 in. Lightweight concrete	0.667	0.1	40.0	0.2	6.67	26.7
C17	8 in. Lightweight concrete block (filled)	0.667	0.08	18.0	0.2	8.34	12.0
C18	8 in. Heavyweight concrete block (filled)	0.667	0.34	53.0	0.2	1.96	35.4
C19	12 in. Lightweight concrete block (filled)	1.000	0.08	19.0	0.2	12.50	19.0
C20	12 in. Heavyweight concrete block (filled)	1.000	0.39	56.0	0.2	2.56	56.0
E0	Inside surface resistance	0.0	0.0	0.0	0.0	0.69	0.0
E1	3/4 in. Plaster or gypsum	0.0625	0.42	100.0	0.2	0.15	6.3
E2	1/2 in. Slag or stone	0.0417	0.83	55.0	0.40	0.05	2.3
E3	3/8 in. Felt and membrane	0.0313	0.11	70.0	0.40	0.29	2.2
E4	Ceiling air space	0.0	0.0	0.0	0.0	1.00	0.0
E5	Acoustic tile	0.0625	0.035	30.0	0.2	1.79	1.9

L = thickness, ft
c_p = specific heat, Btu/lb·°F
k = thermal conductivity, Btu/h·ft·°F
R = thermal resistance, °F·ft^2·h/Btu
ρ = density, lb/ft^3
Mass = unit mass, lb/ft^2

Cooling Load Temperature Differences (continued)

Figure 3.17 (continued)

July Cooling Load Temperature Differences for Calculating Cooling Load from Sunlit Walls 40° North Latitude

Wall Number 1

Wall Face	1	2	3	4	5	6	7	8	9	10	11	12	13	14	15	16	17	18	19	20	21	22	23	24
N	1	0	−1	−2	−3	−1	7	11	11	13	17	21	25	27	29	29	28	29	27	17	11	7	5	3
NE	1	0	−1	−2	−3	2	24	42	47	43	35	28	27	28	29	29	27	24	20	14	10	7	5	3
E	1	0	−1	−2	−2	2	28	51	62	64	59	48	36	31	30	30	28	25	20	14	10	7	5	3
SE	1	0	−1	−2	−3	0	15	32	46	55	58	56	49	39	33	31	28	25	20	14	10	7	5	3
S	1	0	−1	−2	−3	−2	0	4	11	21	33	43	50	52	50	44	34	27	20	14	10	7	5	3
SW	2	0	−1	−2	−2	−2	0	4	8	13	17	25	39	53	64	70	69	61	45	24	13	8	5	3
W	2	1	−1	−2	−2	−2	1	4	8	13	17	21	27	42	59	73	80	79	62	32	16	9	6	3
NW	2	0	−1	−2	−2	−2	0	4	8	13	17	21	25	29	38	50	61	64	55	29	15	9	5	3

Wall Number 2

Wall Face	1	2	3	4	5	6	7	8	9	10	11	12	13	14	15	16	17	18	19	20	21	22	23	24
N	5	3	2	0	−1	−2	−1	3	7	9	11	14	18	21	24	26	27	28	28	27	22	17	12	8
NE	5	3	2	0	−1	−2	2	13	26	36	39	37	33	31	29	29	29	28	26	23	18	14	10	7
E	5	3	2	0	−1	−1	2	15	32	47	55	57	52	44	38	34	32	30	27	23	19	14	11	8
SE	5	3	2	0	−1	−2	0	8	20	33	43	50	53	51	45	39	35	31	28	24	19	14	11	8
S	5	3	2	0	−1	−2	−2	−1	2	7	14	24	33	42	47	48	46	40	33	27	21	15	11	8
SW	7	4	2	1	0	−1	−2	0	2	5	9	13	20	30	41	53	61	65	62	53	39	27	17	11
W	8	5	3	1	0	−1	−2	0	2	5	9	13	17	23	33	46	59	69	73	66	50	34	22	14
NW	8	4	2	1	−1	−2	−2	−1	2	5	9	13	17	21	25	32	41	51	57	54	42	29	19	12

Wall Number 3

Wall Face	1	2	3	4	5	6	7	8	9	10	11	12	13	14	15	16	17	18	19	20	21	22	23	24
N	7	5	3	2	1	0	2	5	7	8	11	14	17	20	23	24	25	26	27	24	20	16	13	10
NE	7	5	3	2	0	0	7	17	26	31	33	31	30	29	29	29	29	28	25	22	18	15	12	9
E	7	5	4	2	1	1	8	21	33	42	47	47	44	40	37	35	33	31	28	24	20	16	13	10
SE	8	5	4	2	1	0	4	12	22	32	39	44	46	44	41	38	35	32	29	24	20	16	13	10
S	8	6	4	2	1	0	0	1	4	9	16	24	31	38	41	42	40	36	31	26	22	17	14	11
SW	12	9	6	4	2	1	1	2	4	6	9	14	21	30	40	49	55	57	54	45	36	28	21	16
W	14	10	7	5	3	1	1	2	4	6	9	13	17	24	34	45	56	63	63	54	43	33	25	19
NW	12	8	6	4	2	1	0	2	3	6	9	13	16	20	25	32	40	48	50	44	35	27	21	16

Wall Number 4

Wall Face	1	2	3	4	5	6	7	8	9	10	11	12	13	14	15	16	17	18	19	20	21	22	23	24
N	11	8	6	4	2	0	0	1	3	5	7	10	13	16	19	22	24	26	27	27	26	22	19	15
NE	10	7	5	3	2	0	0	4	12	21	29	32	33	32	31	30	30	29	28	26	23	20	16	13
E	10	8	5	4	2	1	1	5	15	27	38	45	49	47	44	40	37	34	32	29	25	21	17	14
SE	11	8	6	4	2	1	0	2	8	17	27	36	43	46	46	44	41	37	34	30	26	22	18	14
S	11	8	6	4	2	1	0	−1	0	2	6	13	20	28	35	41	43	42	39	35	30	24	19	15
SW	18	13	9	6	3	2	0	0	0	2	5	8	12	18	27	36	46	53	57	57	51	42	33	25
W	21	15	10	7	4	2	1	0	1	2	5	8	11	15	21	30	40	51	60	64	60	50	40	30
NW	18	13	9	6	3	1	0	0	0	2	4	8	11	15	19	23	30	37	45	49	48	41	33	25

Wall Number 5

Wall Face	1	2	3	4	5	6	7	8	9	10	11	12	13	14	15	16	17	18	19	20	21	22	23	24
N	13	10	8	6	5	3	2	3	5	6	8	9	12	14	17	19	21	23	24	24	23	21	18	15
NE	13	10	8	6	5	3	3	7	14	20	25	27	28	28	28	28	28	27	26	23	21	18	15
E	14	11	9	7	5	4	4	8	17	26	33	39	40	40	38	37	35	34	32	29	26	23	20	17
SE	14	12	9	7	5	4	3	6	11	18	25	32	37	39	39	38	37	35	33	30	27	24	20	17
S	15	12	9	7	5	4	3	2	3	4	8	13	19	25	31	35	36	36	34	32	28	24	21	18
SW	22	18	14	11	8	6	5	4	4	5	6	9	12	17	25	33	40	46	49	48	44	38	32	26
W	25	20	16	13	10	7	5	4	4	5	7	9	11	15	20	28	37	45	52	54	50	44	37	30
NW	21	17	13	10	8	6	4	3	4	4	6	8	11	14	17	21	27	34	40	42	40	35	30	25

Wall Number 6

Wall Face	1	2	3	4	5	6	7	8	9	10	11	12	13	14	15	16	17	18	19	20	21	22	23	24
N	13	11	9	8	6	5	4	5	6	7	8	10	12	14	16	18	20	21	22	23	21	20	17	15
NE	14	12	10	8	6	5	6	10	15	20	23	25	25	26	26	27	27	27	26	25	23	21	18	16
E	16	13	11	9	7	6	7	11	18	25	31	35	36	36	35	35	34	33	31	29	26	24	21	18
SE	16	14	11	9	8	6	6	8	13	18	24	29	33	35	36	35	34	33	32	29	27	24	21	18
S	16	13	11	9	7	6	5	4	4	6	9	13	18	24	28	31	33	33	31	29	27	24	21	18
SW	23	19	16	14	11	9	7	6	6	7	8	10	13	18	24	31	37	42	44	43	40	35	31	27
W	26	22	18	15	13	10	8	7	7	7	8	10	12	15	20	27	35	42	47	48	45	40	35	30
NW	21	18	15	12	10	8	7	6	6	6	8	9	11	14	16	21	26	32	36	38	36	32	28	25

Cooling Load Temperature Differences (continued)

Figure 3.17 (continued)

July Cooling Load Temperature Differences for Calculating Cooling Load from Sunlit Walls 40° North Latitude (*Continued*)

Wall Number 7

Wall Face	\\ Hour 1	2	3	4	5	6	7	8	9	10	11	12	13	14	15	16	17	18	19	20	21	22	23	24
N	13	12	10	9	7	6	6	7	8	8	9	11	12	14	16	17	18	19	20	20	19	18	16	15
NE	15	13	11	10	9	8	9	13	17	20	22	23	23	24	24	25	25	25	24	23	22	20	18	16
E	17	15	13	12	10	9	11	16	21	26	30	32	32	32	32	32	31	30	29	27	25	23	21	19
SE	17	15	13	12	10	9	9	12	16	21	25	28	31	32	32	32	31	30	29	27	25	23	21	19
S	16	14	13	11	10	8	7	7	7	9	12	15	19	23	26	28	29	29	28	26	24	22	20	18
SW	23	20	18	16	13	12	10	10	10	10	11	12	15	20	25	30	35	38	39	37	34	31	28	25
W	25	22	20	17	15	13	12	11	11	11	12	13	14	17	22	28	34	39	42	41	38	34	31	28
NW	20	18	16	14	12	10	9	9	9	9	10	11	13	15	17	21	26	30	33	33	30	28	25	23

Wall Number 9

Wall Face	\\ Hour 1	2	3	4	5	6	7	8	9	10	11	12	13	14	15	16	17	18	19	20	21	22	23	24
N	17	15	13	11	9	7	5	4	4	4	5	7	8	10	12	15	17	19	21	22	23	23	22	20
NE	18	15	13	11	9	7	5	5	6	10	16	20	23	25	26	27	27	28	28	27	26	25	23	20
E	20	17	14	12	10	8	6	5	7	12	19	26	32	36	37	37	37	36	34	33	31	29	26	23
SE	20	17	15	12	10	8	6	5	6	9	13	19	25	31	34	36	37	36	35	34	32	29	26	23
S	21	18	15	12	10	8	6	5	4	3	4	6	10	14	20	25	29	33	34	34	32	30	27	24
SW	31	26	22	18	15	12	9	7	6	5	5	6	8	10	14	19	26	33	39	43	45	44	40	36
W	35	30	25	21	17	14	11	8	7	6	6	7	8	10	12	16	22	30	37	44	48	48	45	41
NW	29	25	21	17	14	11	9	7	5	5	5	6	7	9	11	14	18	22	28	34	37	38	36	33

Wall Number 10

Wall Face	\\ Hour 1	2	3	4	5	6	7	8	9	10	11	12	13	14	15	16	17	18	19	20	21	22	23	24
N	17	15	13	11	9	7	6	5	5	5	6	7	8	10	12	14	17	18	20	22	22	22	21	19
NE	18	16	13	11	9	7	6	6	8	12	16	20	22	24	25	26	27	27	27	26	24	22	20
E	20	17	15	12	10	8	7	7	10	14	20	26	31	34	35	36	36	35	34	33	31	28	26	23
SE	21	18	15	13	10	8	7	6	7	10	15	20	25	30	33	34	35	35	34	33	31	29	26	23
S	21	18	15	13	11	9	7	5	4	4	5	7	11	15	20	24	28	31	32	32	31	29	26	24
SW	31	27	23	19	16	13	10	8	7	6	6	7	8	11	15	20	26	32	38	41	42	41	38	35
W	34	30	26	22	18	15	12	9	8	7	7	7	8	10	13	17	23	30	37	42	45	45	42	39
NW	28	24	21	18	15	12	10	8	6	6	6	6	8	10	12	14	18	23	28	33	35	36	34	31

Wall Number 11

Wall Face	\\ Hour 1	2	3	4	5	6	7	8	9	10	11	12	13	14	15	16	17	18	19	20	21	22	23	24
N	16	14	13	12	10	9	8	7	7	7	8	9	10	11	12	14	15	17	18	19	20	19	18	17
NE	18	17	15	13	12	10	9	9	11	14	17	20	21	22	23	23	24	24	25	25	24	23	21	20
E	21	19	17	16	14	12	11	11	13	17	22	26	29	30	31	31	31	31	31	30	29	27	25	23
SE	21	19	17	16	14	12	11	10	11	14	17	21	24	27	29	30	31	31	30	30	29	27	25	23
S	20	18	16	15	13	11	10	9	8	8	8	10	13	16	19	23	25	27	28	28	27	25	24	22
SW	28	25	23	20	18	16	14	12	11	11	10	11	12	14	17	21	25	30	33	36	36	35	33	30
W	31	28	25	22	20	18	16	14	12	12	11	12	12	13	15	19	23	28	33	37	39	38	36	33
NW	25	23	20	18	16	14	12	11	10	9	9	10	11	12	13	15	18	22	26	29	31	31	29	27

Wall Number 12

Wall Face	\\ Hour 1	2	3	4	5	6	7	8	9	10	11	12	13	14	15	16	17	18	19	20	21	22	23	24
N	16	14	13	12	11	10	8	8	8	8	8	9	10	11	12	14	15	16	17	18	19	19	18	17
NE	18	17	15	14	13	11	10	10	12	14	17	19	21	21	22	23	23	24	24	24	23	22	21	20
E	22	20	18	17	15	13	12	12	14	17	21	25	28	29	30	30	30	30	30	29	28	27	25	24
SE	22	20	18	16	15	13	12	11	12	14	17	21	24	26	28	29	30	30	30	29	28	27	25	23
S	20	19	17	15	14	12	11	10	9	9	9	11	13	16	19	22	24	26	26	26	26	25	23	22
SW	27	25	23	21	19	17	15	14	12	12	12	12	14	17	20	24	28	32	34	34	34	32	30
W	30	28	25	23	21	19	17	15	14	13	13	13	14	16	19	23	27	32	35	37	36	35	33
NW	24	22	20	19	17	15	13	12	11	10	11	11	12	13	15	18	21	25	28	29	29	28	26

Cooling Load Temperature Differences (continued)

Figure 3.17 (continued)

July Cooling Load Temperature Differences for Calculating Cooling Load from Sunlit Walls 40° North Latitude (*Concluded*)

Wall Number 13

Wall Face	Hour																							
	1	2	3	4	5	6	7	8	9	10	11	12	13	14	15	16	17	18	19	20	21	22	23	24
N	15	14	13	12	11	10	9	9	9	9	9	10	10	11	12	14	15	16	17	18	18	18	17	16
NE	18	17	16	15	13	12	11	12	13	16	18	19	20	21	21	22	23	23	23	23	23	22	21	20
E	22	20	19	17	16	15	14	14	16	19	22	25	27	28	29	29	29	29	29	28	27	26	25	23
SE	22	20	19	17	16	14	13	13	14	16	18	21	24	26	27	28	28	28	28	28	27	26	24	23
S	20	18	17	16	14	13	12	11	10	10	11	12	14	16	19	21	23	24	25	25	24	23	22	21
SW	26	25	23	21	19	18	16	15	14	13	13	13	14	15	18	21	24	28	30	32	32	31	30	28
W	29	27	25	23	21	19	18	16	15	15	14	14	15	15	17	20	23	27	31	34	34	34	32	31
NW	23	22	20	18	17	15	14	13	12	12	12	12	12	13	14	16	18	21	24	26	27	27	26	25

Wall Number 14

Wall Face	Hour																							
	1	2	3	4	5	6	7	8	9	10	11	12	13	14	15	16	17	18	19	20	21	22	23	24
N	15	15	14	13	12	11	10	10	10	10	10	10	10	11	12	13	14	15	15	16	17	17	16	16
NE	19	18	17	16	15	14	13	13	14	15	17	18	19	20	20	21	21	22	22	22	22	22	21	20
E	23	22	21	19	18	17	16	15	16	18	21	23	25	26	27	27	28	28	28	28	27	26	25	24
SE	23	21	20	19	18	16	15	15	15	16	18	20	22	24	25	26	27	27	27	27	26	26	25	24
S	20	19	18	17	16	15	14	13	12	12	12	12	14	15	17	19	21	22	23	23	23	23	22	21
SW	26	25	24	22	21	19	18	17	16	15	15	15	15	16	17	19	22	25	27	29	30	30	29	28
W	29	27	26	24	23	21	20	18	17	16	16	16	16	16	17	19	21	24	27	30	32	32	31	30
NW	23	22	21	19	18	17	16	15	14	13	13	13	13	14	14	15	17	19	21	24	25	25	25	24

Wall Number 15

Wall Face	Hour																							
	1	2	3	4	5	6	7	8	9	10	11	12	13	14	15	16	17	18	19	20	21	22	23	24
N	19	18	16	14	12	10	9	7	6	6	6	6	7	8	9	11	13	15	17	19	20	21	21	20
NE	21	19	17	15	13	11	9	8	7	9	11	14	18	20	22	23	25	25	26	26	26	26	25	23
E	25	22	20	17	15	12	10	9	9	10	14	18	23	27	30	32	34	34	34	33	32	31	29	27
SE	25	22	20	17	15	13	11	9	8	8	10	14	18	22	26	30	32	33	34	33	33	31	30	27
S	25	22	20	17	15	13	11	9	7	6	6	6	7	10	13	17	21	25	28	30	30	30	29	27
SW	35	32	28	25	22	18	16	13	11	9	8	8	8	9	11	14	18	23	28	33	37	39	39	37
W	39	35	32	28	24	21	18	15	12	10	9	8	8	9	10	13	16	21	26	32	38	41	42	41
NW	31	28	26	23	20	17	14	12	10	8	7	7	7	8	9	11	13	16	20	25	29	32	33	33

Wall Number 16

Wall Face	Hour																							
	1	2	3	4	5	6	7	8	9	10	11	12	13	14	15	16	17	18	19	20	21	22	23	24
N	18	17	16	14	13	11	10	9	8	7	7	7	8	9	10	11	13	14	16	17	18	19	19	19
NE	21	20	18	16	14	13	11	10	10	11	13	15	17	19	21	22	23	24	24	25	25	24	24	23
E	25	23	21	19	17	15	13	11	11	12	15	19	22	26	28	30	31	31	32	32	31	30	29	27
SE	25	23	21	19	17	15	13	11	10	11	12	15	18	21	25	27	29	30	31	31	31	30	29	27
S	24	22	20	18	16	14	12	11	9	8	8	8	9	11	14	17	20	23	25	27	27	27	27	25
SW	33	30	28	25	23	20	18	15	13	12	11	10	10	11	12	15	18	22	27	30	33	35	35	34
W	36	33	31	28	25	22	20	17	15	13	12	11	11	11	12	14	17	20	25	30	34	37	38	37
NW	29	27	25	23	20	18	16	14	12	11	10	9	9	10	11	12	14	16	19	23	27	29	30	30

Note 1. Direct application of data
- Dark surface
- Indoor temperature of 78°F
- Outdoor maximum temperature of 95°F with mean temperature of 85°F and daily range of 21°F
- Solar radiation typical of clear day on 21st day of month
- Outside surface film resistance of 0.333 (h·ft²·°F)/Btu
- Inside surface resistance of 0.685 (h·ft²·°F)/Btu

Note 2. Adjustments to table data
- Design temperatures

$$\text{Corr. CLTD} = \text{CLTD} + (78 - t_r) + (t_m - 85)$$

where
t_r = inside temperature and
t_m = maximum outdoor temperature $-$ (daily range)/2

- No adjustment recommended for color

Cooling Load Temperature Differences (continued)

Figure 3.17 (continued)

Wall Types, Mass Located Inside Insulation

Secondary Material	R-Factor ft²·°F/Btu	A1	A2	B7	B10	B9	C1	C2	C3	C4	C5	C6	C7	C8	C17	C18
							Principal Wall Material**									
Stucco and/or plaster	0.0 to 2.0	*	*	*	*	*	*	*	*	*	*	*	*	*	*	*
	2.0 to 2.5	*	5	*	*	*	*	*	*	*	5	*	*	*	*	*
	2.5 to 3.0	*	5	*	*	*	3	*	2	5	6	*	*	5	*	*
	3.0 to 3.5	*	5	*	*	*	4	2	2	5	6	*	*	6	*	*
	3.5 to 4.0	*	5	*	*	*	4	2	3	6	6	10	4	6	*	5
	4.0 to 4.75	*	6	*	*	*	5	2	4	6	6	11	5	10	*	10
	4.75 to 5.5	*	6	*	*	*	5	2	4	6	6	11	5	10	*	10
	5.5 to 6.5	*	6	*	*	*	5	2	5	10	7	12	5	11	*	10
	6.5 to 7.75	*	6	*	*	*	5	4	5	11	7	16	10	11	*	11
	7.75 to 9.0	*	6	*	*	*	5	4	5	11	7	*	10	11	*	11
	9.0 to 10.75	*	6	*	*	*	5	4	5	11	7	*	10	11	4	11
	10.75 to 12.75	*	6	*	*	*	5	4	5	11	11	*	10	11	4	11
	12.75 to 15.0	*	10	*	*	*	10	4	5	11	11	*	10	11	9	12
	15.0 to 17.5	*	10	*	*	*	10	5	5	11	11	*	11	12	10	16
	17.5 to 20.0	*	11	*	*	*	10	5	9	11	11	*	15	16	10	16
	20.0 to 23.0	*	11	*	*	*	10	9	9	16	11	*	15	16	10	16
	23.0 to 27.0	*	*	*	*	*	*	*	*	*	*	*	16	*	15	*
Steel or other lightweight siding	0.0 to 2.0	*	*	*	*	*	*	*	*	*	*	*	*	*	*	*
	2.0 to 2.5	*	3	*	*	*	*	*	2	3	5	*	*	*	*	*
	2.5 to 3.0	*	5	*	*	*	2	*	2	5	3	*	*	5	*	*
	3.0 to 3.5	*	5	*	*	*	3	1	2	5	5	*	*	5	*	*
	3.5 to 4.0	*	5	*	*	*	3	2	2	5	5	6	3	5	*	5
	4.0 to 4.75	*	6	*	*	*	4	2	2	5	5	10	4	6	*	5
	4.75 to 5.5	*	6	*	*	*	5	2	2	6	6	11	5	6	*	6
	5.5 to 6.5	*	6	*	*	*	5	2	3	6	6	11	5	6	*	6
	6.5 to 7.75	*	6	*	*	*	5	2	3	6	6	11	5	6	*	10
	7.75 to 9.0	*	6	*	*	*	5	2	3	6	6	12	5	6	*	11
	9.0 to 10.75	*	6	*	*	*	5	2	3	6	6	12	5	6	4	11
	10.75 to 12.75	*	6	*	*	*	5	2	3	6	7	12	6	11	4	11
	12.75 to 15.0	*	6	*	*	*	5	2	4	6	7	12	10	11	5	11
	15.0 to 17.5	*	10	*	*	*	6	4	4	10	7	*	10	11	9	11
	17.5 to 20.0	*	10	*	*	*	10	4	4	10	11	*	10	11	10	11
	20.0 to 23.0	*	11	*	*	*	10	4	5	11	11	*	10	11	10	16
	23.0 to 27.0	*	*	*	*	*	*	*	*	*	*	*	10	*	11	16
Face brick	0.0 to 2.0	*	*	*	*	*	*	*	*	*	*	*	*	*	*	*
	2.0 to 2.5	3	*	*	*	*	*	*	*	*	11	*	*	*	*	*
	2.5 to 3.0	5	11	*	*	*	*	*	6	11	12	*	*	*	*	*
	3.0 to 3.5	5	12	5	*	*	11	*	11	12	12	*	*	12	*	*
	3.5 to 4.0	5	12	6	*	*	12	6	12	12	13	*	*	12	*	*
	4.0 to 4.75	6	13	6	10	*	13	10	12	12	13	*	11	*	*	16
	4.75 to 5.5	6	13	6	11	*	*	11	12	13	13	*	16	*	*	*
	5.5 to 6.5	6	13	6	11	*	*	11	12	13	13	*	*	*	*	*
	6.5 to 7.75	6	13	6	11	*	*	11	13	*	13	*	*	*	*	*
	7.75 to 9.0	6	13	10	16	*	*	11	13	*	13	*	*	*	*	*
	9.0 to 10.75	6	14	10	16	*	*	11	13	*	14	*	*	*	16	*
	10.75 to 12.75	6	14	10	16	*	*	11	13	*	14	*	*	*	16	*
	12.75 to 15.0	6	*	11	16	*	*	12	13	*	*	*	*	*	*	*
	15.0 to 17.5	10	*	11	*	*	*	12	13	*	*	*	*	*	*	*
	17.5 to 20.0	10	*	11	*	*	*	16	*	*	*	*	*	*	*	*
	20.0 to 23.0	11	*	15	*	*	*	16	*	*	*	*	*	*	*	*
	23.0 to 27.0	*	*	*	*	*	*	16	*	*	*	*	*	*	*	*

*Denotes a wall that is not possible with the chosen set of parameters.
**See Table 11 for definition of Code letters.

Cooling Load Temperature Differences (continued)

Figure 3.17 (continued)

Wall Types, Mass Evenly Distributed

Secondary Material	R-Factor ft²·°F/Btu	A1	A2	B7	B10	B9	C1	C2	C3	C4	C5	C6	C7	C8	C17	C18
								*Principal Wall Material***								
Stucco and/or plaster	0.0 to 2.0	1	3	*	*	*	*	*	1	3	3	*	*	*	*	*
	2.0 to 2.5	1	3	1	*	*	2	*	2	4	4	*	*	5	*	*
	2.5 to 3.0	1	4	1	*	*	2	2	2	4	4	*	*	5	*	*
	3.0 to 3.5	1	*	1	*	*	2	2	*	*	*	10	4	5	*	4
	3.5 to 4.0	1	*	1	2	*	*	4	*	*	*	10	4	*	*	4
	4.0 to 4.75	1	*	1	2	*	*	*	*	*	*	10	4	*	*	4
	4.75 to 5.5	1	*	1	2	*	*	*	*	*	*	*	*	*	*	*
	5.5 to 6.5	1	*	2	4	10	*	*	*	*	*	*	*	*	*	*
	6.5 to 7.75	1	*	2	4	11	*	*	*	*	*	*	*	*	*	*
	7.75 to 9.0	1	*	2	4	16	*	*	*	*	*	*	*	*	*	*
	9.0 to 10.75	1	*	2	4	16	*	*	*	*	*	*	*	*	4	*
	10.75 to 12.75	1	*	2	5	*	*	*	*	*	*	*	*	*	4	*
	12.75 to 15.0	2	*	2	5	*	*	*	*	*	*	*	*	*	*	*
	15.0 to 17.5	2	*	2	5	*	*	*	*	*	*	*	*	*	*	*
	17.5 to 20.0	2	*	2	9	*	*	*	*	*	*	*	*	*	*	*
	20.0 to 23.0	2	*	4	9	*	*	*	*	*	*	*	*	*	*	*
	23.0 to 27.0	*	*	*	9	*	*	*	*	*	*	*	*	*	*	*
Steel or other lightweight siding	0.0 to 2.0	1	3	*	*	*	*	*	1	3	2	*	*	*	*	*
	2.0 to 2.5	1	3	1	*	*	2	*	1	3	2	*	*	3	*	*
	2.5 to 3.0	1	4	1	*	*	2	1	2	4	4	*	*	3	*	*
	3.0 to 3.5	1	*	1	*	*	4	1	*	*	*	5	2	4	*	4
	3.5 to 4.0	1	*	1	2	*	*	2	*	*	*	5	2	*	*	4
	4.0 to 4.75	1	*	1	2	*	*	*	*	*	*	10	4	*	*	4
	4.75 to 5.5	1	*	1	2	*	*	*	*	*	*	*	*	*	*	*
	5.5 to 6.5	1	*	1	2	10	*	*	*	*	*	*	*	*	*	*
	6.5 to 7.75	1	*	1	4	11	*	*	*	*	*	*	*	*	*	*
	7.75 to 9.0	1	*	2	4	16	*	*	*	*	*	*	*	*	*	*
	9.0 to 10.75	1	*	2	4	16	*	*	*	*	*	*	*	*	2	*
	10.75 to 12.75	1	*	2	4	*	*	*	*	*	*	*	*	*	4	*
	12.75 to 15.0	1	*	2	5	*	*	*	*	*	*	*	*	*	*	*
	15.0 to 17.5	1	*	2	5	*	*	*	*	*	*	*	*	*	*	*
	17.5 to 20.0	1	*	2	5	*	*	*	*	*	*	*	*	*	*	*
	20.0 to 23.0	2	*	4	9	*	*	*	*	*	*	*	*	*	*	*
	23.0 to 27.0	*	*	*	9	*	*	*	*	*	*	*	*	*	*	*
Face brick	0.0 to 2.0	3	6	*	*	*	*	*	*	*	6	*	*	*	*	*
	2.0 to 2.5	3	10	*	*	*	*	*	5	10	10	*	*	*	*	*
	2.5 to 3.0	4	10	5	*	*	5	*	5	10	11	*	*	10	*	*
	3.0 to 3.5	*	11	5	*	*	10	5	5	11	11	15	10	10	*	10
	3.5 to 4.0	*	11	5	10	*	10	5	5	11	11	16	10	16	*	10
	4.0 to 4.75	*	11	*	11	*	10	5	5	16	11	*	10	16	*	16
	4.75 to 5.5	*	11	*	11	*	10	5	10	16	16	*	10	16	*	16
	5.5 to 6.5	*	16	*	*	*	10	9	10	16	11	*	11	16	*	16
	6.5 to 7.75	*	16	*	*	*	11	9	10	16	16	*	16	16	*	*
	7.75 to 9.0	*	16	*	*	*	15	9	10	16	*	*	15	16	*	*
	9.0 to 10.75	*	16	*	*	*	15	10	10	*	16	*	16	*	10	*
	10.75 to 12.75	*	16	*	*	*	16	10	10	*	*	*	16	*	15	*
	12.75 to 15.0	*	16	*	*	*	16	10	10	*	16	*	*	*	15	*
	15.0 to 17.5	*	*	*	*	*	16	10	15	*	*	*	*	*	16	*
	17.5 to 20.0	*	*	*	*	*	16	15	15	*	*	*	*	*	16	*
	20.0 to 23.0	*	*	*	*	*	*	15	16	*	*	*	*	*	*	*
	23.0 to 27.0	*	*	*	*	*	*	15	*	*	*	*	*	*	*	*

*Denotes a wall that is not possible with the chosen set of parameters.
**See Table 11 for definition of Code letters.

Cooling Load Temperature Differences (continued)

Figure 3.17 (continued)

Wall Types, Mass Located Outside Insulation

Secondary Material	R-Factor ft²·°F/Btu	Principal Wall Material**														
		A1	A2	B7	B10	B9	C1	C2	C3	C4	C5	C6	C7	C8	C17	C18
Stucco and/or plaster	0.0 to 2.0	*	*	*	*	*	*	*	*	*	*	*	*	*	*	*
	2.0 to 2.5	*	3	*	*	*	*	*	2	3	5	*	*	*	*	*
	2.5 to 3.0	*	3	*	*	*	2	*	2	4	5	*	*	5	*	*
	3.0 to 3.5	*	3	*	*	*	2	2	2	5	5	*	*	5	*	*
	3.5 to 4.0	*	3	*	*	*	2	2	2	5	5	10	4	6	*	5
	4.0 to 4.75	*	4	*	*	*	4	2	2	5	5	10	4	6	*	9
	4.75 to 5.5	*	4	*	*	*	4	2	2	5	6	11	5	10	*	10
	5.5 to 6.5	*	5	*	*	*	4	2	2	5	6	11	5	10	*	10
	6.5 to 7.75	*	5	*	*	*	4	2	2	5	6	11	5	10	*	10
	7.75 to 9.0	*	5	*	*	*	5	2	4	5	6	16	10	10	*	10
	9.0 to 10.75	*	5	*	*	*	5	4	4	5	6	16	10	10	4	11
	10.75 to 12.75	*	5	*	*	*	5	4	4	10	6	16	10	10	9	11
	12.75 to 15.0	*	5	*	*	*	5	4	4	10	10	*	10	11	9	11
	15.0 to 17.5	*	5	*	*	*	5	4	4	10	10	*	10	11	10	16
	17.5 to 20.0	*	5	*	*	*	9	4	4	10	10	*	10	15	10	16
	20.0 to 23.0	*	9	*	*	*	9	9	9	15	10	*	10	15	15	16
	23.0 to 27.0	*	*	*	*	*	*	*	*	*	*	*	15	*	15	16
Steel or other lightweight siding	0.0 to 2.0	*	*	*	*	*	*	*	*	*	*	*	*	*	*	*
	2.0 to 2.5	*	3	*	*	*	*	*	2	3	2	*	*	*	*	*
	2.5 to 3.0	*	3	*	*	*	2	*	2	3	2	*	*	*	*	*
	3.0 to 3.5	*	3	*	*	*	2	1	2	4	3	*	*	4	*	*
	3.5 to 4.0	*	3	*	*	*	2	2	2	4	3	5	2	5	*	4
	4.0 to 4.75	*	3	*	*	*	2	2	2	4	3	10	3	5	*	5
	4.75 to 5.5	*	3	*	*	*	2	2	2	5	3	10	4	5	*	5
	5.5 to 6.5	*	4	*	*	*	2	2	2	5	3	10	4	5	*	5
	6.5 to 7.75	*	4	*	*	*	2	2	2	5	4	11	5	5	*	6
	7.75 to 9.0	*	5	*	*	*	2	2	2	5	4	11	5	5	*	6
	9.0 to 10.75	*	5	*	*	*	2	2	2	5	4	11	5	5	4	10
	10.75 to 12.75	*	5	*	*	*	4	2	2	5	5	11	5	5	4	10
	12.75 to 15.0	*	5	*	*	*	4	2	2	5	5	11	5	10	5	10
	15.0 to 17.5	*	5	*	*	*	4	2	4	5	5	16	9	10	9	10
	17.5 to 20.0	*	5	*	*	*	4	4	4	9	5	16	9	10	10	10
	20.0 to 23.0	*	9	*	*	*	4	4	4	9	9	16	10	10	10	11
	23.0 to 27.0	*	*	*	*	*	*	*	*	*	*	16	10	*	10	15
Face brick	0.0 to 2.0	*	*	*	*	*	*	*	*	*	*	*	*	*	*	*
	2.0 to 2.5	3	*	*	*	*	*	*	*	*	11	*	*	*	*	*
	2.5 to 3.0	3	10	*	*	*	*	*	5	10	11	*	*	*	*	*
	3.0 to 3.5	3	11	5	*	*	10	*	5	11	11	*	*	11	*	*
	3.5 to 4.0	3	11	5	*	*	10	5	6	11	11	*	*	11	*	*
	4.0 to 4.75	3	11	5	10	*	10	5	10	11	11	*	10	11	*	16
	4.75 to 5.5	3	12	5	10	*	10	9	10	11	12	*	11	16	*	16
	5.5 to 6.5	4	12	5	10	*	10	10	10	12	12	*	15	16	*	16
	6.5 to 7.75	4	12	5	10	*	11	10	10	12	12	*	16	*	*	16
	7.75 to 9.0	5	12	5	15	*	11	10	10	16	12	*	16	*	*	*
	9.0 to 10.75	5	12	9	15	*	11	10	10	16	12	*	16	*	15	*
	10.75 to 12.75	5	12	10	15	*	11	10	10	*	12	*	16	*	15	*
	12.75 to 15.0	5	*	10	16	*	11	10	11	*	*	*	16	*	15	*
	15.0 to 17.5	5	*	10	16	*	15	10	11	*	*	*	16	*	*	*
	17.5 to 20.0	5	*	10	16	*	16	15	15	*	*	*	*	*	*	*
	20.0 to 23.0	9	*	15	16	*	16	15	15	*	*	*	*	*	*	*
	23.0 to 27.0	*	*	*	*	*	*	15	*	*	*	*	*	*	*	*

*Denotes a wall that is not possible with the chosen set of parameters.
**See Table 11 for definition of Code letters.

Cooling Load Temperature Differences (continued)

Figure 3.17 (continued)

Convection Heat Gain

Convection heat gain is calculated in a manner similar to convection heat loss. Both infiltration and ventilation contribute to heat gain. The formula for convection heat gain is shown below.

$$H_v = 1.1 \text{ cfm } (\Delta T) + 4840 \text{ cfm } (\Delta W)$$
$$1.23 \text{ L/s } (\Delta T) + 3010 \text{ L/s } (\Delta W)$$

Sensible heat gain   Latent heat gain

ΔT = Difference between indoor design and outdoor air (°F) [°C] temperature

ΔW = Difference between design indoor and outdoor air moisture content
(lbs. moisture/lb. dry air)   [kg moisture/kg dry air]

The cfm [L/s] of air entering the building from infiltration is computed in the same way as infiltration affecting heating, except that for cooling, a 7.5 mph [12 km/hr.] wind velocity is interpolated (see Figure 3.12). When

---

Table 34 Cooling Load Temperature Differences (CLTD) for Conduction through Glass

Solar Time, h	CLTD, °F	Solar Time, h	CLTD, °F
0100	1	1300	12
0200	0	1400	13
0300	−1	1500	14
0400	−2	1600	14
0500	−2	1700	13
0600	−2	1800	12
0700	−2	1900	10
0800	0	2000	8
0900	2	2100	6
1000	4	2200	4
1100	7	2300	3
1200	9	2400	2

Corrections: The values in the table were calculated for an inside temperature of 78 °F and an outdoor maximum temperature of 95 °F with an outdoor daily range of 21 °F. The table remains approximately correct for other outdoor maximums 93 to 102 °F and other outdoor daily ranges 16 to 34 °F, provided the outdoor daily average temperature remains approximately 85 °F. If the room air temperature is different from 78 °F and/or the outdoor daily average temperature is different from 85 °F, see note 2, Table 32.

(Copyright 1993 by ASHRAE, from 1993 Fundamentals. Used by permission.)

Footnote: Cross-references in table footnotes pertain to the source from which tables were taken; they do not refer to chapters or pages in this text.

Cooling Load Temperature Differences (continued)

Figure 3.17 (continued)

July Cooling Load Temperature Differences for Calculating Cooling Load from Flat Roofs at 40° North Latitude

Roof No.	1	2	3	4	5	6	7	8	9	10	11	12	13	14	15	16	17	18	19	20	21	22	23	24
1	0	−1	−2	−3	−3	−3	0	7	16	25	33	41	46	49	49	46	41	33	24	14	8	5	3	1
2	1	0	−1	−2	−3	−3	−2	2	9	18	27	34	41	46	48	47	44	39	31	22	14	8	5	3
3	7	4	3	1	0	−1	0	3	7	13	19	26	32	37	40	41	41	37	33	27	21	17	13	9
4	9	6	4	2	1	−1	−2	−2	0	4	9	16	23	30	36	41	43	43	41	37	31	25	19	13
5	12	9	7	4	3	2	1	1	3	7	12	17	23	28	33	37	38	38	36	33	28	23	19	15
8	16	13	12	9	8	7	6	6	7	9	12	16	19	23	27	29	31	32	31	29	27	24	21	18
9	18	14	12	9	7	5	3	2	2	4	7	11	15	20	25	29	33	35	36	35	32	29	25	21
10	21	18	15	13	11	8	7	6	5	6	7	9	13	17	21	24	28	31	32	32	31	29	26	23
13	19	17	16	14	12	11	10	9	9	9	11	13	16	18	21	23	26	27	27	27	26	24	22	21
14	19	18	17	15	14	13	12	11	11	11	12	13	16	18	20	22	23	24	25	25	24	23	22	21

Note 1. Direct application of data
- Dark surface
- Indoor temperature of 25.5 °C
- Outdoor maximum temperature of 35 °C with mean temperature of 29.4 °C and daily range of 11.6 °C
- Solar radiation typical of clear day on 21st day of month
- Outside surface film resistance of 0.059 m² · K/W
- With or without suspended ceiling but no ceiling plenum air return systems
- Inside surface resistance of 0.121 m² · K/W

Note 2. Adjustments to table data
- Design temperatures : Corr. CLTD = CLTD + $(25.5 - t_r) + (t_m - 29.4)$

where

t_r = inside temperature and t_m = mean outdoor temperature

t_m = maximum outdoor temperature − (daily range)/2

- No adjustment recommended for color
- No adjustment recommended for ventilation of air space above a ceiling

Roof Numbers Used in Table

Mass Location**	Suspended Ceiling	R-Factor, m² · K/W	B7, Wood 25 mm	C12, HW Concrete 50 mm	A3, Steel Deck	Attic-Ceiling Combination
Mass inside the insulation	Without	0 to 0.9	*	2	*	*
		0.9 to 1.8	*	2	*	*
		1.8 to 2.6	*	4	*	*
		2.6 to 3.5	*	4	*	*
		3.5 to 4.4	*	5	*	*
		4.4 to 5.3	*	*	*	*
	With	0 to 0.9	*	5	*	*
		0.9 to 1.8	*	8	*	*
		1.8 to 2.6	*	13	*	*
		2.6 to 3.5	*	13	*	*
		3.5 to 4.4	*	14	*	*
		4.4 to 5.3	*	*	*	*
Mass evenly placed	Without	0 to 0.9	1	2	1	1
		0.9 to 1.8	2	*	1	2
		1.8 to 2.6	2	*	1	2
		2.6 to 3.5	4	*	2	2
		3.5 to 4.4	4	*	2	4
		4.4 to 5.3	*	*	*	*
	With	0 to 0.9	*	3	1	*
		0.9 to 1.8	4	*	1	*
		1.8 to 2.6	5	*	2	*
		2.6 to 3.5	9	*	2	*
		3.5 to 4.4	10	*	4	*
		4.4 to 5.3	10	*	*	*
Mass outside the insulation	Without	0 to 0.9	*	2	*	*
		0.9 to 1.8	*	3	*	*
		1.8 to 2.6	*	4	*	*
		2.6 to 3.5	*	5	*	*
		3.5 to 4.4	*	5	*	*
		4.4 to 5.3	*	*	*	*
	With	0 to 0.9	*	3	*	*
		0.9 to 1.8	*	3	*	*
		1.8 to 2.6	*	4	*	*
		2.6 to 3.5	*	5	*	*
		3.5 to 4.4	*	*	*	*
		4.4 to 5.3	*	*	*	*

*Denotes a roof that is not possible with the chosen parameters. **The 50-mm concrete is considered massive and the others nonmassive.

Cooling Load Temperature Differences — In SI Metric Units

Figure 3.17a

Thermal Properties and Code Numbers of Layers Used in Wall and Roof Descriptions

Code Number	Description	Thickness and Thermal Properties					
		L	k	ρ	c_p	R	Mass
A0	Outside surface resistance	0	0.000	0	0.00	0.059	0.00
A1	25 mm Stucco	25	0.692	1858	0.84	0.037	47.34
A2	100 mm Face brick	100	1.333	2002	0.92	0.076	203.50
A3	Steel siding	2	44.998	7689	0.42	0.000	11.71
A4	12 mm Slag	13	0.190	1121	1.67	0.067	10.74
A5	Outside surface resistance	0	0.000	0	0.00	0.059	0.00
A6	Finish	13	0.415	1249	1.09	0.031	16.10
A7	100 mm Face brick	100	1.333	2002	0.92	0.076	203.50
B1	Air space resistance	0	0.000	0	0.00	0.160	0.00
B2	25 mm Insulation	25	0.043	32	0.84	0.587	0.98
B3	50 mm Insulation	51	0.043	32	0.84	1.173	1.46
B4	75 mm Insulation	76	0.043	32	0.84	1.760	2.44
B5	25 mm Insulation	25	0.043	91	0.84	0.587	2.44
B6	50 mm Insulation	51	0.043	91	0.84	1.173	4.88
B7	25 mm Wood	25	0.121	593	2.51	1.760	15.13
B8	62 mm Wood	63	0.121	593	2.51	0.524	37.58
B9	100 mm Wood	101	0.121	593	2.51	0.837	60.02
B10	50 mm Wood	51	0.121	593	2.51	0.420	30.26
B11	75 mm Wood	76	0.121	593	2.51	0.628	45.38
B12	75 mm Insulation	76	0.043	91	0.84	1.760	6.83
B13	100 mm Insulation	100	0.043	91	0.84	2.347	9.27
B14	125 mm Insulation	125	0.043	91	0.84	2.933	11.71
B15	150 mm Insulation	150	0.043	91	0.84	3.520	14.15
B16	4 mm Insulation	4	0.043	91	0.84	0.088	0.49
B17	8 mm Insulation	8	0.043	91	0.84	0.176	0.49
B18	12 mm Insulation	12	0.043	91	0.84	0.264	0.98
B19	15 mm Insulation	15	0.043	91	0.84	0.352	1.46
B20	20 mm Insulation	20	0.043	91	0.84	0.440	1.95
B21	35 mm Insulation	35	0.043	91	0.84	0.792	2.93
B22	42 mm Insulation	42	0.043	91	0.84	0.968	3.90
B23	60 mm Insulation	62	0.043	91	0.84	1.408	5.86
B24	70 mm Insulation	70	0.043	91	0.84	1.584	6.34
B25	85 mm Insulation	85	0.043	91	0.84	1.936	7.81
B26	92 mm Insulation	92	0.043	91	0.84	2.112	8.30
B27	115 mm Insulation	115	0.043	91	0.84	2.640	10.74
C1	100 mm Clay tile	100	0.571	1121	0.84	0.178	113.70
C2	100 mm low density concrete block	100	0.381	609	0.84	0.266	61.98
C3	100 mm high density concrete block	100	0.813	977	0.84	0.125	99.06
C4	100 mm Common brick	100	0.727	1922	0.84	0.140	195.20
C5	100 mm high density concrete	100	1.731	2243	0.84	0.059	227.90
C6	200 mm Clay tile	200	0.571	1121	0.84	0.352	227.90
C7	200 mm low density concrete block	200	0.571	609	0.84	0.352	123.46
C8	200 mm high density concrete block	200	1.038	977	0.84	0.196	198.62
C9	200 mm Common brick	200	0.727	1922	0.84	0.279	390.40
C10	200 mm high density concrete	200	1.731	2243	0.84	0.117	455.79
C11	300 mm high density concrete	300	1.731	2243	0.84	0.176	683.20
C12	50 mm high density concrete	50	1.731	2243	0.84	0.029	113.70
C13	150 mm high density concrete	150	1.731	2243	0.84	0.088	341.60
C14	100 mm low density concrete	100	0.173	641	0.84	0.587	64.90
C15	150 mm low density concrete	150	0.173	641	0.84	0.880	97.60
C16	200 mm low density concrete	200	0.173	641	0.84	1.173	130.30
C17	200 mm low density concrete block (filled)	200	0.138	288	0.84	1.467	58.56
C18	200 mm high density concrete block (filled)	200	0.588	849	0.84	0.345	172.75
C19	300 mm low density concrete block (filled)	300	0.138	304	0.84	2.200	92.72
C20	300 mm high density concrete block (filled)	300	0.675	897	0.84	0.451	273.28
E0	Inside surface resistance	0	0.000	0	0.00	0.121	0.00
E1	20 mm Plaster or gypsum	20	0.727	1602	0.84	0.026	30.74
E2	12 mm Slag or stone	12	1.436	881	1.67	0.009	11.22
E3	10 mm Felt and membrane	10	0.190	1121	1.67	0.050	10.74
E4	Ceiling air space	0	0.000	0	0.00	0.176	0.00
E5	Acoustic tile	19	0.061	481	0.84	0.314	9.27

L = thickness, mm
c_p = specific heat, kJ/(kg·K)
k = thermal conductivity, W/(m·K)
R = thermal resistance, (m²·K)/W
ρ = density, kg/m³
Mass = unit mass, kg/m²

Cooling Load Temperature Differences—In SI Metric Units (continued)

Figure 3.17a (continued)

July Cooling Load Temperature Differences for Calculating Cooling Load from Sunlit Walls 40° North Latitude

Wall Number 1

Wall Face	Hour																							
	1	2	3	4	5	6	7	8	9	10	11	12	13	14	15	16	17	18	19	20	21	22	23	24
N	1	0	−1	−1	−2	−1	4	6	6	7	9	12	14	15	16	16	16	16	15	9	6	4	3	2
NE	1	0	−1	−1	−2	1	13	23	26	24	19	16	15	16	16	16	15	13	11	8	6	4	3	2
E	1	0	−1	−1	−1	1	16	28	34	36	33	27	20	17	17	17	16	14	11	8	6	4	3	2
SE	1	0	−1	−1	−2	0	8	18	26	31	32	31	27	22	18	17	16	14	11	8	6	4	3	2
S	1	0	−1	−1	−2	−1	0	2	6	12	18	24	28	29	28	24	19	15	11	8	6	4	3	2
SW	1	0	−1	−1	−1	−1	0	2	4	7	9	14	22	29	36	39	38	34	25	13	7	4	3	2
W	1	1	−1	−1	−1	−1	1	2	4	7	9	12	15	23	33	41	44	44	34	18	9	5	3	2
NW	1	0	−1	−1	−1	−1	0	2	4	7	9	12	14	16	21	28	34	36	31	16	8	5	3	2

Wall Number 2

Wall Face	Hour																							
	1	2	3	4	5	6	7	8	9	10	11	12	13	14	15	16	17	18	19	20	21	22	23	24
N	3	2	1	0	−1	−1	−1	2	4	5	6	8	10	12	13	14	15	16	16	15	12	9	7	4
NE	3	2	1	0	−1	−1	1	7	14	20	22	21	18	17	16	16	16	16	14	13	10	8	6	4
E	3	2	1	0	−1	−1	1	8	18	26	31	32	29	24	21	19	18	17	15	13	11	8	6	4
SE	3	2	1	0	−1	−1	0	4	11	18	24	28	29	28	25	22	19	17	16	13	11	8	6	4
S	3	2	1	0	−1	−1	−1	−1	1	4	8	13	18	23	26	27	26	22	18	15	12	8	6	4
SW	4	2	1	1	0	−1	−1	0	1	3	5	7	11	17	23	29	34	36	34	29	22	15	9	6
W	4	3	2	1	0	−1	−1	0	1	3	5	7	9	13	18	26	33	38	41	37	28	19	12	8
NW	4	2	1	1	−1	−1	−1	−1	1	3	5	7	9	12	14	18	23	28	32	30	23	16	11	7

Wall Number 3

Wall Face	Hour																							
	1	2	3	4	5	6	7	8	9	10	11	12	13	14	15	16	17	18	19	20	21	22	23	24
N	4	3	2	1	1	0	1	3	4	4	6	8	9	11	13	13	14	14	15	13	11	9	7	6
NE	4	3	2	1	0	0	4	9	14	17	18	17	17	16	16	16	16	14	12	10	8	7	5	
E	4	3	2	1	1	1	4	12	18	23	26	26	24	22	21	19	18	17	16	13	11	9	7	6
SE	4	3	2	1	1	0	2	7	12	18	22	24	26	24	23	21	19	18	16	13	11	9	7	6
S	4	3	2	1	1	0	0	1	2	5	9	13	17	21	23	23	22	20	17	14	12	9	8	6
SW	7	5	3	2	1	1	1	1	2	3	5	8	12	17	22	27	31	32	30	25	20	16	12	9
W	8	6	4	3	2	1	1	1	2	3	5	7	9	13	19	25	31	35	35	30	24	18	14	11
NW	7	4	3	2	1	1	0	1	2	3	5	7	9	11	14	18	22	27	28	24	19	15	12	9

Wall Number 4

Wall Face	Hour																							
	1	2	3	4	5	6	7	8	9	10	11	12	13	14	15	16	17	18	19	20	21	22	23	24
N	6	4	3	2	1	0	0	1	2	3	4	6	7	9	11	12	13	14	15	15	14	12	11	8
NE	6	4	3	2	1	0	0	2	7	12	16	18	18	18	17	17	17	16	16	14	13	11	9	7
E	6	4	3	2	1	1	1	3	8	15	21	25	27	26	24	22	21	19	18	16	14	12	9	8
SE	6	4	3	2	1	1	0	1	4	9	15	20	24	26	26	24	23	21	19	17	14	12	10	8
S	6	4	3	2	1	1	0	−1	0	1	3	7	11	16	19	23	24	23	22	19	17	13	11	8
SW	10	7	5	3	2	1	0	0	1	1	3	4	7	10	15	20	26	29	32	32	28	23	18	14
W	12	8	6	4	2	1	1	0	1	1	3	4	6	8	12	17	22	28	33	36	33	28	22	17
NW	10	7	5	3	2	1	0	0	0	1	2	4	6	8	11	13	17	21	25	27	27	23	18	14

Wall Number 5

Wall Face	Hour																							
	1	2	3	4	5	6	7	8	9	10	11	12	13	14	15	16	17	18	19	20	21	22	23	24
N	7	6	4	3	3	2	1	2	3	3	4	5	7	8	9	11	12	13	13	13	13	12	10	8
NE	7	6	4	3	3	2	2	4	8	11	14	15	16	16	16	16	16	15	14	13	12	10	8	
E	8	6	5	4	3	2	2	4	9	14	18	22	22	22	21	21	19	18	16	14	13	11	9	
SE	8	7	5	4	3	2	2	3	6	10	14	18	21	22	22	21	21	19	18	17	15	13	11	9
S	8	7	5	4	3	2	2	1	2	2	4	7	11	14	17	19	20	20	19	18	16	13	12	10
SW	12	10	8	6	4	3	3	2	2	3	3	5	7	9	14	18	22	26	27	27	24	21	18	14
W	14	11	9	7	6	4	3	2	2	3	4	5	6	8	11	16	21	25	29	30	28	24	21	17
NW	12	9	7	6	4	3	2	2	2	2	3	4	6	8	9	12	15	19	22	23	22	19	17	14

Wall Number 6

Wall Face	Hour																							
	1	2	3	4	5	6	7	8	9	10	11	12	13	14	15	16	17	18	19	20	21	22	23	24
N	7	6	5	4	3	3	2	3	3	4	4	6	7	8	9	10	11	12	12	13	12	11	9	8
NE	8	7	6	4	3	3	3	6	8	11	13	14	14	14	15	15	15	14	14	13	12	10	9	
E	9	7	6	5	4	3	4	6	10	14	17	19	20	20	19	19	19	18	17	16	14	13	12	10
SE	9	8	6	5	4	3	3	4	7	10	13	16	18	19	20	19	19	18	18	16	15	13	12	10
S	9	7	6	5	4	3	3	2	2	3	4	7	10	13	16	17	18	18	17	16	15	13	12	10
SW	13	11	9	8	6	5	4	3	3	4	4	6	7	10	13	17	21	23	24	24	22	19	17	15
W	14	12	10	8	7	6	4	4	4	4	4	6	7	8	11	15	19	23	26	27	25	22	19	17
NW	12	10	8	7	6	4	4	3	3	3	4	5	6	8	9	12	14	18	20	21	20	18	16	14

Cooling Load Temperature Differences—In SI Metric Units (continued)

Figure 3.17a (continued)

July Cooling Load Temperature Differences for Calculating Cooling Load from Sunlit Walls 40° North Latitude (*Continued*)

Wall Number 7

Wall Face	\multicolumn{24}{c}{Hour}
	1	2	3	4	5	6	7	8	9	10	11	12	13	14	15	16	17	18	19	20	21	22	23	24
N	7	7	6	5	4	3	3	4	4	4	5	6	7	8	9	9	10	11	11	11	11	10	9	8
NE	8	7	6	6	5	4	5	7	9	11	12	13	13	13	13	14	14	14	13	13	12	11	10	9
E	9	8	7	7	6	5	6	9	12	14	17	18	18	18	18	18	17	17	16	15	14	13	12	11
SE	9	8	7	7	6	5	5	7	9	12	14	16	17	18	18	18	17	17	16	15	14	13	12	11
S	9	8	7	6	6	4	4	4	4	5	7	8	11	13	14	16	16	16	14	13	12	11	10
SW	13	11	10	9	7	7	6	6	6	6	6	7	8	11	14	17	19	21	22	21	19	17	16	14
W	14	12	11	9	8	7	7	6	6	6	7	7	8	9	12	16	19	22	23	23	21	19	17	16
NW	11	10	9	8	7	6	5	5	5	5	6	6	7	8	9	12	14	17	18	18	17	16	14	13

Wall Number 9

Wall Face	1	2	3	4	5	6	7	8	9	10	11	12	13	14	15	16	17	18	19	20	21	22	23	24
N	9	8	7	6	5	4	3	2	2	2	3	4	4	6	7	8	9	11	12	12	13	13	12	11
NE	10	8	7	6	5	4	3	3	3	6	9	11	13	14	14	15	15	16	16	15	14	14	13	11
E	11	9	8	7	6	4	3	3	4	7	11	14	18	20	21	21	21	20	19	18	17	16	14	13
SE	11	9	8	7	6	4	3	3	3	5	7	11	14	17	19	20	21	20	19	19	18	16	14	13
S	12	10	8	7	6	4	3	3	2	2	2	3	6	8	11	14	16	18	19	19	18	17	15	13
SW	17	14	12	10	8	7	5	4	3	3	3	3	4	6	8	11	14	18	22	24	25	24	22	20
W	19	17	14	12	9	8	6	4	4	3	3	4	4	6	7	9	12	17	21	24	27	27	25	23
NW	16	14	12	9	8	6	5	4	3	3	3	3	4	5	6	8	10	12	16	19	21	21	20	18

Wall Number 10

Wall Face	1	2	3	4	5	6	7	8	9	10	11	12	13	14	15	16	17	18	19	20	21	22	23	24
N	9	8	7	6	5	4	3	3	3	3	3	4	4	6	7	8	9	10	11	12	12	12	12	11
NE	10	9	7	6	5	4	3	3	4	7	9	11	12	13	14	14	15	15	15	14	13	12	11
E	11	9	8	7	6	4	4	4	6	8	11	14	17	19	19	20	20	19	19	18	17	16	14	13
SE	12	10	8	7	6	4	4	3	4	6	8	11	14	17	18	19	19	19	19	18	17	16	14	13
S	12	10	8	7	6	5	4	3	2	2	3	4	6	8	11	13	16	17	18	18	17	16	14	13
SW	17	15	13	11	9	7	6	4	4	3	3	4	4	6	8	11	14	18	21	23	23	23	21	19
W	19	17	14	12	10	8	7	5	4	4	4	4	4	6	7	9	13	17	21	23	25	25	23	22
NW	16	13	12	10	8	7	6	4	3	3	3	3	4	6	7	8	10	13	16	18	19	20	19	17

Wall Number 11

Wall Face	1	2	3	4	5	6	7	8	9	10	11	12	13	14	15	16	17	18	19	20	21	22	23	24
N	9	8	7	7	6	5	4	4	4	4	4	5	6	6	7	8	8	9	10	11	11	11	10	9
NE	10	9	8	7	7	6	5	5	6	8	9	11	12	12	13	13	13	13	14	14	13	13	12	11
E	12	11	9	9	8	7	6	6	7	9	12	14	16	17	17	17	17	17	17	16	15	14	13
SE	12	11	9	9	8	7	6	6	6	8	9	12	13	15	16	17	17	17	17	16	15	14	13
S	11	10	9	8	7	6	6	5	4	4	4	6	7	9	11	13	14	15	16	16	15	14	13	12
SW	16	14	13	11	10	9	8	7	6	6	6	6	7	8	9	12	14	17	18	20	20	19	18	17
W	17	16	14	12	11	10	9	8	7	7	6	7	7	7	8	11	13	16	18	21	22	21	20	18
NW	14	13	11	10	9	8	7	6	6	5	5	6	6	7	7	8	10	12	14	16	17	17	16	15

Wall Number 12

Wall Face	1	2	3	4	5	6	7	8	9	10	11	12	13	14	15	16	17	18	19	20	21	22	23	24
N	9	8	7	7	6	6	4	4	4	4	4	5	6	6	7	8	8	9	9	10	11	11	10	9
NE	10	9	8	8	7	6	6	6	7	8	9	11	12	12	12	13	13	13	13	13	13	12	12	11
E	12	11	10	9	8	7	7	7	8	9	12	14	16	16	17	17	17	17	17	16	16	15	14	13
SE	12	11	10	9	8	7	7	6	7	8	9	12	13	14	16	16	17	17	17	16	16	15	14	13
S	11	11	9	8	8	7	6	6	5	5	5	6	7	9	11	12	13	14	14	14	14	14	13	12
SW	15	14	13	12	11	9	8	8	7	7	7	7	7	8	9	11	13	16	18	19	19	19	18	17
W	17	16	14	13	12	11	9	8	8	7	7	7	7	8	9	11	13	15	18	19	21	20	19	18
NW	13	12	11	11	9	8	7	7	6	6	6	6	6	7	7	8	10	12	14	16	16	16	16	14

Cooling Load Temperature Differences – In SI Metric Units (continued)

Figure 3.17a (continued)

July Cooling Load Temperature Differences for Calculating Cooling Load from Sunlit Walls 40° North Latitude (*Concluded*)

Wall Number 13

Wall Face	Hour																							
	1	2	3	4	5	6	7	8	9	10	11	12	13	14	15	16	17	18	19	20	21	22	23	24
N	8	8	7	7	6	6	5	5	5	5	5	6	6	6	7	8	8	9	9	10	10	10	9	9
NE	10	9	9	8	7	7	6	7	7	9	10	11	11	12	12	12	13	13	13	13	13	12	12	11
E	12	11	11	9	9	8	8	8	9	11	12	14	15	16	16	16	16	16	16	16	15	14	14	13
SE	12	11	11	9	9	8	7	7	8	9	10	12	13	14	15	16	16	16	16	15	15	14	13	13
S	11	10	9	9	8	7	7	6	6	6	6	7	8	9	11	12	13	13	14	14	13	13	12	12
SW	14	14	13	12	11	10	9	8	8	7	7	7	8	8	10	12	13	16	17	18	18	17	17	16
W	16	15	14	13	12	11	10	9	8	8	8	8	8	8	9	11	13	15	17	19	19	19	18	17
NW	13	12	11	10	9	8	8	7	7	7	7	7	7	7	8	9	10	12	13	14	15	15	14	14

Wall Number 14

Wall Face	Hour																							
	1	2	3	4	5	6	7	8	9	10	11	12	13	14	15	16	17	18	19	20	21	22	23	24
N	8	8	8	7	7	6	6	6	6	6	6	6	6	6	7	7	8	8	8	9	9	9	9	9
NE	11	10	9	9	8	8	7	7	8	8	9	10	11	11	11	12	12	12	12	12	12	12	12	11
E	13	12	12	11	10	9	9	8	9	10	12	13	14	14	15	15	16	16	16	16	15	14	14	13
SE	13	12	11	11	10	9	8	8	8	9	10	11	12	13	14	14	15	15	15	15	14	14	14	13
S	11	11	10	9	9	8	8	7	7	7	7	7	8	8	9	11	12	12	13	13	13	13	12	12
SW	14	14	13	12	12	11	10	9	9	8	8	8	8	9	9	11	12	14	15	16	17	17	16	16
W	16	15	14	13	13	12	11	10	9	9	9	9	9	9	9	11	12	13	15	17	18	18	17	17
NW	13	12	12	11	10	9	9	8	8	7	7	7	7	8	8	8	9	11	12	13	14	14	14	13

Wall Number 15

Wall Face	Hour																							
	1	2	3	4	5	6	7	8	9	10	11	12	13	14	15	16	17	18	19	20	21	22	23	24
N	11	10	9	8	7	6	5	4	3	3	3	3	4	4	5	6	7	8	9	11	11	12	12	11
NE	12	11	9	8	7	6	5	4	4	5	6	8	10	11	12	13	14	14	14	14	14	14	14	13
E	14	12	11	9	8	7	6	5	5	6	8	10	13	15	17	18	19	19	19	18	18	17	16	15
SE	14	12	11	9	8	7	6	5	4	4	6	8	10	12	14	17	18	18	19	18	18	17	17	15
S	14	12	11	9	8	7	6	5	4	3	3	3	4	6	7	9	12	14	16	17	17	17	16	15
SW	19	18	16	14	12	10	9	7	6	5	4	4	4	5	6	8	10	13	16	18	21	22	22	21
W	22	19	18	16	13	12	10	8	7	6	5	4	4	5	6	7	9	12	14	18	21	23	23	23
NW	17	16	14	13	11	9	8	7	6	4	4	4	4	4	5	6	7	9	11	14	16	18	18	18

Wall Number 16

Wall Face	Hour																							
	1	2	3	4	5	6	7	8	9	10	11	12	13	14	15	16	17	18	19	20	21	22	23	24
N	10	9	9	8	7	6	6	5	4	4	4	4	4	5	6	6	7	8	9	9	10	11	11	11
NE	12	11	10	9	8	7	6	6	6	6	7	8	9	11	12	12	13	13	13	14	14	13	13	13
E	14	13	12	11	9	8	7	6	6	7	8	11	12	14	16	17	17	17	18	18	17	17	16	15
SE	14	13	12	11	9	8	7	6	6	6	7	8	10	12	14	15	16	17	17	17	17	17	16	15
S	13	12	11	10	9	8	7	6	5	4	4	4	5	6	8	9	11	13	14	15	15	15	15	14
SW	18	17	16	14	13	11	10	8	7	7	6	6	6	6	7	8	10	12	15	17	18	19	19	19
W	20	18	17	16	14	12	11	9	8	7	7	6	6	6	7	8	9	11	14	17	19	21	21	21
NW	16	15	14	13	11	10	9	8	7	6	6	5	5	6	6	7	8	9	11	13	15	16	17	17

Note 1. Direct application of data
- Dark surface
- Indoor temperature of 25.5 °C
- Outdoor maximum temperature of 35 °C with mean temperature of 29.4 °C and daily range of 11.6 °C
- Solar radiation typical of clear day on 21st day of month
- Outside surface film resistance of 0.059 m^2·K/W
- Inside surface resistance of 0.121 m^2·K/W

Note 2. Adjustments to table data
- Design temperatures

$$\text{Corr. CLTD} = \text{CLTD} + (25.5 - t_r) + (t_m - 29.4)$$

where
t_r = inside temperature and
t_m = maximum outdoor temperature − (daily range)/2
- No adjustment recommended for color

Cooling Load Temperature Differences—In SI Metric Units (continued)

Figure 3.17a (continued)

Wall Types, Mass Located Inside Insulation

Secondary Material	R-Factor, m²·K/W	A1	A2	B7	B10	B9	C1	C2	C3	C4	C5	C6	C7	C8	C17	C18
Stucco and/or plaster	0 to 0.35	*	*	*	*	*	*	*	*	*	*	*	*	*	*	*
	0.35 to 0.44	*	5	*	*	*	*	*	*	*	5	*	*	*	*	*
	0.44 to 0.53	*	5	*	*	*	3	*	2	5	6	*	*	5	*	*
	0.53 to 0.62	*	5	*	*	*	4	2	2	5	6	*	*	6	*	*
	0.62 to 0.70	*	5	*	*	*	4	2	3	6	6	10	4	6	*	5
	0.70 to 0.84	*	6	*	*	*	5	2	4	6	6	11	5	10	*	10
	0.84 to 0.97	*	6	*	*	*	5	2	4	6	6	11	5	10	*	10
	0.97 to 1.14	*	6	*	*	*	5	2	5	10	7	12	5	11	*	10
	1.14 to 1.36	*	6	*	*	*	5	4	5	11	7	16	10	11	*	11
	1.36 to 1.59	*	6	*	*	*	5	4	5	11	7	*	10	11	*	11
	1.59 to 1.89	*	6	*	*	*	5	4	5	11	7	*	10	11	*	11
	1.89 to 2.24	*	6	*	*	*	5	4	5	11	11	*	10	11	4	11
	2.24 to 2.64	*	10	*	*	*	10	4	5	11	11	*	10	11	4	11
	2.64 to 3.08	*	10	*	*	*	10	5	5	11	11	*	10	11	9	12
	3.08 to 3.52	*	11	*	*	*	10	5	9	11	11	*	11	12	10	16
	3.52 to 4.05	*	11	*	*	*	10	9	9	16	11	*	15	16	10	16
	4.05 to 4.76	*	*	*	*	*	*	*	*	*	*	*	16	*	15	*
Steel or other low-mass siding	0 to 0.35	*	*	*	*	*	*	*	*	*	*	*	*	*	*	*
	0.35 to 0.44	*	3	*	*	*	*	*	2	3	5	*	*	*	*	*
	0.44 to 0.53	*	5	*	*	*	2	*	2	5	3	*	*	5	*	*
	0.53 to 0.62	*	5	*	*	*	3	1	2	5	5	*	*	5	*	*
	0.62 to 0.70	*	5	*	*	*	3	2	2	5	5	6	3	5	*	5
	0.70 to 0.84	*	6	*	*	*	4	2	2	5	5	10	4	6	*	5
	0.84 to 0.97	*	6	*	*	*	5	2	2	6	6	11	5	6	*	6
	0.97 to 1.14	*	6	*	*	*	5	2	3	6	6	11	5	6	*	6
	1.14 to 1.36	*	6	*	*	*	5	2	3	6	6	11	5	6	*	10
	1.36 to 1.59	*	6	*	*	*	5	2	3	6	6	12	5	6	*	11
	1.59 to 1.89	*	6	*	*	*	5	2	3	6	6	12	5	6	4	11
	1.89 to 2.24	*	6	*	*	*	5	2	3	6	7	12	6	11	4	11
	2.24 to 2.64	*	6	*	*	*	5	2	4	6	7	12	10	11	5	11
	2.64 to 3.08	*	10	*	*	*	6	4	4	10	7	*	10	11	9	11
	3.08 to 3.52	*	10	*	*	*	10	4	4	10	11	*	10	11	10	11
	3.52 to 4.05	*	11	*	*	*	10	4	5	11	11	*	10	11	10	16
	4.05 to 4.76	*	*	*	*	*	*	*	*	*	*	*	10	*	11	16
Face brick	0 to 0.35	*	*	*	*	*	*	*	*	*	*	*	*	*	*	*
	0.35 to 0.44	3	*	*	*	*	*	*	*	*	11	*	*	*	*	*
	0.44 to 0.53	5	11	*	*	*	*	*	6	11	12	*	*	*	*	*
	0.53 to 0.62	5	12	5	*	*	11	*	11	12	12	*	*	12	*	*
	0.62 to 0.70	5	12	6	*	*	12	6	12	12	13	*	*	12	*	*
	0.70 to 0.84	6	13	6	10	*	13	10	12	12	13	*	11	*	*	16
	0.84 to 0.97	6	13	6	11	*	*	11	12	13	13	*	16	*	*	*
	0.97 to 1.14	6	13	6	11	*	*	11	12	13	13	*	*	*	*	*
	1.14 to 1.36	6	13	6	11	*	*	11	13	*	13	*	*	*	*	*
	1.36 to 1.59	6	13	10	16	*	*	11	13	*	13	*	*	*	*	*
	1.59 to 1.89	6	14	10	16	*	*	11	13	*	14	*	*	*	16	*
	1.89 to 2.24	6	14	10	16	*	*	11	13	*	14	*	*	*	16	*
	2.24 to 2.64	6	*	11	16	*	*	12	13	*	*	*	*	*	*	*
	2.64 to 3.08	10	*	11	*	*	*	12	13	*	*	*	*	*	*	*
	3.08 to 3.52	10	*	11	*	*	*	16	*	*	*	*	*	*	*	*
	3.52 to 4.05	11	*	15	*	*	*	16	*	*	*	*	*	*	*	*
	4.05 to 4.76	*	*	*	*	*	*	16	*	*	*	*	*	*	*	*

*Denotes a wall that is not possible with the chosen set of parameters.
**See Table 11 for definition of Code letters.

Cooling Load Temperature Differences—In SI Metric Units (continued)

Figure 3.17a (continued)

Wall Types, Mass Evenly Distributed

Secondary Material	R-Factor, m²·K/W	A1	A2	B7	B10	B9	C1	C2	C3	C4	C5	C6	C7	C8	C17	C18
Stucco and/or plaster	0 to 0.35	1	3	*	*	*	*	*	1	3	3	*	*	*	*	*
	0.35 to 0.44	1	3	1	*	*	2	*	2	4	4	*	*	5	*	*
	0.44 to 0.53	1	4	1	*	*	2	2	2	4	4	*	*	5	*	*
	0.53 to 0.62	1	*	1	*	*	2	2	*	*	*	10	4	5	*	4
	0.62 to 0.70	1	*	1	2	*	*	4	*	*	*	10	4	*	*	4
	0.70 to 0.84	1	*	1	2	*	*	*	*	*	*	10	4	*	*	4
	0.84 to 0.97	1	*	1	2	*	*	*	*	*	*	*	*	*	*	*
	0.97 to 1.14	1	*	2	4	10	*	*	*	*	*	*	*	*	*	*
	1.14 to 1.36	1	*	2	4	11	*	*	*	*	*	*	*	*	*	*
	1.36 to 1.59	1	*	2	4	16	*	*	*	*	*	*	*	*	*	*
	1.59 to 1.89	1	*	2	4	16	*	*	*	*	*	*	*	*	4	*
	1.89 to 2.24	1	*	2	5	*	*	*	*	*	*	*	*	*	4	*
	2.24 to 2.64	2	*	2	5	*	*	*	*	*	*	*	*	*	*	*
	2.64 to 3.08	2	*	2	5	*	*	*	*	*	*	*	*	*	*	*
	3.08 to 3.52	2	*	2	9	*	*	*	*	*	*	*	*	*	*	*
	3.52 to 4.05	2	*	4	9	*	*	*	*	*	*	*	*	*	*	*
	4.05 to 4.76	*	*	*	9	*	*	*	*	*	*	*	*	*	*	*
Steel or other low-mass siding	0 to 0.35	1	3	*	*	*	*	*	1	3	2	*	*	*	*	*
	0.35 to 0.44	1	3	1	*	*	2	*	1	3	2	*	*	3	*	*
	0.44 to 0.53	1	4	1	*	*	2	1	2	4	4	*	*	3	*	*
	0.53 to 0.62	1	*	1	*	*	4	1	*	*	*	5	2	4	*	4
	0.62 to 0.70	1	*	1	2	*	*	2	*	*	*	5	2	*	*	4
	0.70 to 0.84	1	*	1	2	*	*	*	*	*	*	10	4	*	*	4
	0.84 to 0.97	1	*	1	2	*	*	*	*	*	*	*	*	*	*	*
	0.97 to 1.14	1	*	1	2	10	*	*	*	*	*	*	*	*	*	*
	1.14 to 1.36	1	*	1	4	11	*	*	*	*	*	*	*	*	*	*
	1.36 to 1.59	1	*	2	4	16	*	*	*	*	*	*	*	*	*	*
	1.59 to 1.89	1	*	2	4	16	*	*	*	*	*	*	*	*	2	*
	1.89 to 2.24	1	*	2	4	*	*	*	*	*	*	*	*	*	4	*
	2.24 to 2.64	1	*	2	5	*	*	*	*	*	*	*	*	*	*	*
	2.64 to 3.08	1	*	2	5	*	*	*	*	*	*	*	*	*	*	*
	3.08 to 3.52	1	*	2	5	*	*	*	*	*	*	*	*	*	*	*
	3.52 to 4.05	2	*	4	9	*	*	*	*	*	*	*	*	*	*	*
	4.05 to 4.76	*	*	*	9	*	*	*	*	*	*	*	*	*	*	*
Face brick	0 to 0.35	3	6	*	*	*	*	*	*	*	6	*	*	*	*	*
	0.35 to 0.44	3	10	*	*	*	*	*	5	10	10	*	*	*	*	*
	0.44 to 0.53	4	10	5	*	*	5	*	5	10	11	*	*	10	*	*
	0.53 to 0.62	*	11	5	*	*	10	5	5	11	11	15	10	10	*	10
	0.62 to 0.70	*	11	5	10	*	10	5	5	11	11	16	10	16	*	10
	0.70 to 0.84	*	11	*	11	*	10	5	5	16	11	*	10	16	*	16
	0.84 to 0.97	*	11	*	11	*	10	5	10	16	16	*	10	16	*	16
	0.97 to 1.14	*	16	*	*	*	10	9	10	16	11	*	11	16	*	16
	1.14 to 1.36	*	16	*	*	*	11	9	10	16	16	*	16	16	*	*
	1.36 to 1.59	*	16	*	*	*	15	9	10	16	*	*	15	16	*	*
	1.59 to 1.89	*	16	*	*	*	15	10	10	*	16	*	16	*	10	*
	1.89 to 2.24	*	16	*	*	*	16	10	10	*	*	*	16	*	15	*
	2.24 to 2.64	*	16	*	*	*	16	10	10	*	16	*	*	*	15	*
	2.64 to 3.08	*	*	*	*	*	16	10	15	*	*	*	*	*	16	*
	3.08 to 3.52	*	*	*	*	*	16	15	15	*	*	*	*	*	16	*
	3.52 to 4.05	*	*	*	*	*	*	15	16	*	*	*	*	*	*	*
	4.05 to 4.76	*	*	*	*	*	*	15	*	*	*	*	*	*	*	*

*Denotes a wall that is not possible with the chosen set of parameters.
**See Table 11 for definition of Code letters.

Cooling Load Temperature Differences – In SI Metric Units (continued)

Figure 3.17a (continued)

Wall Types, Mass Located Outside Insulation

Secondary Material	R-Factor, m²·K/W	A1	A2	B7	B10	B9	C1	C2	C3	C4	C5	C6	C7	C8	C17	C18
Stucco and/or plaster	0 to 0.35	*	*	*	*	*	*	*	*	*	*	*	*	*	*	*
	0.35 to 0.44	*	3	*	*	*	*	*	2	3	5	*	*	*	*	*
	0.44 to 0.53	*	3	*	*	*	2	*	2	4	5	*	*	5	*	*
	0.53 to 0.62	*	3	*	*	*	2	2	2	5	5	*	*	5	*	*
	0.62 to 0.70	*	3	*	*	*	2	2	2	5	5	10	4	6	*	5
	0.70 to 0.84	*	4	*	*	*	4	2	2	5	5	10	4	6	*	9
	0.84 to 0.97	*	4	*	*	*	4	2	2	5	6	11	5	10	*	10
	0.97 to 1.14	*	5	*	*	*	4	2	2	5	6	11	5	10	*	10
	1.14 to 1.36	*	5	*	*	*	4	2	2	5	6	11	5	10	*	10
	1.36 to 1.59	*	5	*	*	*	5	2	4	5	6	16	10	10	*	10
	1.59 to 1.89	*	5	*	*	*	5	4	4	5	6	16	10	10	4	11
	1.89 to 2.24	*	5	*	*	*	5	4	4	10	6	16	10	10	9	11
	2.24 to 2.64	*	5	*	*	*	5	4	4	10	10	*	10	11	9	11
	2.64 to 3.08	*	5	*	*	*	5	4	4	10	10	*	10	11	10	16
	3.08 to 3.52	*	5	*	*	*	9	4	4	10	10	*	10	15	10	16
	3.52 to 4.05	*	9	*	*	*	9	9	9	15	10	*	10	15	15	16
	4.05 to 4.76	*	*	*	*	*	*	*	*	*	*	*	15	*	15	16
Steel or other low-mass siding	0 to 0.35	*	*	*	*	*	*	*	*	*	*	*	*	*	*	*
	0.35 to 0.44	*	3	*	*	*	*	*	2	3	2	*	*	*	*	*
	0.44 to 0.53	*	3	*	*	*	2	*	2	3	2	*	*	*	*	*
	0.53 to 0.62	*	3	*	*	*	2	1	2	4	3	*	*	4	*	*
	0.62 to 0.70	*	3	*	*	*	2	2	2	4	3	5	2	5	*	4
	0.70 to 0.84	*	3	*	*	*	2	2	2	4	3	10	3	5	*	5
	0.84 to 0.97	*	3	*	*	*	2	2	2	5	3	10	4	5	*	5
	0.97 to 1.14	*	4	*	*	*	2	2	2	5	3	10	4	5	*	5
	1.14 to 1.36	*	4	*	*	*	2	2	2	5	4	11	5	5	*	6
	1.36 to 1.59	*	5	*	*	*	2	2	2	5	4	11	5	5	*	6
	1.59 to 1.89	*	5	*	*	*	2	2	2	5	4	11	5	5	4	10
	1.89 to 2.24	*	5	*	*	*	4	2	2	5	5	11	5	5	4	10
	2.24 to 2.64	*	5	*	*	*	4	2	2	5	5	11	5	10	5	10
	2.64 to 3.08	*	5	*	*	*	4	2	4	5	5	16	9	10	9	10
	3.08 to 3.52	*	5	*	*	*	4	4	4	9	5	16	9	10	10	10
	3.52 to 4.05	*	9	*	*	*	4	4	4	9	9	16	10	10	10	11
	4.05 to 4.76	*	*	*	*	*	*	*	*	*	*	16	10	*	10	15
Face brick	0 to 0.35	*	*	*	*	*	*	*	*	*	*	*	*	*	*	*
	0.35 to 0.44	3	*	*	*	*	*	*	*	*	11	*	*	*	*	*
	0.44 to 0.53	3	10	*	*	*	*	*	5	10	11	*	*	*	*	*
	0.53 to 0.62	3	11	5	*	*	10	*	5	11	11	*	*	11	*	*
	0.62 to 0.70	3	11	5	*	*	10	5	6	11	11	*	*	11	*	*
	0.70 to 0.84	3	11	5	10	*	10	5	10	11	11	*	10	11	*	16
	0.84 to 0.97	3	12	5	10	*	10	9	10	11	12	*	11	16	*	16
	0.97 to 1.14	4	12	5	10	*	10	10	10	12	12	*	15	16	*	16
	1.14 to 1.36	4	12	5	10	*	11	10	10	12	12	*	16	*	*	16
	1.36 to 1.59	5	12	5	15	*	11	10	10	16	12	*	16	*	*	*
	1.59 to 1.89	5	12	9	15	*	11	10	10	16	12	*	16	*	15	*
	1.89 to 2.24	5	12	10	15	*	11	10	10	*	12	*	16	*	15	*
	2.24 to 2.64	5	*	10	16	*	11	10	11	*	*	*	16	*	15	*
	2.64 to 3.08	5	*	10	16	*	15	10	11	*	*	*	16	*	*	*
	3.08 to 3.52	5	*	10	16	*	16	15	15	*	*	*	*	*	*	*
	3.52 to 4.05	9	*	15	16	*	16	15	15	*	*	*	*	*	*	*
	4.05 to 4.76	*	*	*	*	*	*	15	*	*	*	*	*	*	*	*

*Denotes a wall that is not possible with the chosen set of parameters.
**See Table 11 for definition of Code letters.

Cooling Load Temperature Differences – In SI Metric Units (continued)

Figure 3.17a (continued)

an air handling system is used to pressurize the building, the infiltration will equal zero, since no air can infiltrate to a higher pressure. In this situation, outside air is used to both ventilate and pressurize the building; the amount of air required for ventilation is used for the cfm [L/s] value.

Latent Loads (ΔW)

The same cfm [L/s] that causes sensible heat gain also carries moisture that adds to the latent heat load. Change in latent heat, which is measured by the change in moisture content, is the change in moisture measured in pounds of moisture per pound of dry air [grams of moisture per kilogram of dry air]. The moisture content is read from the pychrometric chart (see Figures 3.18 and 3.18a). For example, the amount of latent heat for outdoor air at 88°F [31 °C]/50 percent relative humidity that enters an indoor condition of 78°F [26 °C]/40 percent relative humidity has a ΔW of:

$$0.015 - 0.008 = 0.007 \text{ lbs. moisture/lb. dry air}$$
$$[14.25 - 8.5 = 5.75 \text{ grams moisture/kg dry air}]$$

Cooling Load Temperature Differences (CLTD) for Conduction through Glass

Solar Time, h	CLTD, °C	Solar Time, h	CLTD, °C
0100	1	1300	7
0200	0	1400	7
0300	−1	1500	8
0400	−1	1600	8
0500	−1	1700	7
0600	−1	1800	7
0700	−1	1900	6
0800	0	2000	4
0900	1	2100	3
1000	2	2200	2
1100	4	2300	2
1200	5	2400	1

Corrections: The values in the table were calculated for an inside temperature of 25.5 °C and an outdoor maximum temperature of 35 °C with an outdoor daily range of 11.6 °C. The table remains approximately correct for other outdoor maximums 33 to 39 °C and other outdoor daily ranges 9 to 19 °C, provided the outdoor daily average temperature remains approximately 29.4 °C. If the room air temperature is different from 25.5 °C and/or the outdoor daily average temperature is different from 29.4 °C, see note 2, Table 32.

(Copyright 1993 by ASHRAE, from *1993 Fundamentals, SI Edition.* Used by permission.)

Footnote: Cross-references in table footnotes pertain to the source from which tables were taken; they do not refer to chapters or pages in this text.

Cooling Load Temperature Differences— In SI Metric Units (continued)

Figure 3.17a (continued)

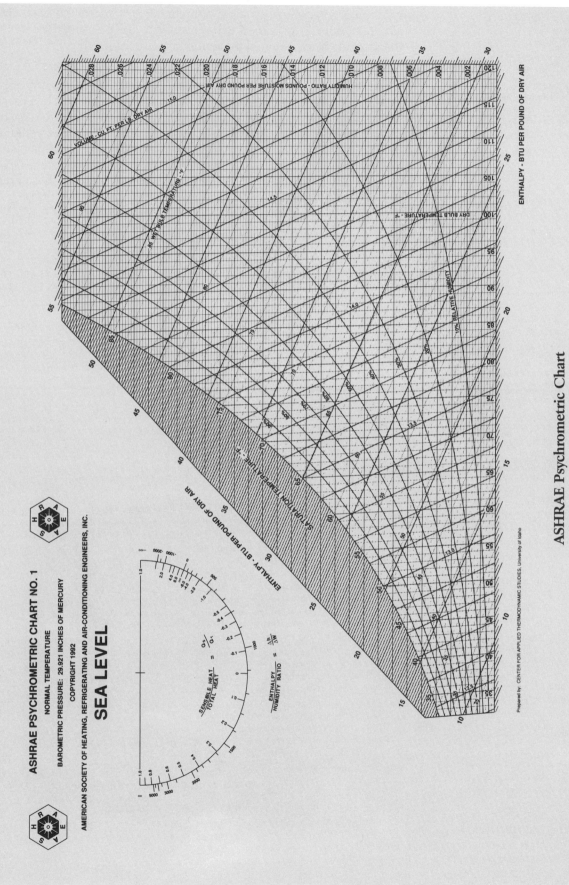

Figure 3.18

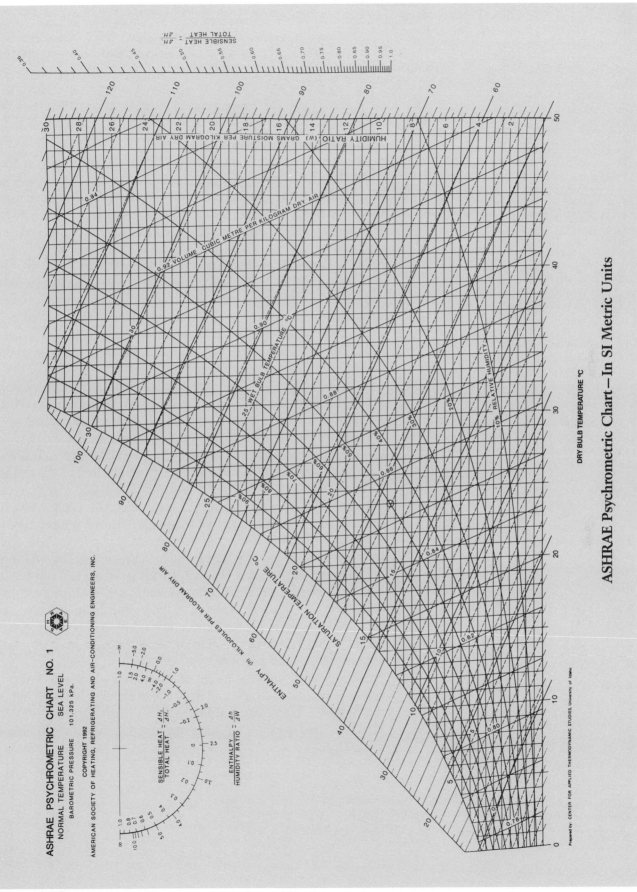

Figure 3.18a

Radiation Heat Gain

Solar heat gain that enters transparent surfaces varies with location, time of day, day of the year, shading, orientation, building color, mass, hours of operation, and daily temperature range. The basic formula for computing radiant energy heat gain is shown below.

H_s = A · (SC) (SCL)
H_s = Radiant heat gain (Btu/hr.)[W]
A = Glass area (s.f.) [m^2]
SC = Shading coefficient (see Figure 3.20)
SCL = Solar cooling load factor with no interior shade, or with shade Btu/hr. s.f. [W/m^2] (see Figure 3.19)

A   Area – This is the glass area for each window, door, skylight, or other opening. The area is measured perpendicular to the ground for windows and doors, and parallel to the ground for skylights and roof openings.

SC   Shading Coefficient – For unshaded single-pane flat glass, the shading coefficient, or SC, equals 1.0. As the type of glass varies – reflective glass or double-pane, for example – less energy can enter into the building and the overall heat gain due to solar radiation is reduced. Similarly, exterior awnings and interior shades or drapes reduce the SC. Figure 3.20 is a table of SC values to be used in the formula for radiant heat gain.

CLF   Cooling Load Factor – These are utilized when calculating sensible heat gains from internal sources such as people, lighting, hooded equipment, and non-hooded equipment. The CLF takes into account all the variables that affect the instantaneous and delayed dissipation of the heat to the space and applies a diversity to the heat gain loads of the people, lights, or the equipment in that space. The CLF can be as high as 1.0 and as low as zero. It is important to note that the CLF's shown in Figure 3.21 assume that the cooling system is on 24 hours per day and that the design internal temperature is 78°F [25.5 °C]. Any alterations will require correction factors and may even require use of a different table. If the cooling system is not on 24 hours per day, stored energy in the structure will not be removed from the building at off-peak hours; a CLF of 1.0 should be used. The CLF is also a function of the construction and type of zone, as outlined in Figure 3.19.

SCL   Solar Cooling Load Factor – This represents the amount of instantaneous heat gain resulting from sunlight radiation at a certain specified time, for a defined structure and interior construction, furnishings, and floor coverings with and without shading. SCL is measured in Btu/hr. $ft.^2$ [W/m^2]. Figures 3.19 and 3.19a list zone types. When zone types are determined, SCL factors can be read.

Zone Types for Use with CLF Tables, Interior Rooms

Room Location	Zone Parameters[a]			Zone Type	
	Middle Floor	Ceiling Type	Floor Covering	People and Equipment	Lights
Single story	N/A	N/A	Carpet	C	B
	N/A	N/A	Vinyl	D	C
Top floor	2.5 in. Concrete	With	Carpet	D	C
	2.5 in. Concrete	With	Vinyl	D	D
	2.5 in. Concrete	Without	b	D	B
	1 in. Wood	b	b	D	B
Bottom floor	2.5 in. Concrete	With	Carpet	D	C
	2.5 in. Concrete	b	Vinyl	D	D
	2.5 in. Concrete	Without	Carpet	D	D
	1 in. Wood	b	Carpet	D	C
	1 in. Wood	b	Vinyl	D	D
Mid-floor	2.5 in. Concrete	N/A	Carpet	D	C
	2.5 in. Concrete	N/A	Vinyl	D	D
	1 in. Wood	N/A	b	C	B

[a] A total of 14 zone parameters is fully defined in Table 20. Those not shown in this table were selected to achieve an error band of approximately 10%.
[b] The effect of inside shade is negligible in this case.

Zone Types for Use with SCL and CLF Tables, Single-Story Building

	Zone Parameters[a]			Zone Type			Error Band	
No. Walls	Floor Covering	Partition Type	Inside Shade	Glass Solar	People and Equipment	Lights	Plus	Minus
1 or 2	Carpet	Gypsum	b	A	B	B	9	2
1 or 2	Carpet	Concrete block	b	B	C	C	9	0
1 or 2	Vinyl	Gypsum	Full	B	C	C	9	0
1 or 2	Vinyl	Gypsum	Half to None	C	C	C	16	0
1 or 2	Vinyl	Concrete block	Full	C	D	D	8	0
1 or 2	Vinyl	Concrete block	Half to None	D	D	D	10	6
3	Carpet	Gypsum	b	A	B	B	9	2
3	Carpet	Concrete block	Full	A	B	B	9	2
3	Carpet	Concrete block	Half to None	B	B	B	9	0
3	Vinyl	Gypsum	Full	B	C	C	9	0
3	Vinyl	Gypsum	Half to None	C	C	C	16	0
3	Vinyl	Concrete block	Full	B	C	C	9	0
3	Vinyl	Concrete block	Half to None	C	C	C	16	0
4	Carpet	Gypsum	b	A	B	B	6	3
4	Vinyl	Gypsum	Full	B	C	C	11	6
4	Vinyl	Gypsum	Half to None	C	C	C	19	−1

[a] A total of 14 zone parameters is fully defined in Table 20. Those not shown in this table were selected to achieve the minimum error band shown in the righthand column for Solar Cooling Load (SCL). The error band for Lights and People and Equipment is approximately 10%.
[b] The effect of inside shade is negligible in this case.

Solar Cooling Load Factors

Figure 3.19

July Solar Cooling Load For Sunlit Glass 40° North Latitude

Zone type A

Glass Face	Hour 1	2	3	4	5	6	7	8	9	10	Solar Time 11	12	13	14	15	16	17	18	19	20	21	22	23	24
N	0	0	0	0	1	25	27	28	32	35	38	40	40	39	36	31	31	36	12	6	3	1	1	0
NE	0	0	0	0	2	85	129	134	112	75	55	48	44	40	37	32	26	18	7	3	2	1	0	0
E	0	0	0	0	2	93	157	185	183	154	106	67	53	45	39	33	26	18	7	3	2	1	0	0
SE	0	0	0	0	1	47	95	131	150	150	131	97	63	49	41	34	27	18	7	3	2	1	0	0
S	0	0	0	0	0	9	17	25	41	64	85	97	96	84	63	42	31	20	8	4	2	1	0	0
SW	0	0	0	0	0	9	17	24	30	35	39	64	101	133	151	152	133	93	35	17	8	4	2	1
W	1	0	0	0	0	9	17	24	30	35	38	40	65	114	158	187	192	156	57	27	13	6	3	2
NW	1	0	0	0	0	9	17	24	30	35	38	40	40	50	84	121	143	130	46	22	11	5	3	1
Hor	0	0	0	0	0	24	69	120	169	211	241	257	259	245	217	176	125	70	29	14	7	3	2	1

Zone type B

Glass Face	Hour 1	2	3	4	5	6	7	8	9	10	Solar Time 11	12	13	14	15	16	17	18	19	20	21	22	23	24
N	2	2	1	1	1	22	23	24	28	32	35	37	38	37	35	32	31	35	16	10	7	5	4	3
NE	2	1	1	1	2	73	109	116	101	73	58	52	48	45	41	36	30	23	13	9	6	5	3	3
E	2	2	1	1	2	80	133	159	162	143	105	74	63	55	48	41	34	25	15	10	7	5	4	3
SE	2	2	1	1	1	40	81	112	131	134	122	96	69	58	49	42	35	26	15	10	8	6	4	3
S	2	2	1	1	1	8	15	21	36	56	74	86	87	79	63	46	37	27	16	11	8	6	4	3
SW	6	5	4	3	2	9	16	22	27	31	36	58	89	117	135	138	126	94	46	31	21	15	11	8
W	8	6	5	4	3	9	16	22	27	31	35	37	59	101	139	166	173	147	66	43	30	21	15	11
NW	6	5	4	3	2	9	16	22	27	31	34	37	37	46	76	108	128	119	51	33	22	16	11	8
Hor	8	6	5	4	3	22	60	104	147	185	214	233	239	232	212	180	137	90	53	37	27	19	14	11

Zone type C

Glass Face	Hour 1	2	3	4	5	6	7	8	9	10	Solar Time 11	12	13	14	15	16	17	18	19	20	21	22	23	24
N	5	5	4	4	4	24	23	24	27	30	33	34	35	34	32	29	29	34	14	10	8	7	6	6
NE	7	6	6	5	6	75	106	107	88	61	49	47	45	43	40	36	31	25	16	13	11	10	9	8
E	9	8	8	7	8	83	130	148	145	124	89	62	56	52	47	43	37	30	20	17	15	13	12	11
SE	9	8	7	6	6	45	82	107	121	121	107	82	59	51	47	42	36	29	19	16	14	13	11	10
S	7	7	6	5	5	12	18	23	36	54	70	79	79	70	54	40	33	26	16	13	12	10	9	8
SW	14	12	11	10	9	15	21	26	29	33	36	57	86	110	124	125	111	80	37	28	23	20	17	15
W	17	15	13	12	11	17	22	27	31	34	36	37	59	98	132	153	156	128	50	35	28	24	21	19
NW	12	11	10	9	8	14	20	25	29	32	34	36	36	44	73	102	118	107	39	26	21	17	15	13
Hor	24	21	19	17	16	34	68	107	144	175	199	212	215	207	189	160	123	83	53	44	38	34	30	27

Zone type D

Glass Face	Hour 1	2	3	4	5	6	7	8	9	10	Solar Time 11	12	13	14	15	16	17	18	19	20	21	22	23	24
N	8	7	6	6	6	21	21	21	24	27	29	31	32	31	30	28	29	32	17	14	12	11	10	9
NE	11	10	9	8	9	63	87	90	77	58	49	48	46	44	42	39	35	29	22	19	17	15	14	12
E	15	13	12	11	11	70	107	123	124	110	85	65	60	57	53	48	43	37	29	25	22	20	18	16
SE	14	13	11	10	10	39	68	90	102	104	95	78	60	55	51	47	42	35	27	24	21	19	17	16
S	11	10	9	8	7	12	17	21	32	46	59	67	69	63	52	41	36	30	22	19	17	15	14	12
SW	21	19	17	15	14	18	22	25	28	31	34	51	74	94	106	109	100	78	45	37	33	29	26	23
W	25	23	20	18	17	21	24	28	30	33	34	35	53	84	112	130	135	116	57	46	39	35	31	28
NW	18	16	15	13	12	17	21	24	27	30	32	33	34	41	64	87	101	94	42	33	29	25	22	20
Hor	37	33	30	27	24	38	64	95	124	150	171	185	191	188	176	156	128	96	72	63	56	50	45	41

Notes:
1. Values are in Btu/h·ft².
2. Apply data directly to standard double strength glass with no inside shade.
3. Data applies to 21st day of July.
4. For other types of glass and internal shade, use shading coefficients as multiplier. See text. For externally shaded glass, use north orientation. See text.

(Copyright 1993 by ASHRAE, from 1993 Fundamentals. Used by permission.)

Footnote: Cross-references in table footnotes pertain to the source from which tables were taken; they do not refer to chapters or pages in this text.

Solar Cooling Load Factors (continued)

Figure 3.19 (continued)

Zone Types for Use with CLF Tables, Interior Rooms

Room Location	Zone Parameters[a]			Zone Type	
	Middle Floor	Ceiling Type	Floor Covering	People and Equipment	Lights
Single story	N/A	N/A	Carpet	C	B
	N/A	N/A	Vinyl	D	C
Top floor	65 mm Concrete	With	Carpet	D	C
	65 mm Concrete	With	Vinyl	D	D
	65 mm Concrete	Without	b	D	B
	25 mm Wood	b	b	D	B
Bottom floor	65 mm Concrete	With	Carpet	D	C
	65 mm Concrete	b	Vinyl	D	D
	65 mm Concrete	Without	Carpet	D	D
	25 mm Wood	b	Carpet	D	C
	25 mm Wood	b	Vinyl	D	D
Mid-floor	65 mm Concrete	N/A	Carpet	D	C
	65 mm Concrete	N/A	Vinyl	D	D
	25 mm Wood	N/A	b	C	B

[a] A total of 14 zone parameters is fully defined in Table 20. Those not shown in this table were selected to achieve an error band of approximately 10%.
[b] The effect of inside shade is negligible in this case.

Zone Types for Use with SCL and CLF Tables, Single-Story Building

Zone Parameters[a]				Zone Type			Error Band	
No. Walls	Floor Covering	Partition Type	Inside Shade	Glass Solar	People and Equipment	Lights	Plus	Minus
1 or 2	Carpet	Gypsum	b	A	B	B	9	2
1 or 2	Carpet	Concrete block	b	B	C	C	9	0
1 or 2	Vinyl	Gypsum	Full	B	C	C	9	0
1 or 2	Vinyl	Gypsum	Half to None	C	C	C	16	0
1 or 2	Vinyl	Concrete block	Full	C	D	D	8	0
1 or 2	Vinyl	Concrete block	Half to None	D	D	D	10	6
3	Carpet	Gypsum	b	A	B	B	9	2
3	Carpet	Concrete block	Full	A	B	B	9	2
3	Carpet	Concrete block	Half to None	B	B	B	9	0
3	Vinyl	Gypsum	Full	B	C	C	9	0
3	Vinyl	Gypsum	Half to None	C	C	C	16	0
3	Vinyl	Concrete block	Full	B	C	C	9	0
3	Vinyl	Concrete block	Half to None	C	C	C	16	0
4	Carpet	Gypsum	b	A	B	B	6	3
4	Vinyl	Gypsum	Full	B	C	C	11	6
4	Vinyl	Gypsum	Half to None	C	C	C	19	−1

[a] A total of 14 zone parameters is fully defined in Table 20. Those not shown in this table were selected to achieve the minimum error band shown in the righthand column for Solar Cooling Load (SCL). The error band for Lights and People and Equipment is approximately 10%.
[b] The effect of inside shade is negligible in this case.

Solar Cooling Load Factors—In SI Metric Units

Figure 3.19a

July Solar Cooling Load For Sunlit Glass 40° North Latitude

Zone type A

Glass Face	Hour 1	2	3	4	5	6	7	8	9	10	11	Solar Time 12	13	14	15	16	17	18	19	20	21	22	23	24
N	0	0	0	0	3	79	85	88	101	110	120	126	126	123	113	98	98	113	38	19	9	3	3	0
NE	0	0	0	0	6	268	406	422	353	236	173	151	139	126	117	101	82	57	22	9	6	3	0	0
E	0	0	0	0	6	293	495	583	576	485	334	211	167	142	123	104	82	57	22	9	6	3	0	0
SE	0	0	0	0	3	148	299	413	473	473	413	306	198	154	129	107	85	57	22	9	6	3	0	0
S	0	0	0	0	0	28	54	79	129	202	268	306	302	265	198	132	98	63	25	13	6	3	0	0
SW	0	0	0	0	0	28	54	76	95	110	123	202	318	419	476	479	419	293	110	54	25	13	6	3
W	3	0	0	0	0	28	54	76	95	110	120	126	205	359	498	589	605	491	180	85	41	19	9	6
NW	3	0	0	0	0	28	54	76	95	110	120	126	126	158	265	381	450	410	145	69	35	16	9	3
Hor	0	0	0	0	0	76	217	378	532	665	759	810	816	772	684	554	394	221	91	44	22	9	6	3

Zone type B

Glass Face	Hour 1	2	3	4	5	6	7	8	9	10	11	Solar Time 12	13	14	15	16	17	18	19	20	21	22	23	24
N	6	6	3	3	3	69	72	76	88	101	110	117	120	117	110	101	98	110	50	32	22	16	13	9
NE	6	3	3	3	6	230	343	365	318	230	183	164	151	142	129	113	95	72	41	28	19	16	9	9
E	6	6	3	3	6	252	419	501	510	450	331	233	198	173	151	129	107	79	47	32	22	16	13	9
SE	6	6	3	3	3	126	255	353	413	422	384	302	217	183	154	132	110	82	47	32	25	19	13	9
S	6	6	3	3	3	25	47	66	113	176	233	271	274	249	198	145	117	85	50	35	25	19	13	9
SW	19	16	13	9	6	28	50	69	85	98	113	183	280	369	425	435	397	296	145	98	66	47	35	25
W	25	19	16	13	9	28	50	69	85	98	110	117	186	318	438	523	545	463	208	135	95	66	47	35
NW	19	16	13	9	6	28	50	69	85	98	107	117	117	145	239	340	403	375	161	104	69	50	35	25
Hor	25	19	16	13	9	69	189	328	463	583	674	734	753	731	668	567	432	284	167	117	85	60	44	35

Zone type C

Glass Face	Hour 1	2	3	4	5	6	7	8	9	10	11	Solar Time 12	13	14	15	16	17	18	19	20	21	22	23	24
N	16	16	13	13	13	76	72	76	85	95	104	107	110	107	101	91	91	107	44	32	25	22	19	19
NE	22	19	19	16	19	236	334	337	277	192	154	148	142	135	126	113	98	79	50	41	35	32	28	25
E	28	25	25	22	25	261	410	466	457	391	280	195	176	164	148	135	117	95	63	54	47	41	38	35
SE	28	25	22	19	19	142	258	337	381	381	337	258	186	161	148	132	113	91	60	50	44	41	35	32
S	22	22	19	16	16	38	57	72	113	170	221	249	249	221	170	126	104	82	50	41	38	32	28	25
SW	44	38	35	32	28	47	66	82	91	104	113	180	271	347	391	394	350	252	117	88	72	63	54	47
W	54	47	41	38	35	54	69	85	98	107	113	117	186	309	416	482	491	403	158	110	88	76	66	60
NW	38	35	32	28	25	44	63	79	91	101	107	113	113	139	230	321	372	337	123	82	66	54	47	41
Hor	76	66	60	54	50	107	214	337	454	551	627	668	677	652	595	504	387	261	167	139	120	107	95	85

Zone type D

Glass Face	Hour 1	2	3	4	5	6	7	8	9	10	11	Solar Time 12	13	14	15	16	17	18	19	20	21	22	23	24
N	25	22	19	19	19	66	66	66	76	85	91	98	101	98	95	88	91	101	54	44	38	35	32	28
NE	35	32	28	25	28	198	274	284	243	183	154	151	145	139	132	123	110	91	69	60	54	47	44	38
E	47	41	38	35	35	221	337	387	391	347	268	205	189	180	167	151	135	117	91	79	69	63	57	50
SE	44	41	35	32	32	123	214	284	321	328	299	246	189	173	161	148	132	110	85	76	66	60	54	50
S	35	32	28	25	22	38	54	66	101	145	186	211	217	198	164	129	113	95	69	60	54	47	44	38
SW	66	60	54	47	44	57	69	79	88	98	107	161	233	296	334	343	315	246	142	117	104	91	82	72
W	79	72	63	57	54	66	76	88	95	104	107	110	167	265	353	410	425	365	180	145	123	110	98	88
NW	57	50	47	41	38	54	66	76	85	95	101	104	107	129	202	274	318	296	132	104	91	79	69	63
Hor	117	104	95	85	76	120	202	299	391	473	539	583	602	592	554	491	403	302	227	198	176	158	142	129

Notes:
1. Values are in W/m².
2. Apply data directly to standard double strength glass with no inside shade.
3. Data applies to 21st day of July.
4. For other types of glass and internal shade, use shading coefficients as multiplier. See text. For externally shaded glass, use north orientation. See text.

(Copyright 1993 by ASHRAE, from 1993 Fundamentals, SI Edition. Used by permission.)

Footnote: Cross-references in table footnotes pertain to the source from which tables were taken; they do not refer to chapters or pages in this text.

Solar Cooling Load Factors — In SI Metric Units (continued)

Figure 3.19a (continued)

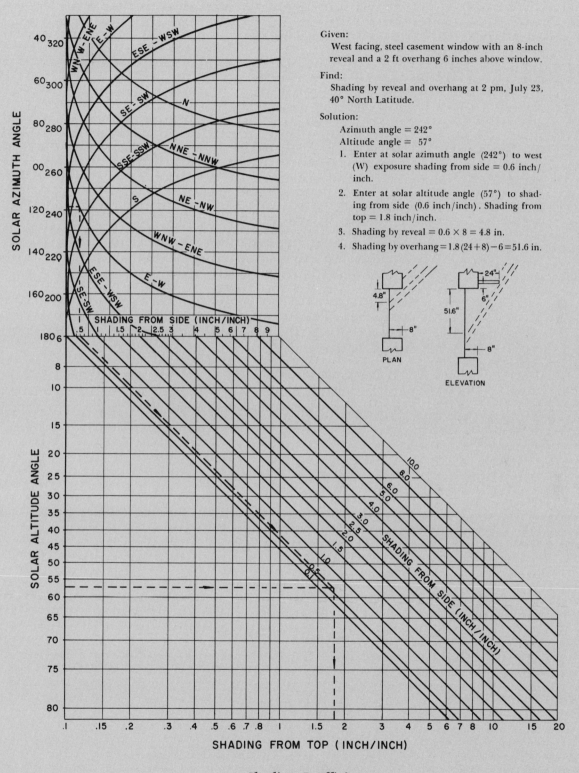

Figure 3.20

Reading CLF Tables

Determination of CLF involves working with two charts. For example, suppose we wish to determine the CLF for people in an office at 1100, 1300, and 1500 hours. Assume the people enter the office at 900 hours and leave at 1700 hours, and the office is in a single story building and has the following characteristics:

- 2 exterior walls
- Vinyl floor covering
- Gypsum partitions
- Full inside shade
- HVAC system operates 24 hours/day

See Figure 3.19; find zone type C for people and equipment. In Figure 3.21, proceed down to the section for zone type C. The left-hand column is for the total hours in the occupied space. The people enter at 900 and leave at 1700; therefore, they occupy the space for 8 hours. The top row is for the number of hours after entry into the space; these numbers for 1100, 1300, and 1500 hours would be 2, 4, and 6 respectively. In zone type C, locate 8 hours in the left-hand column, read across, and find the corresponding CLF numbers for 2, 4, and 6 hours after entry to the space. The correct CLF numbers are 0.69, 0.79, and 0.86.

Cooling load factors for people and unhooded equipment: CLF's for people and unhooded equipment are a function of the total amount of time spent in the space and the time elapsed since entering the space for people, or time elapsed since the equipment was turned on.

Cooling load factors for lights: CLF's for lights are a function of the total amount of time the lights in the space are on, and the time elapsed since the lights were turned on. It is also based on the assumption that

TYPES OF GLASS OR SHADING DEVICES*	Absorptivity (a)	Reflectivity (r)	Transmissibility (t)	Solar Factor†
Ordinary Glass	.06	.08	.86	1.00
Regular Plate, ¼"	.15	.08	.77	.94
Glass, Heat Absorbing	by mfg.	.05	(1 − .05 − a)	---
Venetian Blind, Light Color	.37	.51	.12	.56‡
Medium Color	.58	.39	.03	.65‡
Dark Color	.72	.27	.01	.75‡
Fiberglass Cloth, Off White (5.72 - 61/58)	.05	.60	.35	.48‡
Cotton Cloth, Beige (6.18 - 91/36)	.26	.51	.23	.56‡
Fiberglass Cloth, Light Gray	.30	.47	.23	.59‡
Fiberglass Cloth, Tan (7.55 - 57/29)	.44	.42	.14	.64‡
Glass Cloth, White, Golden Stripes	.05	.41	.54	.65‡
Fiberglass Cloth, Dark Gray	.60	.29	.11	.75‡
Dacron Cloth, White (1.8 - 86/81)	.02	.28	.70	.76‡
Cotton Cloth, Dark Green, Vinyl Coated (similar to roller shade)	.85	.15	.00	.88‡
Cotton Cloth, Dark Green (6.06 - 91/36)	.02	.28	.70	.76‡

*Factors for various draperies are given for guidance only since the actual drapery material may be different in color and texture; figures in parentheses are ounces per sq yd, and yarn count warp/filling. Consult manufacturers for actual values.

†Compared to ordinary glass.
‡For a shading device in combination with ordinary glass.

(Courtesy Carrier Corporation, McGraw-Hill Book Company)

Shading Coefficients (continued)

Figure 3.20 (continued)

the power input and heat dissipated from the light fixture become equal if the lights are left on long enough.

Cooling load factors for hooded equipment: CLF's for hooded equipment are a function of the same variables used for people and equipment; additionally, a portion of the sensible and latent heat is removed directly from the space via the hood.

Internal Heat Gain

All energy consumed in a building eventually results in a form of heat gain. Lights, office equipment, appliances, and computers all consume power that generates heat in the building and adds to the heat gain. Some of this internal heat gain is "dry," which means that it contains no moisture and produces only sensible heat gains. Examples of heat gains that are sensible only are lights, computers, copy machines, and televisions. Other pieces of equipment are "wet," meaning they produce some moisture, which adds approximately 1,000 Btu's of latent heat per pound of moisture [2293 kJ/kg]. Examples of equipment that produce both sensible (dry) and latent (wet) loads include cooking equipment, hair dryers, instrument sterilizers, coffee makers, and indoor fountains. The following formula is used for computing internal heat gain:

H_i = (Sensible load) CLF + (Latent load)
H_i = Internal heat gain (Btu/hr.) [W]
CLF = Cooling load factor

Figure 3.21 lists cooling load factors. Figure 3.22 lists the sensible and latent loads for common internal heat gains. To determine the sensible load of electric appliances and lights not listed, multiply the rated wattage by 3.4 to convert watts to Btu/hour. Add 15 percent if fluorescent fixtures are used to account for energy consumed by the energy-efficient ballasts.

People Heat Gain

The building's occupants represent a special type of internal heat gain. Each person in a building contributes a certain amount to the sensible and latent heat load. The following formula is used for computing heat gained from people:

H_p = $N_o P_s$ (CLF) + $N_o P_L$
H_p = People heat gain (Btu/hr.) [W]
N_o = Number of occupants
P_s = Sensible heat gain per person (Btu/hr.) [W] (see Figures 3.23 and 3.23a)
CLF = Cooling load factor, usually 1.0 (see Figure 3.21)
P_L = Latent heat gain per person (Btu/hr.) [W] (see Figures 3.23 and 3.23a)

The number of occupants, N_o, is determined by the use of each space. For auditoriums, restaurants, theaters, and similar spaces, the occupancy equals the number of seats. Other building uses assign population on a square foot basis. The cooling load factor usually equals 1.0 except in low density applications. Refer to Figure 3.21 or ASHRAE guidelines for a more detailed analysis. Figure 3.23 lists the sensible and latent heat gains for people. The heat gain depends on the amount and type of activity taking place.

Cooling Load Summary

The maximum cooling load is based on the greatest anticipated heat gain during the cooling season. Generally, conditions for 3:00 p.m. on a typical August day are used as one of several conditions tested. Because so many variable conditions are involved, many designers employ sophisticated computer programs that analyze a building hour-by-hour to produce the design conditions for the entire building. This analysis can also be done on

Cooling Load Factors for People and Unhooded Equipment

Number of Hours after Entry into Space or Equipment Turned On

Hours in Space	1	2	3	4	5	6	7	8	9	10	11	12	13	14	15	16	17	18	19	20	21	22	23	24
Zone Type A																								
2	0.75	0.88	0.18	0.08	0.04	0.02	0.01	0.01	0.01	0.01	0.00	0.00	0.00	0.00	0.00	0.00	0.00	0.00	0.00	0.00	0.00	0.00	0.00	0.00
4	0.75	0.88	0.93	0.95	0.22	0.10	0.05	0.03	0.02	0.02	0.01	0.01	0.01	0.01	0.00	0.00	0.00	0.00	0.00	0.00	0.00	0.00	0.00	0.00
6	0.75	0.88	0.93	0.95	0.97	0.97	0.33	0.11	0.06	0.04	0.03	0.02	0.02	0.01	0.01	0.01	0.01	0.00	0.00	0.00	0.00	0.00	0.00	0.00
8	0.75	0.88	0.93	0.95	0.97	0.97	0.98	0.98	0.24	0.11	0.06	0.04	0.03	0.02	0.02	0.01	0.01	0.01	0.01	0.01	0.00	0.00	0.00	0.00
10	0.75	0.88	0.93	0.95	0.97	0.97	0.98	0.98	0.99	0.99	0.24	0.12	0.07	0.04	0.03	0.02	0.02	0.01	0.01	0.01	0.01	0.01	0.00	0.00
12	0.75	0.88	0.93	0.96	0.97	0.98	0.98	0.98	0.99	0.99	0.99	0.99	0.25	0.12	0.07	0.04	0.03	0.02	0.02	0.02	0.01	0.01	0.01	0.01
14	0.76	0.88	0.93	0.96	0.97	0.98	0.98	0.99	0.99	0.99	0.99	0.99	1.00	1.00	0.25	0.12	0.07	0.05	0.03	0.03	0.02	0.02	0.01	0.01
16	0.76	0.89	0.94	0.96	0.97	0.98	0.98	0.99	0.99	0.99	0.99	0.99	1.00	1.00	1.00	1.00	0.25	0.12	0.07	0.05	0.03	0.03	0.02	0.02
18	0.77	0.89	0.94	0.96	0.97	0.98	0.98	0.99	0.99	0.99	0.99	1.00	1.00	1.00	1.00	1.00	1.00	1.00	0.25	0.12	0.07	0.05	0.03	0.03
Zone Type B																								
2	0.65	0.74	0.16	0.11	0.08	0.06	0.05	0.04	0.03	0.02	0.02	0.01	0.01	0.01	0.01	0.00	0.00	0.00	0.00	0.00	0.00	0.00	0.00	0.00
4	0.65	0.75	0.81	0.85	0.24	0.17	0.13	0.10	0.07	0.06	0.04	0.03	0.03	0.02	0.02	0.01	0.01	0.01	0.01	0.00	0.00	0.00	0.00	0.00
6	0.65	0.75	0.81	0.85	0.89	0.91	0.29	0.20	0.15	0.12	0.09	0.07	0.05	0.04	0.03	0.02	0.02	0.01	0.01	0.01	0.01	0.01	0.00	0.00
8	0.65	0.75	0.81	0.85	0.89	0.91	0.93	0.95	0.31	0.22	0.17	0.13	0.10	0.08	0.06	0.05	0.04	0.03	0.02	0.02	0.01	0.01	0.01	0.01
10	0.65	0.75	0.81	0.85	0.89	0.91	0.93	0.95	0.96	0.97	0.33	0.24	0.18	0.14	0.11	0.08	0.06	0.05	0.04	0.03	0.02	0.02	0.01	0.01
12	0.66	0.76	0.81	0.86	0.89	0.92	0.94	0.95	0.96	0.97	0.98	0.98	0.34	0.24	0.19	0.14	0.11	0.08	0.06	0.05	0.04	0.03	0.02	0.02
14	0.67	0.76	0.82	0.86	0.89	0.92	0.94	0.95	0.96	0.97	0.98	0.98	0.99	0.99	0.35	0.25	0.19	0.15	0.11	0.09	0.07	0.05	0.04	0.03
16	0.69	0.78	0.83	0.87	0.90	0.92	0.94	0.95	0.96	0.97	0.98	0.98	0.99	0.99	0.99	0.99	0.35	0.25	0.19	0.15	0.11	0.09	0.07	0.05
18	0.71	0.80	0.85	0.88	0.91	0.93	0.95	0.96	0.97	0.98	0.98	0.99	0.99	0.99	0.99	0.99	1.00	1.00	0.35	0.25	0.19	0.15	0.11	0.09
Zone Type C																								
2	0.60	0.68	0.14	0.11	0.09	0.07	0.06	0.05	0.04	0.03	0.03	0.02	0.02	0.01	0.01	0.01	0.01	0.01	0.01	0.00	0.00	0.00	0.00	0.00
4	0.60	0.68	0.74	0.79	0.23	0.18	0.14	0.12	0.10	0.08	0.06	0.05	0.04	0.04	0.03	0.02	0.02	0.02	0.01	0.01	0.01	0.01	0.01	0.01
6	0.61	0.69	0.74	0.79	0.83	0.86	0.28	0.22	0.18	0.15	0.12	0.10	0.08	0.07	0.06	0.05	0.04	0.03	0.03	0.02	0.02	0.01	0.01	0.01
8	0.61	0.69	0.75	0.79	0.83	0.86	0.89	0.91	0.32	0.26	0.21	0.17	0.14	0.11	0.09	0.08	0.06	0.05	0.04	0.04	0.03	0.02	0.02	0.02
10	0.62	0.70	0.75	0.80	0.83	0.86	0.89	0.91	0.92	0.94	0.35	0.28	0.23	0.18	0.15	0.12	0.10	0.08	0.07	0.06	0.05	0.04	0.03	0.03
12	0.63	0.71	0.76	0.81	0.84	0.87	0.89	0.91	0.93	0.94	0.95	0.96	0.37	0.29	0.24	0.19	0.16	0.13	0.11	0.09	0.07	0.06	0.05	0.04
14	0.65	0.72	0.77	0.82	0.85	0.88	0.90	0.92	0.93	0.94	0.95	0.96	0.97	0.97	0.38	0.30	0.25	0.20	0.17	0.14	0.11	0.09	0.08	0.06
16	0.68	0.74	0.79	0.83	0.86	0.89	0.91	0.92	0.94	0.95	0.96	0.96	0.97	0.98	0.98	0.98	0.39	0.31	0.25	0.21	0.17	0.14	0.11	0.09
18	0.72	0.78	0.82	0.85	0.88	0.90	0.92	0.93	0.94	0.95	0.96	0.97	0.97	0.98	0.98	0.99	0.99	0.99	0.39	0.31	0.26	0.21	0.17	0.14
Zone Type D																								
2	0.59	0.67	0.13	0.09	0.08	0.06	0.05	0.05	0.04	0.04	0.03	0.03	0.02	0.02	0.02	0.01	0.01	0.01	0.01	0.01	0.01	0.01	0.01	0.00
4	0.60	0.67	0.72	0.76	0.20	0.16	0.13	0.11	0.10	0.08	0.07	0.06	0.05	0.05	0.04	0.03	0.03	0.03	0.02	0.02	0.02	0.01	0.01	0.01
6	0.61	0.68	0.73	0.77	0.80	0.83	0.26	0.20	0.17	0.15	0.13	0.11	0.09	0.08	0.07	0.06	0.05	0.05	0.04	0.03	0.03	0.03	0.02	0.02
8	0.62	0.69	0.74	0.77	0.80	0.83	0.85	0.87	0.30	0.24	0.20	0.17	0.15	0.13	0.11	0.10	0.08	0.07	0.06	0.05	0.05	0.04	0.04	0.03
10	0.63	0.70	0.75	0.78	0.81	0.84	0.86	0.88	0.89	0.91	0.33	0.27	0.22	0.19	0.17	0.14	0.12	0.11	0.09	0.08	0.07	0.06	0.05	0.05
12	0.65	0.71	0.76	0.79	0.82	0.84	0.87	0.88	0.90	0.91	0.92	0.93	0.35	0.29	0.24	0.21	0.18	0.16	0.13	0.12	0.10	0.09	0.08	0.07
14	0.67	0.73	0.78	0.81	0.83	0.86	0.88	0.89	0.91	0.92	0.93	0.94	0.95	0.95	0.37	0.30	0.25	0.22	0.19	0.16	0.14	0.12	0.11	0.09
16	0.70	0.76	0.80	0.83	0.85	0.87	0.89	0.90	0.92	0.93	0.94	0.95	0.95	0.96	0.96	0.97	0.38	0.31	0.26	0.23	0.20	0.17	0.15	0.13
18	0.74	0.80	0.83	0.85	0.87	0.89	0.91	0.92	0.93	0.94	0.95	0.95	0.96	0.97	0.97	0.97	0.98	0.98	0.39	0.32	0.27	0.23	0.20	0.17

Note: See Table 35 for zone type. Data based on a radiative/convective fraction of 0.70/0.30.

Cooling Load Factors

Figure 3.21

a room-by-room basis. Cooling loads in buildings consist of conduction, convection, radiation, internal, and people heat gains. This information is summarized in Figure 3.24.

Figure 3.25 lists typical building cooling loads from *Means Mechanical Cost Data* as an overall guide to calculations.

The BIN Method for Estimating Energy Usage

Estimating the energy consumption of a building can be carried out by using simple steady-state (degree-day or BIN) methods. For very complicated and varying situations, a more detailed process may be necessary.

To achieve good results, an energy-estimating method should calculate the energy consumption and varying conditions in different temperature intervals, or BINs. The BIN method follows these criteria, calculating energy consumption for several values of outdoor air temperature and multiplying the consumption by the number of hours within that temperature interval, or BIN.

The degree-day method assumes that both the energy efficiency of the equipment and the building heat loss coefficient remain constant for all outdoor air temperatures. The degree-day is the simplest of the steady-state methods and will give a good indication of the estimated energy consumption of a building.

Cooling Load Factors for Lights

Lights On For	Number of Hours after Lights Turned On																							
	1	2	3	4	5	6	7	8	9	10	11	12	13	14	15	16	17	18	19	20	21	22	23	24
Zone Type A																								
8	0.85	0.92	0.95	0.96	0.97	0.97	0.97	0.98	0.13	0.06	0.04	0.03	0.02	0.02	0.02	0.01	0.01	0.01	0.01	0.01	0.01	0.01	0.01	0.01
10	0.85	0.93	0.95	0.97	0.97	0.97	0.98	0.98	0.98	0.98	0.14	0.07	0.04	0.03	0.02	0.02	0.02	0.02	0.02	0.02	0.01	0.01	0.01	0.01
12	0.86	0.93	0.96	0.97	0.97	0.98	0.98	0.98	0.98	0.98	0.98	0.98	0.14	0.07	0.04	0.03	0.03	0.02	0.02	0.02	0.02	0.02	0.02	0.02
14	0.86	0.93	0.96	0.97	0.98	0.98	0.98	0.98	0.98	0.98	0.99	0.99	0.99	0.99	0.15	0.07	0.05	0.03	0.03	0.03	0.02	0.02	0.02	0.02
16	0.87	0.94	0.96	0.97	0.98	0.98	0.98	0.99	0.99	0.99	0.99	0.99	0.99	0.99	0.99	0.99	0.15	0.08	0.05	0.04	0.03	0.03	0.03	0.02
Zone Type B																								
8	0.75	0.85	0.90	0.93	0.94	0.95	0.95	0.96	0.23	0.12	0.08	0.05	0.04	0.04	0.03	0.03	0.03	0.02	0.02	0.02	0.02	0.02	0.02	0.01
10	0.75	0.86	0.91	0.93	0.94	0.95	0.95	0.96	0.96	0.97	0.24	0.13	0.08	0.06	0.05	0.04	0.04	0.03	0.03	0.03	0.03	0.02	0.02	0.02
12	0.76	0.86	0.91	0.93	0.95	0.95	0.96	0.96	0.97	0.97	0.97	0.97	0.24	0.14	0.09	0.07	0.05	0.05	0.04	0.04	0.03	0.03	0.03	0.03
14	0.76	0.87	0.92	0.94	0.95	0.96	0.96	0.97	0.97	0.97	0.97	0.98	0.98	0.98	0.25	0.14	0.09	0.07	0.06	0.05	0.05	0.04	0.04	0.03
16	0.77	0.88	0.92	0.95	0.96	0.96	0.97	0.97	0.97	0.98	0.98	0.98	0.98	0.98	0.98	0.99	0.25	0.15	0.10	0.07	0.06	0.05	0.05	0.04
Zone Type C																								
8	0.72	0.80	0.84	0.87	0.88	0.89	0.90	0.91	0.23	0.15	0.11	0.09	0.08	0.07	0.07	0.06	0.05	0.05	0.05	0.04	0.04	0.03	0.03	0.03
10	0.73	0.81	0.85	0.87	0.89	0.90	0.91	0.92	0.92	0.93	0.25	0.16	0.13	0.11	0.09	0.08	0.08	0.07	0.06	0.06	0.05	0.05	0.04	0.04
12	0.74	0.82	0.86	0.88	0.90	0.91	0.92	0.92	0.93	0.94	0.94	0.95	0.26	0.18	0.14	0.12	0.10	0.09	0.08	0.08	0.07	0.06	0.06	0.05
14	0.75	0.84	0.87	0.89	0.91	0.92	0.92	0.93	0.94	0.94	0.95	0.95	0.96	0.96	0.27	0.19	0.15	0.13	0.11	0.10	0.09	0.08	0.08	0.07
16	0.77	0.85	0.89	0.91	0.92	0.93	0.93	0.94	0.95	0.95	0.95	0.96	0.96	0.97	0.97	0.97	0.28	0.20	0.16	0.13	0.12	0.11	0.10	0.09
Zone Type D																								
8	0.66	0.72	0.76	0.79	0.81	0.83	0.85	0.86	0.25	0.20	0.17	0.15	0.13	0.12	0.11	0.10	0.09	0.08	0.07	0.06	0.06	0.05	0.04	0.04
10	0.68	0.74	0.77	0.80	0.82	0.84	0.86	0.87	0.88	0.90	0.28	0.23	0.19	0.17	0.15	0.14	0.12	0.11	0.10	0.09	0.08	0.07	0.06	0.06
12	0.70	0.75	0.79	0.81	0.83	0.85	0.87	0.88	0.89	0.90	0.91	0.92	0.30	0.25	0.21	0.19	0.17	0.15	0.13	0.12	0.11	0.10	0.09	0.08
14	0.72	0.77	0.81	0.83	0.85	0.86	0.88	0.89	0.90	0.91	0.92	0.93	0.94	0.94	0.32	0.26	0.23	0.20	0.18	0.16	0.14	0.13	0.12	0.10
16	0.75	0.80	0.83	0.85	0.87	0.88	0.89	0.90	0.91	0.92	0.93	0.94	0.94	0.95	0.96	0.96	0.34	0.28	0.24	0.21	0.19	0.17	0.15	0.14

Note: See Table 35 for zone type. Data based on a radiative/convective fraction of 0.59/0.41.

Cooling Load Factors (continued)

Figure 3.21 (continued)

The building heat loss coefficient is a measure of the heat loss per °F [°C] temperature difference from the balance point temperature t(bal). The balance point temperature t(bal) is the calculated temperature at which all internal heat gains are offset by the heat loss of the building. This balance point temperature is generally accepted as 65°F [18 °C] when used for the degree-day method of energy consumption estimation.

Cooling Load Factors for Hooded Equipment

Hours in Operation	\multicolumn{22}{c}{Number of Hours after Equipment Turned On}

Hours in Operation	1	2	3	4	5	6	7	8	9	10	11	12	13	14	15	16	17	18	19	20	21	23	24
Zone Type A																							
2	0.64	0.83	0.26	0.11	0.06	0.03	0.01	0.01	0.01	0.01	0.00	0.00	0.00	0.00	0.00	0.00	0.00	0.00	0.00	0.00	0.00	0.00	0.00
4	0.64	0.83	0.90	0.93	0.31	0.14	0.07	0.04	0.03	0.03	0.01	0.01	0.01	0.01	0.00	0.00	0.00	0.00	0.00	0.00	0.00	0.00	0.00
6	0.64	0.83	0.90	0.93	0.96	0.96	0.33	0.16	0.09	0.06	0.04	0.03	0.03	0.01	0.01	0.01	0.01	0.00	0.00	0.00	0.00	0.00	0.00
8	0.64	0.83	0.90	0.93	0.96	0.96	0.97	0.97	0.34	0.16	0.09	0.06	0.04	0.03	0.03	0.01	0.01	0.01	0.01	0.01	0.00	0.00	0.00
10	0.64	0.83	0.90	0.93	0.96	0.96	0.97	0.97	0.99	0.99	0.34	0.17	0.10	0.06	0.04	0.03	0.03	0.01	0.01	0.01	0.01	0.01	0.00
12	0.64	0.83	0.90	0.94	0.96	0.97	0.97	0.97	0.99	0.99	0.99	0.99	0.36	0.17	0.10	0.06	0.04	0.03	0.03	0.03	0.01	0.01	0.01
14	0.66	0.83	0.90	0.94	0.96	0.97	0.97	0.99	0.99	0.99	0.99	0.99	1.00	1.00	0.36	0.17	0.10	0.07	0.04	0.04	0.03	0.03	0.01
16	0.66	0.84	0.91	0.94	0.96	0.97	0.97	0.99	0.99	0.99	0.99	0.99	1.00	1.00	1.00	1.00	0.36	0.17	0.10	0.07	0.04	0.04	0.03
18	0.67	0.84	0.91	0.94	0.96	0.97	0.97	0.99	0.99	0.99	0.99	1.00	1.00	1.00	1.00	1.00	1.00	1.00	0.36	0.17	0.10	0.07	0.04
Zone Type B																							
2	0.50	0.63	0.23	0.16	0.11	0.09	0.07	0.06	0.04	0.03	0.03	0.01	0.01	0.01	0.01	0.00	0.00	0.00	0.00	0.00	0.00	0.00	0.00
4	0.50	0.64	0.73	0.79	0.34	0.24	0.19	0.14	0.10	0.09	0.06	0.04	0.04	0.03	0.03	0.01	0.01	0.01	0.01	0.00	0.00	0.00	0.00
6	0.50	0.64	0.73	0.79	0.84	0.87	0.41	0.29	0.21	0.17	0.13	0.10	0.07	0.06	0.04	0.03	0.03	0.01	0.01	0.01	0.01	0.01	0.00
8	0.50	0.64	0.73	0.79	0.84	0.87	0.90	0.93	0.44	0.31	0.24	0.19	0.14	0.11	0.09	0.07	0.06	0.04	0.03	0.03	0.01	0.01	0.01
10	0.50	0.64	0.73	0.79	0.84	0.87	0.90	0.93	0.94	0.96	0.47	0.34	0.26	0.20	0.16	0.11	0.09	0.07	0.06	0.04	0.03	0.03	0.01
12	0.51	0.66	0.73	0.80	0.84	0.89	0.91	0.93	0.94	0.96	0.97	0.97	0.49	0.34	0.27	0.20	0.16	0.11	0.09	0.07	0.06	0.04	0.03
14	0.53	0.66	0.74	0.80	0.84	0.89	0.91	0.93	0.94	0.96	0.97	0.97	0.99	0.99	0.50	0.36	0.27	0.21	0.16	0.13	0.10	0.07	0.06
16	0.56	0.69	0.76	0.81	0.86	0.89	0.91	0.93	0.94	0.96	0.97	0.97	0.99	0.99	0.99	0.99	0.50	0.36	0.27	0.21	0.16	0.13	0.10
18	0.59	0.71	0.79	0.83	0.87	0.90	0.93	0.94	0.96	0.97	0.97	0.99	0.99	0.99	0.99	0.99	1.00	1.00	0.50	0.36	0.27	0.21	0.16
Zone Type C																							
2	0.43	0.54	0.20	0.16	0.13	0.10	0.09	0.07	0.06	0.04	0.04	0.03	0.03	0.01	0.01	0.01	0.01	0.01	0.01	0.00	0.00	0.00	0.00
4	0.43	0.54	0.63	0.70	0.33	0.26	0.20	0.17	0.14	0.11	0.09	0.07	0.06	0.06	0.04	0.03	0.03	0.03	0.01	0.01	0.01	0.01	0.01
6	0.44	0.56	0.63	0.70	0.76	0.80	0.40	0.31	0.26	0.21	0.17	0.14	0.11	0.10	0.09	0.07	0.06	0.04	0.04	0.03	0.03	0.01	0.01
8	0.44	0.56	0.64	0.70	0.76	0.80	0.84	0.87	0.46	0.37	0.30	0.24	0.20	0.16	0.13	0.11	0.09	0.07	0.06	0.06	0.04	0.03	0.03
10	0.46	0.57	0.64	0.71	0.76	0.80	0.84	0.87	0.89	0.91	0.50	0.40	0.33	0.26	0.21	0.17	0.14	0.11	0.10	0.09	0.07	0.06	0.04
12	0.47	0.59	0.66	0.73	0.77	0.81	0.84	0.87	0.90	0.91	0.93	0.94	0.53	0.41	0.34	0.27	0.23	0.19	0.16	0.13	0.10	0.09	0.07
14	0.50	0.60	0.67	0.74	0.79	0.83	0.86	0.89	0.90	0.91	0.93	0.94	0.96	0.96	0.54	0.43	0.36	0.29	0.24	0.20	0.16	0.13	0.11
16	0.54	0.63	0.70	0.76	0.80	0.84	0.87	0.89	0.91	0.93	0.94	0.94	0.96	0.97	0.97	0.97	0.56	0.44	0.36	0.30	0.24	0.20	0.16
18	0.60	0.69	0.74	0.79	0.83	0.86	0.89	0.90	0.91	0.93	0.94	0.96	0.96	0.97	0.97	0.99	0.99	0.99	0.56	0.44	0.37	0.30	0.24
Zone Type D																							
2	0.41	0.53	0.19	0.13	0.11	0.09	0.07	0.07	0.06	0.06	0.04	0.04	0.03	0.03	0.03	0.01	0.01	0.01	0.01	0.01	0.01	0.01	0.01
4	0.43	0.53	0.60	0.66	0.29	0.23	0.19	0.16	0.14	0.11	0.10	0.09	0.07	0.07	0.06	0.04	0.04	0.04	0.03	0.03	0.03	0.01	0.01
6	0.44	0.54	0.61	0.67	0.71	0.76	0.37	0.29	0.24	0.21	0.19	0.16	0.13	0.11	0.10	0.09	0.07	0.07	0.06	0.04	0.04	0.04	0.03
8	0.46	0.56	0.63	0.67	0.71	0.76	0.79	0.81	0.43	0.34	0.29	0.24	0.21	0.19	0.16	0.14	0.11	0.10	0.09	0.07	0.07	0.06	0.06
10	0.47	0.57	0.64	0.69	0.73	0.77	0.80	0.83	0.84	0.87	0.47	0.39	0.31	0.27	0.24	0.20	0.17	0.16	0.13	0.11	0.10	0.09	0.07
12	0.50	0.59	0.66	0.70	0.74	0.77	0.81	0.83	0.86	0.87	0.89	0.90	0.50	0.41	0.34	0.30	0.26	0.23	0.19	0.17	0.14	0.13	0.11
14	0.53	0.61	0.69	0.73	0.76	0.80	0.83	0.84	0.87	0.89	0.90	0.91	0.93	0.93	0.53	0.43	0.36	0.31	0.27	0.23	0.20	0.17	0.16
16	0.57	0.66	0.71	0.76	0.79	0.81	0.84	0.86	0.89	0.90	0.91	0.93	0.93	0.94	0.94	0.96	0.54	0.44	0.37	0.33	0.29	0.24	0.21
18	0.63	0.71	0.76	0.79	0.81	0.84	0.87	0.89	0.90	0.91	0.93	0.93	0.94	0.96	0.96	0.96	0.97	0.97	0.56	0.46	0.39	0.33	0.29

Note: See Table 35 for zone type. Data based on a radiative/convective fraction of 100/0.

(Copyright 1993 by ASHRAE, from 1993 Fundamentals. Used by permission.)

Footnote: Cross-references in table footnotes pertain to the source from which tables were taken; they do not refer to chapters or pages in this text.

Cooling Load Factors (continued)

Figure 3.21 (continued)

In most cases the balance point temperature t(bal), the equipment efficiency, and the heat loss coefficient of the building do not remain constant over a wide range of outdoor air temperatures. The efficiency of boilers and chillers changes indirectly with the outdoor temperature because of the building load change. The efficiency of heat pumps changes dramatically over a wide range of outdoor air temperatures.

Cooling Load Factors for People and Unhooded Equipment

Number of Hours after Entry into Space or Equipment Turned On

Hours in Space	1	2	3	4	5	6	7	8	9	10	11	12	13	14	15	16	17	18	19	20	21	22	23	24
Zone Type A																								
2	0.75	0.88	0.18	0.08	0.04	0.02	0.01	0.01	0.01	0.01	0.00	0.00	0.00	0.00	0.00	0.00	0.00	0.00	0.00	0.00	0.00	0.00	0.00	0.00
4	0.75	0.88	0.93	0.95	0.22	0.10	0.05	0.03	0.02	0.02	0.01	0.01	0.01	0.01	0.00	0.00	0.00	0.00	0.00	0.00	0.00	0.00	0.00	0.00
6	0.75	0.88	0.93	0.95	0.97	0.97	0.33	0.11	0.06	0.04	0.03	0.02	0.02	0.01	0.01	0.01	0.01	0.00	0.00	0.00	0.00	0.00	0.00	0.00
8	0.75	0.88	0.93	0.95	0.97	0.97	0.98	0.98	0.24	0.11	0.06	0.04	0.03	0.02	0.02	0.01	0.01	0.01	0.01	0.01	0.00	0.00	0.00	0.00
10	0.75	0.88	0.93	0.95	0.97	0.97	0.98	0.98	0.99	0.99	0.24	0.12	0.07	0.04	0.03	0.02	0.02	0.01	0.01	0.01	0.01	0.01	0.00	0.00
12	0.75	0.88	0.93	0.96	0.97	0.98	0.98	0.98	0.99	0.99	0.99	0.99	0.25	0.12	0.07	0.04	0.03	0.02	0.02	0.02	0.01	0.01	0.01	0.01
14	0.76	0.88	0.93	0.96	0.97	0.98	0.98	0.99	0.99	0.99	0.99	0.99	1.00	1.00	0.25	0.12	0.07	0.05	0.03	0.03	0.02	0.02	0.01	0.01
16	0.76	0.89	0.94	0.96	0.97	0.98	0.98	0.99	0.99	0.99	0.99	0.99	1.00	1.00	1.00	1.00	0.25	0.12	0.07	0.05	0.03	0.03	0.02	0.02
18	0.77	0.89	0.94	0.96	0.97	0.98	0.98	0.99	0.99	0.99	0.99	0.99	1.00	1.00	1.00	1.00	1.00	1.00	0.25	0.12	0.07	0.05	0.03	0.03
Zone Type B																								
2	0.65	0.74	0.16	0.11	0.08	0.06	0.05	0.04	0.03	0.02	0.02	0.01	0.01	0.01	0.01	0.00	0.00	0.00	0.00	0.00	0.00	0.00	0.00	0.00
4	0.65	0.75	0.81	0.85	0.24	0.17	0.13	0.10	0.07	0.06	0.04	0.03	0.03	0.02	0.02	0.01	0.01	0.01	0.01	0.00	0.00	0.00	0.00	0.00
6	0.65	0.75	0.81	0.85	0.89	0.91	0.29	0.20	0.15	0.12	0.09	0.07	0.05	0.04	0.03	0.02	0.02	0.02	0.01	0.01	0.01	0.01	0.00	0.00
8	0.65	0.75	0.81	0.85	0.89	0.91	0.93	0.95	0.31	0.22	0.17	0.13	0.10	0.08	0.06	0.05	0.04	0.03	0.02	0.02	0.01	0.01	0.01	0.01
10	0.65	0.75	0.81	0.85	0.89	0.91	0.93	0.95	0.96	0.97	0.33	0.24	0.18	0.14	0.11	0.08	0.06	0.05	0.04	0.03	0.02	0.02	0.01	0.01
12	0.66	0.76	0.81	0.86	0.89	0.92	0.94	0.95	0.96	0.97	0.98	0.98	0.34	0.24	0.19	0.14	0.11	0.08	0.06	0.05	0.04	0.03	0.02	0.02
14	0.67	0.76	0.82	0.86	0.89	0.92	0.94	0.95	0.96	0.97	0.98	0.98	0.99	0.99	0.35	0.25	0.19	0.15	0.11	0.09	0.07	0.05	0.04	0.03
16	0.69	0.78	0.83	0.87	0.90	0.92	0.94	0.95	0.96	0.97	0.98	0.98	0.99	0.99	0.99	0.99	0.35	0.25	0.19	0.15	0.11	0.09	0.07	0.05
18	0.71	0.80	0.85	0.88	0.91	0.93	0.95	0.96	0.97	0.98	0.98	0.99	0.99	0.99	0.99	0.99	1.00	1.00	0.35	0.25	0.19	0.15	0.11	0.09
Zone Type C																								
2	0.60	0.68	0.14	0.11	0.09	0.07	0.06	0.05	0.04	0.03	0.03	0.02	0.02	0.01	0.01	0.01	0.01	0.01	0.00	0.00	0.00	0.00	0.00	0.00
4	0.60	0.68	0.74	0.79	0.23	0.18	0.14	0.12	0.10	0.08	0.06	0.05	0.04	0.03	0.02	0.02	0.02	0.01	0.01	0.01	0.01	0.01	0.01	0.01
6	0.61	0.69	0.74	0.79	0.83	0.86	0.28	0.22	0.18	0.15	0.12	0.10	0.08	0.07	0.06	0.05	0.04	0.03	0.03	0.02	0.02	0.01	0.01	0.01
8	0.61	0.69	0.75	0.79	0.83	0.86	0.89	0.91	0.32	0.26	0.21	0.17	0.14	0.11	0.09	0.08	0.06	0.05	0.04	0.04	0.03	0.02	0.02	0.02
10	0.62	0.70	0.75	0.80	0.83	0.86	0.89	0.91	0.92	0.94	0.35	0.28	0.23	0.18	0.15	0.12	0.10	0.08	0.07	0.06	0.05	0.04	0.03	0.03
12	0.63	0.71	0.76	0.81	0.84	0.87	0.89	0.91	0.93	0.94	0.95	0.96	0.37	0.29	0.24	0.19	0.16	0.13	0.11	0.09	0.07	0.06	0.05	0.04
14	0.65	0.72	0.77	0.82	0.85	0.88	0.90	0.92	0.93	0.94	0.95	0.96	0.97	0.97	0.38	0.30	0.25	0.20	0.17	0.14	0.11	0.09	0.08	0.06
16	0.68	0.74	0.79	0.83	0.86	0.89	0.91	0.92	0.94	0.95	0.96	0.96	0.97	0.98	0.98	0.98	0.39	0.31	0.25	0.21	0.17	0.14	0.11	0.09
18	0.72	0.78	0.82	0.85	0.88	0.90	0.92	0.93	0.94	0.95	0.96	0.97	0.97	0.98	0.98	0.99	0.99	0.99	0.39	0.31	0.26	0.21	0.17	0.14
Zone Type D																								
2	0.59	0.67	0.13	0.09	0.08	0.06	0.05	0.05	0.04	0.04	0.03	0.03	0.02	0.02	0.02	0.01	0.01	0.01	0.01	0.01	0.01	0.01	0.01	0.00
4	0.60	0.67	0.72	0.76	0.20	0.16	0.13	0.11	0.10	0.08	0.07	0.06	0.05	0.05	0.04	0.03	0.03	0.03	0.02	0.02	0.02	0.01	0.01	0.01
6	0.61	0.68	0.73	0.77	0.80	0.83	0.26	0.20	0.17	0.15	0.13	0.11	0.09	0.08	0.07	0.06	0.05	0.05	0.04	0.03	0.03	0.03	0.02	0.02
8	0.62	0.69	0.74	0.77	0.80	0.83	0.85	0.87	0.30	0.24	0.20	0.17	0.15	0.13	0.11	0.10	0.08	0.07	0.06	0.05	0.05	0.04	0.04	0.03
10	0.63	0.70	0.75	0.78	0.81	0.84	0.86	0.88	0.89	0.91	0.33	0.27	0.22	0.19	0.17	0.14	0.12	0.11	0.09	0.08	0.07	0.06	0.05	0.05
12	0.65	0.71	0.76	0.79	0.82	0.84	0.87	0.88	0.90	0.91	0.92	0.93	0.35	0.29	0.24	0.21	0.18	0.16	0.13	0.12	0.10	0.09	0.08	0.07
14	0.67	0.73	0.78	0.81	0.83	0.86	0.88	0.89	0.91	0.92	0.93	0.94	0.95	0.95	0.37	0.30	0.25	0.22	0.19	0.16	0.14	0.12	0.11	0.09
16	0.70	0.76	0.80	0.83	0.85	0.87	0.89	0.90	0.92	0.93	0.94	0.95	0.95	0.96	0.96	0.97	0.38	0.31	0.26	0.23	0.20	0.17	0.15	0.13
18	0.74	0.80	0.83	0.85	0.87	0.89	0.91	0.92	0.93	0.94	0.95	0.95	0.96	0.97	0.97	0.97	0.98	0.98	0.39	0.32	0.27	0.23	0.20	0.17

Note: See Table 35 for zone type. Data based on a radiative/convective fraction of 0.70/0.30.

Cooling Load Factors – In SI Metric Units

Figure 3.21a

The building heating energy consumption of a particular temperature interval (BIN) is given by:

$$Q_{BIN} = N_{BIN}(K_{tot}/\text{eff}_h)(t_{bal}-t_o)+$$

where: Q_{BIN} = the energy consumption of a particular BIN [kWh]
N_{BIN} = the number of hours in the BIN
K_{tot} = the total heat loss coefficient (Btu/hr.°F) [W/K]
eff_h = the efficiency of the HVAC system for that BIN
t_{bal} = balance temperature
t_o = outside air temperature
$+$ = only positive values of $t_{bal} - t_o$ will be used for calculating purposes

Energy consumption for the cooling season can also be calculated; however, because of the complexity of the solar gains on all surfaces of a structure and the wide diversity of solar loads from day to day, a more detailed and complicated modified BIN method should be used (see discussion and example below).

BIN data are available from two sources: ASHRAE and the USAF. BINs are normally in 5°F [3 °C] increments and are normally recorded for a temperature range of −10°F [−24 °C] to 105°F [39 °C] and in three 8-hour shifts throughout the day. As there are 23 [20] 5°F [3 °C] BINs between

Cooling Load Factors for Lights

Lights On For	Number of Hours after Lights Turned On																							
	1	2	3	4	5	6	7	8	9	10	11	12	13	14	15	16	17	18	19	20	21	22	23	24
Zone Type A																								
8	0.85	0.92	0.95	0.96	0.97	0.97	0.97	0.98	0.13	0.06	0.04	0.03	0.02	0.02	0.02	0.01	0.01	0.01	0.01	0.01	0.01	0.01	0.01	0.01
10	0.85	0.93	0.95	0.97	0.97	0.97	0.98	0.98	0.98	0.98	0.14	0.07	0.04	0.03	0.02	0.02	0.02	0.02	0.02	0.02	0.01	0.01	0.01	0.01
12	0.86	0.93	0.96	0.97	0.97	0.98	0.98	0.98	0.98	0.98	0.98	0.98	0.14	0.07	0.04	0.03	0.03	0.02	0.02	0.02	0.02	0.02	0.02	0.02
14	0.86	0.93	0.96	0.97	0.98	0.98	0.98	0.98	0.98	0.98	0.99	0.99	0.99	0.99	0.15	0.07	0.05	0.03	0.03	0.03	0.02	0.02	0.02	0.02
16	0.87	0.94	0.96	0.97	0.98	0.98	0.98	0.99	0.99	0.99	0.99	0.99	0.99	0.99	0.99	0.99	0.15	0.08	0.05	0.04	0.03	0.03	0.03	0.02
Zone Type B																								
8	0.75	0.85	0.90	0.93	0.94	0.95	0.95	0.96	0.23	0.12	0.08	0.05	0.04	0.04	0.03	0.03	0.03	0.02	0.02	0.02	0.02	0.02	0.02	0.01
10	0.75	0.86	0.91	0.93	0.94	0.95	0.95	0.96	0.96	0.97	0.24	0.13	0.08	0.06	0.05	0.04	0.04	0.03	0.03	0.03	0.03	0.02	0.02	0.02
12	0.76	0.86	0.91	0.93	0.95	0.95	0.96	0.96	0.97	0.97	0.97	0.97	0.24	0.14	0.09	0.07	0.05	0.05	0.04	0.04	0.03	0.03	0.03	0.03
14	0.76	0.87	0.92	0.94	0.95	0.96	0.96	0.97	0.97	0.97	0.97	0.98	0.98	0.98	0.25	0.14	0.09	0.07	0.06	0.05	0.05	0.04	0.04	0.03
16	0.77	0.88	0.92	0.95	0.96	0.96	0.97	0.97	0.97	0.98	0.98	0.98	0.98	0.98	0.98	0.99	0.25	0.15	0.10	0.07	0.06	0.05	0.05	0.04
Zone Type C																								
8	0.72	0.80	0.84	0.87	0.88	0.89	0.90	0.91	0.23	0.15	0.11	0.09	0.08	0.07	0.07	0.06	0.05	0.05	0.05	0.04	0.04	0.03	0.03	0.03
10	0.73	0.81	0.85	0.87	0.89	0.90	0.91	0.92	0.92	0.93	0.25	0.16	0.13	0.11	0.09	0.08	0.08	0.07	0.06	0.06	0.05	0.05	0.04	0.04
12	0.74	0.82	0.86	0.88	0.90	0.91	0.92	0.92	0.93	0.94	0.94	0.95	0.26	0.18	0.14	0.12	0.10	0.09	0.08	0.08	0.07	0.06	0.06	0.05
14	0.75	0.84	0.87	0.89	0.91	0.92	0.92	0.93	0.94	0.94	0.95	0.95	0.96	0.96	0.27	0.19	0.15	0.13	0.11	0.10	0.09	0.08	0.08	0.07
16	0.77	0.85	0.89	0.91	0.92	0.93	0.93	0.94	0.95	0.95	0.95	0.96	0.96	0.97	0.97	0.97	0.28	0.20	0.16	0.13	0.12	0.11	0.10	0.09
Zone Type D																								
8	0.66	0.72	0.76	0.79	0.81	0.83	0.85	0.86	0.25	0.20	0.17	0.15	0.13	0.12	0.11	0.10	0.09	0.08	0.07	0.06	0.06	0.05	0.04	0.04
10	0.68	0.74	0.77	0.80	0.82	0.84	0.86	0.87	0.88	0.90	0.28	0.23	0.19	0.17	0.15	0.14	0.12	0.11	0.10	0.09	0.08	0.07	0.06	0.06
12	0.70	0.75	0.79	0.81	0.83	0.85	0.87	0.88	0.89	0.90	0.91	0.92	0.30	0.25	0.21	0.19	0.17	0.15	0.13	0.12	0.11	0.10	0.09	0.08
14	0.72	0.77	0.81	0.83	0.85	0.86	0.88	0.89	0.90	0.91	0.92	0.93	0.94	0.94	0.32	0.26	0.23	0.20	0.18	0.16	0.14	0.13	0.12	0.10
16	0.75	0.80	0.83	0.85	0.87	0.88	0.89	0.90	0.91	0.92	0.93	0.94	0.94	0.95	0.96	0.96	0.34	0.28	0.24	0.21	0.19	0.17	0.15	0.14

Note: See Table 35 for zone type. Data based on a radiative/convective fraction of 0.59/0.41.

Cooling Load Factors—In SI Metric Units (continued)

Figure 3.21a (continued)

−10°F [−24 °C] and 105°F [39 °C] and there are three 8-hour shifts per day, there are 69 [60] possible BIN calculations per day. Less exact estimates can be performed on a monthly or yearly basis.

Example:
Estimate the energy consumption of a heat pump in a residence in New York if the design heat loss is 55,000 Btu's/hr. [16 120 W] at a 60°F 15.5 °C] temperature difference. Assume an internal heat gain of 5,500

Cooling Load Factors for Hooded Equipment

Hours in Operation	Number of Hours after Equipment Turned On																						
	1	2	3	4	5	6	7	8	9	10	11	12	13	14	15	16	17	18	19	20	21	23	24
Zone Type A																							
2	0.64	0.83	0.26	0.11	0.06	0.03	0.01	0.01	0.01	0.01	0.00	0.00	0.00	0.00	0.00	0.00	0.00	0.00	0.00	0.00	0.00	0.00	0.00
4	0.64	0.83	0.90	0.93	0.31	0.14	0.07	0.04	0.03	0.03	0.01	0.01	0.01	0.01	0.00	0.00	0.00	0.00	0.00	0.00	0.00	0.00	0.00
6	0.64	0.83	0.90	0.93	0.96	0.96	0.33	0.16	0.09	0.06	0.04	0.03	0.03	0.01	0.01	0.01	0.01	0.00	0.00	0.00	0.00	0.00	0.00
8	0.64	0.83	0.90	0.93	0.96	0.96	0.97	0.97	0.34	0.16	0.09	0.06	0.04	0.03	0.03	0.01	0.01	0.01	0.01	0.01	0.00	0.00	0.00
10	0.64	0.83	0.90	0.93	0.96	0.96	0.97	0.97	0.99	0.99	0.34	0.17	0.10	0.06	0.04	0.03	0.03	0.01	0.01	0.01	0.01	0.01	0.00
12	0.64	0.83	0.90	0.94	0.96	0.97	0.97	0.97	0.99	0.99	0.99	0.99	0.36	0.17	0.10	0.06	0.04	0.03	0.03	0.03	0.01	0.01	0.01
14	0.66	0.83	0.90	0.94	0.96	0.97	0.97	0.99	0.99	0.99	0.99	0.99	1.00	1.00	0.36	0.17	0.10	0.07	0.04	0.04	0.03	0.03	0.01
16	0.66	0.84	0.91	0.94	0.96	0.97	0.97	0.99	0.99	0.99	0.99	0.99	1.00	1.00	1.00	1.00	0.36	0.17	0.10	0.07	0.04	0.04	0.03
18	0.67	0.84	0.91	0.94	0.96	0.97	0.97	0.99	0.99	0.99	0.99	1.00	1.00	1.00	1.00	1.00	1.00	1.00	0.36	0.17	0.10	0.07	0.04
Zone Type B																							
2	0.50	0.63	0.23	0.16	0.11	0.09	0.07	0.06	0.04	0.03	0.03	0.01	0.01	0.01	0.01	0.00	0.00	0.00	0.00	0.00	0.00	0.00	0.00
4	0.50	0.64	0.73	0.79	0.34	0.24	0.19	0.14	0.10	0.09	0.06	0.04	0.04	0.03	0.03	0.01	0.01	0.01	0.01	0.00	0.00	0.00	0.00
6	0.50	0.64	0.73	0.79	0.84	0.87	0.41	0.29	0.21	0.17	0.13	0.10	0.07	0.06	0.04	0.03	0.03	0.01	0.01	0.01	0.01	0.01	0.00
8	0.50	0.64	0.73	0.79	0.84	0.87	0.90	0.93	0.44	0.31	0.24	0.19	0.14	0.11	0.09	0.07	0.06	0.04	0.03	0.03	0.01	0.01	0.01
10	0.50	0.64	0.73	0.79	0.84	0.87	0.90	0.93	0.94	0.96	0.47	0.34	0.26	0.20	0.16	0.11	0.09	0.07	0.06	0.04	0.03	0.03	0.01
12	0.51	0.66	0.73	0.80	0.84	0.89	0.91	0.93	0.94	0.96	0.97	0.97	0.49	0.34	0.27	0.20	0.16	0.11	0.09	0.07	0.06	0.04	0.03
14	0.53	0.66	0.74	0.80	0.84	0.89	0.91	0.93	0.94	0.96	0.97	0.97	0.99	0.99	0.50	0.36	0.27	0.21	0.16	0.13	0.10	0.07	0.06
16	0.56	0.69	0.76	0.81	0.86	0.89	0.91	0.93	0.94	0.96	0.97	0.97	0.99	0.99	0.99	0.99	0.50	0.36	0.27	0.21	0.16	0.13	0.10
18	0.59	0.71	0.79	0.83	0.87	0.90	0.93	0.94	0.96	0.97	0.97	0.99	0.99	0.99	0.99	0.99	1.00	1.00	0.50	0.36	0.27	0.21	0.16
Zone Type C																							
2	0.43	0.54	0.20	0.16	0.13	0.10	0.09	0.07	0.06	0.04	0.04	0.03	0.03	0.01	0.01	0.01	0.01	0.01	0.01	0.00	0.00	0.00	0.00
4	0.43	0.54	0.63	0.70	0.33	0.26	0.20	0.17	0.14	0.11	0.09	0.07	0.06	0.06	0.04	0.03	0.03	0.03	0.01	0.01	0.01	0.01	0.01
6	0.44	0.56	0.63	0.70	0.76	0.80	0.40	0.31	0.26	0.21	0.17	0.14	0.11	0.10	0.09	0.07	0.06	0.04	0.04	0.03	0.03	0.01	0.01
8	0.44	0.56	0.64	0.70	0.76	0.80	0.84	0.87	0.46	0.37	0.30	0.24	0.20	0.16	0.13	0.11	0.09	0.07	0.06	0.06	0.04	0.03	0.03
10	0.46	0.57	0.64	0.71	0.76	0.80	0.84	0.87	0.89	0.91	0.50	0.40	0.33	0.26	0.21	0.17	0.14	0.11	0.10	0.09	0.07	0.06	0.04
12	0.47	0.59	0.66	0.73	0.77	0.81	0.84	0.87	0.90	0.91	0.93	0.94	0.53	0.41	0.34	0.27	0.23	0.19	0.16	0.13	0.10	0.09	0.07
14	0.50	0.60	0.67	0.74	0.79	0.83	0.86	0.89	0.90	0.91	0.93	0.94	0.96	0.96	0.54	0.43	0.36	0.29	0.24	0.20	0.16	0.13	0.11
16	0.54	0.63	0.70	0.76	0.80	0.84	0.87	0.89	0.91	0.93	0.94	0.94	0.96	0.97	0.97	0.97	0.56	0.44	0.36	0.30	0.24	0.20	0.16
18	0.60	0.69	0.74	0.79	0.83	0.86	0.89	0.90	0.91	0.93	0.94	0.96	0.96	0.97	0.97	0.99	0.99	0.99	0.56	0.44	0.37	0.30	0.24
Zone Type D																							
2	0.41	0.53	0.19	0.13	0.11	0.09	0.07	0.07	0.06	0.06	0.04	0.04	0.03	0.03	0.03	0.01	0.01	0.01	0.01	0.01	0.01	0.01	0.01
4	0.43	0.53	0.60	0.66	0.29	0.23	0.19	0.16	0.14	0.11	0.10	0.09	0.07	0.07	0.06	0.04	0.04	0.04	0.03	0.03	0.03	0.01	0.01
6	0.44	0.54	0.61	0.67	0.71	0.76	0.37	0.29	0.24	0.21	0.19	0.16	0.13	0.11	0.10	0.09	0.07	0.07	0.06	0.04	0.04	0.04	0.03
8	0.46	0.56	0.63	0.67	0.71	0.76	0.79	0.81	0.43	0.34	0.29	0.24	0.21	0.19	0.16	0.14	0.11	0.10	0.09	0.07	0.07	0.06	0.06
10	0.47	0.57	0.64	0.69	0.73	0.77	0.80	0.83	0.84	0.87	0.47	0.39	0.31	0.27	0.24	0.20	0.17	0.16	0.13	0.11	0.10	0.09	0.07
12	0.50	0.59	0.66	0.70	0.74	0.77	0.81	0.83	0.86	0.87	0.89	0.90	0.50	0.41	0.34	0.30	0.26	0.23	0.19	0.17	0.14	0.13	0.11
14	0.53	0.61	0.69	0.73	0.76	0.80	0.83	0.84	0.87	0.89	0.90	0.91	0.93	0.93	0.53	0.43	0.36	0.31	0.27	0.23	0.20	0.17	0.16
16	0.57	0.66	0.71	0.76	0.79	0.81	0.84	0.86	0.89	0.90	0.91	0.93	0.93	0.94	0.94	0.96	0.54	0.44	0.37	0.33	0.29	0.24	0.21
18	0.63	0.71	0.76	0.79	0.81	0.84	0.87	0.89	0.90	0.91	0.93	0.94	0.96	0.96	0.96	0.97	0.97	0.56	0.46	0.39	0.33	0.29	

Note: See Table 35 for zone type. Data based on a radiative/convective fraction of 100/0.

(Copyright 1993 by ASHRAE, from 1993 Fundamentals, SI Edition. Used by permission.)

Footnote: Cross-references in table footnotes pertain to the source from which tables were taken; they do not refer to chapters or pages in this text.

Cooling Load Factors—In SI Metric Units (continued)

Figure 3.21a (continued)

Recommended Rate of Heat Gain from Restaurant Equipment Located in Air-Conditioned Area

Appliance	Size	Input Rating, Btu/h Maximum	Input Rating, Btu/h Standby	Recommended Rate of Heat Gain,[a] Btu/h Without Hood Sensible	Without Hood Latent	Without Hood Total	With Hood Sensible
Electric, No Hood Required							
Barbeque (pit), per pound of food capacity	80 to 300 lb	136	—	86	50	136	42
Barbeque (pressurized), per pound of food capacity	44 lb	327	—	109	54	163	50
Blender, per quart of capacity	1 to 4 qt	1550	—	1000	520	1520	480
Braising pan, per quart of capacity	108 to 140 qt	360	—	180	95	275	132
Cabinet (large hot holding)	16.2 to 17.3 ft^3	7100	—	610	340	960	290
Cabinet (large hot serving)	37.4 to 406 ft^3	6820	—	610	310	920	280
Cabinet (large proofing)	16 to 17 ft^3	693	—	610	310	920	280
Cabinet (small hot holding)	3.2 to 6.4 ft^3	3070	—	270	140	410	130
Cabinet (very hot holding)	17.3 ft^3	21000	—	1880	960	2830	850
Can opener		580	—	580	—	580	0
Coffee brewer	12 cup/2 brnrs	5660	—	3750	1910	5660	1810
Coffee heater, per boiling burner	1 to 2 brnrs	2290	—	1500	790	2290	720
Coffee heater, per warming burner	1 to 2 brnrs	340	—	230	110	340	110
Coffee/hot water boiling urn, per quart of capacity	11.6 qt	390	—	256	132	388	123
Coffee brewing urn (large), per quart of capacity	23 to 40 qt	2130	—	1420	710	2130	680
Coffee brewing urn (small), per quart of capacity	10.6 qt	1350	—	908	445	1353	416
Cutter (large)	18 in. bowl	2560	—	2560	—	2560	0
Cutter (small)	14 in. bowl	1260	—	1260	—	1260	0
Cutter and mixer (large)	30 to 48 qt	12730	—	12730	—	12730	0
Dishwasher (hood type, chemical sanitizing), per 100 dishes/h	950 to 2000 dishes/h	1300	—	170	370	540	170
Dishwasher (hood type, water sanitizing), per 100 dishes/h	950 to 2000 dishes/h	1300	—	190	420	610	190
Dishwasher (conveyor type, chemical sanitizing), per 100 dishes/h	5000 to 9000 dishes/h	1160	—	140	330	470	150
Dishwasher (conveyor type, water sanitizing), per 100 dishes/h	5000 to 9000 dishes/h	1160	—	150	370	520	170
Display case (refrigerated), per 10 ft^3 of interior	6 to 67 ft^3	1540	—	617	0	617	0
Dough roller (large)	2 rollers	5490	—	5490	—	5490	0
Dough roller (small)	1 roller	1570	—	140	—	140	0
Egg cooker	12 eggs	6140	—	2900	1940	4850	1570
Food processor	2.4 qt	1770	—	1770	—	1770	0
Food warmer (infrared bulb), per lamp	1 to 6 bulbs	850	—	850	—	850	850
Food warmer (shelf type), per square foot of surface	3 to 9 ft^2	930	—	740	190	930	260
Food warmer (infrared tube), per foot of length	39 to 53 in.	990	—	990	—	990	990
Food warmer (well type), per cubic foot of well	0.7 to 2.5 ft^3	3620	—	1200	610	1810	580
Freezer (large)	73 ft^3	4570	—	1840	—	1840	0
Freezer (small)	18 ft^3	2760	—	1090	—	1090	0
Griddle/grill (large), per square foot of cooking surface	4.6 to 11.8 ft^2	9200	—	615	343	958	343
Griddle/grill (small), per square foot of cooking surface	2.2 to 4.5 ft^2	8300	—	545	308	853	298
Hot dog broiler	48 to 56 hot dogs	3960	—	340	170	510	160
Hot plate (double burner, high speed)		16720	—	7810	5430	13240	6240
Hot plate (double burner, stockpot)		13650	—	6380	4440	10820	5080
Hot plate (single burner, high speed)		9550	—	4470	3110	7580	3550
Hot water urn (large), per quart of capacity	56 qt	416	—	161	52	213	68
Hot water urn (small), per quart of capacity	8 qt	738	—	285	95	380	123
Ice maker (large)	220 lb/day	3720	—	9320	—	9320	0
Ice maker (small)	110 lb/day	2560	—	6410	—	6410	0
Microwave oven (heavy duty, commercial)	0.7 ft^3	8970	—	8970	—	8970	0
Microwave oven (residential type)	1 ft^3	2050 to 4780	—	2050 to 4780	—	2050 to 4780	0
Mixer (large), per quart of capacity	81 qt	94	—	94	—	94	0
Mixer (small), per quart of capacity	12 to 76 qt	48	—	48	—	48	0
Press cooker (hamburger)	300 patties/h	7510	—	4950	2560	7510	2390
Refrigerator (large), per 100 ft^3 of interior space	25 to 74 ft^3	753	—	300	—	300	0
Refrigerator (small), per 100 ft^3 of interior space	6 to 25 ft^3	1670	—	665	—	665	0
Rotisserie	300 hamburgers/h	10920	—	7200	3720	10920	3480
Serving cart (hot), per cubic foot of well	1.8 to 3.2 ft^3	2050	—	680	340	1020	328
Serving drawer (large)	252 to 336 dinner rolls	3750	—	480	34	510	150
Serving drawer (small)	84 to 168 dinner rolls	2730	—	340	34	380	110
Skillet (tilting), per quart of capacity	48 to 132 qt	580	—	293	161	454	218
Slicer, per square foot of slicing carriage	0.65 to 0.97 ft^2	680	—	682	—	682	216
Soup cooker, per quart of well	7.4 to 11.6 qt	416	—	142	78	220	68
Steam cooker, per cubic foot of compartment	32 to 64 qt	20700	—	1640	1050	2690	784
Steam kettle (large), per quart of capacity	80 to 320 qt	300	—	23	16	39	13
Steam kettle (small), per quart of capacity	24 to 48 qt	840	—	68	45	113	32
Syrup warmer, per quart of capacity	11.6 qt	284	—	94	52	146	45

Sensible and Latent Heat Loads for Internal Heat Gains

Figure 3.22

Recommended Rate of Heat Gain from Restaurant Equipment Located in Air-Conditioned Area (*Concluded*)

Appliance	Size	Input Rating, Btu/h Maximum	Input Rating, Btu/h Standby	Recommended Rate of Heat Gain,[a] Btu/h — Without Hood Sensible	Without Hood Latent	Without Hood Total	With Hood Sensible
Toaster (bun toasts on one side only)	1400 buns/h	5120	—	2730	2420	5150	1640
Toaster (large conveyor)	720 slices/h	10920	—	2900	2560	5460	1740
Toaster (small conveyor)	360 slices/h	7170	—	1910	1670	3580	1160
Toaster (large pop-up)	10 slice	18080	—	9590	8500	18080	5800
Toaster (small pop-up)	4 slice	8430	—	4470	3960	8430	2700
Waffle iron	75 in^2	5600	—	2390	3210	5600	1770
Electric, Exhaust Hood Required							
Broiler (conveyor infrared), per square foot of cooking area/minute	2 to 102 ft^2	19230	—	—	—	—	3840
Broiler (single deck infrared), per square foot of broiling area	2.6 to 9.8 ft^2	10870	—	—	—	—	2150
Charbroiler, per square foot of cooking surface	1.5 to 4.6 ft^2	7320	—	—	—	—	307
Fryer (deep fat), per pound of fat capacity	15 to 70 lb	1270	—	—	—	—	14
Fryer (pressurized), per pound of fat capacity	13 to 33 lb	1565	—	—	—	—	59
Oven (large convection), per cubic foot of oven space	7 to 19 ft^3	4450	—	—	—	—	181
Oven (large deck baking with 537 ft^3 decks), per cubic foot of oven space	15 to 46 ft^3	1670	—	—	—	—	69
Oven (roasting), per cubic foot of oven space	7.8 to 23 ft^3	27350	—	—	—	—	113
Oven (small convection), per cubic foot of oven space	1.4 to 5.3 ft^3	10340	—	—	—	—	147
Oven (small deck baking with 272 ft^3 decks), per cubic foot of oven space	7.8 to 23 ft^3	2760	—	—	—	—	113
Range (burners), per 2 burner section	2 to 10 brnrs	7170	—	—	—	—	2660
Range (hot top/fry top), per square foot of cooking surface	4 to 8 ft^2	7260	—	—	—	—	2690
Range (oven section), per cubic foot of oven space	4.2 to 11.3 ft^3	3940	—	—	—	—	160
Range (stockpot)	1 burner	18770	—	—	—	—	6990
Gas, No Hood Required							
Broiler, per square foot of broiling area	2.7 ft^2	14800	660[b]	5310	2860	8170	1220
Cheese melter, per square foot of cooking surface	2.5 to 5.1 ft^2	10300	660[b]	3690	1980	5670	850
Dishwasher (hood type, chemical sanitizing), per 100 dishes/h	950 to 2000 dishes/h	1740	660[b]	510	200	710	230
Dishwasher (hood type, water sanitizing), per 100 dishes/h	950 to 2000 dishes/h	1740	660[b]	570	220	790	250
Dishwasher (conveyor type, chemical sanitizing), per 100 dishes/h	5000 to 9000 dishes/h	1370	660[b]	330	70	400	130
Dishwasher (conveyor type, water sanitizing), per 100 dishes/h	5000 to 9000 dishes/h	1370	660[b]	370	80	450	140
Griddle/grill (large), per square foot of cooking surface	4.6 to 11.8 ft^2	17000	330	1140	610	1750	460
Griddle/grill (small), per square foot of cooking surface	2.5 to 4.5 ft^2	14400	330	970	510	1480	400
Hot plate	2 burners	19200	1325[b]	11700	3470	15200	3410
Oven (pizza), per square foot of hearth	6.4 to 12.9 ft^2	4740	660[b]	623	220	843	85
Gas, Exhaust Hood Required							
Braising pan, per quart of capacity	105 to 140 qt	9840	660[b]	—	—	—	2430
Broiler, per square foot of broiling area	3.7 to 3.9 ft^2	21800	530	—	—	—	1800
Broiler (large conveyor, infrared), per square foot of cooking area/minute	2 to 102 ft^2	51300	1990	—	—	—	5340
Broiler (standard infrared), per square foot of broiling area	2.4 to 9.4 ft^2	1940	530	—	—	—	1600
Charbroiler (large), per square foot of cooking area	4.6 to 11.8 ft^2	16500	510	—	—	—	790
Charbroiler (small), per square foot of cooking area	1.3 to 4.5 ft^2	19700	510	—	—	—	950
Fryer (deep fat), per pound of fat capacity	11 to 70 lb	2270	660[b]	—	—	—	160
Oven (bake deck), per cubic foot of oven space	5.3 to 16.2 ft^3	7670	660[b]	—	—	—	140
Oven (convection), per cubic metre of oven space	7.4 to 19.4 ft^3	8670	660[b]	—	—	—	250
Oven (pizza), per square foot of oven hearth	9.3 to 25.8 ft^2	7240	660[b]	—	—	—	130
Oven (roasting), per cubic foot of oven space	9 to 28 ft^3	4300	660[b]	—	—	—	77
Oven (twin bake deck), per cubic foot of oven space	11 to 22 ft^3	4390	660[b]	—	—	—	78
Range (burners), per 2 burner section	2 to 10 brnrs	33600	1325	—	—	—	6590
Range (hot top or fry top), per square foot of cooking surface	3 to 8 ft^2	11800	330	—	—	—	3390
Range (large stock pot)	3 burners	100000	1990	—	—	—	19600
Range (small stock pot)	2 burners	40000	1330	—	—	—	7830
Steam							
Compartment steamer, per pound of food capacity/h	46 to 450 lb	280	—	22	14	36	11
Dishwasher (hood type, chemical sanitizing), per 100 dishes/h	950 to 2000 dishes/h	3150	—	880	380	1260	410
Dishwasher (hood type, water sanitizing), per 100 dishes/h	950 to 2000 dishes/h	3150	—	980	420	1400	450
Dishwasher (conveyor, chemical sanitizing), per 100 dishes/h	5000 to 9000 dishes/h	1180	—	140	330	470	150
Dishwasher (conveyor, water sanitizing), per 100 dishes/h	5000 to 9000 dishes/h	1180	—	150	370	520	170
Steam kettle, per quart of capacity	13 to 32 qt	500	—	39	25	64	19

[a] In some cases, heat gain data are given per unit of capacity. In those cases, the heat gain is calculated by: q = (recommended heat gain per unit of capacity) * (capacity)

[b] Standby input rating is given for entire appliance regardless of size.

Sensible and Latent Heat Load for Internal Heat Gains (continued)

Figure 3.22 (continued)

Rate of Heat Gain from Selected Office Equipment

Appliance	Size	Maximum Input Rating, Btu/h	Standby Input Rating, Btu/h	Recommended Rate of Heat Gain, Btu/h
Check processing workstation	12 pockets	16400	8410	8410
Computer devices				
Card puncher		2730 to 6140	2200 to 4800	2200 to 4800
Card reader		7510	5200	5200
Communication/transmission		6140 to 15700	5600 to 9600	5600 to 9600
Disk drives/mass storage		3410 to 34100	3412 to 22420	3412 to 22420
Magnetic ink reader		3280 to 16000	2600 to 14400	2600 to 14400
Microcomputer	16 to 640 KByte[a]	340 to 2050	300 to 1800	300 to 1800
Minicomputer		7500 to 15000	7500 to 15000	7500 to 15000
Optical reader		10240 to 20470	8000 to 17000	8000 to 17000
Plotters		256	128	214
Printers				
Letter quality	30 to 45 char/min	1200	600	1000
Line, high speed	5000 or more lines/min	4300 to 18100	2160 to 9040	2500 to 13000
Line, low speed	300 to 600 lines/min	1540	770	1280
Tape drives		4090 to 22200	3500 to 15000	3500 to 15000
Terminal		310 to 680	270 to 600	270 to 600
Copiers/Duplicators				
Blue print		3930 to 42700	1710 to 17100	3930 to 42700
Copiers (large)	30 to 67[a] copies/min	5800 to 22500	3070	5800 to 22500
Copiers (small)	6 to 30[a] copies/min	1570 to 5800	1020 to 3070	1570 to 5800
Feeder		100	—	100
Microfilm printer		1540	—	1540
Sorter/collator		200 to 2050	—	200 to 2050
Electronic equipment				
Cassette recorders/players		200	—	200
Receiver/tuner		340	—	340
Signal analyzer		90 to 2220	—	90 to 2220
Mailprocessing				
Folding machine		430	—	270
Inserting machine	3600 to 6800 pieces/h	2050 to 11300	—	1330 to 7340
Labeling machine	1500 to 30000 pieces/h	2050 to 22500	—	1330 to 14700
Postage meter		780	—	510
Wordprocessors/Typewriters				
Letter quality printer	30 to 45 char/min	1200	600	1000
Phototypesetter		5890	—	5180
Typewriter		270	—	230
Wordprocessor		340 to 2050	—	300 to 1800
Vending machines				
Cigarette		250	51 to 85	250
Cold food/beverage		3920 to 6550	—	1960 to 3280
Hot beverage		5890	—	2940
Snack		820 to 940	—	820 to 940
Miscellaneous				
Barcode printer		1500	—	1260
Cash registers		200	—	160
Coffee maker	10 cups	5120	—	3580 sensible 1540 latent
Microfiche reader		290	—	290
Microfilm reader		1770	—	1770
Microfilm reader/printer		3920	—	3920
Microwave oven	1 ft^3	2050	—	1360
Paper shredder		850 to 10240	—	680 to 8250
Water cooler	32 qt/h	2390	—	5970

[a] Input is not proportional to capacity.

(Copyright 1993 by ASHRAE, from 1993 Fundamentals. Used by permission.)

Footnote: Cross-references in table footnotes pertain to the source from which tables were taken; they do not refer to chapters or pages in this text.

Sensible and Latent Heat Loads for Internal Heat Gains (continued)

Figure 3.22 (continued)

Recommended Rate of Heat Gain from Restaurant Equipment Located in Air-Conditioned Area

Appliance	Size	Input Rating, W		Recommended Rate of Heat Gain,[a] W			
				Without Hood			With Hood
		Maximum	Standby	Sensible	Latent	Total	Sensible
Electric, No Hood Required							
Barbeque (pit), per kilogram of food capacity	36 to 136 kg	88	—	57	31	88	27
Barbeque (pressurized) per kilogram of food capacity	20 kg	210	—	71	35	106	33
Blender, per litre of capacity	1.0 to 3.8 L	480	—	310	160	470	150
Braising pan, per litre of capacity	102 to 133 L	110	—	55	29	84	40
Cabinet (large hot holding)	0.46 to 0.49 m³	2080	—	180	100	280	85
Cabinet (large hot serving)	1.06 to 1.15 m³	2000	—	180	90	270	82
Cabinet (large proofing)	0.45 to 0.48 m³	2030	—	180	90	270	82
Cabinet (small hot holding)	0.09 to 0.18 m³	900	—	80	40	120	37
Cabinet (very hot holding)	0.49 m³	6150	—	550	280	830	250
Can opener		170	—	170	—	170	0
Coffee brewer	12 cup/2 brnrs	1660	—	1100	560	1660	530
Coffee heater, per boiling burner	1 to 2 brnrs	670	—	440	230	670	210
Coffee heater, per warming burner	1 to 2 brnrs	100	—	66	34	100	32
Coffee/hot water boiling urn, per litre of capacity	11 L	120	—	79	41	120	38
Coffee brewing urn (large), per litre of capacity	22 to 38 L	660	—	440	220	660	210
Coffee brewing urn (small), per litre of capacity	10 L	420	—	280	140	420	130
Cutter (large)	460 mm bowl	750	—	750	—	750	0
Cutter (small)	360 mm bowl	370	—	370	—	370	0
Cutter and mixer (large)	28 to 45 L	3730	—	3730	—	3730	0
Dishwasher (hood type, chemical sanitizing), per 100 dishes/h	950 to 2000 dishes/h	380	—	50	110	160	50
Dishwasher (hood type, water sanitizing), per 100 dishes/h	950 to 2000 dishes/h	380	—	56	123	179	56
Dishwasher (conveyor type, chemical sanitizing), per 100 dishes/h	5000 to 9000 dishes/h	340	—	41	97	138	44
Dishwasher (conveyor type, water sanitizing), per 100 dishes/h	5000 to 9000 dishes/h	340	—	44	108	152	50
Display case (refrigerated), per cubic metre of interior	0.17 to 1.9 m³	1590	—	640	0	640	0
Dough roller (large)	2 rollers	1610	—	1610	—	1610	0
Dough roller (small)	1 roller	460	—	460	—	460	0
Egg cooker	12 eggs	1800	—	850	570	1420	460
Food processor	2.3 L	520	—	520	—	520	0
Food warmer (infrared bulb), per lamp	1 to 6 bulbs	250	—	250	—	250	250
Food warmer (shelf type), per square metre of surface	0.28 to 0.84 m³	2930	—	2330	600	2930	820
Food warmer (infrared tube), per metre of length	1.0 to 2.1 m	950	—	950	—	950	950
Food warmer (well type), per cubic metre of well	20 to 70 L	37400	—	12400	6360	18760	6000
Freezer (large)	2.07 m³	1340	—	540	—	540	0
Freezer (small)	0.51 m³	810	—	320	—	320	0
Griddle/grill (large), per square metre of cooking surface	0.43 to 1.1 m²	29000	—	1940	1080	3020	1080
Griddle/grill (small), per square metre of cooking surface	0.20 to 0.42 m²	26200	—	1720	970	2690	940
Hot dog broiler	48 to 56 hot dogs	1160	—	100	50	150	48
Hot plate (double burner, high speed)		4900	—	2290	1590	3880	1830
Hot plate (double burner stockpot)		4000	—	1870	1300	3170	1490
Hot plate (single burner, high speed)		2800	—	1310	910	2220	1040
Hot water urn (large), per litre of capacity	53 L	130	—	50	16	66	21
Hot water urn (small), per litre of capacity	7.6 L	230	—	87	30	117	37
Ice maker (large)	100 kg/day	1090	—	2730	—	2730	0
Ice maker (small)	50 kg/day	750	—	1880	—	1880	0
Microwave oven (heavy duty, commercial)	20 L	2630	—	2630	—	2630	0
Microwave oven (residential type)	30 L	600 to 1400	—	600 to 1400	—	600 to 1400	0
Mixer (large), per litre of capacity	77 L	29	—	29	—	29	0
Mixer (small), per litre of capacity	11 to 72 L	15	—	15	—	15	0
Press cooker (hamburger)	300 patties/h	2200	—	1450	750	2200	700
Refrigerator (large), per 10 m³ of interior space	0.71 to 2.1 m³	780	—	310	—	310	0
Refrigerator (small) per 10 m³ of interior space	0.17 to 0.71 m³	1730	—	690	—	690	0
Rotisserie	300 hamburgers/h	3200	—	2110	1090	3200	1020
Serving cart (hot), per cubic metre of well	50 to 90 L	21200	—	7060	3530	10590	3390
Serving drawer (large)	252 to 336 dinner rolls	1100	—	140	10	150	45
Serving drawer (small)	84 to 168 dinner rolls	800	—	100	10	110	33
Skillet (tilting), per litre of capacity	45 to 125 L	180	—	90	50	140	66
Slicer, per square metre of slicing carriage	0.06 to 0.09 m²	2150	—	2150	—	2150	680
Soup cooker, per litre of well	7 to 11 L	130	—	45	24	69	21
Steam cooker, per cubic metre of compartment	30 to 60 L	214000	—	17000	10900	27900	8120
Steam kettle (large), per litre of capacity	76 to 300 L	95	—	7	5	12	4
Steam kettle (small), per litre of capacity	23 to 45 L	260	—	21	14	35	10
Syrup warmer, per litre of capacity	11 L	87	—	29	16	45	14

Sensible and Latent Heat Loads for Internal Heat Gains—In SI Metric Units

Figure 3.22a

Recommended Rate of Heat Gain from Restaurant Equipment Located in Air-Conditioned Area (*Concluded*)

Appliance	Size	Input Rating, W		Recommended Rate of Heat Gain,[a] W			
				Without Hood			With Hood
		Maximum	Standby	Sensible	Latent	Total	Sensible
Toaster (bun toasts on one side only)	1400 buns/h	1500	—	800	710	1510	480
Toaster (large conveyor)	720 slices/h	3200	—	850	750	1600	510
Toaster (small conveyor)	360 slices/h	2100	—	560	490	1050	340
Toaster (large pop-up)	10 slice	5300	—	2810	2490	5300	1700
Toaster (small pop-up)	4 slice	2470	—	1310	1160	2470	790
Waffle iron	0.05 m^2	1640	—	700	940	1640	520
Electric, Exhaust Hood Required							
Broiler (conveyor infrared), per square metre of cooking area/minute	0.19 to 9.5 m^2	60800	—	—	—	—	12100
Broiler (single deck infrared), per square metre of broiling area	0.24 to 0.91 m^2	34200	—	—	—	—	6780
Charbroiler, per square metre of cooking surface	0.14 to 0.43 m^2	23100	—	—	—	—	970
Fryer (deep fat), per kilogram of fat capacity	7 to 32 kg	820	—	—	—	—	9
Fryer (pressurized), per kilogram of fat capacity	6 to 15 kg	1010	—	—	—	—	38
Oven (large convection), per cubic metre of oven space	0.20 to 0.55 m^3	45900	—	—	—	—	1870
Oven (large deck baking with 15.2 m^3 decks), per cubic metre of oven spacer	0.43 to 1.3 m^3	17300	—	—	—	—	710
Oven (roasting), per cubic metre of oven space	0.22 to 0.66 m^3	28300	—	—	—	—	1170
Oven (small convection), per cubic metre of oven space	0.04 to 0.15 m^3	107000	—	—	—	—	1520
Oven (small deck baking with 7.7 m^3 decks), per cubic metre of oven space	0.22 to 0.66 m^3	28700	—	—	—	—	1170
Range (burners), per 2 burner section	2 to 10 burners	2100	—	—	—	—	780
Range (hot top/fry top), per square metre of cooking surface	0.36 to 0.74 m^2	22900	—	—	—	—	8500
Range (oven section), per cubic metre of oven space	0.12 to 0.32 m^3	40600	—	—	—	—	1660
Range (stockpot)	1 burner	5500	—	—	—	—	2050
Gas, No Hood Required							
Broiler, per square metre of broiling area	0.25 m^2	46600	190[b]	16800	9030	25830	3840
Cheese melter, per square metre of cooking surface	0.23 to 0.47 m^2	32500	190[b]	11600	3400	15000	2680
Dishwasher (hood type, chemical sanitizing), per 100 dishes/h	950 to 2000 dishes/h	510	190[b]	150	59	209	67
Dishwasher (hood type, water sanitizing), per 100 dishes/h	950 to 2000 dishes/h	510	190[b]	170	64	234	73
Dishwasher (conveyor type, chemical sanitizing), per 100 dishes/h	5000 to 9000 dishes/h	400	190[b]	97	21	118	38
Dishwasher (conveyor type, water sanitizing), per 100 dishes/h	5000 to 9000 dishes/h	400	190[b]	110	23	133	41
Griddle/grill (large), per square metre of cooking surface	0.43 to 1.1 m^2	53600	1040	3600	1930	5530	1450
Griddle/grill (small), per square metre of cooking surface	0.23 to 0.42 m^2	45400	1040	3050	1610	4660	1260
Hot plate	2 burners	5630	390[b]	3430	1020	4450	1000
Oven (pizza), per square metre of hearth	0.59 to 1.2 m^2	14900	190[b]	1970	690	2660	270
Gas, Exhaust Hood Required							
Braising pan, per litre of capacity	102 to 133 L	3050	190[b]	—	—	—	750
Broiler, per square metre of broiling area	0.34 to 0.36 m^3	68900	1660	—	—	—	5690
Broiler (large conveyor, infrared), per square metre of cooking area/minute	0.19 to 9.5 m^2	162000	6270	—	—	—	16900
Broiler (standard infrared), per square metre of broiling area	0.22 to 0.87 m^2	61300	1660	—	—	—	5040
Charbroiler (large), per square metre of cooking area	0.43 to 1.1 m^2	51900	1610	—	—	—	2490
Charbroiler (small), per square metre of cooking area	0.12 to 0.42 m^2	62100	1610	—	—	—	2990
Fryer (deep fat), per kilogram of fat capacity	5 to 32 kg	1470	190[b]	—	—	—	100
Oven (bake deck), per cubic metre of oven space	0.15 to 0.46 m^3	79400	190[b]	—	—	—	1450
Oven (convection), per cubic metre of oven space	0.21 to 0.55 m^3	89700	190[b]	—	—	—	2590
Oven (pizza), per square metre of oven hearth	0.86 to 2.4 m^2	22800	190[b]	—	—	—	410
Oven (roasting), per cubic metre of oven space	0.26 to 0.79 m^3	44500	190[b]	—	—	—	800
Oven (twin bake deck), per cubic metre of oven space	0.31 to 0.61 m^3	45400	190[b]	—	—	—	810
Range (burners), per 2 burner section	2 to 10 burners	9840	390	—	—	—	1930
Range (hot top or fry top), per square metre of cooking surface	0.26 to 0.74 m^2	37200	1040	—	—	—	10700
Range (large stock pot)	3 burners	29300	580	—	—	—	5740
Range (small stock pot)	2 burners	11700	390	—	—	—	2290
Steam							
Compartment steamer, per kilogram of food capacity/h	21 to 204 kg	180	—	14	9	23	7
Dishwasher (hood type, chemical sanitizing), per 100 dishes/h	950 to 2000 dishes/h	920	—	260	110	370	120
Dishwasher (hood type, water sanitizing), per 100 dishes/h	950 to 2000 dishes/h	920	—	290	120	410	130
Dishwasher (conveyor, chemical sanitizing), per 100 dishes/h	5000 to 9000 dishes/h	350	—	41	97	138	44
Dishwasher (conveyor, water sanitizing), per 100 dishes/h	5000 to 9000 dishes/h	350	—	44	108	152	50
Steam kettle, per litre of capacity	12 to 30 L	160	—	12	8	20	6

[a] In some cases, heat gain data are given per unit of capacity. In those cases, the heat gain is calculated by: q = (recommended heat gain per unit of capacity) * (capacity)

[b] Standby input rating is given for entire appliance regardless of size.

Sensible and Latent Heat Loads for Internal Heat Gains—In SI Metric Units (continued)

Figure 3.22a (continued)

Rate of Heat Gain from Selected Office Equipment

Appliance	Size	Maximum Input Rating, W	Standby Input Rating, W	Recommended Rate of Heat Gain, W
Check processing work station	12 pockets	4800	2460	2460
Computer devices				
Card puncher	—	800 to 1800	640 to 1410	640 to 1410
Card reader	—	2200	1520	1520
Communication/transmission	—	1800 to 4600	1640 to 2810	1640 to 2810
Disk drives/mass storage	—	1000 to 10000	1000 to 6570	1000 to 6570
Magnetic ink reader	—	960 to 4700	760 to 4220	760 to 4220
Microcomputer	16 to 640 Kbyte[a]	100 to 600	90 to 530	90 to 530
Minicomputer	—	2200 to 6600	2200 to 6600	2200 to 6600
Optical reader	—	3000 to 6000	2350 to 4980	2350 to 4980
Plotters	—	75	37	63
Printers				
Letter quality	30 to 45 char/min	350	175	292
Line, high speed	5000 or more lines/min	1000 to 5300	500 to 2550	730 to 3810
Line, low speed	300 to 600 lines/min	450	225	376
Tape drives	—	1200 to 6500	1000 to 4700	1000 to 4700
Terminal	—	90 to 200	80 to 180	80 to 180
Copiers/Duplicators				
Blue print	—	1150 to 12500	500 to 5000	1150 to 12500
Copiers (large)	30 to 67[a] copies/min	1700 to 6600	900	1700 to 6600
Copiers (small)	6 to 30[a] copies/min	460 to 1700	300 to 900	460 to 1700
Feeder	—	30	—	30
Microfilm printer	—	450	—	450
Sorter/collator	—	60 to 600	—	60 to 600
Electronic Equipment				
Cassette recorders/players	—	60	—	60
Receiver/tuner	—	100	—	100
Signal analyzer	—	25 to 650	—	25 to 650
Mailprocessing				
Folding machine	—	125	—	80
Inserting machine	3600 to 6800 pieces/h	600 to 3300	—	390 to 2150
Labeling machine	1500 to 30000 pieces/h	600 to 6600	—	390 to 4300
Postage meter	—	230	—	150
Wordprocessors/Typewriters				
Letter quality printer	30 to 45 char/min	350	175	292
Phototypesetter	—	1725	—	1520
Typewriter	—	80	—	67
Wordprocessor	—	100 to 600	—	88 to 530
Vending machines				
Cigarette	—	72	15 to 25	72
Cold food/beverage	—	1150 to 1920	—	575 to 960
Hot beverage	—	1725	—	862
Snack	—	240 to 275	—	240 to 275
Miscellaneous				
Barcode printer	—	440	—	370
Cash register	—	60	—	48
Coffee maker	10 cups	1500	—	1050 sensible / 450 latent
Microfiche reader	—	85	—	85
Microfilm reader	—	520	—	520
Microfilm reader/printer	—	1150	—	1150
Microwave oven	28 L	600	—	400
Paper shredder	—	250 to 3000	—	200 to 2420
Water cooler	30 L/h	700	—	1750

[a] Input is not proportional to capacity.

(Copyright 1993 by ASHRAE, from 1993 Fundamentals, SI Edition. Used by permission.)

Footnote: Cross-references in table footnotes pertain to the source from which tables were taken; they do not refer to chapters or pages in this text.

Sensible and Latent Heat Loads for Internal Heat Gains — In SI Metric Units (continued)

Figure 3.22a (continued)

Btu's/hr. [1612 W] and a design internal temperature of 70°F [21 °C]. The heat pump characteristics are outlined in Figure 3.26.

Solution:

See Figure 3.27. The solar heat gain incident on roofs, walls, glazing, and other surfaces is a function of solar heat gain factors, areas of surfaces, shading coefficients, cooling load factors, floor areas, and operating time of the HVAC system. Because solar heat gain is such a widely varying and potentially large value, energy consumption from solar heat gain should not be estimated casually or averaged over a wide period of time. One method for calculating the solar heat gain energy consumption is the modified BIN method.

The modified BIN method of estimating energy consumption is a more detailed and consequently a more accurate method than the degree-day or conventional BIN method. It takes into account details and functions for calculating solar energy as outlined above and can be used for estimating energy for varying load schedules and for specific times of the day. The modified BIN method utilizes diversified heating and cooling loads, whereas the degree-day or conventional BIN method utilizes peak loads as a base

Rates of Heat Gain from Occupants of Conditioned Spaces

Degree of Activity		Total Heat, Btu/h Adult Male	Total Heat, Btu/h Adjusted, M/F[a]	Sensible Heat, Btu/h	Latent Heat, Btu/h	% Sensible Heat that is Radiant[b] Low V	% Sensible Heat that is Radiant[b] High V
Seated at theater	Theater, matinee	390	330	225	105		
Seated at theater, night	Theater, night	390	350	245	105	60	27
Seated, very light work	Offices, hotels, apartments	450	400	245	105		
Moderately active office work	Offices, hotels, apartments	475	450	250	105		
Standing, light work; walking	Department store; retail store	550	450	250	105	58	38
Walking, standing	Drug store, bank	550	500	250	105		
Sedentary work	Restaurant[c]	490	550	275	105		
Light bench work	Factory	800	750	275	105		
Moderate dancing	Dance hall	900	850	305	105	49	35
Walking 3 mph; light machine work	Factory	1000	1000	375	105		
Bowling[d]	Bowling alley	1500	1450	580	105		
Heavy work	Factory	1500	1450	580	105	54	19
Heavy machine work; lifting	Factory	1600	1600	635	105		
Athletics	Gymnasium	2000	1800	710	105		

Notes:
1. Tabulated values are based on 75 °F room dry-bulb temperature. For 80°F room dry bulb, the total heat remains the same, but the sensible heat values should be decreased by approximately 20%, and the latent heat values increased accordingly.
2. Also refer to Table 4, Chapter 8, for additional rates of metabolic heat generation.
3. All values are rounded to nearest 5 Btu/h.

[a] Adjusted heat gain is based on normal percentage of men, women, and children for the application listed, with the postulate that the gain from an adult female is 85% of that for an adult male, and that the gain from a child is 75% of that for an adult male.
[b] Values approximated from data in Table 6, Chapter 8, where V is air velocity with limits shown in that table.
[c] Adjusted heat gain includes 60 Btu/h for food per individual (30 Btu/h sensible and 30 Btu/h latent).
[d] Figure one person per alley actually bowling, and all others as sitting (400 Btu/h) or standing or walking slowly (550 Btu/h).

(Copyright 1993 by ASHRAE, from *1993 Fundamentals*. Used by permission.)

Footnote: *Cross-references in table footnotes pertain to the source from which tables were taken; they do not refer to chapters or pages in this text.*

Sensible and Latent Heat Gains for People

Figure 3.23

for the calculations. The modified BIN method is outlined in detail in the ASHRAE 1993 *Fundamentals Handbook*.

Psychrometrics

Psychrometrics is the analysis of the properties of air as it passes through an HVAC process such as cooling, heating, humidification, dehumidification, or any combination thereof. The psychrometric chart in Figure 3.28 is a convenient and concise summary of the properties of air.

The Psychrometric Chart

Air at any condition can be represented by one point on the psychrometric chart; from this point, seven properties of the air can be read. The seven properties that can be read from any point are as follows (see Figure 3.28a):

- dry bulb temperature (°F) [°C] (line #1)
- wet bulb temperature (°F) [°C] (line #2)
- moisture content or specific humidity (lbs. of moisture per lb. of dry air or grains of moisture per lb. of dry air) [g of moisture per g of dry air] (line #3)
- relative humidity (%) (line #4)
- specific volume (cu.ft./lb.)[m^2/kg] of air (line #5)

Rates of Heat Gain from Occupants of Conditioned Spaces

Degree of Activity		Total Heat, W Adult Male	Total Heat, W Adjusted, M/F[a]	Sensible Heat, W	Latent Heat, W	% Sensible Heat that is Radiant[b] Low V	% Sensible Heat that is Radiant[b] High V
Seated at theater	Theater, matinee	115	95	65	30		
Seated at theater, night	Theater, night	115	105	70	35	60	27
Seated, very light work	Offices, hotels, apartments	130	115	70	45		
Moderately active office work	Offices, hotels, apartments	140	130	75	55		
Standing, light work; walking	Department store; retail store	160	130	75	55	58	38
Walking, standing	Drug store, bank	160	145	75	70		
Sedentary work	Restaurant[c]	145	160	80	80		
Light bench work	Factory	235	220	80	140		
Moderate dancing	Dance hall	265	250	90	160	49	35
Walking 4.8 km/h; light machine work	Factory	295	295	110	185		
Bowling[d]	Bowling alley	440	425	170	255		
Heavy work	Factory	440	425	170	255	54	19
Heavy machine work; lifting	Factory	470	470	185	285		
Athletics	Gymnasium	585	525	210	315		

Notes:
1. Tabulated values are based on 24 °C room dry-bulb temperature. For 27 °C room dry bulb, the total heat remains the same, but the sensible heat values should be decreased by approximately 20%, and the latent heat values increased accordingly.
2. Also refer to Table 4, Chapter 8, for additional rates of metabolic heat generation.
3. All values are rounded to nearest 5 W.

[a] Adjusted heat gain is based on normal percentage of men, women, and children for the application listed, with the postulate that the gain from an adult female is 85% of that for an adult male, and that the gain from a child is 75% of that for an adult male.
[b] Values approximated from data in Table 6, Chapter 8, where V is air velocity with limits shown in that table.
[c] Adjusted heat gain includes 18 W for food per individual (9 W sensible and 9 W latent).
[d] Figure one person per alley actually bowling, and all others as sitting (117 W) or standing or walking slowly (231 W).

(Copyright 1993 by ASHRAE, from *1993 Fundamentals, SI Edition*. Used by permission.)

Footnote: *Cross-references in table footnotes pertain to the source from which tables were taken; they do not refer to chapters or pages in this text.*

Sensible and Latent Heat Gains for People—In SI Metric Units

Figure 3.23a

- enthalpy (Btu/lb. of air) [kJ/kg of dry air] (line #2)
- dew point (°F) [°C] (line #3)

If any two of these properties are known, the other five can be found by plotting the point on the psychrometric chart.

Dry bulb temperature (°F) [°C] is the temperature of the air as read by a normal thermometer.

Wet bulb temperature (°F) [°C] is the temperature registered on a thermometer that has a wet gauze wrapped around its bulb and air blowing across it at 50 fpm [0.25 m/s]. Water evaporates from the gauze as it absorbs heat from the fluid in the bulb. Consequently the wet bulb temperature is lower than the dry bulb temperature unless the air is

---

Heat Gain Formulas

Conduction Heat Gain
$H_C = UA \, (CLTD)$
H_C = Conduction heat gain (Btu/hr.) [W]
U = Overall heat transfer coefficient
  (Btu/hr. s.f. °F) [W.m²/K] (see Figures 3.5 – 3.7)
A = Surface area (s.f.) [m²]
CLTD = Cooling load temperature difference (°F) [°C] (see Figure 3.17)

Convection Heat Gain (Infiltration and Mechanical Ventilation)

$$H_V = \underbrace{1.1 \text{ cfm } [1.23 \text{ L/s}](\Delta T)}_{\text{Sensible heat gain}} + \underbrace{4{,}840 \text{ cfm } [2810 \text{ L/s}] \, (\Delta W)}_{\text{Latent heat gain}}$$

ΔT = Temperature difference between space and outside air (°F)
ΔW = Moisture content difference between inside and outside air
  (lbs. water/lbs. dry air)

Solar Radiation Heat Gain
$H_S = A \, (SC)(SCL)$
H_S = Radiant heat gain (Btu/hr.) [W]
A = Glass area (s.f.) [m²]
SC = Shading coefficient (see Figure 3.20)
SCL = Solar cooling load factor (see Figure 3.19)

Internal Heat Gain
H_i = (sensible load) CLF + (latent load)
H_i = Internal heat gain (Btu/hr.) (see Figure 3.22)
CLF = Cooling load factor (see Figure 3.21)

People Heat Gain
$H_P = N_O P_S CLF + N_O P_L$
H_P = People heat gain (Btu/hr.) [W]
N_O = Number of occupants
P_S = Sensible heat gain per person (Btu/hr.) [W] (see Figure 3.23)
CLF = Cooling load factor, usually 1.0 (see Figure 3.21)
P_L = Latent heat gain per person (Btu/hr.) [W] (see Figure 3.23)

Total Heat Gain
$H_T = H_C + H_V + H_S + H_i + H_P$

All numbers shown in brackets are metric equivalents.

Figure 3.24

Air Conditioning Requirements
BTU's per Hour per S.F. of Floor Area and S.F. per Ton of Air Conditioning

Type of Building	BTU per S.F. [W/m²]	S.F. per Ton [m²/kW]	Type of Building	BTU per S.F. [W/m²]	S.F. per Ton [m²/kW]	Type of Building	BTU per S.F. [W/m²]	S.F. per Ton [m²/kW]
Apartments, Indiv.	26 [82]	450 [12.2]	Dormitory, Rooms	40 [126]	300 [8]	Libraries	50 [158]	240 [6.3]
Corridors	22 [70]	550 [14.3]	Corridors	30 [95]	400 [10.5]	Low Rise Office, Ext.	38 [120]	320 [8.3]
Auditor. & Theaters	40 [126]	300/18* [8]	Dress Shops	43 [135]	280 [7.4]	Interior	33 [105]	360 [9.5]
Banks	50 [158]	240 [6.3]	Drug Stores	80 [252]	150 [4]	Medical Centers	28 [88]	425 [11.4]
Barber Shops	48 [151]	250 [6.6]	Factories	40 [126]	300 [8]	Motels	28 [88]	425 [11.4]
Bars & Taverns	133 [420]	90 [2.4]	Hi-rise Off. —Ext. Rms.	46 [145]	263 [6.9]	Office (small suite)	43 [135]	280 [7.4]
Beauty Parlors	66 [210]	180 [4.8]	Interior Rooms	37 [116]	325 [8.6]	Post Office, Indiv. Off.	42 [132]	285 [7.6]
Bowling Alleys	68 [215]	175 [4.7]	Hospitals, Core	43 [135]	280 [7.4]	Central Area	46 [145]	260 [6.9]
Churches	36 [115]	330/20* [8.7]	Perimeter	46 [145]	260 [6.9]	Residences	20 [63]	600 [15]
Cocktail Lounges	68 [215]	175 [4.7]	Hotel, Guest Rooms	44 [138]	275 [7.2]	Restaurants	60 [190]	200 [5.2]
Computer Rooms	141 [445]	85 [2.3]	Public Spaces	55 [173]	220 [5.8]	Schools & Colleges	46 [145]	260 [6.9]
Dental Offices	52 [165]	230 [6]	Corridors	30 [95]	400 [10.5]	Shoe Stores	55 [173]	220 [5.8]
Dept. Stores, Bsmt.	34 [110]	350 [9.1]	Industr. Plants, Offices	38 [120]	320 [8.3]	Shop'g. Ctrs., Sprmkts.	34 [110]	350 [9.1]
Main Floor	40 [126]	300 [8]	General Offices	34 [110]	350 [9.1]	Retail Stores	48 [151]	250 [6.6]
Upper Floor	30 [95]	400 [10.5]	Plant Areas	40 [126]	300 [8]	Specialty Shops	60 [190]	200 [5.2]

*Persons per ton
12,000 BTU = 1 ton of air conditioning

Figure 3.25

saturated (100% humidity), in which case no evaporation can take place and the dry bulb and wet bulb temperatures are equal. At a given dry bulb temperature the wet bulb temperature is an indication of the relative humidity of the air.

Moisture content or specific humidity (lbs. water/lb. air or grains water/lb. air) [g water/g dry air] is a measure of the amount of water in the air.

Relative humidity (%) is an indication of the degree of saturation of the air.

Specific volume (cu.ft./lb. of dry air) [m^3/kg dry air] of the air is the volume that one pound of air occupies.

Enthalpy (Btu/lb.) [kJ/kg] is a measure of the quantity of energy in the air.

Dew point (°F) [°C] is the temperature at which the air is saturated. At the dew point, the dry bulb temperature is equal to the wet bulb temperature.

Sample Annual Bin Data

City	100/104	95/99	90/94	85/89	80/84	75/79	70/74	65/69	60/64	55/59	50/54	45/49	40/44	35/39	30/34	25/29	20/24	15/19	10/14	5/9	0/4	-5/1	-10/-6	-15/-11	-20/-16	-25/-21
Albuquerque	1	54	191	348	511	617	789	785	816	676	637	720	678	676	560	406	180	101	31	3						
Bismarck		11	68	173	252	320	450	590	625	550	583	506	624	539	626	596	424	399	391	306	364	144	131	43	42	3
Chicago				97	222	362	512	805	667	615	622	585	577	636	720	957	511	354	243	125	66	58	6			
Dallas/Ft. Worth	27	210	351	527	804	1100	947	705	826	761	615	615	523	364	289	57	29									
Denver		3	118	235	348	390	472	697	699	762	783	718	665	758	713	565	399	164	106	65	80	22				
Los Angeles	8	8	9	17	53	194	632	1583	234	2055	1181	394	74	4												
Miami					45	864	1900	2561	1605	871	442	222	105	77	36	12										
Nashville			7	137	407	616	756	1100	866	706	692	650	670	720	582	342	280	107	71	29						
New York City			5	26	170	383	664	820	941	763	699	593	690	765	858	648	377	212	99	20	5					
Seattle					16	62	139	256	450	769	1353	1436	1461	1413	915	358	51	43	15	1						

(Copyright 1993 by ASHRAE, from 1993 Fundamentals. Used by permission.)

BIN Method Heat Pump Table

Figure 3.26

Sample Annual Bin Data

City	39/41	36/38	33/35	30/32	27/29	24/26	21/23	18/20	15/17	12/14	9/11	6/8	3/5	0/2	-3/-1	-6/-4	-9/-7	-12/-10	-15/-13	-18/-16	-21/-19
Chicago				74	176	431	512	960	660	591	780	510	770	686	1671	380	304	125	66	49	11	4
Dallas/Ft Worth	4	170	322	511	922	1100	1077	750	803	870	581	728	418	464	37	3					
Denver			81	217	406	390	570	726	712	902	809	783	750	1467	446	216	106	85	52	44	8
Los Angeles	4	10	9	16	56	194	1016	1874	2280	2208	843	227	23								
Miami				14	648	2147	2581	1852	734	390	202	100	76	14	2						
Nashville		4	82	366	717	756	1291	831	693	801	670	858	639	793	141	89	29				
Seattle				10	88	139	330	497	898	1653	1392	1844	1127	715	40	26	1				

(Copyright 1993 by ASHRAE, from 1993 Fundamentals, SI Edition. Used by permission.)

BIN Method Computation

Figure 3.27

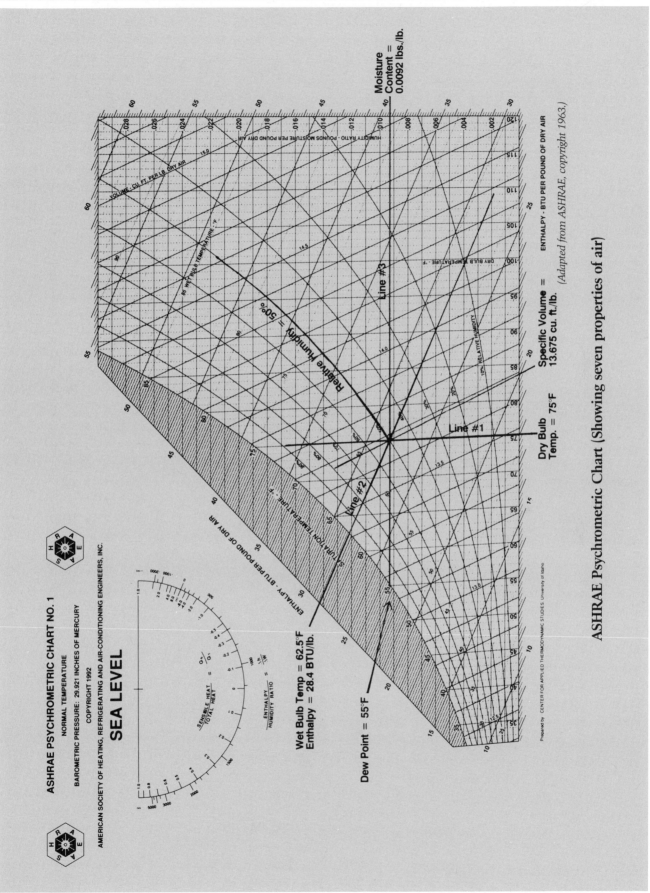

Figure 3.28a

ASHRAE Psychrometric Chart (Showing seven properties of air)

(Adapted from ASHRAE, copyright 1963.)

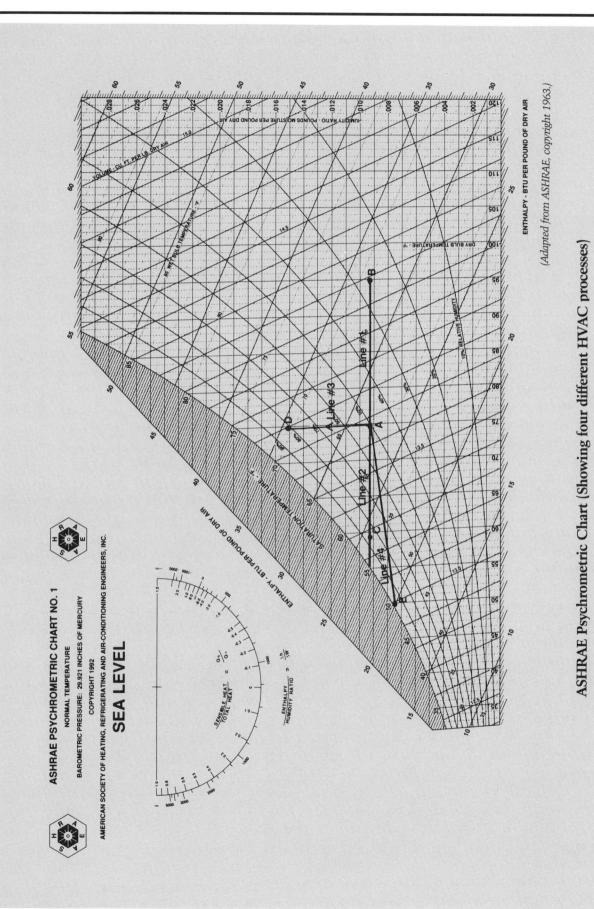

Figure 3.28b

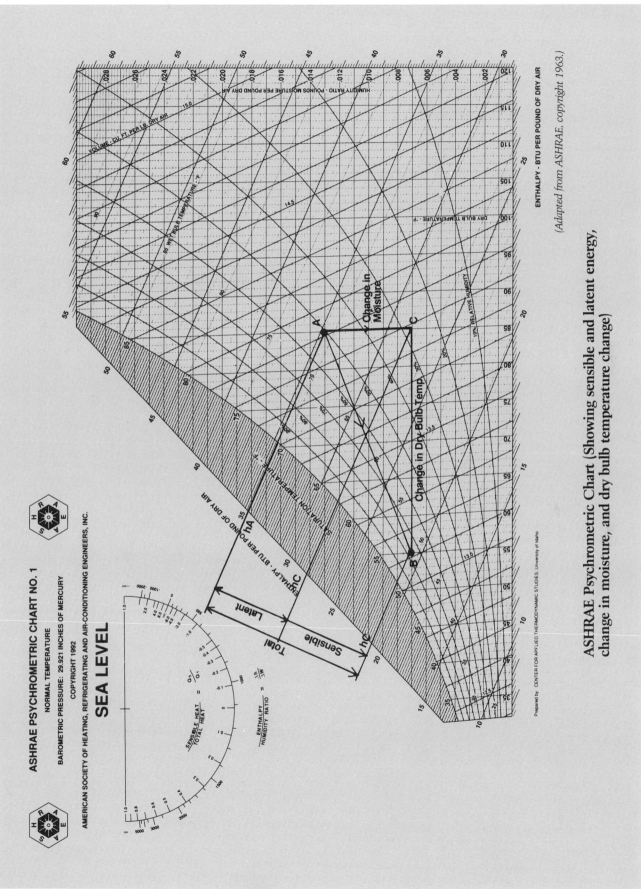

Figure 3.28c

ASHRAE Psychrometric Chart (Showing sensible and latent energy, change in moisture, and dry bulb temperature change)

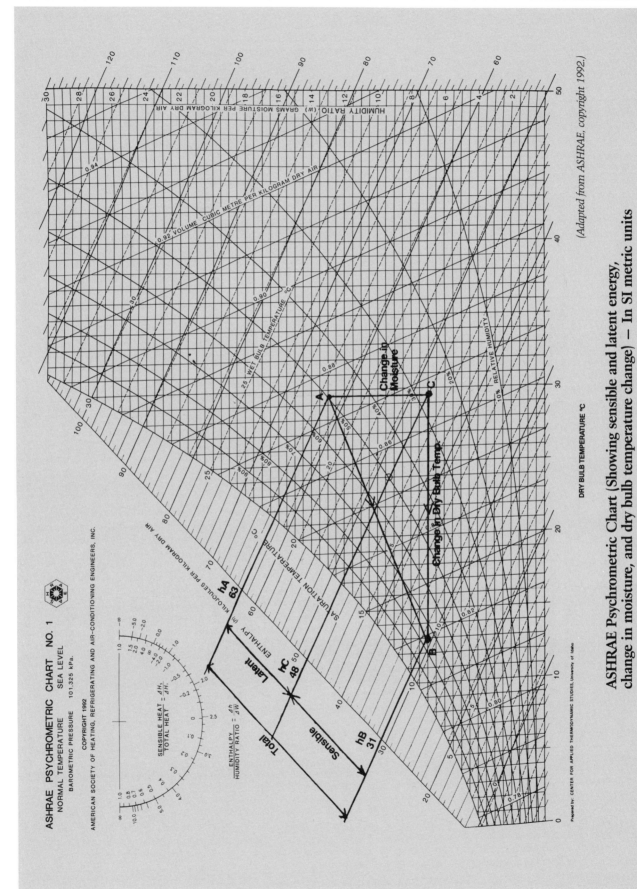

Figure 3.28d

ASHRAE Psychrometric Chart (Showing sensible and latent energy, change in moisture, and dry bulb temperature change) – In SI metric units

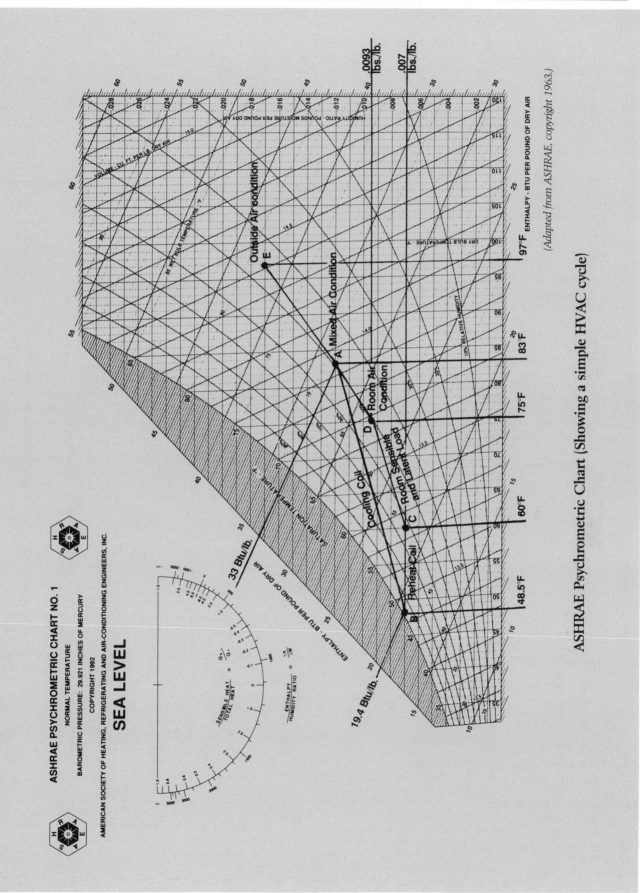

Figure 3.28e

ASHRAE Psychrometric Chart (Showing a simple HVAC cycle)

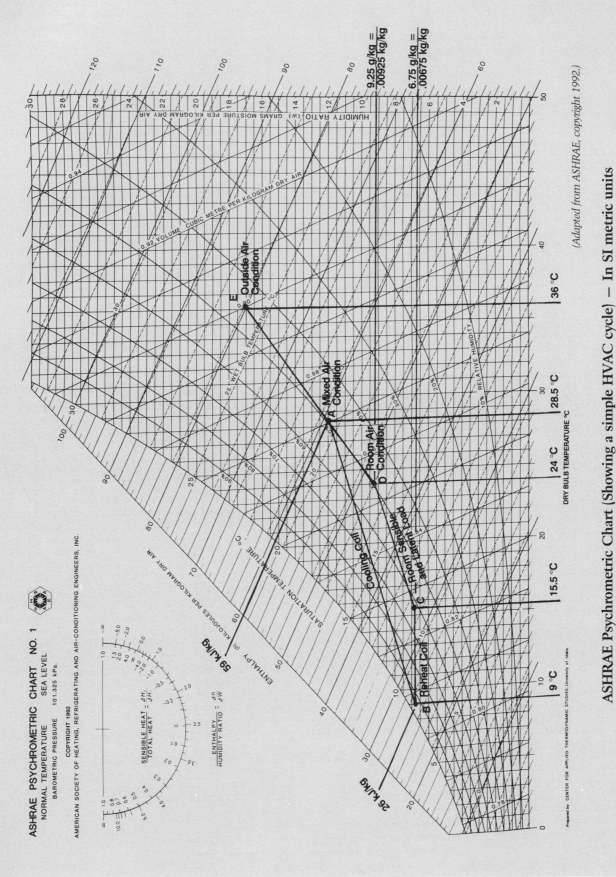

Figure 3.28f

The Psychrometric Process

Air enters HVAC equipment at one set of conditions and leaves at a new set of conditions. Each of these conditions can be represented by a point on the psychrometric chart; a line drawn between these two points represents the HVAC process performed by the equipment. The lines in Figure 3.28b represent different HVAC processes as described below.

Line 1 (A-B) represents a sensible heating process, which may occur as air passes over a hot water or steam heating coil.

Line 2 (A-C) represents a sensible cooling process, which may occur when air passes over a dry cooling coil.

Line 3 (A-D) represents a humidification process, which may occur when air passes over a steam humidifier.

Line 4 (A-E) represents a cooling and moisture reduction process, which would occur as air passes over a cooling coil with a dew point lower than the dew point of the air. This process could also be called a sensible and latent cooling process.

The above processes represent the functions of typical equipment in an HVAC system.

Sensible and Latent Energy in HVAC Processes

An HVAC process requires a change in sensible energy, in latent energy, or in both. Sensible energy is related to changing the dry bulb temperature of the air; latent energy is related to changing the moisture content of the air. Consider the sensible and latent cooling process as indicated by line A-B in Figures 3.28c and 3.28d. Line A-C represents the change in moisture and line C-B represents the dry bulb temperature change.

The change in energy per pound of dry air from point A to point B can be represented by the following equations:

Sensible Load = 1.08 × cfm × dry bulb temperature (°F) [1.23 × L/s × °C] difference

or

Sensible Load = 4.45 × cfm × Enthalpy (Btu/lb.) [1.2 × L/s × kJ/kg] difference

Latent Load = 4782 × cfm × Moisture Content (lbs./lb.) [3010 × L/s × g/g] difference

or

Latent Load = 4.45 × cfm × Enthalpy (Btu/lb.) difference

Total Load = 4.45 × cfm × Enthalpy (Btu/lb.) difference
= Sensible Load + Latent Load

The enthalpy of the air at point A is hA = 34.5 Btu/lb. [63 kJ/kg], and the enthalpy of the air at point B is hB = 20.8 Btu/lb. [31 kJ/kg]. The change in moisture content is given in line A-C; this part of the process relates to the change in latent energy. The change in dry bulb temperature is given in line C-B; this part of the process relates to the change in sensible energy.

The enthalpy of the air at point C must be known in order to determine the latent and sensible loads of the process. The enthalpy at point C is hC = 27.9 Btu/lb. [48 kJ/kg]. Assuming the cfm [L/s] for the process is 1,000 cfm [500 L/s], calculate the loads as follows:

Sensible Load = 4.45 × 1,000 × (27.9 − 20.8) [1.2 × 500 × (48 − 31)]
= 31,595 Btu/hr. [10 200 W]

Latent Load	= 4.45 × 1,000 × (34.5 – 27.9) [1.2 × 500 × (63 – 48)]
	= 29,370 Btu/hr. [9000 W]
Total Load	= 4.45 × 1,000 × (34.5 – 20.8) [1.2 × 500 × (63 – 31)]
	= 60,965 Btu/hr. [19 200 W]
	= Sensible Load + Latent Load

A Simple Psychrometric Cycle

Figures 3.28e and f represent a simple HVAC cycle on a psychrometric chart. The cycle includes a cooling coil and a reheat coil. Assume that there is 500 cfm [250 L/s] of air in the process. Room air at 75°F [24 °C] and 50% RH (point D) is mixed with outside air at 97°F [36 °C] and 45% RH (point E) to produce air at 83°F [28.5 °C] and 49% RH (point A). The mixed air is passed over a cooling coil (process A-B), where there is a reduction in moisture content and dry bulb temperature. The air then passes over a reheat coil (process B-C) to produce the correct humidity in the room. The air enters the room at point C. The air then absorbs the sensible and latent heat gains produced in the room (point C-D); at this point the air is removed from the room, a portion is mixed with the outside air, and the process repeats itself.

In determining the total load of the cooling coil, the load of the reheat coil, and the sensible and latent loads of the room, the following questions must be addressed: Is the load for the reheat coil latent or sensible? In what proportions are the room air and outside air mixed?

Enthalpy at point A is hA	= 33 Btu/lb. [59 kJ/kg]
Enthalpy at point B is hB	= 19.4 Btu/lb. [26 kJ/kg]
Total load of cooling coil	= 4.45 × 500 × (33 – 19.4) [1.2 × 250 × (59 – 26)]
	= 30,260 Btu/hr. [9900 W]
Load of reheat coil	= 1.08 × 500 × (60 – 48.5) [1.23 × 250 × (15.5 – 9)]
	= 6,210 Btu/hr. [2000 W]
Sensible load of room	= 1.08 × 500 × (75 – 60) [1.23 × 250 × (24 – 15.5)]
	= 8,100 Btu/hr. [2614 W]
Latent load of room	= 4782 × 500 × (.0093 – .007) [3010 × 250 × (.0093 – .0068)]
	= 5,500 Btu/hr. [1881 W]

Since the reheat produces a change in dry bulb temperature only and no change in moisture content, its load is sensible.

Dry bulb temperature at point D tD	= 75°F [24 °C]
Dry bulb temperature at point A tA	= 83°F [28.5 °C]
Dry bulb temperature at point E tE	= 97°F [36 °C]
% of outdoor air	= (tA – tD)/(tE – tD)
	= (83 – 75)/(97 – 75)
	[(28.5 – 24)(36 – 24)]
	= 36.4%

Therefore, the air at point A is a mixture of 36.4% outside air and 63.6% room air.

Computer Programs and Available Software

There is a wide selection of computer programs and software targeted to the HVAC designer. These programs include heating and cooling load calculations, energy analysis, system sizing, duct and pipe sizing, psychrometrics, specialized system analysis, cost estimating, electronic catalogs, equipment selections, facility and plant management, and service management.

Most of the software available is very user-friendly and is accompanied by good reference manuals. The many programs are so varied it would be impossible to describe all of them, but brief descriptions of a few programs follow.

Carrier E20-II

The Carrier E20-II software program is a comprehensive set of programs that covers the wide range of HVAC applications. The most critical program is the HAP (Hourly Analysis Program), which utilizes the ASHRAE transfer function method for detailed heating and cooling load calculations. Other programs include engineering economic analysis, commercial and residential load estimating and operating cost analysis, duct design, refrigerant and water piping design, and an electronic catalog for Carrier equipment.

Trane Trace 600 and Load Design 600

Like the Carrier package described above, the Trane Trace 600 and Load Design 600 packages cover a wide range of HVAC design aids and include programs ranging from heating and cooling load analysis to equipment sizing and selection. The Trane CDS selection programs are available to designers to help in sizing, selecting, and specifying equipment.

Customized Software

Customized software is available when an energy management, building management, or direct digital control (DDC) system is installed in a building or facility. These systems are designed to coordinate with existing HVAC systems in the building or facility. They offer programs that aid in preventative maintenance schedules, energy management programs, load shedding, energy-saving control systems, and systems analysis. Generally, these software packages are available only as part of a new DDC- or microprocessor-based control system. The software can be obtained from controls companies such as Johnson Controls, Honeywell, Landys and Gyr Powers, and Andover Controls.

Other HVAC Packages

Other packages are available from local, national, and specialized software companies throughout the country. These packages cover the full range of HVAC system requirements and analysis, from simple U-value calculations to energy consumption estimates and troubleshooting systems in operation.

Manufacturers' Software

With the increased use of CADD (computer aided drafting and design), most equipment manufacturers have developed or are developing software or electronic products that aid the designer in equipment selection and specification. Many of these software packages provide details and drawings compatible with common CADD programs, so information can be imported onto working drawings with relative ease.

Problems

3.1 A space requires 4 tons [14 kW] of cooling in order to maintain its design condition. It is cooled by a refrigeration unit with coefficient of performance (C.O.P.) of 4.0. How many Btu/hr. [W] does the condenser dump to the outside?

3.2 A one-story office building in Boston (see Figure 3.29) has 8-foot [2.5 m] ceilings, has a slab on grade, and is 25 feet [7.6 m] on each side. Each elevation has two 3′ × 5′ [900 mm × 1500 mm] double hung, double glazed windows; the east and west walls each have 3′ × 7′ [900 mm × 2150 mm] solid wood doors.

The exterior walls of the building are constructed of 2″ × 4″ [38 mm × 90 mm] frame construction, wood siding, 5/16″ [7.9 mm] plywood sheathing (see Figure 3.6), and 3/8″ [9.5 mm] gypsum wallboard finish on the interior. The walls are filled with 3″ [75 mm] of fiberglass insulation (see Figure 3.6). The roof is a 1″ [25 mm] wood deck supported on wood framing with a suspended acoustical tile ceiling. There is 3″ [75 mm] of insulation on the wood deck.

Find the building design heat loss.

3.3 Consider the same building you used in problem 3.2 for the heating calculation. Assume there are 10 occupants, one copy machine, and one coffee maker. The general lighting is at 1.5 watts per square foot. Calculate the building heat gain.

3.4 For the three windows in Figure 3.30, find the U value of each. Comment on how much better (if any) the single pane glass window is than the opening that relies principally on the air films only.

3.5 A laboratory takes in outside air at 55°F [13 °C] db and 10% relative humidity and supplies it to a room whose design conditions are 70°F [21 °C] db and 60% relative humidity. How many pounds [kilograms] of water must be added to each pound [kilogram] of fresh outdoor air to maintain the design condition?

3.6 Air at 75°F [24 °C] db and 80% relative humidity is cooled until it is at 55°F [13 °C] fully saturated. What is the change in enthalpy of the air? What % of the total change in enthalpy results from the removal of sensible heat?

3.7 Design a living room window for a house in northern Alaska and Southern New Mexico. Show size, orientation, shading, and the wall construction around the opening.

3.8 Consider the following building elevation and conditions as noted and shown in Figure 3.31:

Location: Providence, RI
Office Building
Summer Cooling: 3 p.m., July 21
All windows 6′ × 6′ [2 m × 2 m] fixed, double glazed
In the winter, what is the heat loss in Btu/hr. [W] through the windows from:

a. conduction
b. convection
c. all other sources
d. total heat loss through the windows shown

In the summer, what is the heat gain in Btu/hr. [W] through the windows from:

a. conduction
b. convection
c. radiation
d. all other sources
e. total heat gain through the windows shown

3.9 Assume that the building shown in problem 3.8 is square in plan and has 100 square feet [10 square meters] per occupant. Each occupant requires 20 cfm [10 L/s] of fresh outdoor air.

a. Calculate the number of building occupants and the total cfm [L/s] required for ventilation.
b. On a particular winter day, the outside temperature is 40°F [5 °C] and the indoor temperature is 68°F [20 °C]. How many Btu/hr. [W] are needed to heat the outside air used for ventilation?
c. On a particular summer day, the outside air is 95°F [35 °C] (db) and 40% relative humidity. The indoor condition is 75°F [24 °C] (db) and also 40% relative humidity. How many Btu/hr. [W] are required to have the fresh air meet the indoor condition?

3.10 True or false?
– The condensor is always warmer than the evaporator when the equipment is running.
– The compressor never takes energy out of a refrigeration loop.
– It is less expensive to cool each Btu [J] of heat gain than it is to heat each Btu [J] of heat loss.
– A square double hung window has more crack length than a casement window of the same overall size.
– Considering their proportion of building area, windows are responsible for a disproportionate share of building heat gain and heat loss.

3.11 A room is at 70°F [21 °C] db and 50% relative humidity. If the inside surface of the glass window is at 45°F [7 °C] db, will water condense on the glass? In the summer, the same room has chilled water piped to the fan coils at 55°F [13 °C]. Will there be any condensation under the coil?

3.12 For wood frame construction, how much insulation is required to obtain the U values listed as code minimums for walls, roofs, and windows? Sketch a wall, roof, and window section showing construction.

3.13 An "adjusted male/female" sleeps for 8 hours, goes to the office for 8 hours, and reads and watches TV for 8 hours. How many calories per day does he or she produce? If the body's efficiency is 25%, how many kilocalories of food should he or she intake?

3.14 A room is set for 75°F [24 °C] and 50% relative humidity. The inside surface temperature of the glass in winter is 50°F [10 °C]. How much water will be removed from each pound of air that comes in contact with the glass?

3.15 Assume you are working on a large assembly hall in Providence, R.I., which is to be designed to the 2-1/2% and 97-1/2% temperature extremes for winter and summer. It contains 5,000 square feet and has a 15-foot ceiling. It can be used for general meetings or for loose seating.

a. Determine the occupancy based on 100 sq. ft. per person and on 7 sq. ft. per person.

b. Calculate the cfm required for outdoor air for each occupancy and the energy required per hour to bring the outdoor air to the room air condition.

c. Calculate the amount of air that the space contains and how long each occupancy could last on a "purge" system of ventilation.

d. Show how you would provide flexibility in the ventilation of the space as the number of people varied depending on the use.

3.16 In the summer, 5,000 cfm [2500 L/s] of outdoor air at 95°F [35 °C] and 40% relative humidity passes over a coil at 50°F [10 °C]. Assuming the coil is 90% efficient, how much energy per hour is needed to condense the water out of the air?

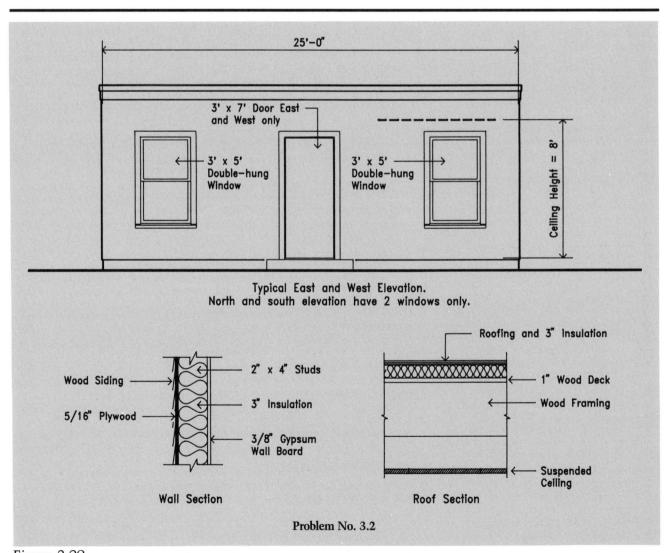

Figure 3.29

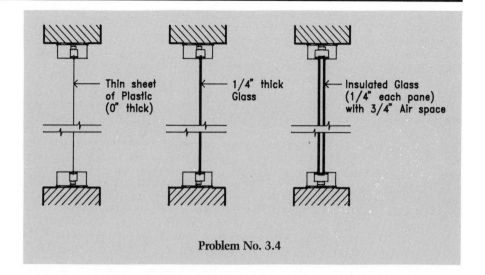

Problem No. 3.4

Figure 3.30

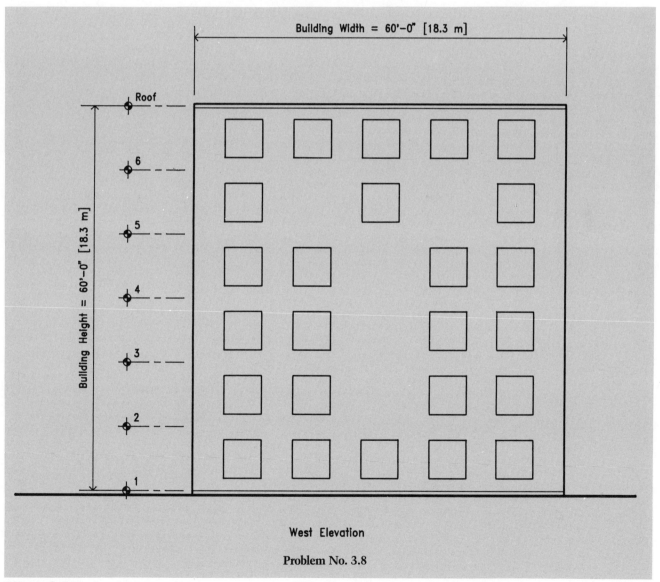

West Elevation

Problem No. 3.8

Figure 3.31

Chapter Four
ENERGY CONSERVATION

People have been building dwellings for thousands of years. Most indigenous building types have evolved to successfully manage environmental forces without having a negative impact on the surroundings. Recent technology and the widespread availability of electricity and fossil fuels have allowed most people to actively control their immediate surroundings, often without regard for the surrounding climate. As cultures have become more mobile, they have transported familiar architectural styles to locations where they might not be ideally suited. Large complexes located far from public transportation are poor conservators of energy, considering the millions of gallons of fuel used by cars to access them. Park Avenue curtainwall buildings have been built in the desert, orientated without regard for the location of the sun.

Buildings consume a significant portion of the energy in the United States. Buildings consume nearly half of all the energy in the country for heat, cooling, and power, and it is estimated that nearly 30 percent of this consumption could be saved by energy conservation and/or sustainable building design and operations. This chapter explores ways in which designers can be more energy conscious and how methods of sustainable design can be incorporated into buildings.

Conservation and Sacrifice

Conservation refers to the protection or preservation of a valuable resource. As we enter the next century, the single greatest contribution that can be made in the field of energy is conservation. This means drawing as much as possible from the natural resources we use, the products we make, and the energy we consume.

Conservation influences what should be made and how we live. If it became common in winter to wear warmer clothes indoors with the result that the indoor design temperature could be reduced, vast quantities of fuel would be conserved, benefiting both the environment and pollution control.

Other basic assumptions need to be reevaluated. Take, for example, a standard downtown office building. The building is probably surrounded by other office towers, and for all practical purposes, the lower floors are always in shade. The building might be rezoned so that the lower floors are not on the same zone as the upper floors for certain portions of the day, promoting both comfort and energy efficiency.

For new facilities, designing for energy conservation means incorporating a wealth of new techniques into building systems. For existing buildings, energy audits can be completed and use patterns analyzed to maximize energy use. For both new construction and renovations there is considerable opportunity to improve the design and performance of buildings.

Sustainable Design

Sustainable design minimizes the impact of the use of materials and energy in the production of a building. Rehabilitating existing structures is more conservative than building new. When buildings are designed with fewer materials, less energy is needed to produce the materials in the first place, and there are fewer materials to eventually discard. In essence, fewer materials means fewer pollutants.

Sustainable design also considers the effect of types of materials chosen. For example, building with appropriate materials can help preserve natural habitats and add to the overall well-being of the environment. *Passive design*, which minimizes energy consumed by burning fuel or using power, is a form of sustainable design. A building that uses only solar energy is an example of passive design.

Because passive systems contribute to environmental conservation, the overall principles of passive design should be understood and used whenever possible. Increasing costs and the negative environmental impact of energy-consumptive buildings make energy-conservative designs more compelling than ever.

Long-forgotten sustainable building systems are being resurrected with success. The principles of Vitrivius – organizing a building around the sun – as well as passive design, revitalization of older buildings, use of recyclable materials, energy scavenger systems, the quest for nuclear fission, and smart building controls are all being brought into the current lexicon of design. Earth-bermed houses have been used by native peoples, modern designers in sustainable design, and Frank Lloyd Wright.

We can no longer afford to ignore the consequences of waste and environmental stress. Designers must reinvigorate interests and knowledge about basic building systems and about the environment we both build in and shield ourselves from.

Indigenous Architecture: Using Natural Energy for Heating, Ventilating, and Air Conditioning

People build dwellings as refuge from the harsh elements of their environment. Virtually all dwellings shield the inhabitants from precipitation. Temperature, and to a lesser extent humidity, are key factors that must be manipulated to provide a comfortable shelter. These "harsh elements" are defined in many ways, depending on the environment. For some climates, comfort means keeping the cold out; for others, it's the heat. For some it is crucial to have a breeze, in others it is deadly. In some areas, a design to accommodate a variety of conditions must be considered.

One of the best ways to understand how to design sustainably and conserve energy in buildings is to revisit indigenous building systems. Early peoples developed building forms over thousands of years to meet the need for shelter in every type of harsh environment without the benefit of large energy-producing machines. There is a basic, sustainable building type for every type of climate; several of these are shown in Figure 4.1. The climates and related indigenous building types are discussed in the following sections.

The Desert

Dry, Hot, and Cold – Pueblo

In what is now the southwestern United States, where hot, dry days are followed by cold nights, native Americans have been building pueblo-style buildings for centuries. These thick-walled adobe dwellings absorb the heat of the day in the thermal mass of the walls, and release this heat at night. In this way the interior is kept cool during the hottest times of the day, and is heated at night. There are deep wall openings that allow breezes but restrict hot summer sun, and the units are built over one another, in apartment-like construction, to minimize surface exposure.

Hot and Dry – Peristyle, Takhtabush, and Mashrabiya

In the hot, dry lands that surround the Mediterranean and Near East, indigenous architecture has turned inward. For centuries, houses in this area have been built with thick, windowless outside perimeter walls. Interior windows look into a central courtyard, or peristyle. The courtyard acts as a cool air trap. As night falls, the warm air of the day rises and is replaced by the cooler night air. This cool air seeps into the rooms of the house, which face the courtyard. When the sun rises, the courtyard remains cool, being shaded on four sides. A pool or fountain often acts as a thermal mass.

A variation on this idea, originating in Arab houses, is the Takhtabush – a covered sitting area located between the courtyard and a larger, less shaded garden. As the larger garden heats up from exposure to the sun, the air rises, drawing the cooler courtyard air through the sitting area.

Another cooling mechanism used in the Middle East is the Mashrabiya, or wooden lattice screen. These screens are traditionally constructed of small wooden pieces, circular in cross section, fastened in a decorative pattern. Small interstitial spaces control the light and air. These spaces, ranging from 1 to 36 square inches, serve several important functions. First, they control the amount of direct sun that comes into the space. Second, they maximize the amount of ventilation area, utilizing whole walls of a space. If a small space does not allow sufficient air to pass through, then the upper portions of the screen, below a roof overhang, are opened up. Third, the screen acts as a humidifier. Because the screen is wooden, it absorbs moisture from the cool night air and releases this moisture to the incoming ventilation air during the day.

Tents

A large population of nomads has always occupied lands from Tibet to Africa, where courtyards and screens are designed for permanent settlements. Another common type of dwelling is the "black tent", which consists of an animal hair membrane stretched over poles and secured with guy ropes. The membrane, woven of goat hair in warmer climates and yak hair in colder ones, breathes easily under the hot sun. In the rain, the membrane contracts tightly to become almost impervious to water. In cold weather, the hair is an excellent insulator. Although black is commonly associated with absorbing heat, the effect of the color in these hot climates is to cast a denser shadow, insulating more effectively against radiation heat. It is common for the poles as well as the membrane to be carried from one site to another.

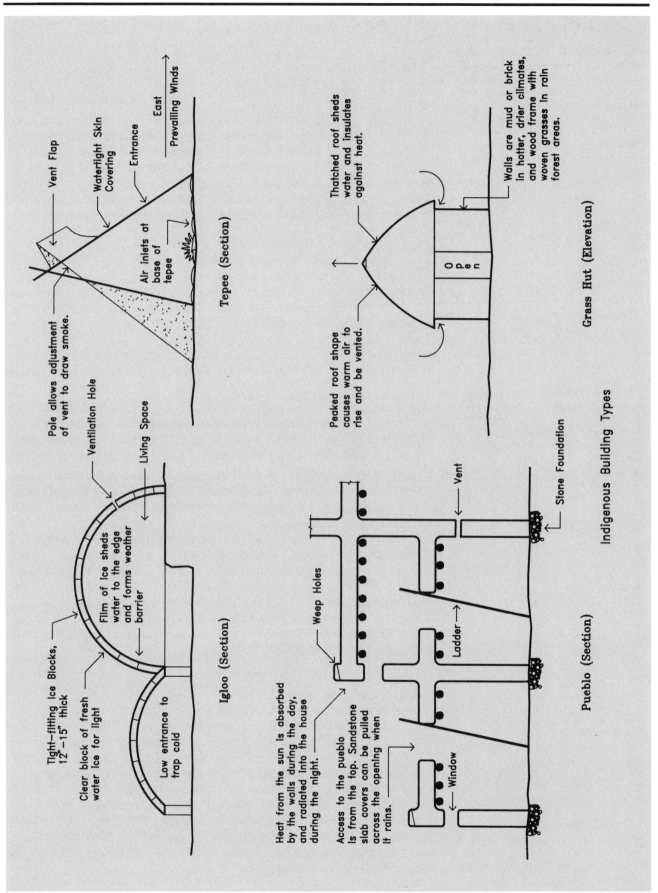

Figure 4.1

The Tropics
Wet and Tropical – Grass Hut

In warm tropical climates, dwellings are elevated above ground to shield them from animals and water. A steeply pitched roof, often grass, is used to shed water. Because these buildings are in or near a thick jungle, the climate is both humid and without breezes. The pitched roof encourages air to move from the low perimeter upward and out the roof opening to create some cooling.

In South America, ancient Indian tribes, such as the Incas, built low mounds from dirt, gravel, or rocks on which to build their dwellings. The mounds, which ranged from three to nine feet high, effectively protected the home against flooding and the buildup of water during heavy rains. The dwellings themselves were made of poles woven with broad leaves or grasses, with a thatched roof.

In Africa, some communities live either in the rain forests or in the grasslands adjacent to them. Their dwellings use many of the same materials and methods as those found in South America. The design of the steeply pitched thatched roof was, and in many cases still is, used in both the grasslands and the forests. In the grasslands, which approach desert-like conditions, walls are usually mud or mud-brick.

The Arctic
Cold and Dry – Igloo

The Americas offer several examples of cultural responses to a variety of conditions. In the north, native Americans build igloos, low domes constructed from blocks of ice 6 – 15 feet in diameter. They are used primarily as winter housing or for temporary shelter during fishing expeditions. When occupied, igloos can be heated with a few candles and the body heat of the occupants despite outdoor temperatures of 30 – 60°F below freezing. A thin shield of ice forms over the inside of the igloo, which, with the thickness of the ice blocks, forms an effective barrier against the wind and cold.

The igloo is accessed via a sunken entrance, which traps the cold air below the living area. Ventilation holes are built in or drilled as necessary. Air quality is an important consideration, because the close, well sealed quarters allow for very little air movement. Candles are kept burning all night, and are closely monitored. If the flame turns orange, a vent is opened or drilled until a white flame is achieved.

Often, a block of clear ice, preferably fresh water, is installed to allow light into the space. A second block of reflective ice can be placed below this window to attain an even higher level of illumination. More permanent dwellings are built out of a combination of stone, sod, skins and bones, depending on available materials.

The Mountains
Temperate and Wet – Wigwam

From the Great Lakes region to the Atlantic coast, the wigwam was a common dwelling prior to the 19th century. These were domed dwellings framed with wooden poles and covered with tightly woven grass mats. These grass mats could be rolled up and transported, leaving the frame. Wood was plentiful in the Northeast, so building a new frame was common. Also, because the climate was wet, materials rotted quickly. In damp weather, the mats would swell to keep the rain out. In the hot, humid summer, the side mats were removed for ventilation.

Temperate Climates
Tepee
The plains region of the northern United States was occupied by many tribes of nomadic or semi-nomadic Indians. Although several types of architecture evolved, the most widely recognized is the tepee. The design of the tepee was very similar to that of the tent just described. Constructed of animal skins stretched around wooden poles, the tepee was easily set up and taken down, thereby well suited to a nomadic lifestyle. When the bottom edge of the tepee was secured to the ground with stakes or stones, it provided an effective barrier against the cold, being easily heated with body heat and a central fire that was vented through a flap in the top. In the summer, the skin could be loosened and partially pulled back to allow air to travel up through the space to the vent hole above.

Self-regulation of Body Systems in Animals
People have devised ways to regulate their environment in ways that are external to their bodies. Animals, however, still depend on heating and cooling methods that are integral with their body functions and structure. Through evolution, nature has produced an enormous variety of animals who respond in very different ways to similar, or even the same, environments. Some systems used by animals for environmental control are shown in Figure 4.2.

\	How Animals Control Their Environment	
Animal	**Heat Response**	**Cold Response**
Dogs	Shed fur from coat. Pant to increase evaporation and ventilation within chest. Sleep to reduce metabolic rate.	Grow heavier fur coat. Bristle fur to increase insulation with trapped air. Increase activity. Sleep in curled position to reduce exposed surfaces.
Fish	Seek colder, usually deeper water. No internal control. Wide range of tolerable body temperature.	Seek warmer, shallower water. No internal control. Wide range of tolerable body temperature.
Birds	Bathe in water for evaporative cooling. Molt feathers for reduced insulation.	"Fluff" feathers for increased insulation. Perch facing wind to reduce air circulation within feathers.
Polar Bears	Swim for evaporative cooling. Increase respiration to increase internal ventilation and cooling. Shed fur.	Fur has no pigment, which allows solar heat to penetrate coat and be trapped. Increase thickness of coat. Oils in coat prevent water saturation.
Elephants and Rabbits	Extend ears as 'radiators' to increase heat loss.	Keep ears tight to body to minimize heat loss.
Humans	Perspire to promote evaporative cooling. Reduce activity to decrease metabolic rate. Dilate surface capillaries to give up core heat at the skin surface.	Body hair bristles to entrap air at skin surface. Blood circulation is shunted away from the extremities in favor of the vital organs. Shivering response increases metabolic rate. Perspiration decreases.

Figure 4.2

Energy Conservation Techniques

While considering the principles of passive systems noted previously, we must realize that buildings consume energy for a wide range of uses. There are techniques for actively minimizing the energy consumed in buildings; these methods are discussed in the following sections.

Energy-efficient Lighting

Lighting systems can account for up to 40 percent of a building's energy costs. In recent years there has been an effort by both code and utility companies to ensure that electrical systems are made more energy-efficient. Some of the requirements and recommendations to achieve these objectives are outlined below.

Local and State Codes

Most local and state codes now have maximum allowable lighting levels and energy usage per square foot. These values range from state to state, but 2 watts per square foot is average. Decorative lighting needs careful attention, and certain rooms (hospital operating rooms, for example) are not as affected as general office landscapes.

Use of Natural Daylight

Daylight is by far the most efficient method of lighting. When there is adequate window area, natural daylight can be used as a primary lighting source. In these instances, strategic placement of the furniture in the room and use of reflective surfaces and directional blinds can aid in efficient natural lighting. Sensors may be used to control the level of electrical lighting and compensate for lowered levels.

Incandescent to Fluorescent Lighting

Incandescent lighting is not as efficient as fluorescent. Incandescent lighting fixtures use three to four times the power of fluorescent lighting for the same lighting levels. The payback for replacing incandescent fixtures or bulbs with fluorescent fixtures or bulbs is usually only a few months. Incandescent lighting should be replaced wherever possible.

Energy-saving Ballasts and Bulbs

All fluorescent fixtures require a small transformer or ballast to boost the voltage so lamps will operate. Most ballasts currently manufactured and installed are energy-saving ballasts. Older ballasts typically consumed energy equal to approximately 20 to 30% of the wattage of the lamps. New ballasts consume energy equal to approximately 10 to 15% of the wattage of the lamps.

Fixtures should also be equipped with energy-saving lamps. Energy-saving lamps typically save an additional 20% of energy costs for a lighting fixture. Energy-saving ballasts and lamps are more expensive than the non-energy-saving type, but the benefits outweigh the expense and payback is realized very quickly.

Note: There is a potential hazardous waste issue when replacing old ballasts with new. Old ballasts may contain PCBs and should be disposed of in conformance with the applicable regulations for hazardous waste.

Occupancy Sensors

Occupancy sensors are an effective means of controlling lights for rooms that are not used consistently, such as smoking rooms, conference rooms, or storage rooms. Such rooms waste energy if the lighting is on all day. Occupancy sensors use an infrared detector or motion sensor to turn the lights on when the room is occupied. They can be used instead of or in conjunction with a conventional wall switch. Occupancy sensors usually have an adjustable time delay (15 minutes, for example) before turning the lights off.

Lighting Control Panels
For areas or buildings where the occupancy pattern is constant and predictable from day to day, the entire lighting system in the building can be controlled with a central control panel that switches the lights on and off as scheduled. Control systems usually have local manual override buttons to accommodate exceptions.

Energy-saving Incentives
Most electric utility companies offer rebates or grant programs to consumers for energy-conservative measures. These programs act as an incentive for consumers to install energy-saving systems. Many utility companies will pay all or a substantial part of the cost to change fixtures from incandescent to fluorescent, or to change the ballasts and lights to the high-efficiency types.

Electrical energy (watts) is obtained by multiplying the instantaneous voltage by the instantaneous amperage. The maximum voltage does not always occur at the same instant as the maximum amperage.

The efficiency of electrical systems can be improved by adjusting the *power factor*. Power factor is the ratio of true power (watts) to the apparent power (volt-amperes) and occurs when the current in a circuit leads or lags the voltage as a result of inductance or capacitance. Inductance causes the current to lag the voltage and capacitance causes the current to lead the voltage. Power factor is expressed as a decimal or a percentage, and can range from 0 to 1.0 or 0 to 100%. When the current leads or lags the voltage, the true power (watts) delivered is less than the apparent power (volt-amperes). The excess power is wasted as heat.

In essence, the power factor can be thought of as the efficiency of an electrical system. Most electrical utilities penalize users for having a system with low power factor (below 0.80). The power factor for buildings usually lags because of inductive loads such as fluorescent lighting, induction motors, small heating appliances, and small transformers. Low power factor can be improved by 1) replacing underloaded induction motors with smaller motors; 2) replacing induction motors with synchronous motors; 3) installing static capacities across the line to adjust the peak of the current to be more coincident with the voltage peak.

Lower Cooling Loads
When energy-saving lighting systems are utilized instead of normal lamps, there is a reduction not only in the lighting energy consumed but also in the cooling load. Heat gain from lighting is a large percentage of the total cooling load; lowering this load will lower fuel costs and may result in smaller, more efficient equipment being adequate for the air conditioning system.

Variable Frequency Drive Motors
In Chapter 8, the laws governing pumps and fans are discussed in detail. Pumps and fans are the primary motors used throughout HVAC systems. The power consumed by a motor is proportional to the cube of the speed of the motor; the speed of the motor is proportional to the frequency of the electric current feeding the motor. Therefore, the power consumed by a motor will vary as the frequency of the electric current varies.

In the United States, electrical power is delivered at 60 cycles per second, or 60 hertz (Hz); elsewhere power is delivered at 50 Hz. Converters (variable frequency drives) are now available to vary the frequency to the motor from the power company. This results in lower motor speeds and hence lower power consumption in accordance with the fan and pump laws. The speed of a motor is modulated by its variable frequency drive in response to actual demand on the equipment. Consequently, the volume of air or water being blown or pumped through the fan or pump follows the building needs more accurately and saves energy at the same time. This is the basis of variable frequency/speed motors.

Variable speed motors are economical to use for applications such as variable air volume (VAV) air conditioning systems or for variable pumping of chilled or hot water in fan coil systems. In a VAV system, the volume of air required decreases as the cooling load of the spaces in the building drops and the VAV boxes close to a small percentage of their full flow. The lower air volume blown by the fan can be achieved by lowering the frequency of the electricity supply to the motor, hence lowering the speed and reducing the volume of air flow to the required level.

When the cooling or heating loads drop in a fan coil system there are two alternatives: one is to vary the temperature of the supply water to the units, and the other is to vary the flow of the water supplied to the units. Varying the flow of the water to the units very often will save more energy than varying the temperature of the water supplied to the units, because boilers and chillers do not have many stages of load control and do not respond as quickly as a variable frequency motor.

Variable frequency/speed motors can save considerable energy. As stated above, the power consumed by a motor is proportional to the cube of the speed of the motor. In most systems it is not necessary for the equipment to run above 50% capacity for more than 1/3 of the time. Thus, if the cooling load of a system requires the speed of the motor to reduce to half its full load speed, the power consumed by the motor would reduce to $(1/2)^3 = 1/8$ of the original power consumed.

Variable frequency motors and their associated controls are more expensive than conventional equipment, but the payback for variable speed motors larger than 5 or 10 H.P. [3.75 or 7.5 kW]* can be as early as two years.

Cogeneration

Cogeneration refers to providing two forms of energy from one piece of equipment, such as an electrical generator. Fuel or energy from a single primary source is used to produce two useful energy forms, usually heat and power. These systems are designed to capture excess energy that would ordinarily be lost when heat and power are generated in separate processes.

Figure 4.3 illustrates some cogeneration principles. Cogeneration systems can be either bottoming cycles or topping cycles. In bottoming cycles, thermal energy is produced and used to satisfy thermal loads, and power is generated from the recaptured energy. In topping cycles, power is generated first, and the recaptured thermal energy is used to satisfy thermal loads. The primary components of a cogeneration system are the prime mover, generators, waste heat recovery systems, controls, electrical transmission and thermal distribution system, and connections to building services.

*Throughout the book, SI metric equivalents are provided for most imperial units. The SI units appear in brackets [ ].

The Prime Mover

The prime mover converts the fuel to mechanical energy. The prime mover can come in a variety of forms, from various types of turbines to internal combustion engines. The designer's choice of prime mover governs the overall operation of the system.

There are two principles to be considered when looking at prime movers. The first is overall system efficiency. Since it is at the first stage that fuel is used, it is the system's first and last opportunity to extract the maximum possible energy out of the fuel. Second, the needs of the facility should be considered. Some prime movers have relatively little heat output, but a very high power output. In other systems, where the ratio between the two is closer, power production may be insufficient to meet facility demands.

The Generator

The generator converts the mechanical, or shaft, energy to power. The main concerns in generator selection are efficiency and matching the generator to facility needs. Because generators are capable of providing a wide range of options in a variety of situations, they are often packaged with a prime mover.

Waste Heat Recovery Systems

As stated above, the primary advantage to a cogeneration system is that waste heat is recovered to meet facility demands for that heat. Heat recovery systems vary with the type of prime mover selected. The waste heat can come in the form of steam or exhaust gases. Applications include boilers, absorption chillers, steam turbines, and other heating or heat exchanger systems. Often, these "supplementary" systems that must be satisfied are complex or numerous, so the designer must look at the availability and type of this excess heat in different systems to match demand. Some

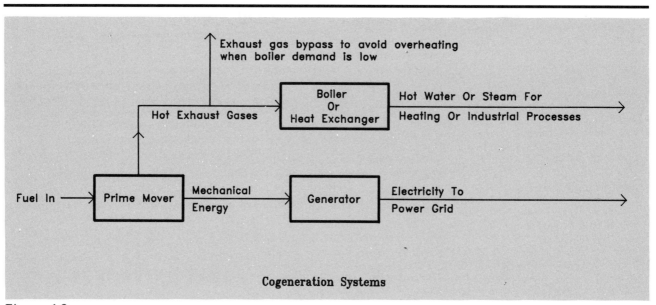

Figure 4.3

exhaust gases contain more oxygen than others, and thus are able to be used with burners in the exhaust, to feed a boiler. If waste steam is used to feed an absorption chiller, the available heat must be carefully sized and balanced to ensure that the correct amount of heat is available at the correct time.

Another source of heat recovery is the cooling system. The prime mover and the generator produce a significant amount of heat, which is often carried away by coolants such as oil or water. The coolants can produce hot water or steam for use by people or industrial processes. An additional heat rejection method may be required to dissipate heat if heat is not called for by the facility.

Controls

Control systems for cogeneration plants serve to coordinate a complex array of functions. They must regulate the prime mover and the generator while coordinating the cooling systems with the heat recovery systems for both efficiency and safety. The main pieces of equipment must be maintained at a steady temperature to prevent thermal stress. Systems fed by the power and heat must be kept supplied. The system also should be kept from cycling too often, which can cause excessive wear. A complete system of alarms, overrides, and safety checks is imperative to preserve the life and use of the system.

Transmission and Distribution

Once generated, the power or heat needs to be transported to its end use. There are different methods to distribute the energy depending on the design concerns. Electricity is subject to voltage drop over long runs. It can be installed in a grid system or by a direct feed to the user. Steam and hot water should be well insulated and placed below the frost line. Hot gases should maintain a high enough temperature to prevent condensation. A variety of supply, return exhaust, and other system parts must be designed to provide the maximum benefit to the user.

Solar Energy

Solar energy can be used in a variety of ways to generate hot water or hot air for heating, and hot water for domestic uses. Solar energy also has applications in absorption cooling.

The primary interest in solar energy is heating. For heating, the overall estimate of the proportion of solar collectors to floor area must be determined. For a rough approximation, consider that the sun shines each day on average for 8 hours on a collector that on average receives 100 Btu/hr./s.f. [312 W/m^2]. Thus allowing for some cloudy days, each day 800 Btu [240 Wh] are collected, which equals 800 Btu/day × 250 sunny days/yr., or 200,000 Btu/yr./s.f. [650 kWh/yr./m^2] Solar systems on average are 40% efficient; thus of the 200,000 Btu/s.f./yr. incident upon the collector, roughly 80,000 Btu/yr./s.f. will be available to the building – 60,000 in winter and 100,000 in summer. These numbers are gross approximations, which as indicated vary from summer to winter. Comparing these figures with the energy code objective of a building with an energy consumption of 55,000 Btu/s.f./yr., it is clear that if a collector in winter is capable of 60,000 Btu/s.f., the building area and solar collector area should be approximately equal.

Solar systems require a backup system in the event of a long cloudless period, so the cost of solar systems tends to be high. Because large areas of collectors are necessary for space heating and because solar systems are complicated, they have longer paybacks and are less frequently used. One application for solar panels has been domestic hot water systems, which, on average, require two to three panels (50 s.f.) [4.7 m^2] of collector surface for a family unit. The panels preheat the incoming cold water and reduce the cost of domestic water heating.

Solar energy can be provided by passive and active systems.

Passive Solar Energy

When solar energy is used directly, without any mechanically operated equipment, it is said to use a passive system. Passive solar energy principles are highly effective when made an integral part of a building. Examples of passive systems include:

- Windows that let in sunlight for heat in winter
- Use of a dark masonry floor to capture and store radiated heat
- Placing a building in a berm or beneath an earth roof to take advantage of the natural cooling and insulating effect of the earth
- Use of a massive wall of drums filled with water to store and re-radiate heat

Figure 4.4 illustrates some passive solar energy systems.

Passive design captures the incident solar radiation on a building. The amount of solar radiation varies from day to day and with latitude and orientation. South walls and the roof are the most effective places to catch solar radiation. In Chapter 3, Figure 3.19 shows the amount of radiation at 40 degrees latitude. More exact tables are used in actual calculations.

A central feature of many passive systems is a south facing exterior wall, which receives the most direct sunlight. The wall is often glazed, so as the sun strikes the furniture, floor, or other thermal masses such as a dark floor or rear wall, the heat is trapped inside and gradually released to the space. Spaces can actually become unusually warm using this method, so shades, such as venetian blinds, are often employed to control the amount of sun that enters the space. Curtains at night serve to keep the heat from escaping back to the night air. If the system uses hand-operated shades it is still considered a passive system.

In solar systems, the roof on the south side should overhang the wall so that the high summer sun is blocked out while low winter sun is allowed to reach the space. Figure 3.20 illustrates shading on windows. The designer should check the position of the sun at the solstices to determine the proper depth of the overhang. Sun control in summer is best achieved with awnings, shades, or blinds. In winter, heavy curtains are effective in insulating the window area and keeping heat in the building – hence the expression "summer shades and winter drapes".

Another passive system places a large thermal mass directly behind and inside a glass wall. The mass has openings on both the top and bottom. The cooler lower air is heated by the wall and naturally convected out of the top openings to the building.

Active Solar Energy

Active solar energy systems can be used to heat or preheat domestic hot water effectively. Active solar systems also work ideally with absorption

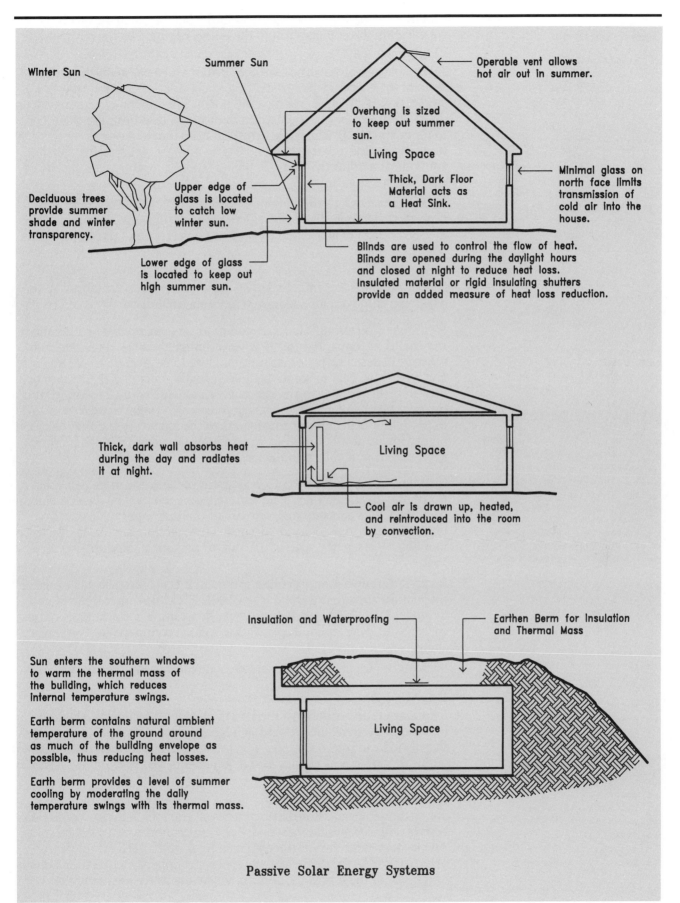

Figure 4.4

cooling systems, since the heat load in the building increases with the solar load in direct proportion to the energy captured by the solar system to run the absorber.

Active solar systems can be used with hydronic or air systems. In a hydronic system, the collectors heat water or a glycol solution and return it to a storage tank. At night the pumps will not operate and the system may drain down. Heat exchangers are used to transfer the heat from the solar loop to the heating loop, which takes the heat from the storage tank. In this way heat collected during the day can be used at night. Air systems use the sun to heat air directly, which is then distributed to the space directly or sent over for storage.

In active systems, the design of the storage tanks and heat exchangers depends on latitude, heat loads, orientation, glazing area, and other factors. Detailed design of active and passive systems is beyond the scope of this book.

Most active solar systems use either flat plate or concentrating solar collectors, which are connected to an active pumping or fan control system. They may also contain a storage tank or thermal mass.

Flat Plate Collectors: Flat plate collectors are constructed of a glazing material that transmits the short-wave (infrared) solar irradiation onto collector plates. The plates, in turn, produce long-wave thermal radiation, and "give" this heat to some medium (usually water or air) via tubes or fins, to be distributed to the space. Glass is usually used as the glazing material, because it allows a very high percentage of short-wave irradiation to pass while being almost impervious to the long-wave thermal radiation needed for heating. Some designs use double-wall glass facing to minimize loss of the captured heat to the air.

Concentrating Collectors: Concentrating collectors can be parabolic troughs or dishes, or a flat plate with reflective wings. The direct radiation from the sun is concentrated onto an absorber or water tube, which brings the heat to the location desired. The collectors are usually designed to follow, or track, the path of the sun, to maximize the gathering of heat.

Because the concentrating collector moves, it is able to take advantage of more of the time that the sun is available. Also, the concentrator is able to produce very high-temperature water or steam. The advantages of this collector are lower cost and the ability to collect more of the sun's available energy.

Some examples of active solar systems are illustrated in Figure 4.5.

Ice Storage

Ice storage is an economical method of providing cooling for a building. Ice storage systems are charged at night, utilizing off-peak utility rates, thereby reducing electrical bills. The ice storage system then discharges during the day at peak cooling load requirements. Figure 4.6 shows a typical ice storage system.

Ice was the first cooling medium; it was harvested on lakes and ponds during the winter and stored in ice houses all year long. Ice was a commodity and was shipped around the world. Early refrigerators were called iceboxes. The iceman came two or three times a week and loaded ice on the top of the icebox. As the ice melted, the cool water dripped around the channels in the side of the box to the bottom, where the water was removed.

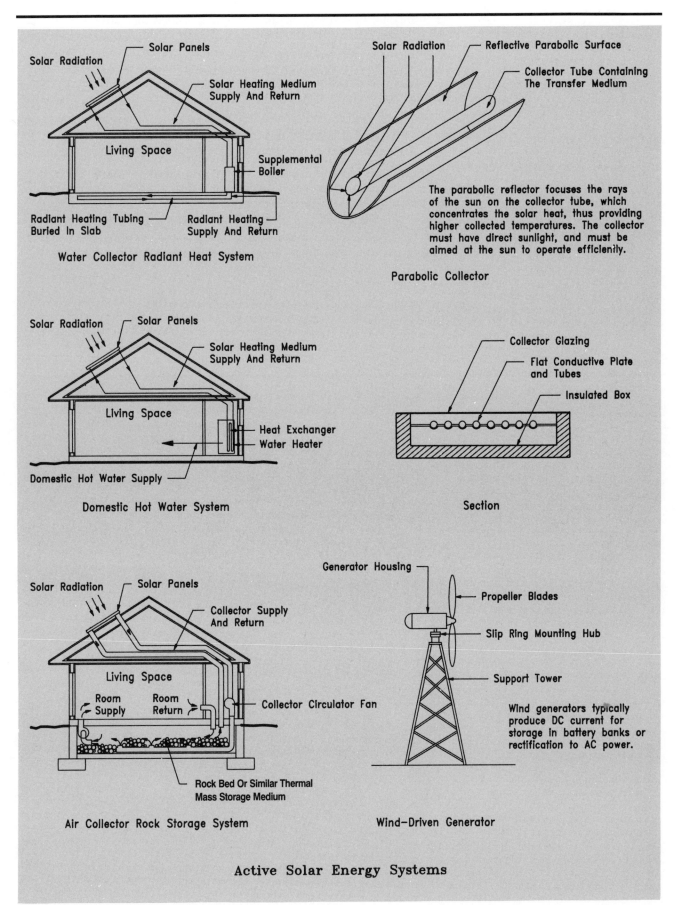

Figure 4.5

Ice storage for buildings generates cool air in much the same way as an icebox. There are two basic types of ice storage systems: encapsulated and harvester. Encapsulated systems make the ice on a daily basis during off-peak hours in a closed-loop system. Harvester systems make large quantities of ice in winter, in a manner similar to snow-making machines, in open-loop systems. They spray a mist that is frozen by the air and placed in pits for use in the summer.

Ice can be stored efficiently, and the cost of the added tanks, piping, chiller capacity, and controls is often easily offset by the savings in electric rates.

The heat transfer medium for ice storage and thermal storage systems varies. In harvester systems, water is commonly used because the system utilizes melted ice and the temperature of the liquids is above freezing. In encapsulated systems, a mixture that forms a slush or eutectic is often "frozen". The eutectic is usually a mixture of water and glycol or a mixture of inorganic salts (mainly sodium sulfate) and water with nucleating

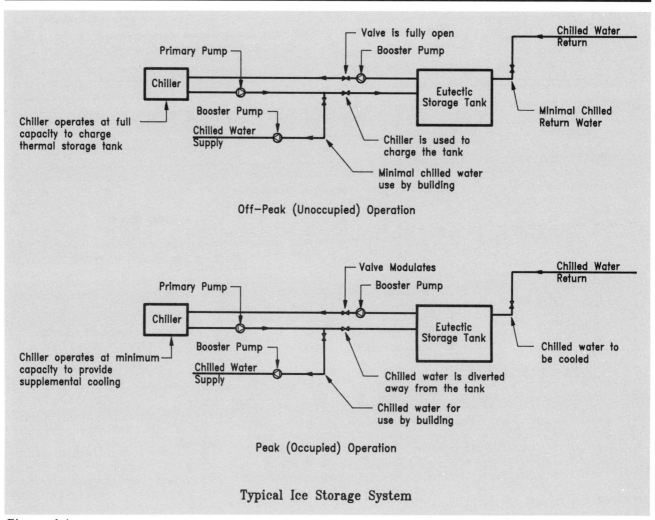

Figure 4.6

and stabilizing agents. Eutectics are engineered to melt and freeze at approximately 44–48°F [6.5–8.7 °C], which means that a conventional chiller set to operate at 40°F [4°C] can be used to charge the system and make the "ice".

Some cooling storage systems utilize water instead of ice. In this way, existing chillers do not need to be modified as drastically as they do for making ice. However, the latent heat of fusion of ice is 144 Btu/lb. [33 kJ/kg]; because of this, ice storage systems are theoretically 144 times smaller than cold water storage systems.

Brine is also used in thermal storage systems, but since it does not freeze, it is technically not an ice storage system. Larger storage tanks are necessary in brine systems because they do not have the advantage of latent heat released during a phase change, but the piping is free of freezing problems. Otherwise the principles of thermal storage and system layout are similar.

In designing ice storage systems, the following factors should be considered:
- The utility rate structure now and in the future for on- and off-peak hours
- The building loads as profiled during the year
- The type of ice storage system and fluids to be used
- The size of storage tanks, building diversity, differential electric rates, and cost of equipment
- The system layout; in particular, will the ice storage or chiller be primary?

Suppose a building is to be used for an ice storage system. It has a design load of 1,500 tons [5276 kW] and operates 10 hours a day with an average diversity of 75%. A chiller is to be sized for ice storage such that it will make ice at night while running at 70% capacity and run at 100% capacity during the day to assist the ice system.

During the day the equipment is required to produce:

1,500 tons [5276 kW] × 10 hours × .75 (diversity) = 11,250 ton hours [3957 W]

The chiller capacity to meet this load =

$$\frac{11,250 \text{ ton hours [3957 W]}}{[(24-10) \text{ hr.} \times .70 + (10 \text{ hr.} \times 1.00)]} = 568 \text{ tons (say 600 tons) [1998 kW]}$$

Using a 600-ton chiller with an ice storage system could compare favorably to a 1,500-ton chiller for a particular installation. Ice storage has demonstrated paybacks of less than five years on large installations.

Existing installations can be converted to ice storage; however, those that utilize chilled water are more easily and cheaply retrofitted to accept chilled water storage systems than ice storage systems. Some systems using a brine mixture may provide pumping and heat exchanger complications, but the fact that brine temperatures are colder than chilled water may offset these problems.

Positive displacement centrifugal chillers are more easily converted for ice storage. A centrifugal chiller can be equipped for ice storage by changing the compressor wheel to provide the higher pressure ratio required; this may also involve providing a new motor to accommodate the increase in compressor power. In some cases it is more economical to replace the entire cooling plant with an ice storage plant than to convert an existing system.

Modular and Dual Fuel Boilers

Heating loads vary considerably over the heating season. For most climates the heating load is less than 50% of the maximum design load more than two-thirds of the time, so most boilers are grossly oversized for most of the heating season. Boilers run most efficiently when fully operative; their efficiency ratings drop significantly when not under full load.

Modular boilers are smaller package units connected in parallel across the supply and return mains. Water that reaches the design set temperature passes through successive boilers that remain off. As the temperature of the supply water decreases, the next boiler in the series will come on line to maintain temperature. The boilers are piped to permit bypassing in case one of them is down for repairs. An example of a modular boiler system is shown in Figure 4.7.

In a bank of modular boilers, the lead module operates when heat is called for. The second module does not operate until the lead boiler has reached full load, the third module operates when the second module has reached full load, and so on. The modules shut off in reverse order of firing – that is, the last one on is the first one off.

As mentioned above, boilers run less efficiently on part load than on full load. For example, compare an installation of a 1,500 MBH [440 kW] gas-fired three-stage boiler with a 5-module bank of 300 MBH [88 kW] modular boilers. At a 60% load of 900 MBH [264 kW], the large boiler is firing well below its peak load and therefore well below its potential efficiency – say at about 85% efficiency. In the bank of modular boilers, modules 1, 2, and 3 are operating at full load and at an efficiency in the high 90's. What this means is that having more than one boiler will allow for multiple boilers to be brought on line as the load requires. Multiple boilers allow for overall greater efficiency and provide some additional reserve in the event of a boiler failure.

For larger systems, using two boilers sized at one-third and two-thirds of the design capacity can achieve better overall efficiency. Larger boilers can also be staged at different firing rates; a high and low setting is common in larger boilers. During most of the year the smaller boiler will run. When it cannot make the load, the first stage of the larger boiler will activate, followed by both boilers on full load for those few days of the year when the outside temperature is critical.

It is also common to equip boilers with the capacity to burn more than one fuel when such fuels are available to the building. Dual fuel boilers can burn gas or oil, thus offering the consumer an opportunity to choose the cheaper fuel at any given time. Dual fuel boilers provide a backup fuel should anything happen to the primary source, such as utility company strikes, leaks in the oil tank, and long periods between the ordering and delivery of oil. Dual fuel boilers are an asset in hospitals, homes, and health care facilities.

Dual fuel boilers do involve a higher initial installation cost. In some cases, a gas company may demand a contract that obliges the consumer to use a minimum amount of gas in a given year. However, having a backup fuel source and the option to use either source at any time is a significant advantage.

Controls

Controls are the single most effective way to cut the cost of operating a system. A few energy-effective control systems are described below.

- **Time clocks** turn equipment off when not needed.

- **Outside reset** adjusts the boiler temperature to respond to changes in outside temperature.
- **Zone systems** allow one zone only to respond.
- **Economizers** bring in cool outdoor air in spring and fall for cooling and turn off refrigeration equipment.
- **Energy management control systems (EMCS)** integrate all building

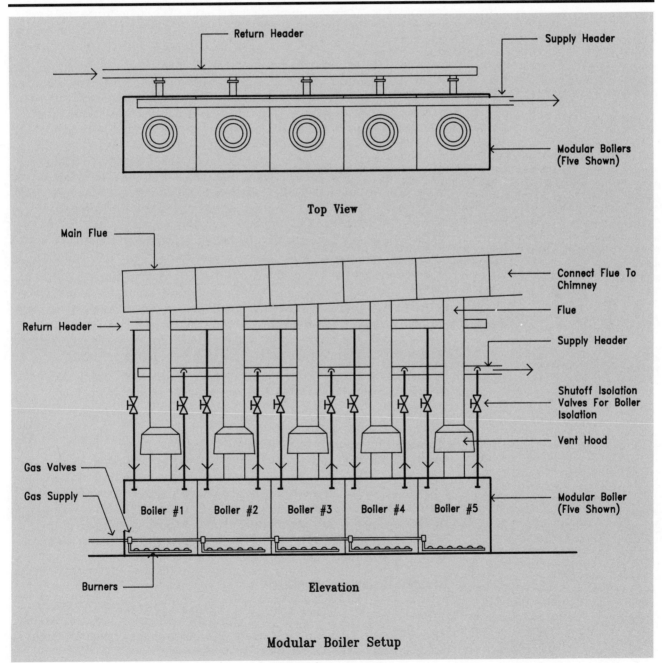

Figure 4.7

energy uses – HVAC, elevators, lighting – and operate them for minimum energy use. They also control equipment to minimize electric utility demand charges and cycle equipment automatically.

Chapter 12 covers controls and control systems in more detail.

Hospital and College Facilities

Energy consumption for large campus facilities ranges from 80 to 250 MBH per s.f. [250 – 800 kW/m^2]. This wide range is attributed to the efforts expended on building maintenance and upkeep. Facilities that have reduced consumption to low levels have generally upgraded boiler and chiller plants and modernized control systems. They also have an operation and maintenance program that regularly maintains equipment, including maintaining steam traps, cleaning boilers, replacing filters, adjusting belts, and keeping control systems such as dampers, economizers, and control settings in calibration and good repair. With current technology, consumptions of under 100 MBH per square foot [3786 kW/m^2] are readily achievable.

Large facilities are often operated by a team of architects, engineers, and maintenance staff. The cost of new equipment is balanced against the costs of maintenance and repair. The single greatest maintenance cost to facilities after housekeeping is repair of HVAC equipment. In buildings equipped with an effective central control system, much staff time is saved by being able to diagnose problems or adjust a system from a central control point. Central control systems also allow for space control in facilities where user patterns vary. In college campuses, for example, classrooms, lecture halls, and laboratories are not in use all the time, and other spaces are activated for programmed activities such as special lectures and concerts. A central control system allows these spaces to be set back when not in use and activated when necessary.

Most large campus facilities operate from a central plant that generates medium pressure steam, which is distributed to the buildings. Steam is an efficient transfer of energy, but maintenance of steam equipment such as traps is time consuming; energy losses on a poorly maintained building are considerable. Consequently, steam distribution can be limited once it enters a building to the mechanical equipment room, with a possible run of main air handlers on the roof or in equipment rooms at each floor.

Steam radiation throughout a building is expensive to maintain. Many facilities convert the steam to hot water once it enters the building.

A classic problem is the need to heat and cool different portions of a building simultaneously. In the spring and fall, the east side of a hospital or college wing requires some cooling, while the west exposure may need heating to continue from the previous cool night. Simultaneous heating and cooling capability can be provided by four-pipe fan coils, heat pumps, dual duct systems, and reheat systems. Problems in effective building control result when only single systems, such as two-pipe fan coil systems, are provided, since they permit only heating or cooling throughout their loop. It is not feasible to switch between heating and cooling in the same loop on a daily basis.

Energy Recovery Systems

In addition to the measures already discussed, there are many ways to conserve energy in buildings. Most involve a modest installation cost, but once implemented will provide energy savings that are a benefit to both the owner and to the community at large.

Steam Reclamation

When steam is condensed, it is at 220°F [103 °C] and still able to provide additional useable energy. At minimum, it should be returned to the boiler rather than dumped. When steam is returned to the utility company, additional energy can be extracted by utilizing the 220°F [103 °C] condensate to preheat incoming domestic water.

Since steam charges are often metered as they exit the building, condensate can also be used when permitted to provide makeup water to cooling towers, which reduces both the steam and water consumption.

Backfeeding Steam

In many large facilities or campuses, or in district heating applications, medium-pressure steam is distributed to individual buildings, where it is reduced to low pressure through a pressure-reducing station. The pressure-reducing station often produces significant volumes of medium-pressure condensate. This medium-pressure condensate very often is not returned directly to the central plant but is first fed to a flash tank, where it is reduced to low-pressure condensate. The low-pressure condensate is then pumped back to the central plant or sometimes even dumped into the sewer.

Condensate is produced when steam is applied in a heating application and the temperature of the steam falls below the boiling point of water at a given pressure. In the flash tank, as much as 15% of the medium-pressure condensate will flash to produce low-pressure steam; this low-pressure steam is very often vented to the outside atmosphere as a by-product. The low-pressure steam produced in the flash tank need not be vented; it could be backfed into the local low-pressure steam system in the building, thus saving considerable amounts of energy. In addition, if the low-pressure condensate is dumped into the sewer for one reason or another, this condensate (at temperatures over 200°F [92 °C] could be used to preheat domestic hot water or boiler makeup water. At the very least, the condensate should be returned to the boiler plant to avoid wasting water being used as makeup water.

Preheating with Condensate

Low-pressure condensate can be used to preheat domestic hot water and/or boiler makeup water. In residential buildings or dormitories, where a large volume of hot water is required, considerable energy savings are possible. Preheating the boiler makeup water is recommended not only to save energy but also to save wear on the boiler, as cold water being introduced to the boiler will "shock" the boiler and can cause some cracking. In applications that use heavy oils (such as #6 oils), which can be sluggish in cold weather, the condensate could be used to warm the oil so it flows better and avoids clogs.

Water Conservation

HVAC equipment often utilizes makeup water. Boilers, cooling towers, and water loops all have uses for potable water. Drinking fountains, steam condensate, and cooling coil condensate are all possible sources of makeup water. Using these sources reduces fresh water usage. Low-flush toilets, water-flow restrictors, and leak detection and correction also reduce fresh water usage.

Heat Recovery

Air systems are a source of potential heat recovery. Exhaust air from toilets, kitchens, and general exhaust contains heat (or cool air in summer) that can be reclaimed without contaminating incoming fresh air.

Energy transfer equipment such as rotary wheels, fixed plate exchangers, multiple tower exchangers, dual ducts, finned coil exchangers, and runaround coil exchangers or heat recovery loops are all available to scavenge energy from exhaust systems for reuse. Figure 4.8 illustrates some common heat recovery systems.

Thermal Heat Wheels

Thermal heat wheels are energy recovery devices that transfer energy from an exhaust airstream to a supply or makeup airstream. The supply air and exhaust air ducts must be side by side or on top of each other. The wheel is installed in both airstreams. The wheel has an energy transfer surface which, when slowly rotated, absorbs both sensible and latent energy from the warm airstream and transfers this total energy to the cooler airstream in the second half of its rotation. The rotation of a typical wheel is about 25 RPM. Manufacturers claim 75% to 85% effectiveness for their thermal wheels.

Thermal wheels work well in rooms with a large constant volume of exhaust and makeup air, such as function rooms. Thermal wheels can be used in both summer and winter for heating and cooling applications. Care must be taken in their application, as contaminants from the exhaust airstream may also get transferred to the supply airstream.

Heat Pipes

Heat pipes serve the same purpose as thermal wheels in that they transfer energy between two ducts that are side by side; however, heat pipes transfer only sensible energy. Heat pipes are installed halfway in both supply and exhaust ducts and contain an energy absorbing fluid with a low boiling point. As the fluid absorbs energy in the warm airstream, it boils and tends to move along the outer wall of the two-walled pipe to the cooler airstream, where the energy is absorbed by the cooler airstream and the fluid condenses and returns to the inner wall of the pipe. The cycle is purely thermodynamic and no moving parts are incorporated.

Heat pipes can be used in rooms with a constant volume of exhaust and makeup air, such as kitchens. Because there is no direct surface area contact between the transfer medium and the airstreams, there is no chance of contamination between the exhaust and supply airstreams.

Runaround Coils

Runaround coils are a method of transferring energy between two airstreams that are not side by side. A coil in each airstream contains the heat transfer medium, which is pumped from one coil to the other. The heat transfer fluid absorbs energy in the warm airstream and is pumped to the cold airstream, where the energy is absorbed by the cooler air. The pump is controlled by thermostats in both airstreams. The length of piping run between the coils should be limited to about 50 feet [15 m] one way.

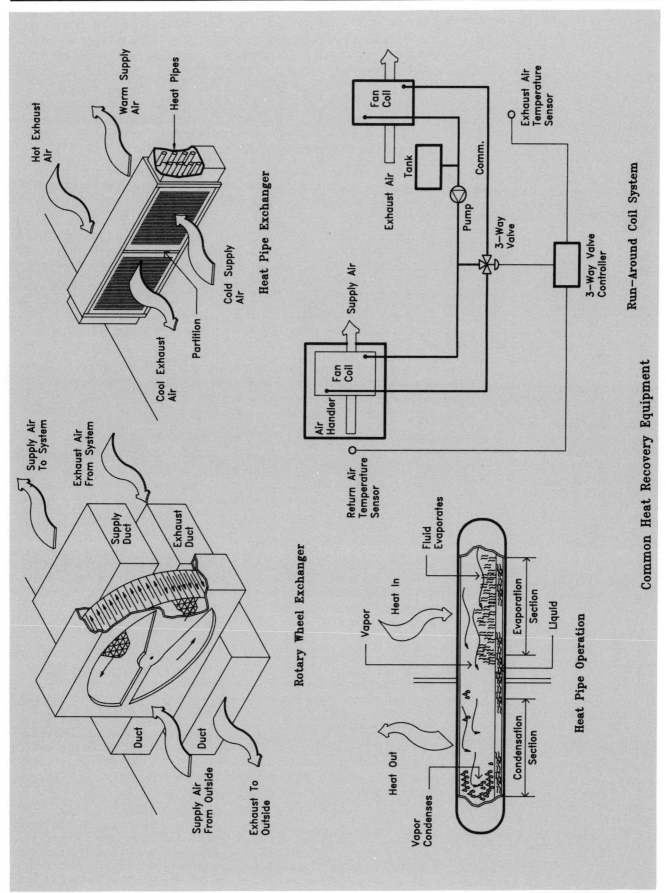

Figure 4.8

Economizer Sections on AHU's

In a typical air handling unit (AHU) application, a minimum volume of outside fresh air is introduced, normally 10–15% depending on the occupancy. An equal or slightly lower volume of air is exhausted to compensate and the remaining recirculated air is mixed with the fresh air, treated at the AHU, and distributed. On a design cooling day the outside air temperature (say 90°F [32 °C]) will be well above the temperature of indoor or return air (say 78°F [25 °C]); in this case more energy is required to cool the outside than to cool the same volume of return air down to the supply air temperature (say 60°F [15 °C]).

Since fresh air must be introduced for ventilation and to improve the indoor air quality, it can also be used for cooling on cool days. For example, assume that the building requires cooling on a non-design cooling day when the outside temperature is at 70°F [21 °C]. In this case less energy is required to cool the outside than to cool the same volume of return air down to the supply air temperature. (Be careful to check the humidity to ensure that the total enthalpy of the air is less than the exhaust air.) In this case it is more economical and energy efficient if all the air returning from the building is exhausted and 100% fresh air is supplied to the space. An economizer section on an AHU provides the necessary sensors and automatic dampers to provide this type of control of the outside and return air.

For an economizer section to operate correctly, an additional fan is required to either exhaust the air or provide the fresh air, depending on the configuration of the supply air fan. If the supply air fan is a draw-through type, then an exhaust or return fan will be required.

In more humid regions of the country, enthalpy sensors, rather than temperature sensors, are used in the economizer section. This is because cooler, more humid outside air may have more enthalpy (energy) than warmer, dryer return air. In this case 100% outside air must not be introduced until the enthalpy of the outside air is less than the enthalpy of the return air.

Refrigeration Condensate

In addition to heat recovery from exhaust air, refrigeration equipment produces hot gases between the compressor and the condenser. The heat that the condenser would otherwise release to the atmosphere can be reclaimed from refrigeration systems for preheating incoming air or domestic water.

Gas-fired Equipment

Gas-fired equipment is currently available for heat pumps and chillers. These machines are designed to produce cooling utilizing gas in lieu of an electrically driven compressor, very much like the original gas refrigerators.

Expert Systems

HVAC equipment is currently available with self-diagnostic modules, similar to those in automobiles. They aid in rapid analysis and correction of deficiencies and provide energy savings. Chapter 12 discusses in more detail the full range of control benefits.

Energy Audits

Since buildings consume large quantities of energy, building owners or managers are constantly trying to improve overall efficiency and performance. An *energy audit* is a comprehensive energy conservation study of a building or facility. For an existing facility, an energy audit is an exceptionally successful money-saving tool.

The audit report describes the building's existing characteristics, energy systems, operating procedures, maintenance routines, and energy consumption patterns. It examines the building envelope, weather tightness, and equipment efficiencies and conditions. Based on existing building conditions, the report outlines all appropriate energy conservation opportunities, estimates their costs, and calculates their energy-saving potential and estimated payback period. The report then highlights and recommends for implementation those energy-saving measures that are most effective in terms of a comprehensive payback. Recommendations with low cost or a payback of less than one year are called *operation and maintenance measures*, or O&Ms. Recommendations with a longer payback are termed *energy conservation measures*, or ECMs.

Energy audits inform building owners or managers where the building's energy is being used and what measures they can take to increase the energy efficiency of the building. The report is also very often used as part of an application for a state or federal grant to finance some or all of the recommended measures. These grants are commonly available to schools, hospitals, and institutions.

The energy audit report is normally completed by a registered architect or professional engineer who is experienced with the required procedures of an energy audit and who is registered in the state in which the premises being audited are located.

An Outline of the Energy Audit Report

A typical energy audit analysis and report consists of the following sections.

1. Executive summary
2. Site investigation (all systems, building condition, and use patterns)
3. HVAC investigation (building systems and heating and cooling load calculations)
4. Energy consumption data (utility bills)
5. Current energy prices
6. Annual energy consumption
7. O&Ms and ECMs
8. Implementation of O&Ms and ECMs
9. Combustion efficiency report
10. Fuel consumption justification
11. Scoring of report

Each of these elements is described in the following paragraphs.

1. Executive Summary

The executive summary generally reviews the findings, summarizes the recommendations, and indicates the expected savings and costs to implement the recommendations. It serves as an overall introduction and building description; it outlines the intent of the energy audit, and the authority being applied to if the audit is a part of a grant application. The executive summary also lists the energy-saving measures to be implemented.

The report should begin with a brief, nontechnical building description, followed by an outline of the building type, use, location, and structure. Finally, the mechanical and electrical systems should be described, from generation equipment to terminal equipment and controls.

2. Site Investigation

The site investigation is a thorough analysis of the building and its energy systems. Information to be determined includes:
- Type of HVAC systems in the building
- HVAC and energy operation and control systems
- Types of all fuel consumed, including for heating and for hot water
- Occupancy patterns of the building
- Type of electrical service
- Number and types of lighting
- Age and efficiencies of the equipment

The building is examined to determine that all systems are insulated and tight. The construction and envelope of the building is also evaluated to determine materials, cracks, and types and condition of windows. It is also checked for the presence of insulation in the walls and on the roof. When possible, the building's original plans are obtained and reviewed. The building occupancy and use patterns are also evaluated. The building may have only partial use, which may mean the building's schedule can be reprogrammed for more energy-efficient utilization.

3. HVAC Investigation

The HVAC system is completely evaluated. From the fuel source and generation equipment to the driving system, distribution system, and terminal units and control system, every component is examined for condition, effectiveness, efficiency, and improvements.

Heating: Heating plants need special attention. Boilers are the most common piece of equipment, and the most likely to be in poor condition. In multi-story or large-zoned buildings, there might be two or more boilers per building or zone. Each boiler should be examined thoroughly.

Figure 4.9 shows a checklist for heating plants. In some cases information may not be readily available; the building custodian or engineer is often helpful.

The boiler should be visually inspected for leak stains and rust buildup; the floor around the boiler should be checked for water stains; and the boiler blow-off and drain piping should be checked for drips and leakage. If the boiler is a steam boiler it will probably have a condensate receiver with pumps, which should be tested. All gauges and thermostats should be checked to ensure correct operation.

Boiler investigation is more difficult if the system is not in operation; it should be carried out during a typical heating day if possible.

The distribution system piping should be checked for leaks at joints, and valves should be randomly checked for tightness and ease of operation. If the system is a steam system the traps should be randomly tested to see if they leak steam into the condensate lines. Finally, all piping should be inspected for age, rust, presence of insulation, and general condition.

The terminal units should be visually inspected, counted, and randomly tested for operation. If units such as floor-mounted fan coils with a fresh air damper are being used, the damper and sealing around the wall louver should be checked for tightness. Radiators should be inspected for leaks and large buildup of dirt. All valves on terminal units should be tested for operation and wear.

Ventilation and Air Conditioning: All air handling units should be inspected for condition of the fan and coils, insulation tightness, operation of automatic outside air dampers, and condition of the motor and belts. The unit should be turned on and off to see if the belts screech, the bearings rub harshly, or the motor gets unusually warm. The unit casing and all flexible connections should be inspected for holes, seals, and insulation. Filters should be checked for clogging and to determine how often they are replaced.

The distribution ductwork should be traced and inspected for tight seals and condition. In general, supply ductwork should be insulated. Automatic dampers should be inspected and tested randomly for ease of operation. Finally, all supply and return/exhaust diffusers, registers, and grilles should be checked to ensure that they operate as they should. A piece of light paper will tell which way the air is flowing and will give an experienced surveyor an indication of the strength of airflow. If VAV (variable air volume) boxes with reheat are used, they should be checked to ensure that the reheat coil does not operate as a result of overcooling of the space.

HVAC Controls: The controls of the HVAC system need to be closely analyzed for the following:

- Type of control system installed
- Condition of control components (i.e., compressor thermostats, tubing, relays, wiring, and valves)
- How the control system is operated and who operates it
- What the control system time and temperature settings are for heating and cooling

\multicolumn{4}{c}{**Heating Plant Checklist**}			
Type of Boiler or Furnace	#1	#2	#3
Age in Years			
Burner Type			
Burner Age			
Fuel Type			
Capacity			
Steam Pressure or Hot Water Temperature			
Electric Ignition?	___ yes ___ no	___ yes ___ no	___ yes ___ no
Automatic Flue Damper?	___ yes ___ no	___ yes ___ no	___ yes ___ no
Variable Firing Rate?	___ yes ___ no	___ yes ___ no	___ yes ___ no
Testing Hole in Flue?	___ yes ___ no	___ yes ___ no	___ yes ___ no
Describe Controls			

Figure 4.9

This information is then used to determine whether the existing controls are being used effectively and what recommendations will save energy.

DHW Investigation: The DHW (domestic hot water) system should be inspected for the following:
- Size and capacity of DHW heater and tank
- Fuel/source of heat for heating water (Is the water heated by the building heating system?)
- Age and condition of the tank
- Alternative fuels locally available
- Is there dual temperature water control?
- Water temperature setting

This information is used to determine the fitness of fuel type and temperature settings, the condition of the existing heater, and the capacity of the tank and heater for peak conditions in the building.

Electrical System Investigation: The investigation of the electrical system is principally concerned with analyzing the lighting system and determining the number and characteristics of the electrical services.

The size (voltage, phases, and amperage) and capacity of the transformers and switchboard should be noted. Use of circuit breakers or fuses should also be determined.

When analyzing the lighting system, each and every light fixture should be counted and noted as to its type, wattage, and location. Building personnel should be consulted about the hours of operation of the lights and the availability of any night lights or lights with emergency power. A sample log of lighting fixtures is shown in Figure 4.10.

Building Envelope Investigation: In addition to examining the relevant mechanical and electrical systems, the building envelope must be studied in preparation for the heating and cooling load analysis. The envelope should be examined for the following:

\multicolumn{7}{c}{Lighting Fixture Survey Form}
Area	Lamp Fixture Type	Watts/Fixture	Number of Fixtures	Total Watts	Hours/Week	kWh/Week

Figure 4.10

- The types, sizes, and condition of all windows, including their orientation, number, and any weatherstripping and/or storm windows (see Figure 4.11)
- The types, sizes, and condition of all doors, including their orientation, number, and any weatherstripping and/or storm doors (see Figure 4.11)
- The presence of insulation in walls
- The presence of insulation in the roof
- Any leaky stairwells or loose doors and windows

Contact with the Building Owner: During the site investigation, the building owner should be interviewed for all relevant information about the building that is not available during the site investigation. The occupancy pattern of the building should be confirmed with the building owner. See Figure 4.12 for a typical occupancy pattern chart.

In addition, copies of the building's original plans, and copies of all utility bills dating back at least three years, should be obtained. The plans will aid in understanding the building layout and are key for heating and cooling load calculations. The utility bills will be needed to estimate past energy consumption and for the fuel verification calculation.

Door and Window Survey Form							
Type	Description	Shading	Glazing	Size	Number	Weatherstrip	Remarks

Figure 4.11

Building Occupancy Pattern Chart								
Area or Space	Weekdays		Saturdays		Sundays		Holidays	
	No.	Hours	No.	Hours	No.	Hours	No.	Hours

No. indicates number of occupants in spaces.
Hours indicates period of time of occupancy.

Figure 4.12

Building Heating and Cooling Load Calculations: The building heating and cooling load calculations should be made using the information obtained from the site investigation and the floor plans. The calculation procedures outlined in Chapter 3 should be followed closely. Few assumptions will need to be made, since information such as condition of windows, number of occupants, time of occupancy, type of activities, lighting types and lighting schedule, and internal heat gains will have been obtained from the site investigation. This is in contrast with design work on a new facility, when this information is very often assumed.

The heating and cooling loads calculated for an energy audit are more accurate and customized than the calculations carried out for a design project, because there are fewer assumptions as to what will be built and how the building will be used. This energy consumption will be further justified in the fuel consumption justification computations section.

4. Energy Consumption Data

The utility bills obtained during the site investigation are gathered for the previous three years and tabulated in order of fuel type, month of consumption, cost of monthly consumption, and yearly consumption and cost. Figure 4.13 shows a typical table with fuel consumption data.

Annual Fuel Consumption Chart

Year 19__	Electric		Electric Demand		# ____ Oil		Natural Gas		Other	
Month	kWh	$	kWh	$	Gal.	$	CCF	$	Unit	$
Jan.										
Feb.										
March										
April										
May										
June										
July										
Aug.										
Sept.										
Oct.										
Nov.										
Dec.										
Total			N/A							
Average $/Unit										
Total MBTU			N/A							
MBTU/s.f.			N/A							

Average $/Unit = average price per unit.   Building _____   Owner _____
Total MBTU = total MBTU's consumed per year.
MBTU/s.f. = average MBTU's per square foot.   Building Gross Square Footage of Heated Space _____ s.f.

Figure 4.13

Tables of annual fuel consumption should be generated for a minimum of three years to give a good overall picture of the building's energy demands. Using a three-year span will also enable analysis of the energy consumption trend from year to year and point out unusually high or low consumption rates at particular times. These tables give a bottom line energy consumption total for a given year and will be used for generating base year consumption data and for comparison to consumptions after all recommended energy-saving measures are implemented.

For facilities with more than one building on one meter, fuel consumption must be divided according to technical analysis of each building's heating and cooling load calculations, occupancy pattern, and equipment. A less accurate method would be to divide the fuel consumption on a square foot basis. Dividing the energy consumption by the square footage ignores building occupancy and thermal conditions; this method will probably yield incorrect results and complications later in the energy audit calculations.

5. Current Energy Prices

Current energy prices are required for an energy audit because they will form the basis of the cost of operating the building and the savings possible with various energy-saving recommendations. They are obtained from the fee schedules of suppliers or local utility companies. Electrical costs per kWh, demand charges, and costs of therms of natural gas will usually be on the monthly electric bills; the latest bills can be used to calculate current prices. Because of the fluctuating price of oil, suppliers should be contacted for the latest price; however, in the case of a large consumer of oil there may be a yearly contract issued to the lowest bidder. The current contract price can then be used.

The average price per unit, marginal cost, and the date of the price should be tabulated (see Figure 4.14). The marginal cost of a fuel applies to fuels that are billed on a declining block rate. The marginal cost of electricity is calculated by taking the cost of the last unit purchased and adding any per-unit charges. For example, a building may have a declining block rate for its electrical bill: $0.12/kWh for the first 1000 kWh, $0.10/kWh for the next 1000 kWh, $0.08/kWh for the remaining usage, and an additional charge of $0.04/kWh for fuel costs on top of the block rate. Assume a monthly usage of 2500 kWh. The average fuel bill would be approximately $0.10/kWh, plus the $0.04 for fuel cost, or $0.14/kWh. However, the cost of the last block of 500 kWh is $0.08/kWh. and the marginal cost is $0.08 + $0.04 = $0.12/kWh. The marginal cost, not the average cost, must be used in the energy audit calculations to obtain meaningful payback analysis. Electrical per-unit and demand costs should also be separate line items.

6. Annual Energy Consumption

This section calculates the costs of previous annual energy usage and estimates the future energy consumption of the building for a typical year after the recommended energy-saving measures have been implemented. The backup data for this information, such as the specifics on operation and maintenance measures (O&Ms) and energy conservation measures (ECMs) are discussed in more detail in the next section.

The annual energy consumption lists the overall bottom line totals of energy consumed during the following times:
- The base year (previous year) (see Figure 4.13)
- Adjusted base year (see Figure 4.15)

- Adjusted base year with all O&M measures implemented (see Figure 4.16)
- Adjusted base year with all O&M and ECM measures implemented (see Figure 4.16)

In Figure 4.15, for **base year**, list actual energy consumption for a 12-month period selected as the base year. Utilize data from Annual Fuel Consumption Data (Figure 4.13).

For the **adjusted base year**, complete the chart if any of the following apply:
- Adjust for an O&M or ECM implemented in the base year.
- Adjust if there is a significant change in the future planning of the building, equipment, or use patterns.
- Adjust for an atypical heating or cooling season or situation during the base year.
- Adjust for a fuel delivery schedule that does not reflect annualized consumption (i.e., topping off tanks at the end of a season in one year for the next calendar year).

If an adjusted base year chart is applicable, provide an explanation of the adjustments.

Current Energy Price Form				
Energy Type	Units	Average Price/Unit	Marginal Price/Unit (if applicable)	Energy Price Date or Period
Electricity	kWh			
Electrical Demand	kW			
# _____ oil				
Natural Gas				
Other				

Figure 4.14

Annual Energy Consumption Chart					
Energy Type	Quantity	Units	MBTU's	MBTU/S.F.	Cost
Electricity		kWh			
Electric Demand	N/A	kW	N/A	N/A	
# _____ oil		gal.			
Natural Gas		ccf			
Other					
_____ Base Year			_____ Adjusted Base Year		

Figure 4.15

The information concerning annual energy consumption in Figures 4.15 and 4.16 is a summary of the totals of the energy and cost of the different fuels consumed for one year. A comparison of these tables establishes the effectiveness of the recommended energy-saving measures.

Explanation of O&M and ECM energy- and cost-saving procedures and calculations is discussed in the following section.

7. O&Ms and ECMs

All energy-saving recommendations are divided into either Operating & Maintenance routines (O&Ms) or Energy Conservation Measures (ECMs). For each O&M and ECM, cost estimates to implement, estimates of energy savings, and calculations of payback periods are made.

O&Ms: Operation and maintenance procedures (O&Ms) are measures that do not involve any initial capital cost, but do require good "housekeeping", staff cooperation, awareness, and dedication. O&M measures can be as simple as raising the thermostat setting for air conditioning or cleaning out the tubes in a heat exchanger. A typical O&M checklist is outlined in Figure 4.17.

O&Ms have the greatest rate of return because of the low implementation costs and potentially high energy savings of each measure.

ECMs: Energy-conservation procedures and measures (ECMs) involve a capital investment and have a simple payback of greater than one or two years (see discussion below for more details about payback). ECMs usually involve hiring an architect, engineer, and contractor and can be looked at as a capital improvement. Some examples of ECMs are replacing a boiler, insulating a roof, providing new controls for an HVAC system, or replacing a motor with a variable frequency drive motor.

Many authorities and grant-awarding boards regard ECMs with a payback of less than one or two years as being O&M procedures, because the owner will have a cash reserve at the end of the short payback. Therefore, most organizations do not merit grant monies for these types of ECMs.

See Figure 4.18 for a list of typical O&Ms and ECMs prioritized by payback.

Estimated Energy Consumption Chart					
Energy Type	**Quantity**	**Units**	**MBTU's**	**MBTU/S.F.**	**Cost**
Electricity		kWh			
Electric Demand	N/A	kW	N/A	N/A	
# ____ oil		gal.			
Natural Gas		ccf			
Other					
____ After Implementing O&Ms			____ After Implementing ECMs		

Figure 4.16

O&M Checklist

Indicate A, B, or C	Item	Date (Mo./Yr.) to be Done	O&M or ECM No.	Frequency (Date)
	Envelope			
	1. Install or replace caulking around window and door frames and other penetrations.			
	2. Repair or replace window weatherstripping.			
	3. Repair or replace door weatherstripping.			
	4. Repair or replace broken or cracked glazing.*			
	5. Repair doors and windows that do not fit or that do not operate properly.*			
	6. Install, repair, or replace attic access hatches, weatherstripping, and insulation.			
	7. Repair or fill in penetrations in top plates (in attic at wall perimeter).			
	8. _____			
	Heating, Ventilating, and Air Conditioning (HVAC)			
	1. Test and adjust burners, air/fuel ratios, and tune up.			
	2. Clean boiler tubes and heat transfer surfaces.			
	3. Repair/recalibrate HVAC controls.			
	4. Inspect boiler water for rust and other contaminants. Add correct water treatment if necessary (to improve boiler tube heat transfer and keep steam traps from clogging).*			
	5. Clean sediment screens in boilers and hydronic lines.*			
	6. Reduce boiler blowdown via correct water treatment.*			
	7. Lower indoor temperature during heating season to 65°F or other appropriate setting.			
	8. Raise indoor temperature during cooling season to 78°F or other appropriate setting.			
	9. During unoccupied hours, lower indoor temperature to 55°F or less and eliminate unnecessary use of the HVAC system (setback program).			
	10. Adjust thermostats for local radiators, fan coil units, and univents.			
	11. Raise temperature of chilled water supply.			
	12. Lower hot water supply temperature for heating coils and room radiation and check operation of reset controls.			
	13. Adjust temperature of discharge air from air handling units.			
	14. Repair existing locking thermostats and lock unguarded thermostats.*			
	15. Lower boiler steam pressure.			
	16. Reduce outside air intake to minimum required levels.			
	17. Shut down HVAC system in unoccupied areas.			
	18. Clean coils in air handling units.*			
	19. Clean A/C condenser tubes and coils.*			
	20. Replace air filters.*			
	21. Reduce fan speeds and adjust belt drives.			

* The item may be recommended *without* providing a separate description and analysis if it is apparent that such an analysis would be more expensive than the expected savings.

A = Not applicable to this building.   C = Applicable, not in place, recommended for implementation.
B = Applicable and in place.

Figure 4.17

O&M Checklist (continued)

Indicate A, B, or C	Item	Date (Mo./Yr.) to be Done	O&M or ECM No.	Frequency (Date)
	22. Check operation of steam traps, and retrofit or replace traps (with correct capacity units) as necessary.			
	23. Repair or replace distribution pipe and/or duct insulation.			
	24. Repair steam leaks.			
	25. Repair air leaks in supply ducts, especially in suspended ceilings and in passages through unconditioned spaces.*			
	26. Clean registers and ensure that air louvers are directing air properly.*			
	27. Ensure that dampers and actuators operate properly.			
	28. _____			
	29. _____			
Domestic Hot Water				
	1. Repair leaks in domestic hot water system.			
	2. Lower domestic hot water temperature.			
	3. Install low-flow showerheads and aerators.			
	4. Reduce the flow by reducing water pressure.			
	5. Repair or replace DHW pipe insulation.			
	6. _____			
Lighting				
	1. Remove unnecessary lamps.			
	2. Disconnect unnecessary ballasts.			
	3. Turn off lights when area is unoccupied.*			
	4. Turn off lights near windows when appropriate. Maximize daylight by opening drapes, shades, etc.*			
	5. Reduce or eliminate evening activities such as building cleaning.			
	6. Reduce/eliminate unnecessary exterior lighting.			
	7. Initiate a program to clean fixtures on a regular maintenance basis.*			
	8. Use lower wattage or higher efficiency lamps.			
	9. Use one high-wattage lamp rather than several smaller ones.			
	10. Replace burned out ballasts with more efficient ballasts and matching lamps.			
	11. Replace high usage incandescent bulbs with screw-in compact fluorescent.			
	12. _____			

* The item may be recommended *without* providing a separate description and analysis if it is apparent that such an analysis would be more expensive than the expected savings.

A = Not applicable to this building.
B = Applicable and in place.
C = Applicable, not in place, recommended for implementation.

Figure 4.17 (continued)

8. Implementation of O&Ms and ECMs

The objective of an energy audit is to obtain a useful list of energy-saving recommendations. Any O&M or ECM with a payback of less than five years should be done. The simple payback is the cost of implementing a measure divided by the annual cost it saves. For example:

$$\text{Payback in years} = \frac{\text{cost to implement O\&M or ECM in dollars}}{\text{annual energy savings in dollars}}$$

The cost should include the price of labor and materials to implement the measure, any design fees, related costs such as disposal or containment of asbestos (if a boiler is being removed) or PCBs (if a transformer or older lighting ballast is being removed), plus any administrative costs for phasing, renting other space during construction, or related project costs.

The annual energy savings is the amount of money saved each year by the implementation of each O&M or ECM. These savings must be computed as an *interactive* payback, not as a simple payback. For example, suppose that for a particular building you recommend that the roof have more insulation, which costs $5,000, and that the boiler be changed to a high-efficiency boiler, which costs $7,500. The roof insulation would decrease building heat loss and by itself would result in an annual savings of $1,000 and a simple payback of ($5,000/$1,000) 5 years. The more efficient boiler would also decrease energy consumed per year and would save $1,200 per year for a simple payback of ($7,500/$1,200) 6.25 years, if it were installed by itself without the roof insulation.

Clearly, if they were implemented in the order of their simple payback – roof first, with a 5-year simple payback, and boiler second, with a 6.25-year simple payback – the real boiler payback would be more than its simple

---

O&M Checklist (continued)

Indicate A, B, or C	Item	Date (Mo./Yr.) to be Done	O&M or ECM No.	Frequency (Date)
	Other			
	1. Raise chilled drinking water temperatures.*			
	2. Put all appropriate equipment on timers.			
	3. Check and adjust all time clocks, especially at changes in daylight savings time.			
	4. Defrost refrigerators and freezers regularly and clean condenser coils.*			
	5. Replace refrigerator and freezer door gasket seals if necessary.			
	6. _____			
	7. _____			

* The item may be recommended *without* providing a separate description and analysis if it is apparent that such an analysis would be more expensive than the expected savings.
A = Not applicable to this building.
B = Applicable and in place.
C = Applicable, not in place, recommended for implementation.

Figure 4.17 (continued)

Typical O&M and ECM Paybacks

O&M No.	Description	Cost ($)	Fuel Type Saved	Annual Amount Saved	MBTU Saved/Year	MBTU/S.F. Saved/Year	1st Year Energy Cost Savings	Payback (Years)	Useful Life (Years)	Date
1.	Reduce occupied space temperature.							Immediate		
2.	Reduce unoccupied space temperature.							Immediate		
3.	Reduce DHW temperature.							Immediate		
4.	Install flow restrictions on faucets.							0.1 – 0.2		
5.	Replace incandescent bulbs with high efficiency fluorescent tubes.							0.5 – 1.5		
6.	Replace broken panels of glass.							1 – 2		
7.	Replace fluorescent tubes.							1 – 3		
8.	Insulate attic.							1 – 2		
9.	Insulate roof.							8 – 10		
10.	Replace windows.							15+		
ECM No.										
1.	Replace steam traps.							1 – 3		
2.	Replace boiler, burner/warm air furnace.							2.5 – 5		
3.	Insulate DHW heater.							3 – 3.5		
4.	Replace DHW heater.							4		
5.	Replace incandescent fixtures with high efficiency fluorescent lamps.							3 – 5		
6.	Weatherstrip doors/windows.							1 – 4		
7.	New temperature zoning valves, and controls.							2 – 6		
8.	Upgrade thermostats.							3 – 7		
9.	New storm windows.							7 – 8		

Building _____ Owner _____

Figure 4.18

payback. After the roof is installed, the building will burn less fuel, so the savings produced by the roof insulation would interact with the savings from the boiler. Actual calculations would typically show a 15–25% increase in the boiler payback period for a real interactive boiler payback of 8 years.

Thus in determining the recommended O&Ms and ECMs, the implications of all recommended energy-saving measures must be analyzed, including cost estimates, calculation of fuel savings, prioritizing measures, and calculating interactive paybacks. When the site investigation is complete and all of the checklists have been completed, the appropriate O&Ms and ECMs are usually apparent. Before an O&M or ECM can be recommended, the following procedures should be followed (see Figure 4.19):

- Determine the type of measure and the fuel consumption that will be affected; assign a title to the measure.
- Estimate the implementation cost of the measure.
- Estimate the energy savings and the cost savings of the measure.
- Calculate the simple payback period of the measure.
- Estimate the useful life of the project.
- Estimate the change in operating and maintenance costs.
- Estimate the salvage value or disposal cost of the implementation.
- Determine whether the implementation has any environmental effects (release of CFCs, for example); notify the client of the correct requirements and regulations; determine who should implement the measure.
- Recalculate all O&Ms and ECMs in the order of the simple payback to determine their interactive payback.

O&M Example: As part of an energy audit, the following was observed during the site investigation: the heating thermostat was faulty (the setpoint was 68°F [20 °C] but the actual temperature in the space was 70°F [21 °C]); the heating load at 70°F [21 °C] was 550,000 Btu/hr. [161 kW], based on an outside design temperature of 10°F [−12 °C]; and the annual consumption of fuel oil was 32,000 gallons [121 000 L]. It was recommended that the thermostat be replaced and the setpoint be maintained at 68°F [20 °C]. Assume the cost of oil is $.75 per gallon [$.20 per liter].

Temperature differential at 70°F [12 °C]	= (70 − 10) [21 − (−12)] 60°F [33 °C]
Temperature differential at 68°F [20 °C]	= (68 − 10) [20 − (−12)] 58°F [30 °C]
% reduction in heat load and fuel consumption	= $\frac{(60-58)}{60} \times 100$ = 3.33%
Reduction in fuel consumed	= (32,000 [121 000 L] × 0.0333) = 1,070 gals [4033 L]
Cost of fuel saved	= (1,070 × .75) [4033 × .20] = $802
Cost of new thermostat	= $100
Simple payback	= 0.125 yrs

ECM Example: During the same site investigation and subsequent analysis, it was discovered that the 25-year-old gas-fired boiler serving the building was sized at 200% of the estimated building peak heating load and consequently was operating very inefficiently – at about 25% load or less for most of the heating season. The historical annual fuel consumption was estimated to be 32,000 gallons [121 000 L] per year at 75 cents per gallon [20 cents per liter]. On average, five repair visits per year were required as a result of the age of the boiler; each visit cost $300.

O&M/ECM Detail Sheet

Building _____   O&M No. _____   ECM No. _____

Owner _____

1. Name of measure: _____

2. Type of measure: Envelope   Lighting   Mechanical
                     DHW   Renewable Other

3. Estimated total project implementation cost for this measure including design services, acquisition, installation, and other costs. (Attach detail breakdown.) $ _____

4. Estimated energy/cost savings. (Attach detail breakdown.)

Fuel Type	Annual Fuel Savings	Units	Annual Energy Savings (MBTUs)	First-Year Energy Cost Savings
Totals				

5. Simple payback period:
   Total project cost: $ _____ = _____ simple payback/(years).
   First year energy savings: $ _____

6. Estimated useful life of this measure. (Must be more than payback period.)   _____ years

7. Estimated annual increase/decrease (+/−) in operating or maintenance costs resulting from this measure. (Attach explanation.) $ _____

8. Estimated salvage value (+) or disposal cost (−) of the measure, if any, at the end of its useful life. (Current not discounted costs.) $ _____

9. Would the implementation of this measure require the preparation of an environmental notification form? Include costs in project cost above.   ___ Yes ___ No

Figure 4.19

It was determined that a new boiler, sized at 115% of the building peak load with a high and low firing electronic spark burner, would consume only 22,000 gallons [83 270 L] of oil per year (the calculation was carried out using the fuel consumption equation shown later). Only one maintenance check per year would be necessary, at a cost of $200 per visit. The demolition of the existing boiler would cost $2,000 and the installation of the new boiler and associated work would cost $25,000.

Total project cost = $27,000

Fuel saved = 10,000 gallons [37 850 L]

Fuel cost savings = 10,000 × 0.75 [37 850 × .20] = $7,500

Maintenance savings = $1,500 − $200 = $1,300

Total savings = $8,800

Simple payback = 27,000/8,800 = 3.07 yrs

A payback of 3.07 years for replacing a boiler is very attractive. However, schools and institutions often have difficulty obtaining the initial $27,000 and consequently submit the energy audit report as part of an application for a grant or subsidy.

Simple Payback versus Interactive Payback: The idea of a simple payback gives us a first indication of the effectiveness of the proposed energy-saving measure. However, in a typical analysis there will be many different proposed measures. Each measure will have an interactive effect on the others.

Regarding the two previous examples, replacing the thermostat reduces the heat load and fuel consumption by 3.33%. This implementation has priority over replacing the boiler because of its low cost and short payback. After implementing the O&M measure the new fuel consumption is 32,000 − 1,070 = 30,930 gallons [121 000 − 4033 = 116 967 L]. The estimated consumption of the new boilers is 22,000 gallons [83 270 L] and the true fuel savings is 30,930 − 22,000 = 8,970 gallons [116 967 − 83 270 = 33 697 L] – not 10,000 gallons [37 850 L] as calculated above. The corresponding savings calculate to $7,545 instead of $8,800, thus increasing the payback number from 3.07 years to 3.57 years. This increase in the payback is very slight because only one O&M and no ECMs were implemented; however, it is not uncommon for a simple payback to increase a full year or more as a result of preceding O&M and ECM measures.

Figure 4.18 gives examples of typical O&Ms and ECMs that are often recommended in energy audit reports and give their expected payback.

9. Combustion Efficiency Report

To determine the effectiveness and efficiency of the boiler in a building, a combustion efficiency test must be conducted. The efficiency of the existing boiler is needed for two reasons: to get a general indication of how well the boiler is running and what condition it is in, and to obtain the fuel consumption verification calculation (see below).

The combustion efficiency test could be carried out during the site investigation, but this is a specialized and time-consuming procedure. It should be carried out at a different time, by a contractor or engineer who is experienced in the procedure. The importance of the combustion efficiency report cannot be overemphasized – it verifies the fuel consumption, which is the basis for most of the payback calculations for the O&M and ECM measures.

The combustion efficiency report should include information on all test readings and information on the boiler, such as boiler type, age, capacity, and fuel burned (see Figure 4.20).

10. Fuel Consumption Justification

A valid energy audit should compare the actual fuel consumed with the calculated values. Calculating the fuel consumption is necessary to verify that the assumptions and numbers used in the energy savings and heat load calculations are correct. The heat losses of the building and the related efficiencies of the boiler and the system are the basis for the validity of the proposed energy-saving measures.

The objective of the fuel consumption justification is to calculate the fuel consumption so that the result is within ±5% of the estimated fuel consumption, based on fuel bills for the adjusted base year.

The fuel consumed or required for consumption is given by the equation:

$$F = \frac{24 \, (DD) \, H_t C}{e \, (T_i - T_o) V}$$

This formula is explained further in Chapter 6.

If the value calculated is substantially different, then the actual building thermal values (R values), infiltration assumptions, boiler efficiencies,

Combustion Efficiency Report			
Identification of Heating Units			
Unit Designation	#1	#2	#3
Boiler Furnace Type			
Capacity (MBH)			
Age (Years)			
Fuel			
Combustion Efficiency Test Circumstances			
Date of most recent cleaning and adjusting			
Date of this reporter test			
Test Readings			
(a) Stack Temperature			
(b) Boiler Room Temperature			
Net Stack Temperature (a − b)			
% CO_2 or % O_2			
Smoke Density			
Draft			
Combustion Efficiency			
Comments			

Figure 4.20

occupancy patterns, and other assumptions need to be reexamined and corrected so that the computed fuel values are in line with the actual fuel consumed by the building.

11. Scoring of Report

Once the owner has approved the completed report, it is often submitted to the state or federal energy office as part of an application for a grant or subsidy. The accuracy of the report, its completeness, the overall average interactive payback of measures for which funding is being applied, and the projected annual savings all influence whether or not the owner will receive a grant.

Rebate Programs

Utility companies throughout the country have instituted energy-savings programs that seek to help existing buildings become more energy-efficient and encourage energy-efficient design in new construction. In residential buildings, owners have been encouraged to change light bulbs to fluorescent, add door and window weatherstripping, and install water flow restrictors and domestic hot water insulating covers. The cost of implementing these measures, which generally fall into the O&M category, have been borne in large measure by utility companies. In fact, all utility consumers are eligible for some type of energy-saving rebate program.

There are also programs available for new construction. Utility companies in some communities will pay all or a portion of the cost of implementing energy measures that exceed minimum code requirements. The basis used for calculating the energy saved, and hence the amount of rebate, are normally the local or state energy code requirements. The difference between the code maximum allowable and the actual energy consumed as a result of the consumer's energy-saving systems will determine the amount of rebate. Incentive programs like these demand a higher capital expense but will result in lower energy bills.

Energy audits are one step in the conservation of energy. As a detailed and exhaustive process, audits identify real areas where significant savings can be made. Energy audits make building owners aware of energy use and techniques where consumption can be reduced. This process often leads to better conservation, overall programs such as carpooling and better building utilization, and generally an improved, more sustainable facility.

Problems

4.1 Find an example of an indigenous building for humans (e.g., tepee, adobe, mud hut, lean-to, igloo) and a habitat for animals (e.g., rabbit warren, termite mound, beaver hutch), and sketch a plan and section of each. Describe and illustrate how the building types serve to mitigate the environmental stress of the climate and provide heating, ventilating, and air conditioning.

4.2 Examine standard references for building materials and find examples of two energy-efficient lighting fixtures.

4.3 Sketch the form of a one-family residence in your climate region and show how the form responds to climatic conditions, orientation, passive solar techniques, and wind direction.

4.4 A hotel has a heat load of 1,000 MBH [293 kW]. From manufacturers' literature, select the boilers (make and model number) for this load as follows:

   a. One large boiler
   b. Two boilers sized for 1/3 and 2/3 of the load
   c. 10 modular boilers

   Assume the boilers have a load of 350 MBH [103 kW]. For the efficiencies listed at this load, determine the overall efficiency of each boiler setup.

4.5 For your residence, complete the O&M checklist.

Chapter Five
HOW TO SELECT A SYSTEM

The four previous chapters were designed to familiarize the reader with the principles behind heating and cooling, methods of calculating loads, and codes and regulations that should be followed in designing HVAC systems. This chapter is an introduction to system selection and contains guidelines for using the information presented in the remaining chapters of the book.

Selection Criteria

Certain criteria must be determined for every building in order to select the appropriate HVAC system. Many of these criteria are established by the owner. They influence what components the building will have and the range and quality of the selected components. Basic system selection criteria include building life, cost, technical feasibility, overall fit, and whether the building is new or is being remodeled. Each of these criteria is described in the following sections.

Building Life

The anticipated life of a building is an important factor to consider when designing an HVAC system. Commercial office buildings, for example, have a financial life of ten to twenty years. Many developers typically sell a property within seven years of building it. Institutional buildings are often expected to have a life span greater than fifty years. Hospitals have certain components that are upgraded every three to ten years. Clearly, the effort and cost for a short-lived project are different from those for a building that should last over fifty years. The useful life of the equipment in the building should, accordingly, be selected to match the anticipated building life span.

Cost

The costs for a project can be evaluated in three ways: initial cost, life cycle cost, and maintenance cost. First, the cost of *initial construction* must be considered. This is the cost to complete the project, including land, legal, design, marketing, and construction costs. Most projects are approved or not approved on the basis of the anticipated initial cost. This is also the easier cost to establish. The *life cycle cost* of the facility is not as easy to anticipate. This is the total cost of a building over its entire useful life. It is a more accurate, but more complex, financial analysis. Projects are not paid for all at once. In reality, money is borrowed for a project's development costs and paid off over the term of the note. In addition, *operating and maintenance costs* are added to the annual principal and interest charges along with all other expenses, such as taxes, replacement costs, and broker fees. If the overall life cycle costs are less than the income the building is expected to produce, the project is considered profitable. A careful life cycle analysis will typically recommend more efficient, but more expensive, equipment than an initial cost analysis, because over each year of the building's life, the maintenance and operation of such equipment is less costly than the initially less expensive items.

Technical Feasibility

In selecting a system, it is useful to determine as early as possible any technical limitations. What fuels are available? Are gas lines in the street? Can the building have a basement? Will the structure hold the added weight of the intended equipment? Do local environmental considerations prohibit certain types of equipment? Is the existing plant large enough for the new load? Can the roof support the mechanical equipment? Are the spaces adequate? What are the local prohibitions?

Many initial choices are eliminated because they are not possible or permitted. If these limitations can be identified early, there will be more time to design for the appropriate choices.

Overall Fit

An HVAC system must be designed to fit into the building. The actual arrangement of a system that threads its way through a building takes a considerable amount of time and is often one of the major challenges a designer faces. Some items that should be determined early in a project include:

- Space requirements for mechanical equipment, including the size and availability of a mechanical equipment room.
- Location and sizes for chimney and shafts.
- Routes and space requirements for pipes and ducts.
- Places where mechanical systems cross other systems and may require larger chases or lower ceilings.

Close cooperation between design team members is essential to produce an efficiently designed building.

Renovation Projects

The charts and recommendations in this book can be used as a starting point for new construction. In existing buildings, however, the generation equipment is usually replaced after fifteen to forty years of use. For existing buildings, the decisions regarding new equipment are guided by the following additional considerations:

- **Energy audit recommendations:** An energy audit is usually performed by a professional engineer and results in recommended changes based on an evaluation of existing systems. The cost of replacement is weighed against the annual savings it will bring. Paybacks under three years are usually implemented.
- **Condition of existing equipment:** Worn out or inefficient equipment is often replaced as part of a capital improvement program. Savings in repairs and tax benefits are part of the economic benefits.
- **Operation and maintenance cost of existing equipment:** Equipment that is costly to run or repair may be completely replaced.

- **Cost and disruption of any replacement:** When replacing equipment, it is important to maintain continuity of service. This may require off-hour premium time, which can increase the replacement cost.
- **Increased income generated by the improvements:** New equipment, however expensive, may open the building to new, higher-paying markets. Rent projections may show a positive cash flow for improvements.
- **Feasibility:** The overall ability to implement the work must be verified. Considerations such as the capacity of the roof to support a cooling tower and access to adequate electrical power are crucial to any plan.

Each building type has a typical set of general requirements for common heating and cooling systems. For most building types, there is more than one choice for a suitable HVAC system. Selecting an HVAC generating system follows a logical decision-making process, illustrated in Figure 5.1. There are four basic selections to make in order to establish the type of HVAC system:

1. Select the type of system (heating and/or cooling and ventilating).
2. Select the fuel.
3. Select the distribution system and the terminal units.
4. Select the generation equipment and accessories.

Once the above steps are completed, the basic characteristics of the entire system can be established.

The following discussion applies to both new and renovated systems. However, when a worn out or inefficient piece of equipment is to be replaced, the new equipment must be compatible with the systems that remain.

The four basic selections are discussed in the following sections. Each section demonstrates how the selection is made. Following the discussion, three separate examples – one for heating only, one for cooling only, and one combined system – are presented. Once the basic characteristics of a system are selected, the remaining portions of the book should be used to size each separate part of the entire system, including generation, distribution, terminal units, controls, and accessories.

Figure 5.2 lists building types, along with comments on the typical types of heating and cooling systems available. This chart also notes whether the heating and cooling systems are separate or combined.

Selecting the Type of System

The decision process shown in Figure 5.1 illustrates the basic choices to be made when selecting an HVAC system. Heating is mandatory in most buildings in temperate climates for reasons of human health and comfort as well as for public safety (for example, to prevent sprinkler systems from freezing). The decision to be made is whether or not the building requires cooling. Cooling is generally optional. However, most new and renovated buildings are expected to provide cooling.

Figure 5.2 lists common HVAC systems for particular building types. For example, for a one-story commercial building, a combined HVAC rooftop unit is listed as the common system. In addition to the recommendations provided in this chart, current practice in the area for similar buildings should be considered. Another factor to be considered in choosing the HVAC system is the effect that the type of cooling system chosen will have on the overall marketing of the property.

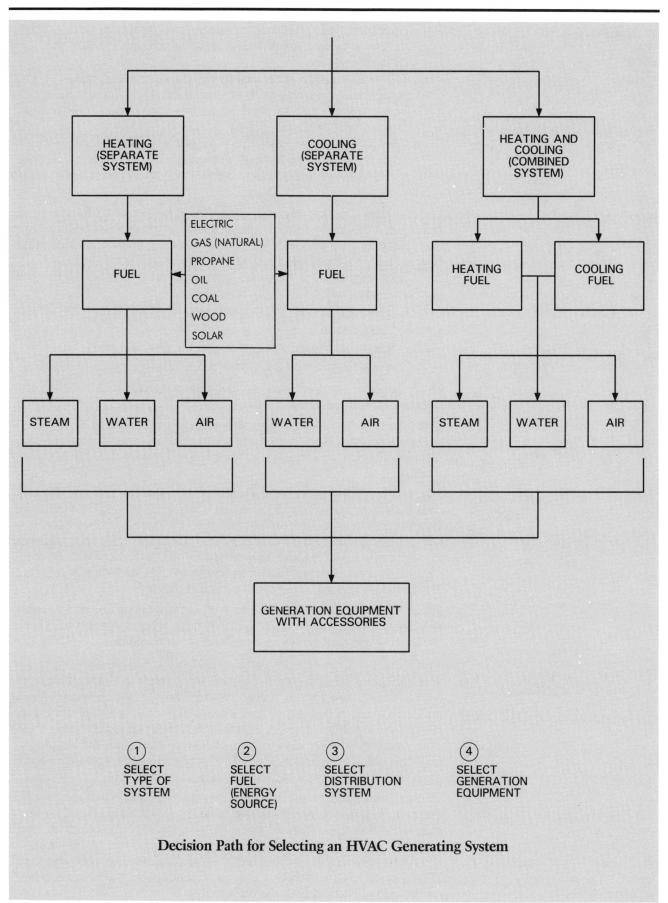

Figure 5.1

Types of Heating and Cooling Systems for Buildings

Building Type	Heating — Separate (Fig. 1.6)	Cooling — Separate (Fig. 1.11)	Heating and Cooling Combined (Fig. 1.23)	Remarks
Assembly				
Movie Theaters, Cinemas	—	—	All air central system	Large make-up air consideration. Check local codes.
Nightclubs, Restaurants	—	—	All air central system	Smoke and humidity removal requires large make-up air consideration. Negative pressure in kitchens for odor removal.
Libraries, Museums	—	—	All air central system	Humidity control.
Churches	Perimeter radiation	DX	All air central system	—
Schools	Thru-the-wall fan coils Perimeter radiation	—	—	Local codes regulate outdoor air make-up.
Swimming Pools	—	—	All air central system	Humidity control.
Business				
Offices	Perimeter radiation	Central air	Package-Multizone	—
Bank	Perimeter radiation	—	All air central system	—
Courthouse	Perimeter radiation	Local units	All air central system	—
Multi-story	Perimeter-hot water	—	All air central system	—
Lobby, Hallways	—	—	Separate units	Separate hours.
Typical Office Floor	Reheat coils, electric or hot water	Fan coils, VAV units	Central air with economizer, heat pumps	VAV, induction, dual duct, fan coil.
Computer Rooms	—	Tenant supplied unit	—	Tie to cooling tower.
Lobby	—	—	—	Set 5°-8° buffer.
Commercial/ Retail				
Department Store	Reheat coils as required	—	All air central system	—
One-story Stores	Supplemental radiation at perimeter	—	Rooftop HVAC unit	—
Supermarkets	—	—	Rooftop HVAC unit	Reclaim heat from refrigeration compressors
Factory	Electric, gas-fired space heaters or hydronic	Special conditions only	All air central system	—
Hazardous	Hot water	Chilled water	Hydronic	Non-sparking motors and controls.
Institutional				
Hospitals	Steam-combined with hot water	Central	All air-dual duct	Special ventilation standards.
Penal	Hydronic, concealed or remote	—	—	Special security standards.

Figure 5.2

Based on these factors, select from among the following types of HVAC systems:

- separate heating,
- separate cooling, and
- combined heating and cooling.

The possible systems for each of these choices are illustrated on the heating ladder (Figure 1.6), the cooling ladder (Figure 1.11), and the air conditioning ladder (Figure 1.23) presented in Chapter 1.

Types of Systems

Heating, ventilating, air conditioning, and humidity control can be provided in many different ways. Not all systems can provide all four functions. The purpose of this section is to provide basic information about the more common system types, with advantages and disadvantages (see Figure 5.3). When determining the type of system to use, it is important to compare options, brainstorm about possibilities, and then select the system that will best fit the space and serve the user. Chapter 14 discusses in some detail the particular requirements of special systems, such as high-rises, atriums, covered malls, laboratories, and hospitals.

The following sections describe some commonly used systems. Each has three major components: a generator, a distribution system, and terminal units. Some systems are capable of other configurations, using a different type of one or more of these three components.

Types of Heating and Cooling Systems for Buildings (continued)				
Building Type	**Heating — Separate (Fig. 1.6)**	**Cooling — Separate (Fig. 1.11)**	**Heating and Cooling Combined (Fig. 1.23)**	**Remarks**
Residential				
1 & 2 Family	Warm air system or hydronic with radiators	Split system or window units	Heat pump (air-to-air) requires booster or with electric reheat	Operable windows and bathroom exhaust fan.
Mid-rise Multi-family Residences and Elderly Housing	Electric or hydronic	Window units	Heat pump (water-to-air) fan coil units	Fresh air to corridors — bathroom and kitchen exhaust.
Hotels	—	—	Hydronic or electric fan coil units	Same as above.
Storage Warehouses	Electric unit, gas or hydronic unit heaters	Walk-in coolers Built-up direct expansion systems	— —	Check requirements for material stored.
Existing Buildings Renovation	Note the type of system presently in the building and compare with the above table. Review the recommendations in the book. See discussion on installing new systems versus renovation. When the renovated use is different from the existing system, check for overall feasibility.			

Figure 5.2 (continued)

System Type	System Diagram	Spaces Commonly Served	System Description	Advantages	Disadvantages	H	V	AC	HU
Baseboard System	Boiler / Baseboard / Steam and Chilled Water Supply and Return	Storage areas, Residences, Utility rooms. Supplement to cooling-only systems. Perimeter heat.	Baseboard terminal units fed by boiler with either steam or hot water. Heating only.	Low cost. No ducts required. Flexible control. Minimum space requirements.	Baseboard is exposed. May interfere with furniture. No ventilation, air conditioning, or humidity control.	●			
4-Pipe Fan Coil System	Fan Coil Unit	Areas that need isolated control. Hospitals and Colleges.	Four-pipe fan coil units fed by district hot and chilled water. Heating and air conditioning only.	Low cost. Individual room control. Minimal space required.	Noisy. Limited size range. No ventilation. Air handler requires space.	●	*	●	O
Packaged Unit System	Packaged HVAC Unit	Open office, Retail, and Patient rooms.	Self-contained unit. Serves all HVAC functions.	Quiet. Easily designed and installed. Energy costs can be individually billed.	Requires maintenance. Extensive ductwork.	●	●	●	O
Split System	Air Handling Unit / Condenser Unit	Rooms with limited access to outside, Conference rooms, and multiple tenancy	Air handler with expansion coil and optional electric heat with condenser unit. Serves all HVAC functions except ventilation.	Small indoor component. Minimal ductwork. Quiet. DX—no freezing.	Multiple pieces of equipment to maintain.	●	*	●	O
Variable Air Volume System	Main Air Handler / VAV Box	Areas with many spaces, where indirect control is required: Offices/Clinics.	Central air handler/air conditioner with VAV boxes. Serves all HVAC functions. Variable speed fan optional.	Individual room control. Good for large applications. Can work with economizer system.	Reheat required.	●	●	●	
Induction System	Induction Unit with Coils / Exhaust/Return Air System / Supply Air AHU	Spaces with ventilation requirements: Offices and Patient rooms.	Central air handler/air conditioner with individual room induction units. Serves all HVAC functions.	No ductwork. Minimal amount of equipment to maintain. Individual room control. Provides ventilation air.	High airflow in rooms. Noisy.	●	●	●	O
Heat Pump System	Heat Pump / Cooling Tower / Boiler	Offices, high-end Residences, Nursing homes.	Heat pumps with cooling tower and boiler. Heating and air conditioning only.	Energy efficient. Energy moved around building.	Will not operate in low air temperatures. No ventilation or humidity control. Units require maintenance.	●	*	●	

* Ventilation can be achieved by supplying tempered air into return air plenum with separate system.

O Humidification possible by adding in-line humidifier.

System Selection Chart

Figure 5.3

Hot Water Boiler, Piping, and Baseboard Heat

Baseboard heat is the simplest, least expensive heating system that can be installed. It is usually comprised of fin-tube baseboard units, distribution piping, and the hot water boiler with pump. The boiler generates hot water, which is then distributed by the pump. The piping for a baseboard system is straightforward and easily zoned. The fin-tube units are wall-mounted close to the floor, where the air is cooler. As the air is heated by the fin-tube units, it rises, and in doing so draws in more cool air from below. The system can be controlled by room thermostats, zone thermostats, or outside air temperature sensors.

Baseboard heat does have disadvantages. It does not ventilate or provide air conditioning. Further, baseboards can exacerbate dry air conditions during the already dry winter season. The baseboard units are exposed and can interfere with furniture in the space. Also, while cast iron baseboard is rugged, fin-tube units with sheet metal covers are easily dented.

Options: A steam boiler can be used. In this system, a condensate return pump is substituted for the circulation pump. Electric baseboard is also available, but it is more expensive to operate.

Steam Boiler, Chiller, Piping, Ductwork, and Fan Coil Units

Fan coil units consist of a fan and heating and/or cooling coils, as well as the associated distribution ductwork and terminal units. Distribution piping is run to a boiler and chiller (each with an associated pump, which supplies hot and chilled water).

Fan coil systems can be either two-pipe or four-pipe. In a two-pipe system, one pipe is for the supply of both the heating and the cooling medium, and one pipe is for the return of both the heating and the cooling medium. A two-pipe system works well in simple systems where there would not be any chance for heating to be called for in one area while cooling is called for in another. In a four-pipe system, the cooling and heating loops are independent. A four-pipe system allows many zones to operate independently, which is often necessary during the spring and fall, when temperatures can vary radically from one zone to another.

Fan coil units are convenient because they can be located in tight spaces. They can be noisy, however, and care should be taken to either provide adequate sound insulation or place the unit in or above unoccupied areas. Fan coil units are usually used for taking in large quantities of fresh outdoor air, because of the possibility of freezing the coils. When fresh air is required, a separate system should be used to temper the air before it is distributed through the fan coil unit. Fan coil units come in smaller sizes than other types of units, and are not normally designed to serve multiple zones or large areas.

Options: Fan coil units are often tied into existing distribution mains from central hot and chilled water systems. A small booster pump may be required if there is insufficient pressure in the line to push the liquid through the lines.

Packaged Units, Ductwork, and Terminal Units

Packaged units are available in a great variety of sizes and capabilities. Most of these systems use direct expansion cooling. Packaged units have several advantages over fan coil units. They are not subject to freezing problems, so they are more versatile as to where they can be placed (e.g., rooftops and other unheated spaces). They are also able to take in fresh air, which is usually required, directly. This fresh air intake can be tied to an economizer cycle, which allows the unit to use cooler outdoor air to relieve internal heat gain. The unit can be provided with heat in the form of either electric or gas.

These units are often roof-mounted, and can be either single-zone or multi-zone. The latter is designed to have several supply ducts leading directly from the unit to separate zones.

Packaged units are very convenient for designers because they contain all the components with controls and are usually considerably less expensive than a custom assembly of the component parts. All four seasons can be handled with one piece of equipment.

Because the packaged unit is centralized, more ductwork is required to distribute the air than for other systems. Operation and maintenance costs are higher than most systems, because the unit is exposed to the weather. Roof structure should be checked, and reinforced if necessary, to accommodate the units since they can be quite heavy.

Options: Another type of packaged unit commonly used is the terminal through-the-wall unit, designed for individual room control.

Split Systems, Ductwork, and Terminal Units

Split systems usually operate by direct expansion. As the name implies, there are two components, one outside and one inside. The outside component contains the compressor and condenser, which is either water- or air-cooled. A water-cooled condenser will require a cooling tower. Refrigerant supply and return lines run to an evaporator coil and a supply air fan in the air handling section (the indoor component), from which the air is distributed through the zone. Less ductwork is required for split systems than for packaged air conditioners.

Because split systems often serve multiple zones, the units have to be oversized to serve the 100% load of all spaces. When demand is not high, special controls should be considered to prevent the refrigerant from freezing up. Also, when the unit is satisfying the load of a large space, reheat may be required in smaller, ancillary spaces to prevent them from getting too cold.

Options: Many manufacturers can provide a heating coil (steam or hot water) in the air handling section. Small units, which only recirculate air, are available from some manufacturers. These can be very convenient, but they do not supply the fresh air required by most codes.

Variable Air Volume Systems (Cooling only)

Variable air volume (VAV) systems are comprised of a central air handler, distribution ductwork, and the VAV boxes themselves. The central air handler is provided with a cooling medium (refrigerant or chilled water) and supplies cool air to each room (or zone) VAV box. A thermostat controls a damper in each box, adjusting the amount of air that enters the space, thus controlling the temperature.

VAV systems are used for larger scale installations, wherein one air handler might serve a whole floor or even a whole building. Because the amount of air required by the system fluctuates, the amount of air supplied by the central fan must be controlled. There are four methods by which this can be accomplished. The first, and least expensive, is to simply allow the friction caused by the closing VAV box dampers to create friction in the system. This is the least energy-efficient method. The second is to install a fan for which the pitch of the blades is adjustable: airflow is then controlled by sensors that adjust these blades to maintain system air pressure in response to demands from the system. The third method is to adjust the speed of the fan. The fourth is to provide a damper in front of the fan to adjust the amount of air available to the fan.

VAV systems operate well to regulate the temperature in many small spaces. Also, the cooling source (i.e., chiller or cooling tower) need not be designed for the 100% load of the system, because over the course of the day, different areas will peak at different times. VAV systems are not capable of heating and cooling at the same time, so a separate heating system is often used.

Induction Systems

Induction systems bring fresh air, tempered as necessary, to induction units. The induction unit is similar to a fan coil unit in that it can be two- or four-pipe. The induction air blows over the fins and induces room air to flow and be heated or cooled as necessary. The induction system has no fan. The air is generally supplied constantly to provide ventilation air; the valves of the induction unit modulate the hot and chilled water supply to achieve room design conditions. A main air handler with a high pressure fan, and cooling and reheat coils, are necessary to generate the air for the induction piping, which is normally at high pressure and velocity.

Induction systems have no ducts except the small induction air. They have a limited amount of breakable equipment and they avoid having to have two systems (perimeter and interior) for a building. A separate exhaust system is required, however, and the induction air can sometimes be noisy as the high velocity air is introduced into the unit.

Heat Pump Systems

Water source heat pumps provide an economical heating and cooling system for buildings that can be zoned. They are basically small air handlers that distribute treated air to a zone via a duct system. Each zone contains a heat pump, usually suspended from the ceiling. A water loop, which is actually the condensing line for each heat pump, flows to each heat pump, which either takes heat out of the loop for heating or dumps heat into the loop if cooling utilizing the refrigeration cycle. The water temperature in the loop varies from 60° to 90°F [15.3 – 32 °C],* which allows plastic piping to be used. Heat pump systems are unusually efficient since they can take heat from one section of a building and deliver to another section by the water loop, resulting in an overall reduction in energy operating costs of approximately 30%.

Heat pumps have two major disadvantages. Approximately 3 to 6 percent of heat pumps require repair at any one time. They are relatively complicated, being an entire refrigeration unit as compared to a simple box with a damper as in a VAV system, and require active maintenance. The second disadvantage is that they do not provide ventilation air. Ventilation air is usually delivered to the ceiling plenum for distribution by the air handler of the heat pump.

*Throughout the book, SI metric equivalents are provided for most imperial units. The SI units appear in brackets [ ].

Computer Room Design

As a specialized example of system selection, consider the design for a computer room. Figure 5.4 illustrates four choices that can be made for the layout of a computer room, with the advantages and disadvantages of each. HVAC systems benefit from a review of particular requirements and the establishment of alternatives for an overall design that meets client needs as well as budgetary restrictions.

Selecting the Type of Fuel

Before selecting any HVAC equipment, the type of fuel must be chosen. The fuel for heating systems is selected independent of cooling systems and is usually chosen based on economy and availability. Figure 5.5 lists the basic properties and costs of common fuels.

The basic fuels are natural gas, propane gas, oil, steam, electricity, solar energy, and coal. The following sections contain comments on the use of each type.

Electricity

Electricity is universal to all buildings because it is necessary for lighting and power. It is the most commonly used fuel for air conditioning because the majority of refrigeration is achieved with compressors, which require electricity to operate. When compared with other fuels, electricity has a high cost and alternatives are generally sought. There are, however, some particular advantages to using electricity that often outweigh its high cost.

- Electrical systems can be individually metered, making them preferable for multi-tenant buildings such as residential condominiums.
- Electrical systems are free of freezing problems and leakproof. As long as plumbing lines are heat-traced electrically, the heating system can be completely turned off, a particular advantage for spaces that may be unoccupied for long periods of time. (Air systems have the similar advantage of being freezeproof and leakproof.)
- Electrical systems can be regulated by individual room control thermostats. This means that unused rooms can be turned down or off, thus compensating for higher fuel costs.
- Electrical systems do not need chimneys, air intake louvers, fuel tanks, or pumps. They have few (if any) parts that wear out and are generally less expensive than hydronic or air systems to install and maintain.
- The efficiency of electric heat is 100 percent. All the consumed electrical energy is delivered to the space as heat.

Steam

When steam is available from the utility in the street or from a central plant, it has the particular advantage of not requiring a boiler or chimney. Pressure-reducing stations are used to reduce the high-pressure steam, as required, to low or medium pressure. Heat exchangers are utilized in producing hot water from the utility steam.

Oil

Fuel oil is one of the most commonly used heating fuels. Fuel oil is graded from #1, highly refined, to #6, crude. The refined oils are more expensive and have a lower heat content, but are chosen because they flow at room temperatures; #2 oil is commonly used in heating applications with small burners; #4 oil is commonly used for larger burners; #5 and #6 are nearly solid at room temperature and must be preheated to flow.

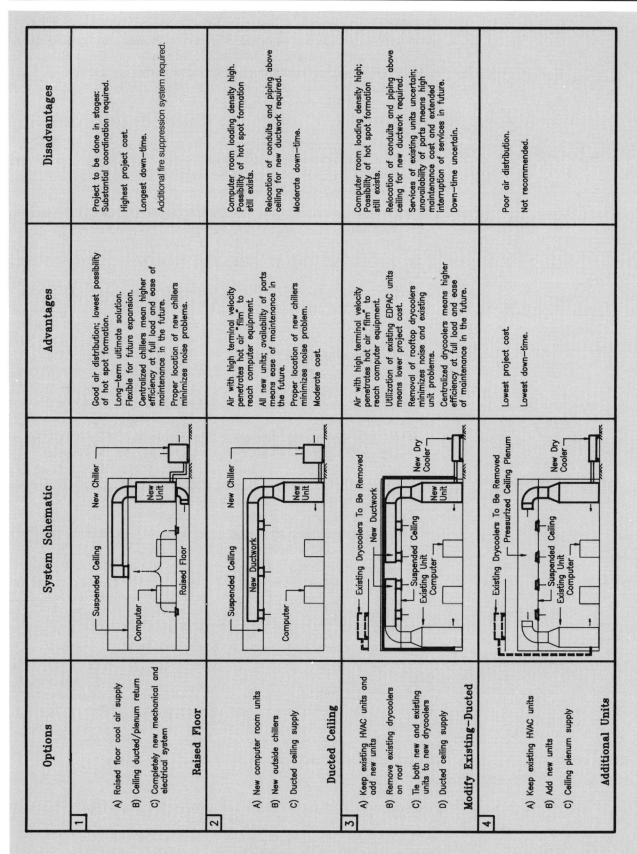

Figure 5.4

Oil is normally stored in tanks, which may be exposed or buried in the ground. Current laws may require containment and/or leak detection systems. Tanks may be made of single- or double-wall fiberglass or steel. Transfer pumps that supply oil to the burner reservoir or day tank are usually required on long runs.

Gas

There are several gases available for heating. They include natural gas, propane (often called bottled gas), butane, methane, hydrogen, and carbon monoxide.

Natural gas and propane are the most common types of fuel used. When available from the street, natural gas has the advantage of not requiring a storage tank. Natural gas is manufactured from coke or oil and obtained directly from the earth. Propane is most often used in rural areas where natural gas lines are not available. Gas can also be used for cooking and is often selected for this reason.

Coal

Coal is a solid fuel and requires special consideration to feed the boiler (stoker) and remove the ash. It is no longer practical to perform these procedures manually; larger coal furnaces and boilers have been developed with automatic stoking and ash removal. Nevertheless, the practical considerations of delivery, storage, and handling of solid fuel, in addition to its sulphur content (which has been associated with environmental problems), have restricted its use to large industrial or campus plants.

Basic Properties and Costs of Fuels

Fuel Type	Unit	Heating Value Btu/Unit [kWh/Unit]	Cost per Unit (cents)*	Overall System Efficiency	Net Cost per MBH* (cents)	Remarks
Electricity	Kilowatt-hours	3,412 [1]	12–15	100	3.5–4.4	
Steam	Pounds (at atmospheric pressure)	1,000 [.3]	—	—	—	
Oil #2	Gallon	138,000 [40.6]	90–130	60–88	0.7–1.5	
Oil #4	Gallon	145,000 [42.5]	90–130	60–88	0.74–1.57	
Oil #6	Gallon	152,000 [44.5]	—	—	—	Preheat
Natural Gas	CCF (100 C.F.)**	103,000 [30.2]	0.5–0.8	65–92	2–5.5	
Propane	Gallon	95,500 [28]	80–110	65–90	—	
Coal	Pound	13,000 [3.8]	—	45–75	—	
Solar	S.F. of collector—varies with location and collector					

All numbers shown in brackets are metric equivalents.
*Northeastern U.S., 1991
**Note: 1 therm = 1.013 CCF

Figure 5.5

Coal is classified as either anthracite or bituminous – hard or soft – in a series of ratings with a relatively even heat value (12,000 – 14,000 Btu/lb.) [7.680 – 8.960 kW/kg].

Solar

Solar energy can be used to generate domestic hot water, warm air, and hot water for absorption chillers. The costs vary with each application. Among other issues, the roof structure must be checked to ensure that it can support the solar panels for both gravity and uplift forces.

Selection

The selection of fuel should be based not only on the cost per MBH, but also on any special installation costs or fuel conversion charges. Oil and propane, for example, require fuel tanks, while steam, electricity, and natural gas involve utility connection charges.

In urban areas, steam is a preferred heating fuel because it brings noncombustible fuel into the building – a great asset from the viewpoint of the fire department and insurance companies – and it requires no boiler or chimney. Therefore, steam systems occupy less of a building's valuable floor space – an asset to owners and developers. Steam heat is almost always used when it is available in urban high-rise buildings.

With the exception of steam, gas, or hot water-fed absorption chillers, virtually all cooling equipment runs by electricity.

In existing buildings, the practicality of continuing with the fuel currently used should be examined. In many cases, conversion to a different fuel is more economical despite the costs of demolition and installation of a new system. Where gas is less expensive than oil, the cost of removing an existing oil tank and boiler and replacing it with a new gas system may result in lower monthly fuel bills, which can offset the replacement cost in less than five years.

Selecting the Distribution System

For a complete description of distribution systems, refer to Chapter 10, "Distribution and Driving Systems." The following is a brief discussion on the distribution system selection process for renovation projects.

Renovation Projects

It is common to retain and reuse existing distribution systems when the same type of system (separate, or combined heating and/or cooling) will be used. Other considerations are whether the existing distribution system is in good condition, and if it can serve the same overall layout or use after renovation. When it is practical to retain the existing distribution system, the result is often considerable cost savings when compared with the cost of the demolition and installation of a new system.

When the type of system or use of the renovated building will change, the general guidelines for new construction are followed to select the type of distribution system.

When the condition of the distribution system is poor, it should be replaced with a new system. Repairs to existing systems – particularly older steam or water systems – can be very difficult and often never-ending.

Figure 5.6 illustrates the distribution system selection process. This chart can be used to determine the appropriate type of distribution system for the building.

Selecting the Generation Equipment

Once the choices for the type of overall building system, the type of fuel, and the type of distribution system have been made, the type of generation equipment and its accessories can be determined. Figure 5.7 shows portions of the heating ladder, the cooling ladder, and the air conditioning ladder, which can be used to determine the type of generating equipment required. (See Figures 1.6, 1.11, and 1.23 in Chapter 1 for more detail regarding the heating ladder, the cooling ladder, and the air conditioning ladder, respectively.)

Equipment Selection Examples

The following examples illustrate the selection process for separate heating, separate cooling, and combined heating and cooling systems. The principles previously explained are illustrated in the following three examples.

Selection of a Separate Heating System

For this example, the building is a mid-rise housing complex in a suburban area. The heat load for the complex is 1,500 MBH [440 kW]. According to the selection process shown in Figure 5.1, decisions are made in the following order.

Type of System: From Figure 5.2, the residential building, select a separate heating system.

Type of Fuel: From Figure 5.5, #2 oil is selected. In a suburban area, steam and natural gas may not be available. Electricity, depending on cost, could have been selected, especially if individual heat is to be metered and paid by each tenant.

Preliminary Selection of Distribution Systems		
Type of Building System	**Type of Distribution System Conditions**	**Select**
Separate Heating	Individual room control and/or tenant billing required	Electric
	Steam heating and other steam uses—industrial	Steam
	Continuous use buildings	Hot water
	Intermittent use with freezing potential	Warm air
Separate Cooling	One- and two-story buildings	Air
Heating and Cooling	High-rise office — fan coil units — induction units/central air — Variable air volume	Water Air and reheat Air and reheat

Figure 5.6

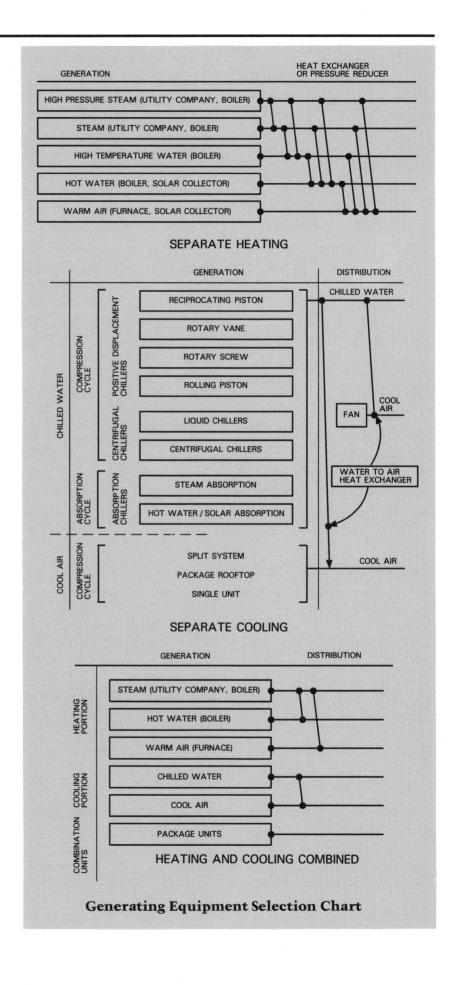

Figure 5.7

Type of Distribution System: From Figure 5.6, select hot water.

Type of Generating Equipment (with accessories): The main heating generating equipment is an oil-fired hot water boiler (#2 oil). (The Means Division number for this equipment is 155-120. Equipment data sheets for generation equipment appear at the end of Chapter 6.) The path of this example is traced in Figure 5.8, part of the heating ladder. Some designers recommend installing two boilers in each building. Because the maximum design load occurs only a few days each year, sizing one boiler for one-third and the second boiler for two-thirds of the load maximizes overall efficiency, cuts down on wear and tear, and provides a backup for heating when one boiler is not operative. For simplicity, one boiler is used for this example.

Selection of a Separate Cooling System

The building used for this example is an existing four-story office building. It has an existing centralized steam heating system using perimeter convectors/radiators in good condition. Cooling is to be provided. The cooling load is 40 tons [141 kW]. According to the selection process shown in Figure 5.1, decisions are made in the following order.

Type of System: Using Figure 5.2, check the "existing buildings" entry. Because the new use will be the same as the old use and the existing heating system is in good condition, leave the heating alone and design a separate "stand-alone" cooling system.

Type of Fuel: From Figure 1.11 in Chapter 1, it can be seen that electricity is the common fuel for cooling except where cheap steam, gas, or solar energy is available to run large absorption machines.

Type of Distribution System: Because radiation already lines the perimeter of the building, cooling is best provided from the ceiling. Since existing (older) buildings generally have high ceilings, an air system is the first choice because it will not interfere with the steam system; it will fit into the high ceiling; and it will provide distribution in the ceiling, the ideal location for cooling. Since the heating system is centralized, a centralized cooling system is also selected. (A different choice would be made if each floor or tenant were paying individually for electricity.) A central system requires a piece of equipment to reject the heat from the building to the outside. This device might be an air-cooled condenser, a cooling tower, an evaporative condenser, or a rooftop unitary piece of equipment. The choice of this equipment should be reviewed by a structural engineer to make sure that the weight of the equipment can be supported by the particular type of roof framing.

Type of Generating Equipment (with accessories): The main cooling generating equipment is an electric package unit mounted on the roof and zoned to each tenant. Figure 5.9, part of the cooling ladder, traces the selection path for this example.

Selection of a Combined Heating and Cooling System

In this example, the building is a 15-story office building with 8,000 square feet [743 m^2] per floor. It includes a ground-floor retail area and is located in a downtown section of the city. Hook-ups to all major services and utilities are available, including steam. The heating design load is 2,500 MBH [733 kW] and the cooling load is 300 tons [1055 kW]. Cooling is required year-round for the interior of the building. According to the selection process shown in Figure 5.1, decisions are made in the following order.

Type of System: From Figure 5.2, select a combined heating and cooling system.

Type of Fuel: For combined systems, the heating and cooling fuels are selected independently. Steam is selected as the heat source because it is available; is reasonably competitive (see Figure 5.5); generates savings; and because boilers, tanks, and chimneys are not necessary.

Electricity is selected as the energy source for cooling because the building has sufficient electrical capacity available. (An absorption unit is also worthy of consideration in this case.)

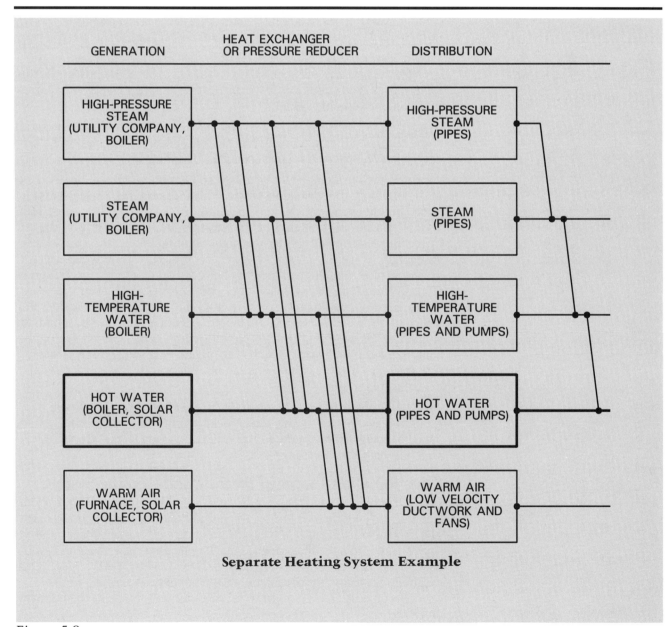

Figure 5.8

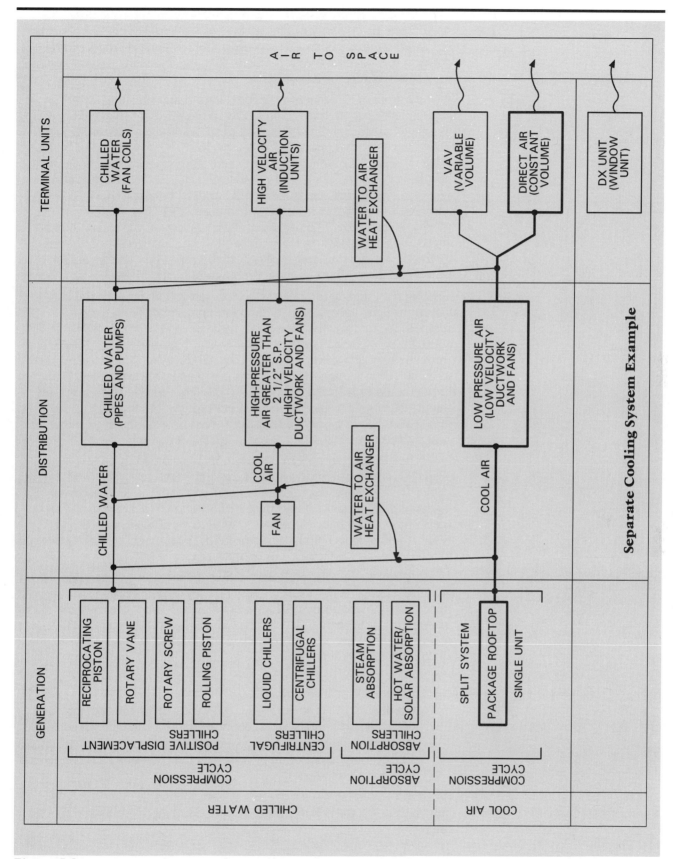

Figure 5.9

Type of Distribution System: The possible mediums for the distribution system are air and water. The basic choices are a centralized plant that services the entire building or a decentralized system of smaller units, down to one or more for each tenant. Usually, individual tenant systems may be separately metered.

A building's HVAC system is selected after many meetings with the owner, developer, architect, and HVAC engineer. For this building, it is assumed that after several trial runs and cost estimates, a centralized system is selected. For other projects in different cities with varying costs, building designs, marketing strategies, and tenant requirements, the choices will be different. To make the best possible choice, it is necessary to combine experience with trial selections of full systems. For each trial selected, compare the advantages and disadvantages, including the cost of the mechanical system. The impact of these choices on the overall building should also be considered. With the centralized system selected, the possible mediums for heating and cooling can be analyzed. The layout of centralized distribution systems is shown in Figure 5.10.

Cooling the building interior is required year-round, because the heat gain from lights, people, and equipment contributes to an excess temperature that must be removed from the interior, even in winter. By contrast, because the perimeter of a typical high-rise building loses heat through the window wall, heating of the window wall surfaces is required in winter, even though the interior zone is usually always cooled. The need to simultaneously heat and cool office buildings in winter and cool the interior during the cooler spring and fall months has resulted in the development of two major energy conservation systems: free cooling and heat pumps (see Chapter 6, "Generation Equipment").

Now the distribution system is chosen. It is assumed that there is an available central core in the interior for supply and return air, and that the owner wishes to be responsible for and maintain as few pieces of equipment as possible. Based on the chart shown in Figure 5.10, an economizer cycle, low-velocity air system to variable air volume (VAV) boxes is selected for cooling the interior. This system will provide ventilation air and individual control for interior spaces.

Heating and cooling the perimeter can be achieved in several ways. Fan coils are selected for this example because they do not occupy any more floor or ceiling space for air ducts; they can be placed under windows, an ideal location for heat and condensation control; they can be individually controlled; and they can accommodate the different volumes of air required for heating and cooling.

Generation Equipment: The primary generating equipment for this combined system is as follows:
- **Cooling:** an electric 300-ton [1055-kW] chiller and cooling tower tied to an air handler with economizer and perimeter fan coils.
- **Heating:** utility steam to pressure-reducing station, heat exchangers for hot water distribution to perimeter fan coils, and freeze protection coils for air handler.

Figure 5.11 illustrates the path this system follows on the cooling ladder, and Figure 5.12, the heating ladder, for this combined heating and cooling system.

Summary

By using the basic outline presented, the overall system can be selected. Equipment data sheets are provided at the end of the chapters on generation, distribution, terminal units, controls, and accessory equipment to further

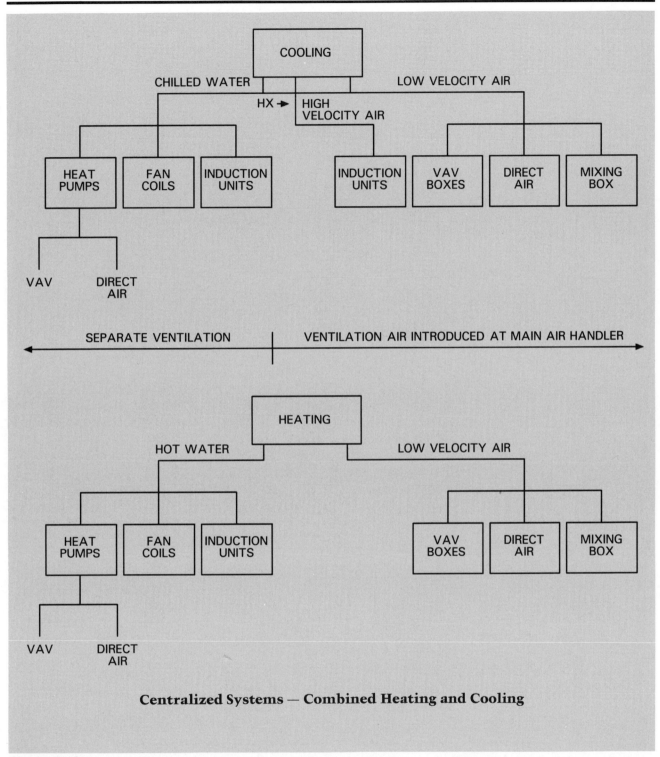

Figure 5.10

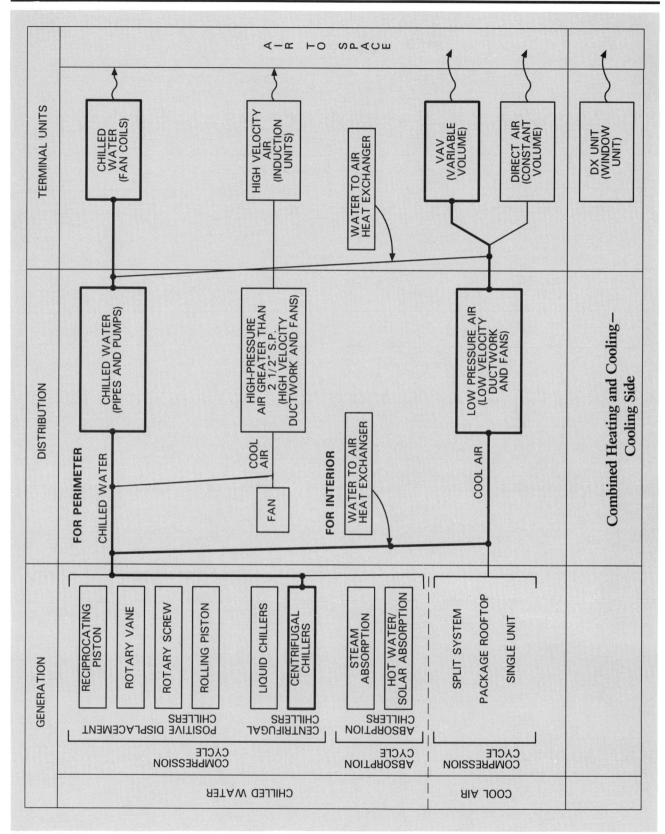

Figure 5.11

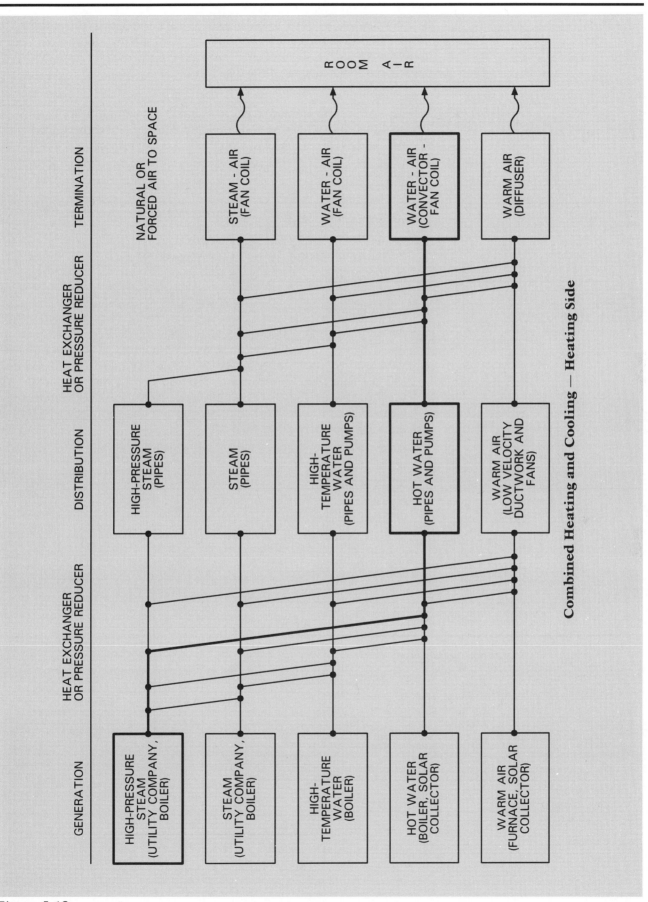

Figure 5.12

Problems

5.1 For each of the building types in Figure 5.13, state if heating only, cooling only, or heating and cooling combined is recommended and list the type of system you would select. Give reasons.

5.2 Rank the following in order of shortest to longest anticipated useful life:
   a. Hospital
   b. Suspension bridge
   c. Commercial 3-story office building
   d. Moderate income 100-unit housing complex
   e. Restaurant
   f. Library
   g. Warehouse

5.3 An existing nursing home is considering a new HVAC system. It has 2'-6" [.76 m] between the suspended ceiling and the floor above with steel bar joists in the space, which are 16" [.41 m] deep and 3'0" [.91 m] on center. Compare a heat pump system with a fan coil system. Prepare a chart that shows advantages and disadvantages of each. Diagrammatically sketch each system and show how ventilation air would be obtained in each case.

5.4 Select one of the systems below and sketch the system for a 3-story office building.
   a. Baseboard
   b. VAV
   c. Heat pump
   d. Split system
   e. Fan coil
   f. Induction system

Problem No. 5.1			
Building Type	**H, C, or H&C**	**System Type**	**Reasons**
Low-rent Housing			
High-income Condominiums			
Elementary School			
College Laboratory			
Storehouse, Dry Goods			
Mid-rise Office Tower			
Computer Room			
Sports Arena			
Operating Room			

Figure 5.13

Part Two

EQUIPMENT SELECTION

In Part Two, the equipment selection process is discussed, including layouts, sizing instructions, and other special considerations for generation equipment, distribution and driving systems, coils and heat exchangers, terminal units, controls, and accessories. Equipment data sheets are included at the end of each chapter, listing capacities, advantages and disadvantages, and other considerations for each piece of equipment. Means line numbers are included for easy reference to the annual cost guide, *Means Mechanical Cost Data*.

Chapter Six
GENERATION EQUIPMENT

Generation equipment includes all major pieces of equipment that generate heating or cooling and their accessories, including boilers, furnaces, heat exchangers, expansion tanks, chillers, and cooling towers, as well as associated pumps, fans, and accessories. Generation equipment is divided into systems that provide heating only, systems that provide cooling only, and systems that provide both heating and cooling.

Hydronic Systems – Heating Only

Hydronic systems that provide heating only are described in this section, which includes discussions of boilers, burners, storage tanks, and expansion tanks.

Boilers

Heating boilers are designed to produce steam or hot water. The water in the boilers is heated by either coal, oil, gas, wood, or electricity. Some boilers have dual fuel capabilities. Boilers are manufactured from cast iron, steel, or copper.

Cast iron sectional boilers may be assembled in place or shipped to the site as a completely assembled package. These boilers can be made larger on site by adding intermediate sections. The boiler sections may be connected by push nipples, tie rods, and gaskets. Cast iron boilers are noted for their durability.

Steel boilers are usually shipped to the site completely assembled. Large steel boilers may be shipped in segments for field assembly. The components of a steel boiler consist of tubes within a shell and a combustion chamber. If the water being heated is inside the tubes, the unit is called a water tube boiler. If the water is contained in the shell and the products of combustion pass through tubes surrounded by this water, the unit is called a fire tube boiler. Water tube boilers may be manufactured with steel or copper tubes.

Electric boilers have electric resistance heating elements immersed in the water and do not fall into either category of tubular boilers. Steel boilers in the larger sizes are often slightly more efficient than cast iron and are generally constructed to be more serviceable, which with proper maintenance adds to their useful life.

Several types of boilers are available to meet the hot water and heating needs of both residential and commercial buildings. Some different boiler types are shown in Figure 6.1.

Low-pressure hot water boilers operate at less than 30 psig [300 kPa]*. Low-pressure steam boilers operate at less than 15 psig [200 kPa]. Above 30 psig [300 kPa], high-pressure boilers must conform to stricter requirements. Water is lost from a heating system through minor leaks and evaporation; therefore, a makeup water line must feed the boiler with fresh makeup water.

Heating boilers are rated by their hourly output. The output available at the boiler supply nozzle is referred to as the gross output. The gross output in Btu per hour divided by 33,475 indicates the boiler horsepower rating. [The gross output in kilowatts divided by 9.8 indicates the boiler horsepower rating.] The net rating of a boiler is the gross output less allowances for the piping tax, the pickup load, and uncertainties that are typically limited by codes to 25 percent. The net load should match the actual building heat load. Another term often used in sizing boilers is *square feet of radiation* (also known as equivalent direct radiation (EDR)); one square foot of radiation equals 240 Btu/hour. (For hydronic systems one square foot of radiation equals 150 Btu/hour.)

Because of the high cost of fuels, efficiency of operation is a prime consideration when selecting a boiler. Recent innovations in the manufacturing field have led to more efficient and compact boiler designs. Figure 6.2 shows a typical boiler installation.

Throughout the book, SI metric equivalents are provided for most imperial units. The SI units appear in brackets [ ].

Boiler Selection Chart

Boiler Type	Output Capacity Range—MBH [kW] Efficiency Range	Fuel Types	Uses
Cast Iron Sectional	80 – 14,500 [23.5 – 4250] 80 – 92%	Oil, Gas, Coal, Wood/Fossil	Steam/Hot Water
Steel	1,200 – 18,000 [350 – 5275] 80 – 92%	Oil, Gas, Coal, Wood/Fossil, Electric	Steam/Hot Water
Scotch Marine	3,400 – 24,000 [1000 – 7000] 80 – 92%	Oil, Gas	Steam/Hot Water
Pulse Condensing	40 – 150 [12 – 44] 90 – 95%	Gas	Hot Water
Residential/Wall Hung	15 – 60 [4.5 – 17.5] 90 – 95%	Gas, Electric	Hot Water

All numbers shown in brackets are metric equivalents.

Efficiencies shown are averages and will vary with specific manufacturers.
For existing equipment, efficiencies may be 60-75%.

Figure 6.1

The Department of Energy has established test procedures to compare the "Annual Fuel Utilization Efficiency" (AFUE) of comparably sized boilers. Better insulation, heat extractors, intermittent ignition, induced draft, and automatic draft dampers contribute to the near 90-percent efficiencies claimed by manufacturers today.

In the search for higher efficiency, a new concept in gas-fired water boilers has been introduced. This innovation is the pulse condensing boiler, which relies on a sealed combustion system rather than on a conventional gas burner. The AFUE ratings for pulse-type boilers are in the low- to mid-90-percent range. Pulse-type boilers cost more initially than conventional types, but savings in other areas help to offset this added cost. Because these units vent through a plastic pipe to a side wall, no chimney is required. The pulse-type boiler also takes up less floor space, and its high efficiency saves on fuel costs.

Another innovation in the boiler field is the introduction from Europe of wall-hung, residential-size boilers. These gas-fired, compact, efficient (up to 80-percent AFUE) boilers may be directly vented through a wall or to a conventional flue. Combustion make-up air is directed to a sealed combustion chamber similar to that of the pulse-type boiler. Storage capacity is not needed in these boilers because the water is heated instantaneously as it flows from the boiler to the heating system. The boiler material consists mostly of steel. Heat-exchanger water tubes are

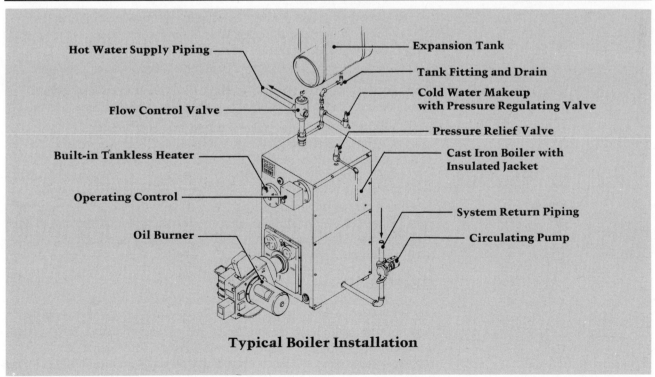

Typical Boiler Installation

Figure 6.2

usually made of copper or stainless steel, although some manufacturers use cast iron.

Conditions to consider when selecting a boiler include the following:
- **Accessibility:** Both for installation and for future replacement, cast iron boiler sections can be delivered through standard door or window openings. Some steel fire-tube boiler replacement tubes can be installed through strategically located door or window openings. Some steel water-tube boilers are made long and narrow to fit through standard door openings.
- **Economy of installation:** Truly packaged boilers have been factory-fired and tested and arrive on the job ready to be rigged or manhandled into place. Connections required are minimal – fuel, water, electricity, supply and return piping, and a flue connection to a stack or breeching.
- **Economy of operation:** In addition to the AFUE ratings, boiler output should be matched as closely as possible to the building heat loss. Installation of two or more modular boilers, piped and controlled to step-fire to match varying load conditions, should be carefully evaluated. Only on maximum design load would all boilers be firing at once. This method of installation not only increases boiler life, but also provides for continued heating capacity in the event that one boiler should fail.

Selecting the Boiler

When selecting boilers, a primary decision is the choice between steam and hot water. For small buildings, hot water is often selected because the costs of steam generation equipment and distribution piping are higher than for an all-water system. However, steam is an excellent medium because:

1. The high latent heat of steam vaporization permits large quantities of heat to be transmitted from boiler to terminal units with little loss of temperature.
2. Steam promotes its own circulation. It flows from naturally higher pressures (in boiler) to lower pressures (steam lines). Steam does not require pumps or fans.
3. Boiler output is easily modulated by varying steam pressure.

While hot water is the appropriate choice for many situations, steam should be considered when the budget allows and applications are suitable.

When selecting boilers, the general output size and fuel tend to limit the choices. Figure 6.1 shows that the types selected follow a general size relationship. Cast iron, steel, and scotch marine boilers are generally acceptable in the 3,000 to 14,000 MBH [875 – 4100 kW] range. Other factors, such as cost, availability, and personal preference based on experience, are also considerations. After a boiler type has been selected from Figure 6.1, the generation equipment data sheets at the end of this chapter can be used to make additional determinations and select features to complete the systems.

Boiler Sizing

The main features in sizing a boiler are its type and output. The boiler type is selected from Figure 6.1 and the generation equipment data sheets. Boiler output is designated as both *net* and *gross*: net output is the actual heat delivered by the boiler to the terminal units; gross or nozzle output is always larger than the net output and represents the heat content of the fuel consumed by the boiler. The boiler efficiency equals the net output,

H_n, divided by the gross output, H_g.

$$\text{Boiler efficiency} = \frac{\text{boiler net output (H}_n\text{)}}{\text{boiler gross output (H}_g\text{)}} = \frac{\text{(fuel output from boiler)}}{\text{(fuel value input to boiler)}}$$

To determine the net output desired for the boiler, the heat load, H_t, is used and increased for piping tax and pickup load.

H_n = H_t + 10 – 25% contingency reserve (for possible future loads and design uncertainties)
    + 5 – 15% pipe tax (5% if pipes are insulated)
    + 0 – 15% pickup load (0 if boiler runs 24 hours)

Generally, the required net output of the boiler is approximately 1.25 H_t.

Safety Features

Boilers have many incorporated safety features: (1) *Pressure relief valves* must be used to relieve excess pressure and prevent an explosive buildup of pressure. (2) A *low water cut-off* senses the water level and turns the burner off if the water level drops too low, preventing overheating of the metal. Other safety features incorporated into the system include (3) *fusible elements*, (4) *water feeders*, (5) *alarms*, (6) *thermal controls*, (7) *Fire-o-matic*, (8) *backflow prevention*, (9) *local shut-off switches*, and (10) *gauge glass*.

Accessories

Accessories, such as breeching, flues, and fuel piping, are also required to operate a boiler and must be considered when selecting a system. (For more information on accessories, see Chapter 13.)

Fuel Consumption

Annual fuel cost is determined using the following formula (see Figure 6.3):

$$F = \frac{24(DD)H_tC}{e(\Delta)V} \quad \text{i.e.,} \quad \frac{24 \text{ hours/day} \times \text{heating degree days/year} \times H_tC}{e \times \Delta T \text{ Design} \times V \text{ fuel}}$$

F = Fuel consumed per year in units (for unit, see Figure 5.3)
DD = Degree days per year (see Figure 3.9) (In metric applications, degree days are based on °C)
H_t = Total heat load (in MBH)
C = Correction factor for oversizing and outdoor temperature (see Figure 6.3)
e = System efficiency (see Figure 6.1)
ΔT = Design indoor minus design outdoor temperature (°F) [°C] ($T_i - T_o$)
V = Heating value of fuel per unit (MBH/unit) [kW/unit] (see Figure 5.3)

As an example, the fuel consumed for a building in Casper, Wyoming, burning #2 oil, using a boiler with an output of 600 MBH for a building heat load of 500 MBH, uses the values listed below.

DD = 7,400 (see Figure 3.9)
H_t = 500 MBH (given)
C = See Figure 6.3

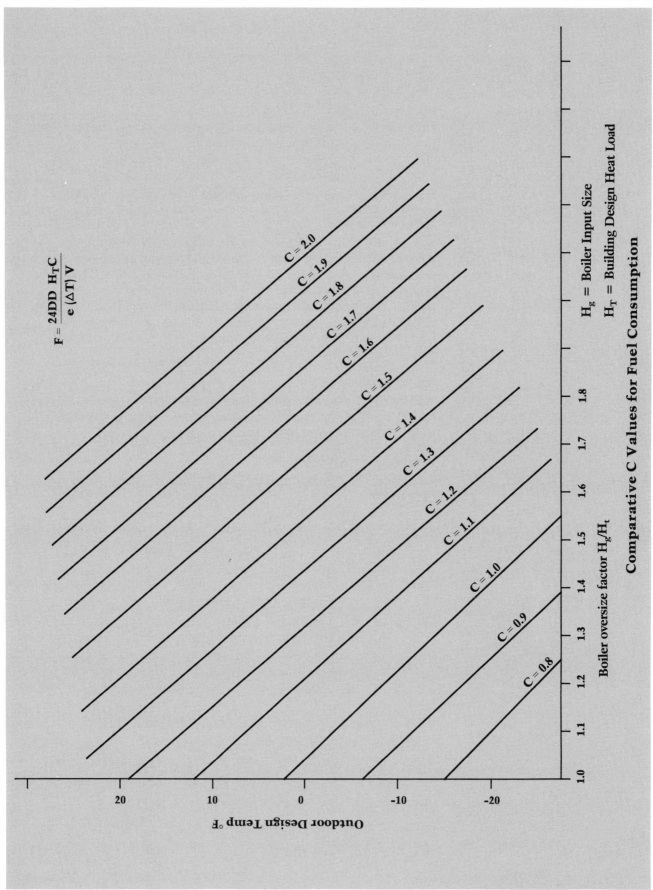

Figure 6.3

H_i/H_t = 600/500 = 1.2
T_o = −5°F
C = 1.05
e = .85 (verify w/boiler selected) (see Figure 6.1)
ΔT = 68 − (−5) = 73°F
V = 138.5

Thus, for this example, the fuel consumption is calculated as follows:

$$F = \frac{24 \times 7,400 \times 500 \times 1.05}{0.85 \times 73 \times 138.5} = 10,850 \text{ gal./yr.}$$

Cost = F × unit price of fuel

Burners

Fuel for boilers is injected and burned in the combustion chamber by the burner. Burners may be fueled by oil, gas, or a combination of the two. The combined burner has the advantage of switching to the more economical fuel. When oil is used, the burner atomizes the oil into a fine mist, which is then ignited. The finer the mist, the more complete and efficient the combustion. In a gun-type burner, a nozzle is used to mist the oil, which is fed to it under pressure. Rotary burners vaporize the oil as it leaves a rotating cup while mixing with the combustion air. Gas burns directly without any mechanical systems to atomize it.

Burners are controlled to shut down under several conditions for safety. The stack temperature is monitored to ensure that combustion has taken place. Failure to cause a rise in stack temperature within a few minutes due to loss of fuel or ignition will stop the burner from operating.

Burner flames are ignited using one of two methods: pilot or spark. In pilot ignition, a pilot light burns continually and the oil mist or gas injected over it on call is ignited. With spark ignition (sometimes called electronic ignition), a spark is passed across the oil mist or gas. Some electronic combustion controls have the ability to continuously monitor the combustion air and adjust the quantity for optimum conditions.

Tanks

There are two common tanks used with HVAC equipment: fuel storage tanks and expansion tanks. Both are shown in the generation equipment data sheets (Figure 6.11) at the end of this chapter.

Fuel Storage Tanks

Oil and gas may be stored on site for a building's HVAC system. Oil is the most common fuel stored in tanks. Fuel tanks may be above ground or below ground. Above-ground tanks are subject to safety provisions, such as containment areas to hold a volume equal to the liquid volume of the tank, so that in the event of a spill or major leak, the fuel can be contained. Walls are used to create the containment area.

Underground tanks may be of single- or double-wall construction. The double-wall tanks offer better protection against leaks. Leak detection systems are required for underground tanks in some communities. Cathodic protection for underground tanks to prevent electrolytic deterioration as well as bituminous coatings are routinely used to extend the tank life.

Buried tanks must be held down because in the spring, when the tanks are likely to be empty, the spring rains flood the surrounding earth. The resulting buoyant force can lift the tank unless it is secured to a concrete pad. The pad is usually sized to weigh 1.5 times the buoyant force of the tank and should be approximately 1'-6" [.5 m] larger than the tank overall. The buoyant force is calculated as follows:

$$B = \frac{62.5 \text{ lbs.}}{\text{ft.}^3} \cdot \frac{0.133 \text{ ft.}^3}{\text{gal.}} = 8.33T \left[\frac{100 \text{kg}}{\text{m}^3} \cdot \frac{0.001 \text{ m}^3}{\text{L}} = T \right]$$

B = Buoyant force

T = Tank capacity (gal.) [L]

Concrete used for this purpose should weigh 150 pounds per cubic foot; therefore, the volume of the concrete pad required is calculated as follows:

$$V_p(\text{ft.}^3) = \frac{8.33T \text{ (gal.)} \cdot 1.5}{150 \text{ lbs./ft.}^3} = 0.0833T$$

$$\left[\frac{T \text{ Liters} \cdot 1.5}{2400 \text{ kg/m}^3} = 6.25 \times 10^{-4} T \right]$$

Example

The concrete pad required to hold down a 5,000 gallon [19 000 liter] fuel tank 23 feet [7 m] long and 7.5 feet [2.25 m] in diameter is calculated below.

$V_p \text{ft.}^3 = 0.0833T$ $[6.25 \times 10^{-4} \times T = 6.25 \times 10^{-4} \times 19\ 000]$
$\qquad = 0.0833 \cdot 5,000$ $\qquad\qquad\qquad\qquad\qquad = 11.9 \text{ m}^3$
$\qquad = 416.5 \text{ ft.}^3$

Pad length $\qquad 23 + 1.5 = 24.5$ ft. $[7 + 0.5 = 7.5$ m$]$
 width $\qquad\quad 7.5 + 1.5 = 9$ ft. $[2.25 + .5 = 2.75]$
 thickness try 2'-0" [0.6 m]
 $24.5 \cdot 9 \cdot 2 = 441$ ft.^{3} > 416.5 ft.^{3} [$7.5 \times 2.75 \times 0.6 = 12.375$ m^3
 $\qquad\qquad\qquad\qquad\qquad\qquad\qquad\qquad\qquad\qquad > 11.9$ m^3]

Therefore, 441 cubic feet [12.375 cubic meters] of concrete would be adequate to secure the tank.

A more exact analysis that uses the weight of earth above the pad or a low water table can be used to reduce the above requirements, which are generally conservative. Hold-down straps or angles and rods must be provided to secure the tank. They should be anchored in the concrete and the straps should be passed over the top of the tank to hold it down.

Expansion Tanks

All hydronic systems undergo changes in temperature that cause the water to expand and contract. An expansion tank is always provided on each closed loop piping system, because the tank allows the water to expand into it as the water volume increases with the temperature.

There are two formulas for computing the expansion volume required: one for hot water and one for cold water.

For hot water (160–280°F) [70–140 °C]

$$V_t = \frac{(0.00041T - 0.0466)}{\dfrac{P_a}{P_f} - \dfrac{P_a}{P_o}} V_3 \qquad \left[V_t \left(\frac{0.000739t - 0.0336}{\dfrac{P_a}{P_f} - \dfrac{P_a}{P_o}} \right) V_s \right]$$

For chilled water (40 – 100°F) [5 – 40 °C]

$$V_t = \frac{(0.006) V_s}{\dfrac{P_a}{P_f} - \dfrac{P_a}{P_o}}$$

V_t = expansion tank size (gal.) [L]
V_s = total gallons [liters] in system
t = average water temperature (°F) [°C]
P_a = atmospheric pressure (pressure in tank before water fill) (psia) [kPa]
P_f = design fill pressure in tank (psia) [kPa]
P_o = final pressure at tank (psia) [kPa]

Expansion tanks may permit the liquid and air to be in contact or they may be separated by a diaphragm. Properly sized expansion tanks have the ability to maintain a set pressure range at the tank.

Heat Exchangers

If steam or hot water service is available to a building from a remote source of supply, the need for a boiler is eliminated. If the proposed system is forced hot water and the remote source of supply is steam or high-temperature hot water, a shell and tube-type heat exchanger (converter) must be provided. A heat exchanger is also required if the building is served by low-temperature chilled water or brine for building cooling.

Warm Air Systems – Heating Only

Warm air systems heat and distribute warm air without using water. They are convenient because they avoid some problems associated with freezing, but they do require ductwork, which is larger than the piping used in hydronic systems.

Furnaces

Forced warm air furnaces utilize self-contained fans and combustion chambers. Forced warm air furnaces use burners similar to hydronic boilers. The furnace operates by drawing cool air around the combustion chamber and blowing this heated air into the distribution system, utilizing the fan. The net output of the furnace should equal the building design heat loss plus the allowances for losses in a manner similar to boilers.

H_n = H_t + 10 – 25% contingency reserve
       + 2 – 10% duct loss
       + 0 – 5% pickup load

Generally, the required net output, H_n, of the furnace is approximately 1.15 H_t.

Furnaces are specified by the net output and/or by the cfm [L/s] of air and static pressure the fan delivers. The fan static pressure must be verified to ensure that it has enough pressure to deliver the air through the ductwork. For a supply air temperature of 130°F [55 °C] and a room air temperature of 70°F [21 °C] (a 60°F [34 °C] temperature difference), the quantity of air needed to compensate for heat loss can be found using the following equation:

H_t = 1.1 cfm ΔT

$$\text{cfm} = \frac{H_t}{1.1 \cdot \Delta T} = \frac{H_t}{1.1 \cdot 60} \left[\frac{H_t}{1.23 \cdot 34} \right]$$

Figure 6.4 illustrates a typical warm air heating system.

Duct Heaters

Duct heaters are frequently used to reheat cool air that needs local tempering before it is delivered to a space. Duct heaters are a convenient device because they can easily be individually controlled independent of the main heating system.

Many applications require electric duct heaters. A typical example is a fresh air supply intake. Sometimes mounted on the roof, remote from the main distribution system, an electric coil easily heats incoming cold air to 55 – 60°F [12 – 16 °C] without threat of freezing.

Makeup Air Units

When the fresh air requirements are large, special makeup air units are employed. A common example is a commercial kitchen, where large quantities of air must be introduced to make up for the air exhausted by the range hood. The quantity of air should be ten to twenty percent larger than the exhaust air to keep spaces under positive pressure. The following

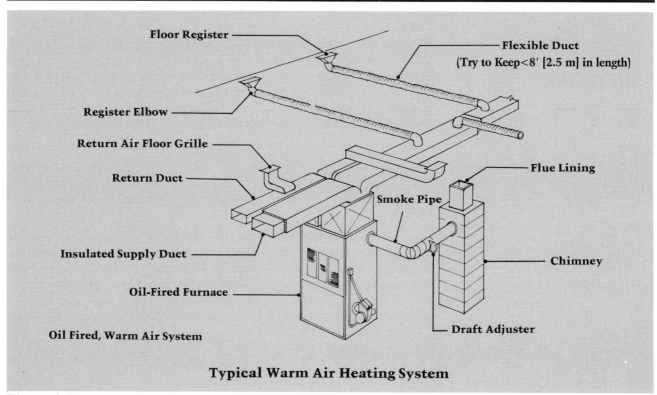

Typical Warm Air Heating System

Figure 6.4

formula is used to determine the rating (in Btu) [W] of the makeup air system:

H	= 1.1 cfm ΔT [1.23 L/s ΔT]
cfm [L/s]	= exhaust air quantity
ΔT	= $T_i - T_o$ °F [°C]
T_i	= indoor design temperature
T_o	= outdoor design temperature

Hydronic Systems – Cooling Only

The central component of any chilled water air conditioning system is the water chiller. Chillers produce water from 42 – 55°F [5.5 – 13 °C], which is then distributed to coils, fan coil units, or central station air handling units for cooling. This is done by passing water through the evaporator section, where refrigerant gas absorbs the heat and chills the water. The chilled water goes out through the distribution system. The refrigerant gas from the evaporator goes to the compressor section. The hot gas is cooled by a condenser and the refrigerant is then returned to the evaporator.

Chillers are sized to meet the total building heat gain in addition to an allowance for piping tax, pickup load, and contingency. Because cooling often is not done twenty-four hours a day, the pickup load can be large or the equipment can be activated two to four hours before being occupied.

Packaged water chillers are available in three basic designs: the reciprocating compressor, direct-expansion type; the centrifugal compressor, direct-expansion type; and the absorption type generator (see Figures 1.16 and 1.17).

Chillers

Chillers and other cooling apparatus are sized by the ton [kW]. One ton of cooling equals the melting rate of one ton of ice in a twenty-four hour period, or 12,000 Btu's per hour. The three types of chillers vary significantly in their cooling power, as well as in their operation.

The *reciprocating compressor chiller*, which generates cooling capacities in the range of 10 to 200 tons [35 to 700 kW], is usually powered by an electric motor. These machines often have multiple compressors, which allow for staged cooling.

The *centrifugal compressor*, which generates cooling capacities ranging from one hundred to several thousand tons, is also commonly powered by an electric motor, but it may be designed for a steam-turbine drive as well. It is possible to power a centrifugal compressor or reciprocating compressor from an internal combustion engine. Some engines for this purpose run on natural gas.

Absorption-type chillers provide cooling capacities ranging from 3 to 1,600 tons [10.5 to 5600 kW]. Because they use water as a refrigerant and lithium bromide or other salts as an absorbent, this system consumes only about 10 percent of the electrical power required to operate the conventional reciprocating and centrifugal direct-expansion chillers. This low consumption advantage is particularly desirable in buildings where an electrical power failure triggers an emergency backup system, such as in hospitals, data processing centers, electronic switching system locations, and other buildings that must continue to function on auxiliary power. Absorption-type chillers require a heat source to maintain the absorption process, which is economically advantageous in areas where electric power is scarce or costly, where gas rates are low, or where waste or process steam or hot water is available during the cooling season. As discussed in Chapter 1, absorption chillers are also free of CFCs. Solar power may also be used in some areas to generate the heat required for the absorption process. Figure 6.5 shows a typical chiller, condenser, and cooling tower installation.

Condensers

The hot refrigerant gas from the chillers may be air- or water-cooled. The heat absorbed by the chiller must be removed, which is done by rejecting the heat to the outside. Very small chillers are available with an air-cooled condenser built into the package. For larger systems and for systems that may create too much noise for their location, air-cooled condensers are installed at a distance from the chiller, and the two units are connected with refrigerant piping.

The basic process for air conditioning systems using refrigerants as the cooling medium is the cooling and condensing back to liquid form of the refrigerant gas that was heated during the evaporation stage of the cycle. This condensation is achieved by cooling the gas with air, water, or a combination of both.

Air-cooled condensers cool the refrigerant by blowing air directly across the refrigerant condenser coil; evaporative condensers use the same method of cooling, but with the addition of a spray of water over the coil to expedite the process.

The condenser for water-cooled chillers is piped into a remote water source, such as a cooling tower, pond, or river, via the "condenser water system". Completely packaged chiller systems may include built-in chilled water pumps and all interconnecting piping, wiring, and controls. All of the components of the completely packaged unit are factory-installed and tested prior to shipment to the installation site for connection to the chilled water and condenser water systems.

When water is used as the condensing medium and is abundant enough that recycling is not required, it may be piped to a drain after performing its cooling function and returned to its source. The source may be a river,

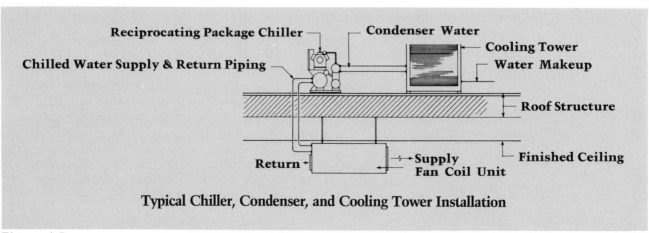

Typical Chiller, Condenser, and Cooling Tower Installation

Figure 6.5

a pond, or the ocean. If the water supply is limited, expensive, or regulated by environmental restrictions, a water-conserving or recycling system must be employed. Several types of systems may be installed to conform to these limitations. For example, a water-regulating valve, a spray pond, a natural draft cooling tower, or a mechanical draft cooling tower may be used.

In very small cooling systems, a temperature-controlled, water-regulating valve may be used, provided that such a system is permitted by local environmental and/or building codes. The regulating valve system functions by allowing cooling water to flow when the condenser temperature rises, and conversely, by stopping the flow as the temperature falls. The problem with this system, however, is that the heated condenser water cannot be recycled and is therefore wasted during the flow cycle.

Cooling Towers

Although both are viable and available methods of cooling, *spray pond* and *natural draft* cooling tower systems are not commonly used for building air conditioning. Because of water loss caused by excessive drift and the large amount of space required for their installation and operation, spray pond and natural draft cooling tower systems are less desirable than *mechanical draft* cooling tower systems.

Mechanical draft cooling tower systems are classified in two basic designs: *induced draft* and *forced draft*. In an induced-draft tower, a fan positioned at the top of the structure draws air upward through the tower as the warm condenser water spills down. A cross-flow induced-draft tower operates on the same principle, except that the air is drawn horizontally through the spill area from one side. The air is then discharged through a fan located on the opposite side. In a forced-draft tower, the fan is located at the bottom or on the side of the structure. Air is forced by the fan into the water spill area, through the water, and then discharged at the top. All designs of mechanical draft towers are rated based on the tons of cooling; three gallons of condenser water per minute per ton is an approximate tower sizing method.

After the water has been cooled in the tower, it cycles back through a heat exchanger, or condenser, in the refrigeration unit. Here, it again picks up heat and is pumped back to the cooling tower. The piping system is called the condenser water system. Figure 6.6 shows the layout of a typical condenser water system.

The actual process of cooling within the mechanical draft cooling tower takes place when air is moved across or counter to a stream of water falling through a system of "fill" to the tower basin. After the cooled water reaches the basin, it is piped back to the condenser. Some of the droplets created by the fill are carried away by the moving air as "drift" and some of the droplets evaporate. This limited loss of water is to be expected as part of the operation of the tower system.

Because of the loss of water by drift, evaporation, and bleed-off, replenishment water must be added to the tower basin to maintain a predetermined level and to ensure continuous operation of the system. To prevent scale buildup, algae, bacterial growth, or corrosion, tower water should be treated with chemicals or ozone applications.

The materials used in constructing mechanical draft cooling towers include redwood (which is most common), other treated woods, and various metals, plastics, concrete, or ceramic materials. The fill, which is the most important element in the tower's construction, may be manufactured from the same wide variety of materials used in the tower structure. Factory-assembled, prepackaged towers are available and are usually preferable to built-in-place units. Multiple tower installations are now being used for large systems.

The location of cooling towers is an important consideration for both practical and aesthetic reasons. They may be located outside of the building on the roof or on the ground. If space permits, a cooling tower may be installed indoors by substituting centrifugal fans for the conventional noisy propeller type, and adding air intake and exhaust ductwork. The tower discharge should not be directed into the prevailing wind or toward doors, windows, or building fresh air intakes. In general, common sense should be used when determining tower placement so that the noise, heat, and humidity the system creates do not interfere with building operation and comfort. The manufacturer's guidelines for installation should be strictly followed, especially the sections that address clearances for maximum air flow, maintenance, and future unit replacement.

Economical operation of a cooling tower system may be achieved through effective control and management of several critical aspects of its operation, including careful monitoring of water treatment, selecting and

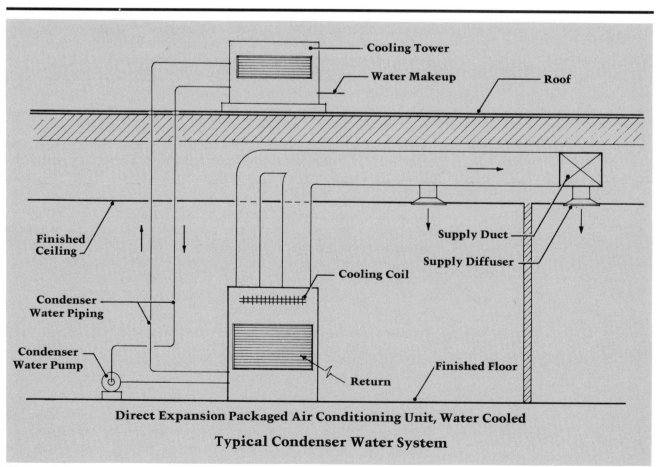

Figure 6.6

maintaining the most efficient condensing temperature, and controlling water temperature with fan cycling and two-speed fans. Variable frequency drives are being installed on power fans as an energy-conservation measure.

A recent development in tower water system operation allows the tower to substitute for the water chiller under certain favorable climatic conditions. This method cannot be implemented in all cases, but in situations where it can be employed, substantial savings result in the reduced cost of chiller operation. A plate coil-type heat exchanger is used to good advantage in this type of system.

Cooling towers are shown in the generation equipment data sheets (Figure 6.11) at the end of this chapter. Cooling towers are sized to match the load of the chiller (the chiller net output plus the mechanical load of the compressor).

Split Systems

Split systems are used both for cooling only and for combined heating and cooling systems. Split systems are discussed in the following section.

Combined Heating and Cooling Systems

Central Station Air Handling Units

An air handling unit consists of a filter section, a fan section, and a coil section on a common base. For large installations, each component is selected individually. Smaller units, often called package units, contain all components in a single prefabricated enclosure. They are used to distribute clean, cooled or heated air to the occupied building spaces. These units are available in a wide range of capacities, from 200 cubic feet per minute [95 L/s] to tens of thousands of cubic feet per minute [L/s]. The units also vary in complexity of design and versatility of operation. Small units tend to have relatively simple coil and filter arrangements and a modest-sized fan motor. Larger, more sophisticated units usually require remote placement. Because of the need to overcome resistance losses caused by intake and supply ductwork and by complex coil, filter, and damper configurations, the fan motor horsepower must be dramatically increased. Figure 6.7 shows a typical rooftop air handling system.

Small air handling units may be located and mounted in a variety of settings and by different methods. They may be mounted on the floor or hung from walls or ceilings with no discharge ductwork required in the room they service. Small air handling units require supply and return piping for heating and/or cooling. If these units are used for cooling, a drain pan and piping connection are also required to run the condensation from the cooling coil to waste.

Determining the proper size, number, capacity, type, and configuration of coils in the unit is a prime consideration when selecting and/or designing an air handling unit. As a general rule, the amount of air (in cubic feet per minute) to be handled by the unit determines the size and number of the various coils.

Electric or hydronic coils are used for heating; chilled water or direct expansion coils are used for cooling. As the units increase in size and complexity, the coil configurations and arrangements become limitless. A simple heating and cooling unit, for example, may use the same coil for either hot or chilled water. A large unit usually demands different types of coils to perform many separate functions. In humid conditions, the air temperature may be intentionally lowered to remove moisture. In this case, a reheat coil is added to bring the air temperature up to its desired level.

Conversely, in dry conditions, a humidifier component is built into the unit. If outside air is introduced to the unit at subfreezing temperatures, then a preheat coil is placed in the outside air intake duct. Precautions must be taken to prevent icing or freeze-up of this coil.

Certain precautions should be taken to prevent damage and to ensure the efficiency of the unit's coils and other components. To protect the coil surfaces from accumulating dust and other airborne impurities, a filter section is a necessary addition to the unit. If a unit is designed to cool air, a drain pan must be included beneath the coiling section. This pan is then piped to an indirect drain to dispose of the unwanted condensation.

Another protection precaution involves the internal insulation of the fan coil casing. If this precaution is not taken to protect the cooling coil section and all other sections "downstream," corrosive or rust-causing condensation will damage the casing and discharge ductwork. Insulating this casing also helps deaden the noise of the fan. Noise can be further reduced by installing flexible connections between the unit and its ductwork, as well as by mounting the unit on vibration-absorbing hangers.

Free Cooling

In free cooling, or economizer cycle and enthalpy control, fresh outdoor air is used to cool in spring, fall, and winter, instead of running the chiller whenever the outside temperature is cooler than the inside design temperature and it is not too humid. As mentioned previously, the cooling tower can also be used "in reverse" for producing cool water on mild days for hydronic systems. In air systems, the savings are similar. No compressors need to be run for free cooling. Except for the cost of running

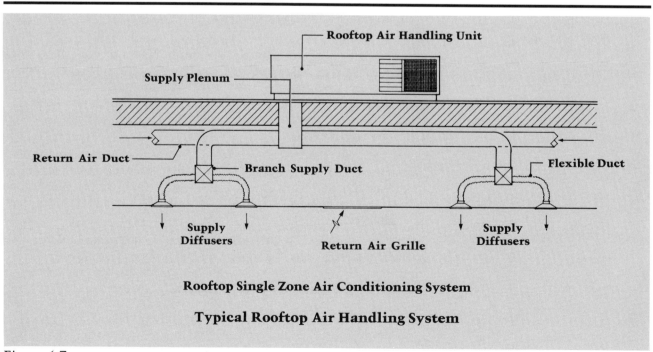

Figure 6.7

the fans, the cooling is "free." As the outside air gets colder than 65°F [13 °C], it is mixed with the building return air (typically 70°F) [21 °C] to keep the supply air from being too cold. This method automatically provides large quantities of outside air, which also satisfies the ventilation requirements.

As the outside air approaches 32°F [0 °C], some heat is provided to prevent freezing of the coils. If the free cooling is tied to an automatic control system that measures indoor and outdoor temperatures and controls supply, return, and exhaust air dampers, the building is said to have an economizer cycle. If, in addition to the economizer, the control system also measures the humidity of the air and computes the latent and sensible energy of the air in modulating the dampers, the building is said to have enthalpy control.

Figure 6.8 illustrates an economizer system. The economizer system monitors mixed air temperature and outdoor temperature and controls outdoor-air, return-air, and exhaust-air dampers. If the system is properly designed there should be no need to provide heat for the outdoor air as it approaches 32°F. Naturally, there must be a heating coil downstream of the mixed air plenum.

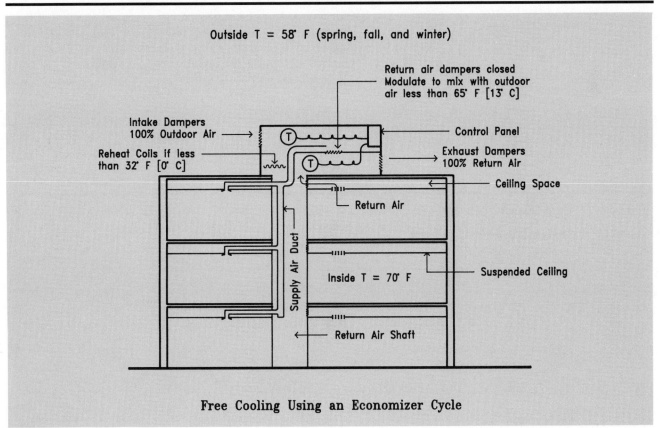

Figure 6.8

Packaged Units

For loads under 100 tons [350 kW], packaged units are a convenient option. In one assembly, the fan, refrigeration equipment, and heating equipment are assembled and ready to operate with minimal field connections. They are most often air-cooled and mounted on a roof or through the wall where outside air can cool the condenser coils. Most manufacturers have a wide range of packaged units suitable for various installation situations and fuel types.

Rooftop Units

For single-story buildings or spaces with an accessible flat roof, economical rooftop packaged units are often preferred. They can provide both heating and cooling, supply the necessary fresh outdoor air, and have economizer options. In addition, they should generally have a local supply of replacement parts and service.

Through-the-Wall Units

Self-contained through-the-wall units are similar to rooftop packaged units, except they are placed inside a building, usually near or through an exterior wall.

Split Systems

For locations where the space to be cooled is remote from an available wall or roof, split systems are used. In split systems, the condenser and compressor are placed outdoors on a pad or roof, and the evaporator and supply fan with evaporator coil is placed in the space, often in a closet or above the ceiling. The refrigerant gas from the evaporator in the space is piped to the outdoor condensing unit by copper tubing, where it rejects the heat and returns as a liquid to the evaporator. This system removes the noise of the compressor from the conditioned or occupied space, and is therefore very desirable. It also allows particular tenants, such as first-floor retail or individual residential units, to have cooling and pay for it individually. Figure 6.9 illustrates the layout of a split system.

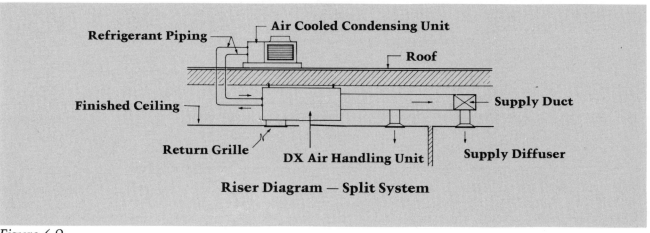

Figure 6.9

Heat Pumps

A method of saving energy when buildings are simultaneously heated and cooled is with water source heat pumps, illustrated in Figure 6.10. A heat pump is a complete package refrigeration unit that can "pump" heat in or out of the water passing through it. Also available are air-to-air heat pumps for milder climates where only minimal heating is required.

In the case of a water source heat pump system, water is continuously pumped in a loop, which connects a boiler, a cooling tower, and a series of heat pumps. The boiler and cooling tower are activated as necessary to maintain the temperature as close to 75°F [24 °C] as possible (always above 60°F [15.5 °C] and below 90°F [41.5 °C]). By "pumping" the heat absorbed from the room into the water, the heat pump raises the temperature of the water. This is possible because the heat pump is able to reverse the direction of the refrigerant in its loop. When heating a space, the air stream coil is used as the condenser and the water stream coil acts as the evaporator. When cooling a space, an internal valve in the heat pump reverses the flow of the refrigerant. Now the air stream is used as the evaporator, the air is cooled, and the water loop is warmed as it flows over the condenser. By themselves, heat pump systems do not provide ventilation, but they do take advantage of situations where heating and cooling occur simultaneously. Because there is considerable heat inertia in the water of the heat pump loop and because some heat pumps are adding heat from the building to the water loop while other heat pumps are removing heat from the water loop, it is possible to move heat from one portion of the building to another without using a boiler or cooling tower. This saves considerable energy on the total fuel bill.

Both heat pumps and economizer systems offer energy savings. Both use fans, but the compressors in the heat pumps consume energy, which can make them slightly more expensive than economizer systems, depending on the specific applications. On the other hand, all of the heat pump energy (fans and compressors) can be individually run, metered, and paid for by the tenant, who can expect to pay only for what is used and to share with other tenants in the benefits of simultaneous heating and cooling during the year.

Heat pumps supply air that can be sent into a constant volume ductwork system or further connected to variable air volume (VAV) boxes. When one heat pump is connected to several spaces or offices, VAV boxes allow a better measure of individual control.

Generation Equipment Data Sheets

Figure 6.11 (a and b) on the following pages contains equipment data sheets for generation equipment. Each of these sheets includes efficiency ratings, useful life, typical uses, capacity ranges, advantages and disadvantages, special considerations, accessories, and additional equipment needed for each piece of generating equipment.

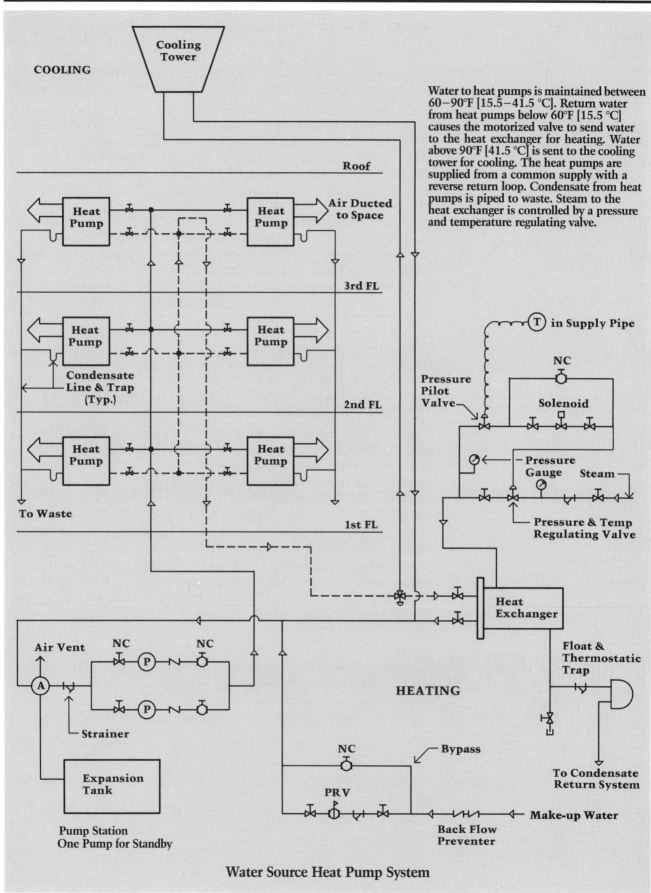

Figure 6.10

Problems

6.1 Despite their high cost, steam and electricity are commonly selected as fuels when available in high density urban areas. Why?

6.2 A cabin in the woods near Portsmouth, New Hampshire is to be heated with propane. It has a 40,000 Btu/hr. [11 725 W] design heat loss; the furnace is oversized by 25% and is 80% efficient. How many gallons [liters] of fuel will be needed each year?

6.3 A hospital in San Francisco, California has a design heat load of 10,000 MBH [2930 kW].
   a. If #4 oil is used for fuel, calculate the amount of fuel used per year.
   b. For the amount of fuel, what size tanks would be needed if delivery in winter can be made weekly?
   c. For a buried tank, calculate the hold-down pad, and sketch the tank. Show devices for oil containment. Comment on the tank size and suggest an alternative.

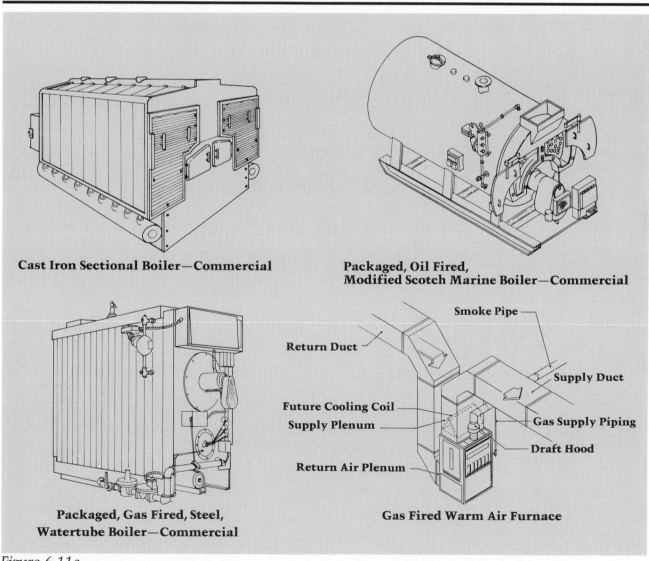

Cast Iron Sectional Boiler—Commercial

Packaged, Oil Fired, Modified Scotch Marine Boiler—Commercial

Packaged, Gas Fired, Steel, Watertube Boiler—Commercial

Gas Fired Warm Air Furnace

Figure 6.11a

Generation Equipment—Heating

Heating	Type and R.S. Means No.	Capacity (MBH) [kW]	Efficiency (%)	Advantages	Disadvantages	Related Equipment and Accessories	Considerations
Cast Iron Sectional Boilers	Gas-fired 155-115	80 – 7,000 [24 – 2050]	80 – 92	25 years (plus) useful life.	Heavy. Hard to clean. High standby losses.	Insulated jacket. Type B vent required for gas. Push nipples or gaskets required at sections. Flue and breeching. Safety devices. Two-stage firing available. May be shipped as a package or broken down for field assembly.	Gas service required.
	Oil-fired 155-120	100 – 7,000 [30 – 2050]	80 – 92	Long life. Durability. Expandable. Wide range of sizes. Ease of replacement. Cracked section may be isolated rather than shut down entire boiler.			Fuel oil storage and pumping required.
	Gas/oil combination 155-125	720 – 13,500 [210 – 4000]	80 – 92				Allows for use of least expensive fuel available.
	Solid fuel 155-130	148 – 4,600 [43 – 1350]	N/A				Allows for use of inexpensive fuel or burning of waste by-products.
Steel Boilers	Solid fuel 155-130	1,500 – 18,000 [440 – 5275]	N/A	15 to 25 years useful life. Less expensive to purchase. Tubes easily accessed for cleaning or replacement. Leaking tubes may be plugged for future replacement.	Regular maintenance required to extend boiler life and retain efficiency. Space allowance to pull or punch tubes.	Insulated jacket. Type B vent required for gas. Flue and breeching. Safety devices. Two-stage firing available. Must be shipped as one piece (two-piece available on special order).	Allows for use of inexpensive fuel or burning of waste by-products.
	Oil or gas 155-115/135	144 – 23,435 [42 – 6868]	80 – 92				Fuel service and/or storage required.
	Electric 155-100	6 – 2,500 [2 – 750]	100				No flue or chimney required. No fuel storage required. Large electric service required.
Novel Residential Type Boilers	Pulse/condensing gas-fired 155-115	44 – 134 [13 – 40]	90 – 95	No chimney required. Less floor space. Ease of installation.	Noisy. Acid waste.	PVC through-wall flue. Plastic drain lines.	Gas service required.
	Wall hung gas-fired 155-115	44 – 64 [13 – 19]	85	No chimney required. Less floor space. Ease of installation.	Structural support considerations.	Through-wall flue.	Gas service required. Available in cast iron or steel.
Furnaces — Warm Air	Electric 155-420	30 – 141 [9 – 41]	100	Quick response. No freeze-up of system. Air cleaning feature. Cooling option at approximately 1/3 of heating output. 15 to 30 years useful life.	Space considerations for ductwork. Humidification required. Large temperature swings. Noisy. Air filtering required.	Insulated jacket. Type B vent required for gas. Safety devices. Shipped as a package.	No flue or chimney required. No fuel storage required. Large electric service required.
	Gas 155-420	42 – 400 [12.5 – 120]	80 – 92				Gas service required.
	Oil 155-420	55 – 400 [16 – 120]	80 – 92				Fuel oil storage and pumping required.
	Solid fuel 155-420	112 – 170 [33 – 50]	N/A				Allows for use of inexpensive fuel or burning of waste by-products, etc.

Figure 6.11a (continued)

Generation Equipment – Heating (continued)

Heating	Type and R.S. Means No.	Capacity (MBH) [kW]	Efficiency (%)	Advantages	Disadvantages	Related Equipment and Accessories	Considerations
Makeup Air Unit	Gas 155-461	168 – 6,275 [50-1840]	80 – 92	Quick response. No freeze-up of system. Air cleaning feature. Cooling option at approximately 1/3 of heating output. 15 to 30 years useful life.	Space consideration for ductwork. Humidification required. Large temperature swings. Noisy. Air filtering required.	Insulated jacket. Type B vent required for gas. Safety devices. Shipped as a package.	Makeup air for kitchens or other large volume considerations. May be roof mounted or indoors.
Burners	Gas 155-230	35 – 5,760 [10 – 1690]	80 – 92	Clean burning. Available fuel.	Danger of leakage.		Usually furnished with boiler or furnace.
	Oil 155-230	.5 – 12 (GPH) [5.25 × 10⁻⁴ –63 × 10⁻⁴ L/s]	80 – 92		Oil spill hard to contain or clean up. Oil supply vulnerable.		Usually furnished with boiler or furnace.
	Gas/Oil combination 155-230	400 – 6,300 [120 – 1850]	80 – 92	Choice of available fuel.	High original cost.		May be furnished as original equipment. Takes advantage of least expensive fuel available.
	Coal stoker 155-230	1,000 – 7,300 [293 – 2140]	N/A	Inexpensive fuel.	Dirty storage and exhaust. Needs coal storage space.	Conveyor or manual feed.	Uses inexpensive coal or waste by-products.
Fuel Oil Storage Tanks	Fiberglass underground 155-671	550 – 48,000 gal. [2130 – 186 000 L]	N/A	Cannot rust. Ease of handling. Dielectric material.	Leak detection difficult.	Fill vent and sound piping to grade or higher. Hold-down pads and anchors.	Double wall available. Extreme caution in backfilling.
	Steel underground 155-671	500 – 30,000 gal. [1950 – 116 250 L]	N/A	Inherent strength of steel prevents crushing.	Leak detection difficult. Subject to electrolytic action.		Double wall available. Normal caution in backfilling.
	Steel above ground	275 – 5,000 gal. [1065 – 19 375 L]	N/A	Ease of replacement. No excavation or backfill required.	Takes up valuable space.		Visual leakage control.
Expansion Tanks	Liquid expansion	15 – 400 gal. [58 – 1550 L]	N/A	Can be atmospherically recharged.	Subject to flooding. Pressures in system can vary.	Diverter fittings. Drain valve.	20 – 30 years useful life. Used on hot and chilled water systems.
	Diaphragm (captive air)	19 – 528 gal. [7.5 – 2050 L]	N/A	Cannot waterlog. Constant pressure possible.	Diaphragm can rupture or lose seal.		20 years (plus) useful life. Used on hot and chilled water systems.

All numbers shown in brackets are metric equivalents.

Figure 6.11a (continued)

Generation Equipment — Cooling

Packaged Water Chillers R.S. Means No. 157-110 157-190	Typical Uses: Produce chilled water for cooling coils, fan coils Useful Life: 15 – 30 years Capacity Range: 2 – 250 tons [7 – 880 kW]	
Advantages	**Disadvantages**	**Remarks**

Direct Expansion
  Chilled water is piped throughout the building and available on demand.
  No environmental concern.
	Reciprocating type can be noisy.	Most chillers operate on electricity.

Absorption
  Very competitive where heat source is inexpensive.
  Few mechanical parts.
  Use little electrical power for operating.
	High initial cost.
	Requires frequent maintenance.
	Disposal of lithium bromide is an environmental concern.
		Energy source is steam or hot water. Gas is also used frequently.

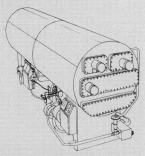

Absorption, Gas-Fired, Air Cooled

Condensers R.S. Means No. 157-225	Typical Uses: Cools refrigerant gases leaving compressor — air cooled or water cooled Useful Life: 15 – 30 years Capacity Range: 20 – 100 tons [70 – 350 kW] (air cooled)	
Advantages	**Disadvantages**	**Remarks**
Air Cooled Cools refrigerant gas directly. Fewer overall components to maintain. Can be grouped for larger capacities.	Limited in size but can be grouped as modules. Noisy.	Air cooled preferred when available. Heavy loads — check with structural engineer.
Water Cooled Generally integral with hydronic chiller.	Cooling tower or evaporative condenser required. Freezing of water lines a problem in northern climates if winter operation is needed.	Evaporative condensers are very efficient under certain conditions and can be used for the cooling options. Local codes regulate their use due to heavy water consumption.

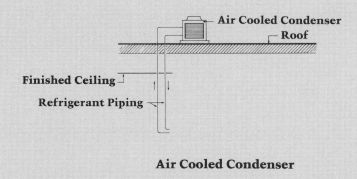

Air Cooled Condenser

All numbers shown in brackets are metric equivalents.

Figure 6.11b

Generation Equipment – Cooling (continued)

Rooftop Air Conditioning Units R.S. Means No. 157-180	Typical Uses: Commercial applications Useful Life: 10 – 20 years Capacity Range: 3 – 100 tons [10.5 – 350 kW]	
Advantages	**Disadvantages**	**Remarks**
Single package, low initial cost. Ease of installation. No flue required. Valuable floor space not taken up for a mechanical room. Very serviceable. Minimum field connections.	Noisy. May have to be screened or shielded for aesthetic purposes.	Electric for cooling. Gas or electric for heating. Multi-zone capability with some units. Economizer option recommended.

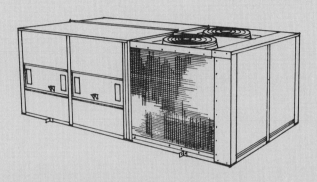

Packaged Rooftop Air Conditioner

Self-contained Air Conditioning Units R.S. Means No. 157-185	Typical Uses: Commercial applications Useful Life: 10 – 20 years Capacity Range: 3 – 60 tons [10.5 – 17.5 kW]	
Advantages	**Disadvantages**	**Remarks**
No major remote pieces of generating equipment necessary other than a condenser. Can be free-blow without ducts if located in conditioned space.	Ventilation must be provided separately. Takes up valuable floor space.	Heating coil may be added to these units. Units typically air cooled. Fire regulations may require smoke detector in ductwork to shut down unit upon detection.

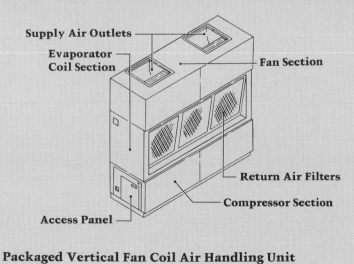

Packaged Vertical Fan Coil Air Handling Unit

All numbers shown in brackets are metric equivalents.

Figure 6.11b (continued)

Generation Equipment — Cooling (continued)

Cooling Towers R.S. Means No. 157-240	Typical Uses: Cools condenser water from refrigerant condenser. Useful Life: 25 – 40 years Capacity Range: 60 – 1,000 tons [210 – 3500 kW]	
Advantages	**Disadvantages**	**Remarks**

Natural Draft
  Low initial cost.
  Few mechanical parts.
  Quiet operation.
Mechanical Draft
  Energy savings possible by modulating fan speed.
  Acts like natural draft with fans off.
  High capacity is possible.
	All cooling towers require anti-bacterial water treatment.
	
	High initial cost.
	Noisy.
		Provide makeup water to replace evaporated water and to wash out (blow down) system as water is fouled.
		
		Mechanical draft towers are available as forced draft or as induced draft. In northern climates, if winter operation is a requirement, closed circuit cooling towers containing antifreeze solution are an alternative.

Induced Draft, Double Flow, Cooling Tower

Central Station Air Handling Units R.S. Means No. 157-125	Typical Uses: HVAC of building Useful Life: 15 – 30 years Capacity Range: 1300 – 60,000 cfm [615 – 28 320 L/s]	
Advantages	**Disadvantages**	**Remarks**

Combines all HVAC components into one system.
Energy savings via economizers possible.
Can be coordinated with VAV systems.
	Ducts and shafts require space.
	Must protect coils from freezing in the event of a power failure.
		Heating:
		  steam
		  electricity
		  gas
		  hot water
		Cooling:
		  electric
		  chilled water
		  direct expansion
		Check structural loads for floor or roof mounting.

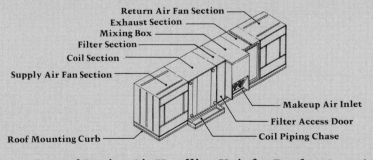

Central Station Air Handling Unit for Rooftop Location

All numbers shown in brackets are metric equivalents.

Figure 6.11b (continued)

Generation Equipment—Cooling (continued)

Split Systems R.S. Means No. 157-150/187 157-230	Typical Uses: Individual space conditioning remote from mechanical equipment. Useful Life: 10 – 20 years Capacity Range: 1 – 10 tons [3.5 – 35 kW]	
Advantages	**Disadvantages**	**Remarks**
Individual cooling coil for each tenant. Provides separate cooling for spaces remote from condensing unit location. No noise in conditioned space.	Separate ventilation air required. Long runs of refrigerant piping.	Distance between two sections is limited to approximately 60'. Refrigerant lines between both units must be insulated. Optional heating coil.

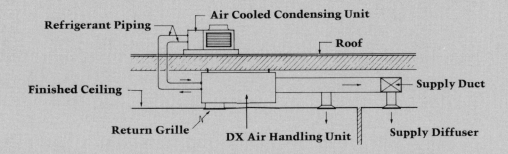

Heat Pumps R.S. Means No. 157-160	Typical Uses: Heating and cooling of spaces. Useful Life: 10 – 20 years Capacity Range: 1.5 – 50 tons [5.25 – 175 kW]	
Advantages	**Disadvantages**	**Remarks**
Individual metering of space possible for electricity. Energy-efficient, as some heat and others cool. Boiler and chiller off for portions of spring and fall.	Compressor noise in space. For winter cooling, a glycol solution in a closed circuit cooling tower or air cooled condenser is necessary.	Boiler must be controlled to limit supply water to 90°F maximum. Low operating temperatures are conducive to plastic piping.

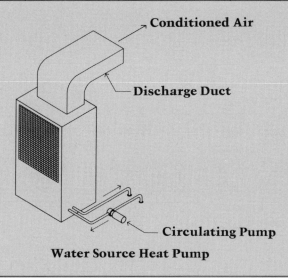

All numbers shown in brackets are metric equivalents.

Figure 6.11b (continued)

Chapter Seven

GENERATING EQUIPMENT ASSEMBLIES

The previous chapter describes and illustrates individual generation equipment components. Now those components are put together to create some common generation equipment assemblies. Illustrations and discussion in this chapter show how generation equipment systems operate, depending on the type of fuel available and the configuration and purpose of the building to be heated or cooled.

Generation equipment for buildings is usually comprised of *primary* and *secondary* equipment. Primary equipment "makes" the heated or chilled water. Secondary equipment, usually located in the machine room, takes the hot or cool medium through a secondary or auxiliary piece of equipment such as a pump, fan, or heat exchanger. The following paragraphs are a review of these basic categories.

Primary Equipment

Primary equipment for heating systems includes boilers and furnaces. Boilers are used with hot water and steam systems. Located in mechanical rooms, boilers require fuel lines, flues, and combustion air. Fire safety regulations also require heat detectors and fire-rated walls. Furnaces are used with air systems and require similar ancillary equipment as boilers, in addition to the necessary space for ductwork.

Primary equipment for cooling systems includes compression cycle and absorption cycle refrigeration equipment.

- **Compression cycle refrigeration equipment**
  - Through-the-wall direct expansion units (common window air conditioners) are noisy, provide poor humidity control, and block windows. These units are routinely used in residences and small offices.
  - Chillers, condensers, and cooling towers are large pieces of equipment and are usually located in a mechanical room or on the roof. Installation of these components requires the attention of a structural engineer, who must ensure adequate provisions for loads and vibration. Noise control must also be provided.
  - Split system units are composed of two components installed remotely from one another. The fan and evaporator are usually installed in the ceiling in the conditioned space, and the

condenser or condensing unit placed outdoors. The condensate line from the evaporator coil is then run to a convenient open waste.
- **Absorption cycle refrigeration equipment** is typically cumbersome and heavy. Installation of these components requires the attention of a structural engineer. Also required are vibration isolation and noise control.

Secondary Equipment

Examples of secondary equipment include pumps, fans, and heat exchangers.
- **Pumps** include circulating pumps for hot or chilled water in the building and pumps between the chiller and cooling tower.
- **Fans** include supply, return, and exhaust fans for buildings.
- **Heat exchangers** for main equipment include steam to hot water; steam to warm air; hot water to warm air; and chilled water or refrigerant to cool air.

Typical Assemblies

Figures 7.1 through 7.7 illustrate several common generating equipment assemblies. These illustrations can be used as references in the selection and pricing process since they show the major components of various systems and provide Means line numbers for each component.

Figure 7.1 shows a supplied steam assembly. This type of system is commonly used in large urban areas where the local utility has waste steam available, but it may also be used in any large facility that is comprised of a group of buildings serviced by a central steam-generating power plant. The central power source is the primary equipment. It is connected to mechanical rooms that house the secondary equipment in the serviced buildings. In the supplied steam system, the steam pressure is reduced and can then be used to heat a building, either directly or following conversion to hot water. Among the advantages of this system is the fact that it requires neither a chimney nor fuel storage.

An oil-fired (boiler) steam generation system is shown in Figure 7.2. When steam is the medium chosen for heating, an oil-fired boiler is typically chosen as the generation equipment. In this case, the system is located in the boiler room of the building it serves. The steam is distributed through convectors, radiators, coils, and air handling units. The oil for this type of steam generation system is stored in a tank in the building or underground.

The system shown in Figure 7.3 differs from the system shown in Figure 7.2 in that it is fired by gas and uses hot water rather than steam. The gas is supplied by a utility in the street, and the hot water is distributed by radiators and convectors. This is a very common heating system for both residential and commercial applications. For rural areas not served by a local gas utility, propane gas is an option and can be stored in a tank on the site.

Figure 7.4 is an illustration of an oil-fired warm air furnace system. The primary generation equipment consists of a furnace, rather than a hydronic boiler as shown in the previous systems. The heat is distributed through ductwork to grilles or registers. This system requires fuel storage on the site. When the system is gas-fired, the need for a tank is eliminated. It is a popular choice for residential and small commercial applications.

A primary chilled water generating source is shown in Figure 7.5. This is a centrifugal-type, water-cooled chiller. It supplies chilled water to a

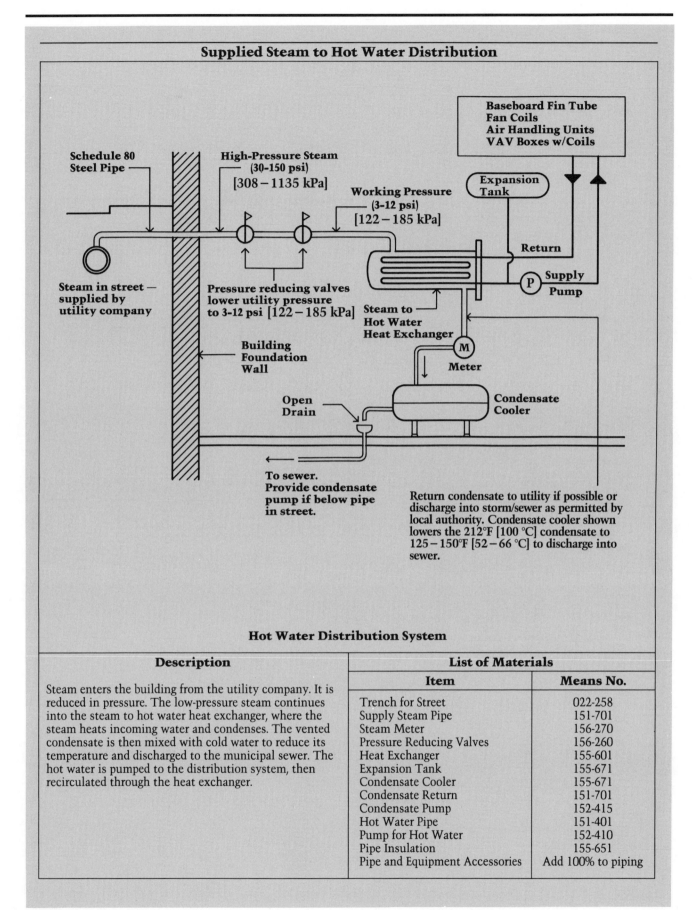

Figure 7.1

Steam Boiler/Two-Pipe Vapor System

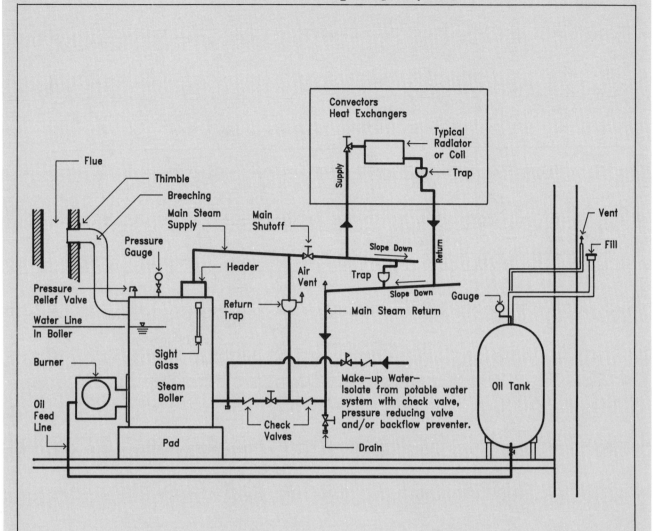

Steam Distribution System

Description	List of Materials	
	Item	Means No.
Oil from the tank is fed to the oil burner, which mixes air with an oil mist, burns it, and injects the flame at the base of the steam boiler. Condensed steam, which is returned from the system, is mixed with cold make-up water, heated by the burner into steam, and sent to the distribution system. The combusted air flows up the breeching pipe and into the flue.	Steam Boiler — Oil	155-120
	Breeching	155-680
	Steam Piping	151-701
	Condensate Piping	151-701
	Fittings and Accessories	Add 50% to piping
	Oil Tank	155-671
	Oil Transfer Pump	155-250

Figure 7.2

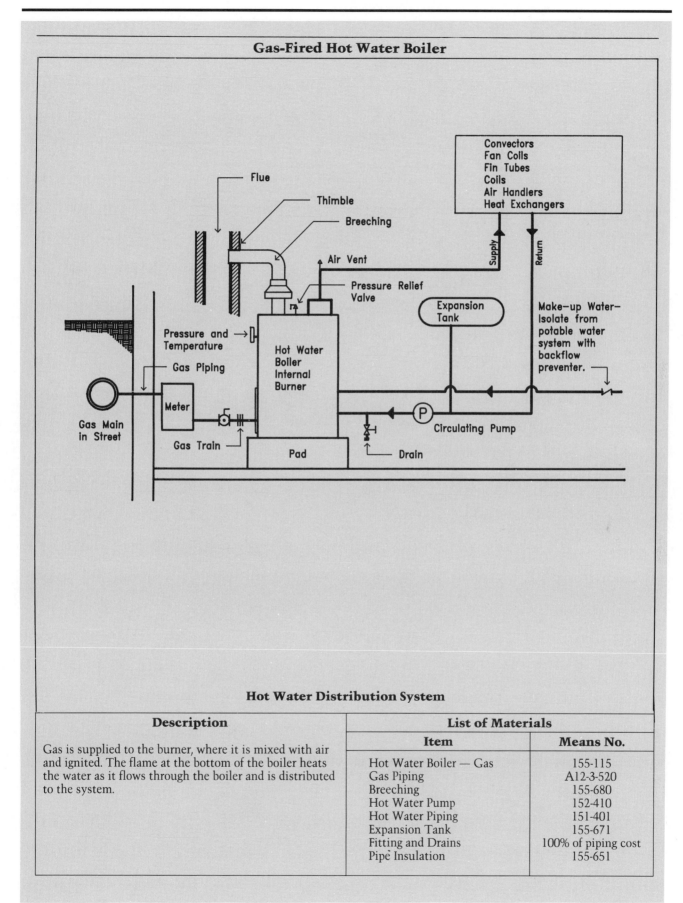

Figure 7.3

Warm Air Furnace

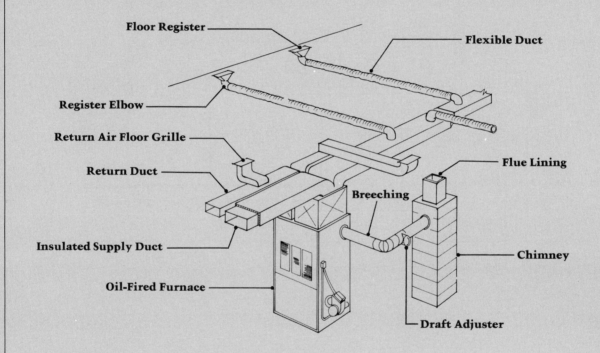

Description	List of Materials	
Oil from the tank is fed to the oil burner, which mixes air with an oil mist, burns it, and injects the flame into the combustion chamber of the warm air furnace. Return air is reheated in the furnace and recirculated via the fan to the duct distribution system. The exhaust gases leave the furnace through the breeching and flue.	**Item**	**Means No.**
	Warm Air Furnace	155-430
	Breeching	157-250
	Ductwork	157-250
	Duct Insulation	155-651
	Oil Tank	155-671
	Floor Registers	157-470
	Fittings and Accessories	Add 15% to ductwork

Figure 7.4

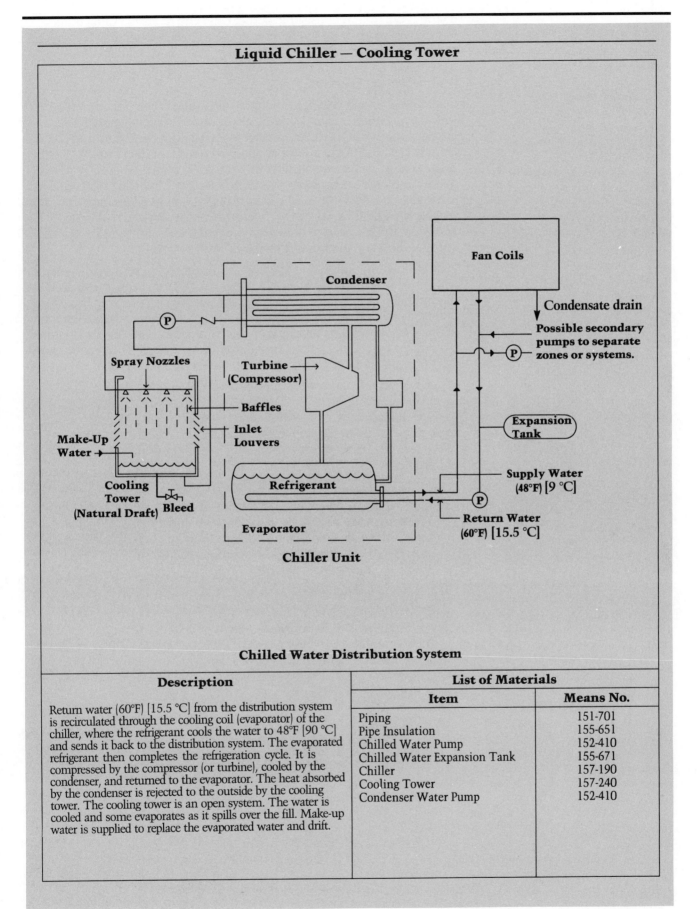

Figure 7.5

distribution system consisting of duct coils and/or air handling units. Condenser water gives up the heat rejected from the building through an outdoor cooling tower. Cooled condenser water is recirculated back to the chiller. This type of cooling system is commonly used for large commercial applications.

Figure 7.6 shows a packaged rooftop multi-zoned unit. This is a self-contained heating and cooling assembly. The only required field connections for this system are electricity to power the fans and the compressor, and ductwork for supply and return air distribution (cooling). If the heating is also supplied by this unit, a gas line would be required to feed a gas burner or, less commonly, electricity supplied to an electric resistance coil. The multi-zone feature of this unit provides individual zone control for specific areas of the building through a system of dampers and controllers. This type of system is popular for commercial applications such as shopping malls and low-rise office buildings.

A built-up air handling unit is shown in Figure 7.7. This assembly is comprised of several modules: a cooling coil section, heating coil section, filtering section, pre-heat coil section, fan section, and mixing box. This particular model combines primary and secondary generation equipment in one location. Hot or cold air is circulated via the supply air distribution system, fed by a chiller assembly and a hydronic source. This type of system may be used for large commercial and industrial applications.

All generation equipment, both primary and secondary, can be priced using the latest edition of *Means Mechanical Cost Data*. This cost reference provides installed prices, not just material costs. Using Means cost data together with the design criteria presented in this book, the designer can select the most economical system or component to fit a predetermined budget.

Problems

7.1 A building is to be heated with steam from the utility in the street. The steam is to be converted to hot water and distributed to a four-pipe fan coil system. Cooling is achieved by an air-cooled chiller located on the roof. The major pieces of equipment are shown in Figure 7.8. Complete the piping diagram.

7.2 For the hot water distribution system shown in Figure 7.1, illustrate as many ideas as possible to save, conserve, or scavenge energy.

7.3 For the hot water distribution system shown in Figure 7.3, what is the estimated construction cost of the mechanical system in the boiler room if the system is sized for a 150 MBH [44 kW]* boiler and a 30-gallon [115-liter] expansion tank? There are 100 feet [30 m] of piping in the boiler room.

*Throughout the book, SI metric equivalents are provided for most imperial units. The SI units appear in brackets [ ].

Packaged Rooftop Multizone Air Conditioning Unit

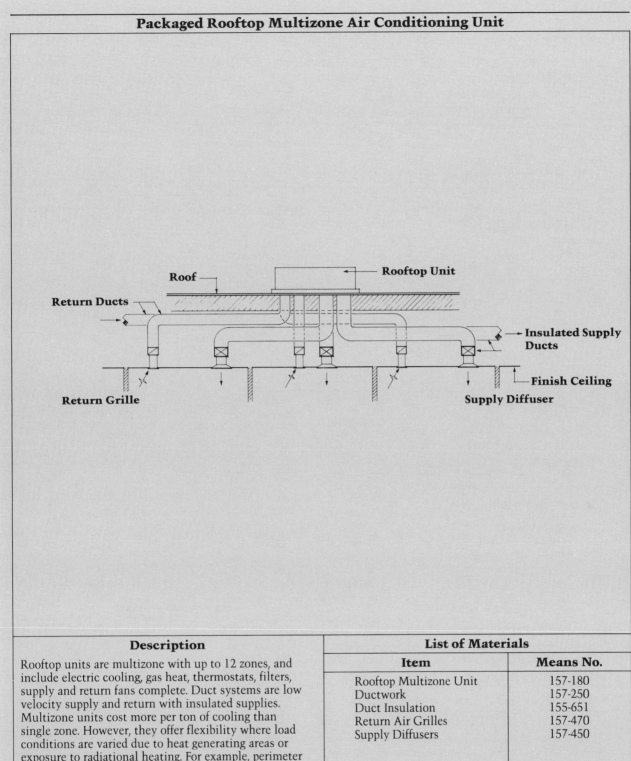

Description

Rooftop units are multizone with up to 12 zones, and include electric cooling, gas heat, thermostats, filters, supply and return fans complete. Duct systems are low velocity supply and return with insulated supplies. Multizone units cost more per ton of cooling than single zone. However, they offer flexibility where load conditions are varied due to heat generating areas or exposure to radiational heating. For example, perimeter offices on the "sunny side" may require cooling at the same time "shade side" or central offices may require heating. It is possible to accomplish similar results using duct heaters in branches of the single zone unit. However, heater location could be a problem and total system operating energy efficiency could be lower.

List of Materials

Item	Means No.
Rooftop Multizone Unit	157-180
Ductwork	157-250
Duct Insulation	155-651
Return Air Grilles	157-470
Supply Diffusers	157-450

Figure 7.6

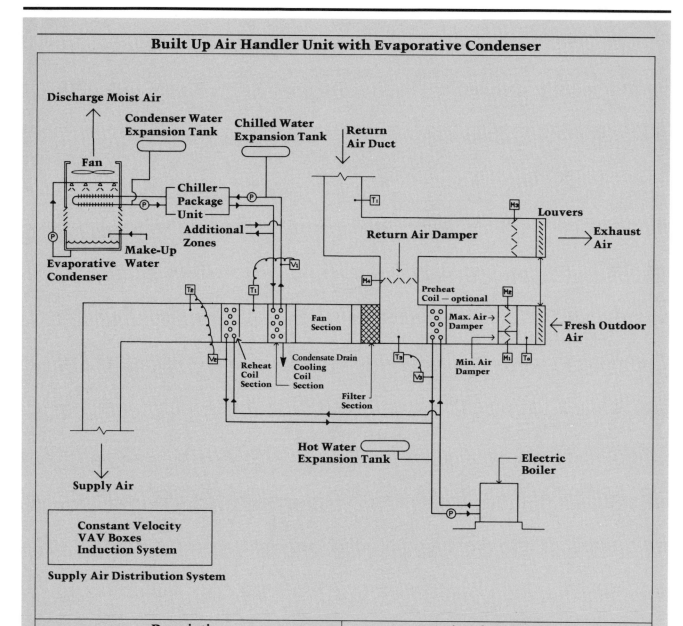

Built Up Air Handler Unit with Evaporative Condenser

Description

Depending on outside conditions, some return air may be exhausted to the outside and some may be recirculated. In the spring and fall, the chiller may be shut down and 100% outside air may be used and all return air exhausted (an economizer cycle). Modulating dampers are operated by motors (M_1, M_2, M_3, and M_4), which are controlled by sensors (T_1 and T_2) to regulate optimum flow of air. The air can be preheated (controlled by T_3, V_3); cooled (T_1, V_1); and tempered (T_2, V_2). The cooling coil produces condensate which is piped to waste. The evaporative condenser sprays water over the condenser water that is in the fin tubes. Note: There are three expansion tanks for this system and four pumps.

List of Materials

Item	Means No.
Piping	151-701
Pipe Insulation	155-651
Electric Boiler	155-110
Expansion Tanks (3)	155-671
Hot Water Pump	152-410
Central Station Air Unit	157-125
Ductwork	157-250
Duct Insulation	155-651
Dampers	157-480
Controls	157-420
Chiller Unit	157-190
Chilled Water Pump	152-410
Evaporative Condenser	157-225
Condenser Pump	152-410
Louvers	157-482

Figure 7.7

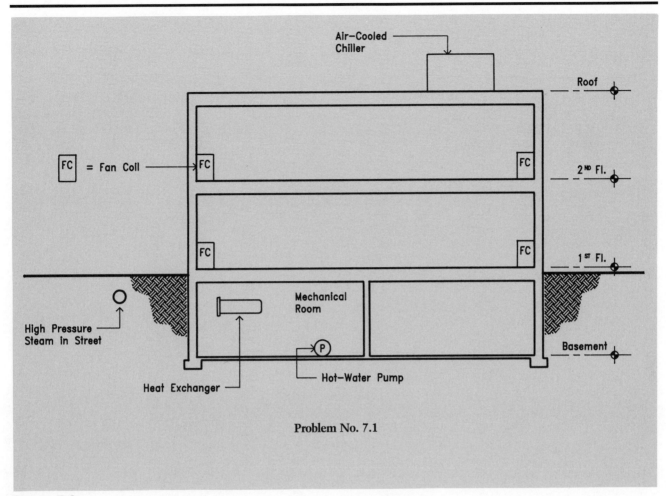

Figure 7.8

Chapter Eight

DRIVING SYSTEMS

Driving systems consist primarily of pumps and fans, the equipment used to "drive" water or air through pipes or ducts. These pumps and fans are normally located in the mechanical room and drive air or water across or through the generation equipment, then throughout the building. In order to properly select a driving system, the following factors must be determined.

- **Type:** There are several choices of pumps and fans for each specific system.
- **Capacity:** The capacity is measured in gallons of liquid per minute (gpm) [liters of liquid per second (L/s)]* for pumps and cubic feet of air per minute (cfm) [liters of air per second (L/s)] for fans.
- **Head or pressure:** Head or pressure is measured as feet of head (H) [kilopascals (kPa)] for pumps and inches of water (wg) [Pascals (Pa)] for fans. One foot of head is the pressure at the bottom of a column of water one foot high. One inch of water is the pressure at the bottom of a column of water one inch high. One kilopascal (1000 Pascals) is the pressure created when the force of one kilogram (measured in Newtons, or N) is applied over an area of one square meter.

These pressure designations are illustrated in Figure 8.1.

Pumps

Pumps are generally used on *closed loop systems*, where the fluid is continuously recycled and contained within a piping system. The pump must be sized to provide the proper quantity of water (gpm), and must be rated large enough to provide the required pressure. Pressure from a pump is used to overcome the friction, or drag, that exists between the fluid and the pipe and within itself as it moves through the piping system. (This pressure/friction relationship is discussed in Chapter 10, "Distribution and Driving Systems.")

Some pumps are used in *open systems*. In open systems, the fluid leaves the pipe and is exposed to the atmosphere. Water pumps for evaporative cooling towers are used in open systems, because during the circuit, the water leaves the pipe as it enters "open" air while it spills over the cooling

---

Throughout the book, SI metric equivalents are provided for most imperial units. The SI units appear in brackets [ ].

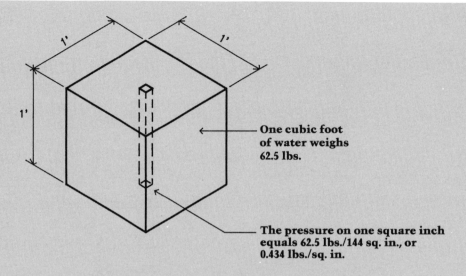

One foot of head — Used for Hydronic Systems

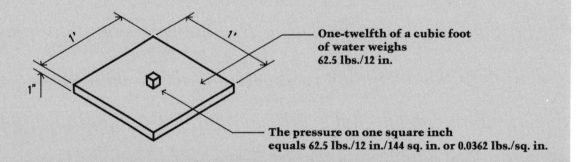

One inch of pressure (Hg) — Used for Gaseous Systems

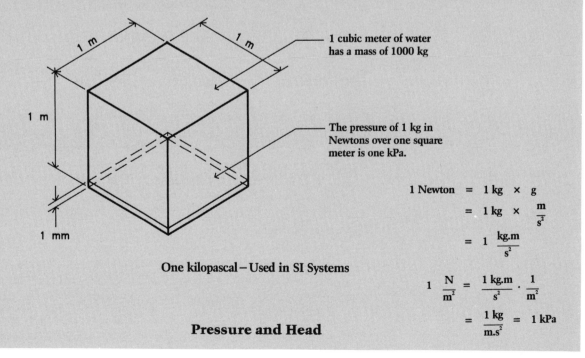

One kilopascal — Used in SI Systems

Pressure and Head

Figure 8.1

tower fill. Pumps in an open system must have a pump head large enough to compensate for friction losses, and must have the additional capacity to lift the liquid from one elevation to another. This application for the evaporative condenser pump is shown in Figure 7.7.

Centrifugal Pumps

The most commonly used pumps in HVAC systems are centrifugal pumps, which are driven by an electric motor. Most pump bodies are cast iron. Bronze bodies are available for potable water systems at additional cost.

Centrifugal pumps are comprised of an impeller, which rotates and drives the fluid; the casing, which surrounds the impeller; and the electric motor, which turns the impeller. Pumps over 3/4 horsepower (H.P.) should be connected with flexible hoses and mounted on vibration isolators or absorbers on concrete inertia pads in order to avoid transmission of noise and vibrations to the piping system.

Figure 8.2 lists the types of pumps that are commonly used in HVAC systems. This chart can be used as a guide for selection of pumps, in conjunction with manufacturers' literature and space considerations. (See also the equipment data sheets in Figure 8.11 at the end of this chapter.)

Pump Curves

Pumps provide less water flow (in gpm) [in L/s] as the head increases. Pump curves are diagrams that plot how a particular pump performs over

Selection Guide for Pump Types		
Centrifugal Pump	**Use**	**Remarks**
Circulator	Small systems and residential heating and domestic hot water recirculation.	In line.
Close Coupled	Hot water circulation (heating and domestic hot water). Cold water booster (domestic water).	High temperature seals.
Base Mounted (single stage double suction)	Primary and secondary chilled water	
Base Mounted (multi stage)	Condenser water (cooling tower).	Check suction pressure, strainer on supply side.
In Line (vertical)		
Condensate Pump	Steam (condensate) return.	
Most pumps listed above are single-suction, volute, flexible-coupled with a single impeller except where noted. Multiple impellers are available with most pump types.		

Figure 8.2

the range of normal use. Centrifugal pumps, for example, have a predictable pattern, or curve, throughout their range. Typical performance curves for a specific type of pump are shown in Figure 8.3.

Pump curves plot *head* versus *flow* for each pump. Pump size on the curves refers to the nominal size of the pump connection. As a pump delivers more gallons per minute (gpm) [L/s], the head, or pressure, imparted to the water declines. The reverse is also true. A pump connected to a system with a low head requirement will deliver more gpm [L/s] than when connected to a system with a high head. Generally, centrifugal pump curves drop rapidly and are considered nearly vertical. Examine the "ideal" pump curve shown in Figure 8.4 and compare it with the actual curves from Figure 8.3 shown for 20 gpm at a 6.5 foot head.

Whenever a pump is selected, it is rare that the actual head and flow of the pump exactly match the head and flow requirements of the system design. It is better to slightly oversize the pumps to allow for future connections and site balancing. Figure 8.4 shows the system curve passing through the system requirements of 6.5 feet of head and 20 gpm. It also intersects the two pump curves for 1-1/4" and 1-1/2" pumps. The 1-1/4" pump is too small, since it would, if connected to the system, deliver only 18 gallons per minute, a rate that would "starve" some of the equipment. The 1-1/2" pump, on the other hand, delivers 21 gpm – slightly more than necessary – and it should be selected. A perfect 20 gpm can be achieved by "throttling down" slightly, although this is rarely necessary.

Efficiency

Another consideration in pump selection is efficiency. Usually more than one type of pump is suitable. When this situation occurs, the pump with the highest efficiency in the design range is selected. The precise diameter of the impeller may also be specified.

When selecting a pump, the design engineer should choose a type that will avoid cavitation. Cavitation occurs when the pressure on the suction side of the pump falls below the vapor pressure, thus causing vapor (gas) pockets within the pump. Each pump has a rating for the required net positive suction head (NPSH), which must be met. Meeting this rating is particularly important for hot water systems, where the vapor pressure is most critical.

Pump Laws

The relationships between flow, head, speed, horsepower, impeller diameter, and fluid are shown in Figure 8.5. For known conditions, the pump laws are used to predict pump characteristics when one condition, such as pump speed, is changed.

Arrangement of Pumps

For small installations, such as residences or pumps smaller than 3/4 H.P., one constant speed pump is normally selected. For larger installations, a variety of pump arrangements are used. These are shown in Figure 8.6.

Generally, whenever a pump is greater than 3/4 H.P., it is good practice to provide an auxiliary or back-up pump so that there will be no interruption in service should one pump be out of service. Thus two pumps, a main and an auxiliary, are usually used in large systems.

In some cases, a particular system may have an odd characteristic system curve. A system that requires a high head at low flow, for example, can be accommodated by placing pumps in series. Where a low head but high

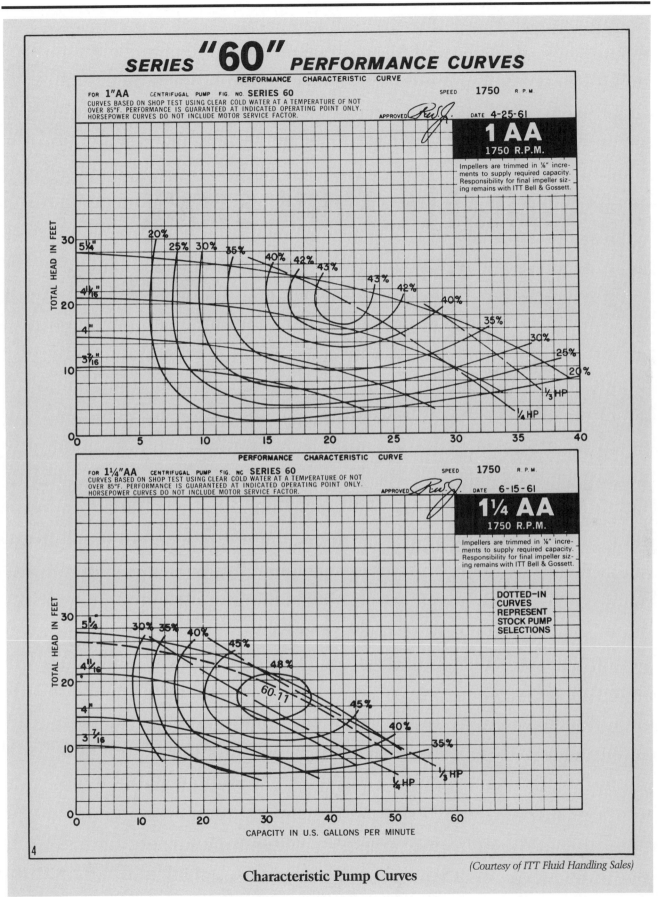

Characteristic Pump Curves

(Courtesy of ITT Fluid Handling Sales)

Figure 8.3

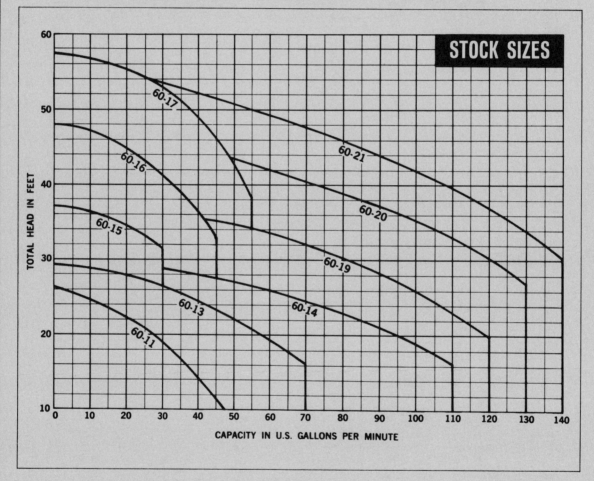

Figure 8.3 (continued)

flow is required, the pumps can be arranged in parallel as shown in Figure 8.6. These two pump arrangements are provided when it is determined that one pump is not practical for the required flow and head.

Staging

Another consideration in pumping is that two speeds (or stages), high and low, may need to be provided. Staging may be needed for systems that have variable demands on pumps. For example, low flow may be needed in winter for hot water, and high flow in summer for chilled water on a two-pipe fan coil system. In such a situation, two-speed pumps or additional pumps are brought on line, or staged, as necessary to pick up the greater cooling load.

Primary and Secondary Pumping

Pumps can be arranged for primary and secondary loops, as illustrated in Figure 8.6. The primary loop will circulate continuously; it is sized for

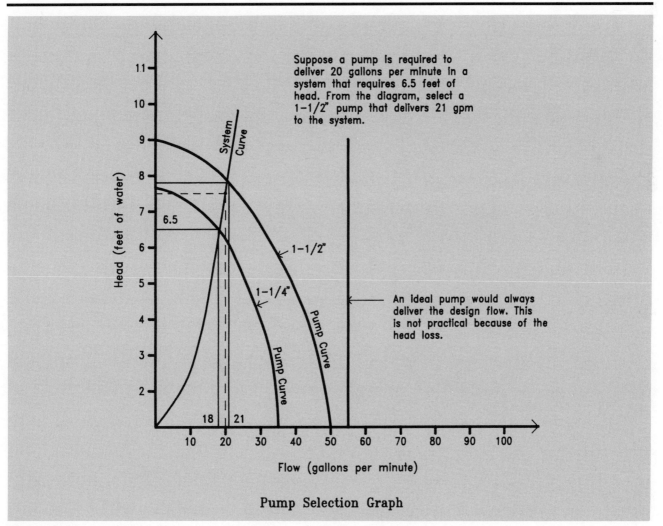

Figure 8.4

the flow of the main loop. The secondary pumps are tied to zone control devices and are sized for the flow in each zone.

Fans

Fans are used most often in open systems. The air from the fan is transmitted through ducts and discharged into or exhausted from the building spaces. When air is recirculated within a system, it is still not a tight enough system to be called closed. Leakage, infiltration, exhaust air, makeup air, and room losses mean that the air is supplied and returned by separate open systems.

Size and Location

Fans must be of an appropriate size to overcome the friction losses in the ducts and fittings. Because air is so light relative to water (0.075 pounds per cubic foot [1.2 kg/m^3] for air versus 62.5 pounds per cubic foot [1000 kg/m^3] for water), the pressure head to lift the air from one height to another is small and usually ignored. The weight of the water cannot be overlooked in pumping systems where the density of water is so much more

Pump Laws

Variable	Constant	No.	Law	Formula
Speed of impeller (rpm) (N)	Fluid specific gravity (W)	1	Capacity varies as the speed	$\dfrac{Q_1}{Q_2} = \dfrac{N_1}{N_2}$
	Pump impeller diameter (D)	2	Pressure varies as the square of speed	$\dfrac{P_1}{P_2} = \left\{\dfrac{N_1}{N_2}\right\}^2$
		3	Horsepower varies as the cube of speed	$\dfrac{HP_1}{HP_2} = \left\{\dfrac{N_1}{N_2}\right\}^3$
Pump impeller diameter (D)	Fluid specific gravity (W)	4	Capacity varies as the diameter	$\dfrac{Q_1}{Q_2} = \dfrac{D_1}{D_2}$
	Speed of impeller (N)	5	Pressure varies as the square of the diameter	$\dfrac{P_1}{P_2} = \left\{\dfrac{D_1}{D_2}\right\}^2$
		6	Horsepower varies as the cube of the diameter	$\dfrac{HP_1}{HP_2} = \left\{\dfrac{D_1}{D_2}\right\}^3$
Fluid specific gravity	Speed of impeller (N) Pump impeller diameter (D)	7	Horsepower varies as the specific gravity	$\dfrac{HP_1}{HP_2} = \dfrac{W_1}{W_2}$

Q = Flow through pump (gpm) [L/s]
N = Speed of impeller (rpm) [rpm]
P = Pressure (feet of head) [kPa]
HP = Pump horsepower
D = Impeller diameter (in.) [cm]
W = Fluid specific gravity

Figure 8.5

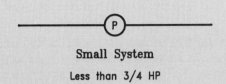

Small System

Less than 3/4 HP

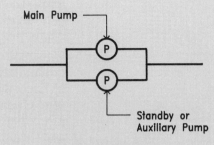

Typical Installation

3/4 HP and above

(Pumps may be programmed to alternate daily or weekly operation.)

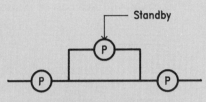

Pumps in Series

Low flow, high head conditions

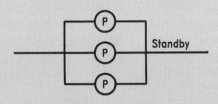

Pumps in Parallel

High flow, low head conditions

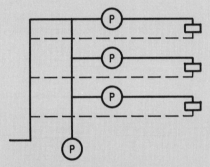

Primary and Secondary Pumping

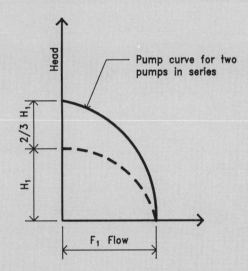

Second pump head is approximately 2/3 effective.

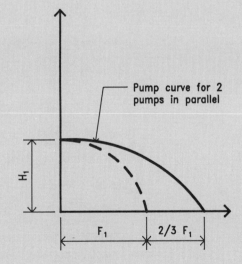

Pump Arrangements

Figure 8.6

than that of air. As with pumps, fans are sized to provide the proper flow (cfm) [L/s] and pressure (in wg) [Pa] for the particular system. There are performance curves for fans and fan laws to regulate their use, just as there are for pumps.

Fans may be exposed in the area served, as in a bathroom ceiling exhaust, or remotely located, either on a roof or in a mechanical room. Fans may be direct drive or belt driven. A direct drive fan is less expensive, but might send objectionable noise directly into the ductwork. Belt drives contain a belt that wraps around two pulleys, or sheaves – one on the motor drive and the other around the fan drive. Belt drives offer more flexibility in performance since the diameter of the sheaves can be changed to regulate the fan speed.

Fan Types

There are two broad categories of fans commonly used in driving systems: centrifugal fans and axial flow fans (see Figures 8.7 and 8.8). Centrifugal fans take in air at the side (circular opening) and discharge it radially into the supply stream (rectangular opening). Axial flow fans are mounted in the air stream and boost the pressure of the air. Figure 8.7 may be used to determine the basic fan type. Fans supply air from as low as 50 cfm [25 L/s] to more than 500,000 cfm [236 000 L/s] and at pressures from 0.1" wg [30 Pa] to 6" wg [1775 Pa]. Fan speeds can be adjusted by using two-speed motors, variable pitch sheaves or blades, or variable frequency motors. Noise control is a particular concern in fan design; noise ratings are taken

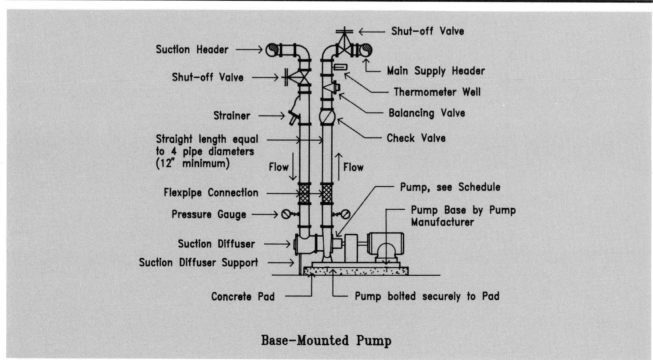

Base-Mounted Pump

Figure 8.6 (continued)

into account in fan selection. When fans are near combustible fumes, nonsparking materials are usually specified for blades and housing.

Fan Curves

Each type of fan has a characteristic performance curve. This information is supplied by the manufacturer. As with pumps, the fan curves plot pressure versus flow and indicate other characteristics, such as efficiencies and horsepower over the range of operation. Typical fan performance characteristics are shown in Figure 8.8 for forward, backward, and radial pitched centrifugal fans.

As with pumps, it is important to select a fan large enough to handle the pressure and flow required for the system. Oversizing slightly (by 10 – 15 percent) is common practice. Comparisons between various manufacturers and models should be made to select for overall durability, ease of maintenance, cost, and efficiency.

Selection Guide for Fan Types

Fan Type	Type and Impeller Design	Use	Remarks
Centrifugal	**Airfoil:** Airfoil-contour blades curve away from direction of rotation.	General HVAC. Usually for larger systems, where cost is justified by energy savings.	Highest efficiency of any centrifugal type. Power reaches maximum near peak efficiency.
	Backward-Curved: Blades curve away from direction of rotation.	Same as airfoil; used in corrosive environments that might damage airfoil blades. Also used in high-speed applications.	Similar to airfoil; slightly less efficient.
	Radial: Blades are not curved.	Industrial materials handling. Medium speed applications.	High pressure characteristics. Easy to repair. Not common in HVAC. Least efficient of centrifugal types.
	Forward-Curved: Blades curve toward direction of rotation.	Low-pressure HVAC applications. Small residential and packaged units. Low-speed applications.	Fans in series.
Axial	**Propeller:** 2 – 5 blades on a small hub.	Large volume at low pressure. Ventilation and make-up air applications.	Low efficiency, noisy. Best when not ducted.
	Tubeaxial: 4 – 8 blades on a medium-sized hub. Blades are either airfoil or single thickness.	Heavy duty propeller fans used where space is tight. Industrial exhaust. Low and medium pressure HVAC applications.	More efficient than propeller type, and capable of developing more static.
	Vaneaxial: Many blades, large hub. Blades can be adjustable.	General HVAC and industrial applications. Used where space is tight and costs permit.	Most efficient propeller type, with high-pressure capability. Medium flow rate. Good downstream air distribution, more compact than centrifugal fans.

Figure 8.7

Fan Laws

As with pumps, fan laws may be used to predict fan performance under conditions different from the conditions listed in a fan curve. The fan laws, shown in Figure 8.9, are used to determine the proper fan speed for a system. Slowing the fan to a lower speed is always less expensive than running it at high speed and partially closing dampers (throttling).

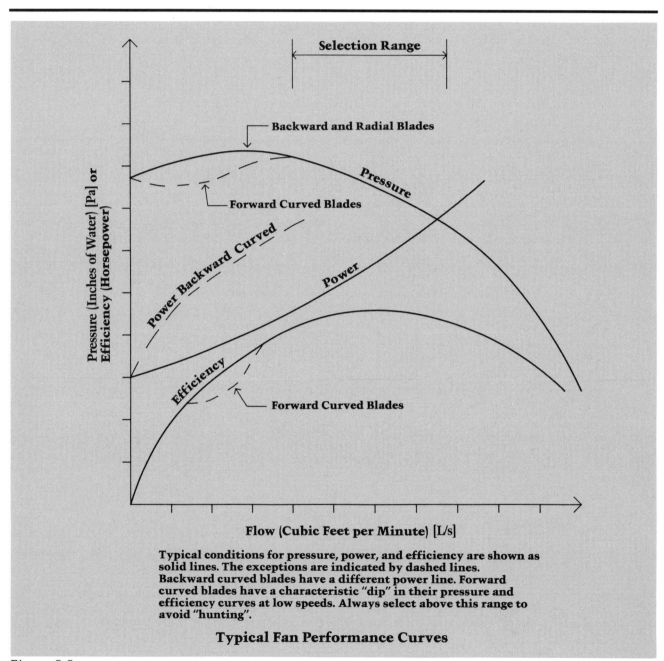

Figure 8.8

Fan Arrangement

Fans may be placed in series or parallel in a manner similar to pumps. In practice, however, this approach is rarely taken because fans come in a wide range of capacities suitable for most installation conditions. Also, the compressibility of air (as compared to water) makes it more difficult to properly pair fans in series or parallel.

LAWS OF FAN PERFORMANCE

Fan laws are used to predict fan performance under changing operating conditions or fan size. They are applicable to all types of fans.

The fan laws are stated in *Table 5*. The symbols used in the formulas represent the following quantities:

- Q — Volume rate of flow thru the fan.
- N — Rotational speed of the impeller.
- P — Pressure developed by the fan, either static or total.
- Hp — Horsepower input to the fan.
- D — Fan wheel diameter. The fan size number may be used if it is proportional to the wheel diameter.
- W — Air density, varying directly as the barometric pressure and inversely as the absolute temperature.

FAN LAWS

VARIABLE	CONSTANT	NO.	LAW	FORMULA
SPEED	Air Density, Fan Size, Distribution System	1	Capacity varies as the Speed.	$\frac{Q_1}{Q_2} = \frac{N_1}{N_2}$
		2	Pressure varies as the square of the Speed.	$\frac{P_1}{P_2} = \left(\frac{N_1}{N_2}\right)^2$
		3	Horsepower varies as the cube of the Speed.	$\frac{Hp_1}{Hp_2} = \left(\frac{N_1}{N_2}\right)^3$
FAN SIZE	Air Density, Tip Speed	4	Capacity and Horsepower vary as the square of the Fan Size.	$\frac{Q_1}{Q_2} = \frac{Hp_1}{Hp_2} = \left(\frac{D_1}{D_2}\right)^2$
		5	Speed varies inversely as the Fan Size.	$\frac{N_1}{N_2} = \frac{D_2}{D_1}$
		6	Pressure remains constant.	$P_1 = P_2$
	Air Density, Speed	7	Capacity varies as the cube of the Size.	$\frac{Q_1}{Q_2} = \left(\frac{D_1}{D_2}\right)^3$
		8	Pressure varies as the square of the Size.	$\frac{P_1}{P_2} = \left(\frac{D_1}{D_2}\right)^2$
		9	Horsepower varies as the fifth power of the Size.	$\frac{Hp_1}{Hp_2} = \left(\frac{D_1}{D_2}\right)^5$
AIR DENSITY	Pressure, Fan Size, Distribution System	10	Speed, Capacity and Horsepower vary inversely as the square root of Density.	$\frac{N_1}{N_2} = \frac{Q_1}{Q_2} = \frac{Hp_1}{Hp_2} = \left(\frac{W_2}{W_1}\right)^{1/2}$
	Capacity, Fan Size, Distribution System	11	Pressure and Horsepower vary as the Density.	$\frac{P_1}{P_2} = \frac{Hp_1}{Hp_2} = \frac{W_1}{W_2}$
		12	Speed remains constant.	$N_1 = N_2$

(Courtesy Carrier Corporation, McGraw–Hill Book Company)

Footnote: Cross-references in table footnotes pertain to the source from which tables were taken; they do not refer to chapters or pages in this text.

Figure 8.9

When design conditions in a system vary, it is more common to select a two-stage fan or, in larger installations, to utilize variable pitched blades. Advancements have recently been made in varying the frequency of power to motors from the normal 60 hertz. Further developments are likely in this area.

Equipment Data Sheets

Figures 8.10a and 8.10b at the end of this chapter contain equipment data sheets for fans and pumps. These sheets describe the efficiency, useful life, typical uses, capacity range, special considerations, accessories, and additional equipment needed for driving system components.

These sheets can be used in the selection of each piece of equipment. A Means line number, where applicable, is included. Costs can be determined by looking up the line numbers in the current edition of *Means Mechanical Cost Data*.

Problems

8.1 Consider the series 60 1" AA and 1-1/4" AA pumps shown earlier in the chapter. The 3-7/16", 4", 4-11/16", and 5-1/4" refer to the impeller sizes and the 10%, 20%, 30%, etc. refer to pump efficiencies.
   a. What do the 1" and 1-1/4" represent?
   b. Which pump and impeller are most efficient to deliver a flow of 20 gpm at 15 feet of head?
   c. Assume that for a 20 gpm flow and 15 feet head design condition you have selected a 1" AA with a 5-1/4" impeller (this is not the correct answer to (b) above). Estimate the actual flow and pressure on the system and show how you obtained it. Hint: review system curve.

8.2 A 5 H.P. pump provides 100 gpm [6.5 L/s] at a speed of 1,750 rpm at 60 hertz.
   a. If the variable frequency drive drops the power to 50 hertz, what will be the flow rate of the pump?
   b. At 50 hertz, how much will the horsepower of the pump decrease?
   c. Comment on the variable speed drive's effectiveness in this case.

8.3 What type of pump would be used for each of the following?
   a. Hot water system at boiler
   b. Condensate pump for chiller
   c. To send steam condensate back to boiler

8.4 What type of fan would be used for each of the following?
   a. Toilet exhaust
   b. Kitchen exhaust
   c. At an inline duct heating coil
   d. Supply air fan for main air handler

Driving Equipment Data Sheets for Pumps

Driving Equipment	In-Line Centrifugal Pumps	Means No. 152-410
In-Line Centrifugal Pump	Typical Uses: Residential, Light Commercial Hydronic Heating	Capacity Range: 1/40 – 1-1/2 H.P.
	Comments: Common close coupled pump for small, residential applications. No base required. Pump often supported directly by piping. Easily repaired or replaced without disturbing the piping system. Non-ferrous models available for domestic water recirculation systems.	
	Close Coupled Centrifugal Pumps	Means No. 152-410
Close Coupled Centrifugal Pump	Typical Uses: Commercial Hot and Chilled Water Systems	Capacity Range: 1-1/2 – 25 H.P./40 – 1,500 gpm [2.5 – 95 L/s]
	Comments: Factory assembled for longer life but difficult to repair. Entire unit can be readily replaced. Vibration connections available.	
	Base Mounted Pumps – Single Stage	Means No. 152-410
Base Mounted Centrifugal Pump	Typical Uses: Hot and Chilled Water Systems	Capacity Range: 1-1/2 – 25 H.P./40 – 1,500 gpm [2.5 – 95 L/s]
	Comments: Most common for intermediate and large commercial installations. Many configurations available for higher capacities. Vibration bases and/or couplings usually required. Chill water pumps may require a drip ledge base and drain tapping. Units available as factory assembled or with motor furnished by others for field assembly.	
	Condensate Return Pumps	Means No. 152-415
Condensate Return Pump	Typical Uses: Return Condensate to Boiler or Utility	Capacity Range: Modular – No Limit
	Comments: Condensate return pumps may be single or duplex and are available with cast iron or steel receivers. A vent line from the receiver relieves any pressure buildup and prevents steam from entering the pump itself. On larger systems, with the addition of sophisticated controls, a boiler feed system capable of handling several boilers can be evolved. A duplex alternator adds to the pump life expectancy.	

All numbers shown in brackets are metric equivalents.

Figure 8.10a

Driving Equipment Data Sheets for Fans

Centrifugal Ceiling or Cabinet Exhaust Fan	**Right angle ceiling or cabinet exhaust fan**	**Means No.: 157-290**
	Typical use: Exhaust	**cfm Range: 50 – 2,500** [25 – 1200 L/s]
	Comments: Usually direct drive. Larger sizes are used for conference rooms and offices. Smaller models are used in toilets, and often switched with lights. Check for compatible voltages. Typically low pressure, short duct run to exterior or plenum.	
Centrifugal Sidewall Exhaust Fan	**Centrifugal exhaust fan**	**Means No.: 157-290**
	Typical use: Exhaust	**cfm Range: 100 – 7,000** [47 – 3300 L/s]
	Comments: Direct or belt drive. Mounted on the exterior, where interior mounting space is limited, and higher flow rate is required.	
Centrifugal In-line Fan	**Centrifugal in-line fan**	**Means No.: 157-290**
	Typical use: Supply or exhaust	**cfm Range: 100 – 1,200** [47 – 5665 L/s]
	Comments: Direct or belt drive. Greater performance than cabinet exhausters, usually concealed above ceiling. Not very attractive.	
Centrifugal Utility Fan	**Centrifugal utility fan**	**Means No.: 157-290**
	Typical use: Exhaust	**cfm Range: 300 – 180,000** [140 – 85 000 L/s]
	Comments: Direct or belt drive. Roof mounted, very efficient. Typically for exhaust in low- to medium-pressure applications. Typically uses airfoil blades.	
Centrifugal Tubular Fan	**Centrifugal tubular fan**	**Means No.: 157-290**
	Typical use: Exhaust	**cfm Range: 1,600 – 120,000** [750 – 56 650 L/s]
	Comments: Similar to the centrifugal utility fan, but designed for in-line type of mounting.	
Centrifugal Roof Fan	**Centrifugal roof fan**	**Means No.: 157-290**
	Typical use: Exhaust	**cfm Range: 1,100 – 37,000** [520 – 17 500 L/s]
	Comments: Direct or belt drive. Typically for exhaust, but can be used for supply. Commonly used at the top of toilet, kitchen or dryer stacks in commercial buildings.	
Centrifugal Fans – Variable Pitch Blade	**Variable pitch blade**	**Means No.: 157-290**
	Typical use: VAV systems	
	Comments: Activated on pressure. Efficient.	
Centrifugal Fan – Variable Speed	**Variable speed**	**Means No.: 157-290**
	Typical use: VAV systems	
	Comments: Frequency of current to motor is varied from 60 Hz (cycles/second) to adjust speed of blades. Economical.	

All numbers shown in brackets are metric equivalents.

Figure 8.10b

Driving Equipment Data Sheets for Fans (continued)

Propeller Paddle Blade Air Circulator	**Propeller air circulator**	Means No.: 157-290
	Typical use: Air circulation at ceiling of space.	cfm Range: 100 – 7,000 [47 – 3 300 L/s]
	Comments: Usually direct drive. Weak pressure, good airflow. Typically used in high spaces to prevent stratification. Rotation can be reversed for heating or cooling applications.	
Propeller Sidewall	**Propeller sidewall fan**	Means No.: 157-290
	Typical use: Exhaust	cfm Range: 200 – 90,000 [95 – 42 500 L/s]
	Comments: Direct or belt drive. Typically not ducted. Used for cooling and ventilation in the summer. Often used to keep mechanical and electrical room from overheating. In smaller, residential applications, this can be used to draw cooler basement air to the upper floors.	
Propeller Tube Axial	**Propeller tube axial fan**	Means No.: 157-290
	Typical use: Supply or exhaust	cfm Range: 4,000 – 100,000 [1900 – 47 500 L/s]
	Comments: Designed for the industrial environment, such as high temperatures and exhausting gases and particulate matter. Typically roof mounted.	
Propeller Vane Axial Supply or Exhaust Fan	**Propeller vane axial**	Means No.: 157-290
	Typical use: Supply or exhaust	cfm Range: 300 – 72,000 [140 – 34 000 L/s]
	Comments: The best flow and pressure characteristics for any propeller fan.	
Propeller Roof Exhaust or Supply Fan	**Propeller roof fan**	Means No.: 157-290
	Typical use: Supply or exhaust	cfm Range: 300 – 72,000 [140 – 34 000 L/s]
	Comments: Similar to the centrifugal roof exhaust fan.	

All numbers shown in brackets are metric equivalents.

Figure 8.10b (continued)

Chapter Nine

HEAT EXCHANGERS

A heat exchanger is required whenever there is a change in the type of heating or cooling medium. In a boiler room, for example, a shell and tube heat exchanger transfers heat from steam to hot water. This hot water is eventually converted to warm air through a fan coil heat exchanger. Figures 1.6, 1.11, and 1.23 in Chapter 1 show the locations of heat exchangers within a system. In these figures, a diagonal line designates a change in medium and a transfer of heat, requiring a heat exchanger.

There are two basic types of heat exchangers. *Shell and tube* heat exchangers are used when converting from a liquid to a gas and vice versa. *Fin and tube* heat exchangers (coils) are used to warm or cool air. This is accomplished with steam, hot water, or chilled water contained within the tubes, and air passing through the fins. Types of heat exchangers are illustrated in Figure 9.1.

Shell and Tube Heat Exchangers

In a shell and tube heat exchanger, the shell surrounds the tubes and carries the primary heating or cooling medium (steam or water). The tubes contain the secondary heated or cooled liquid. There is typically one shell and several tubes. When steam is used, a steam trap keeps steam in the heat exchanger until it has been converted to condensate, which improves the efficiency of the system.

The important selection criteria for shell and tube heat exchangers are:

Transfer load: The MBH [kW]* of heating or cooling that is to be delivered to the system by the exchanger. All heat exchangers are rated in MBH [kW].

Primary fluid rate: The flow rate of the primary fluid through the shell, expressed in gpm [L/s] or pounds of steam per hour [kilograms per hour].

Entering primary fluid condition: The temperature and pressure of the fluid supplied to the shell side of the heat exchanger.

Leaving primary fluid condition: The temperature and pressure of the fluid leaving the shell, typically 20°F [12 °C] lower than the primary entering fluid temperature for hot water systems.

---

*Throughout the book, SI metric equivalents are provided for most imperial units. The SI units appear in brackets [ ].

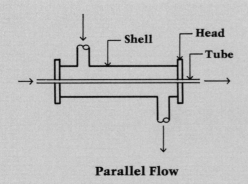

Parallel Flow

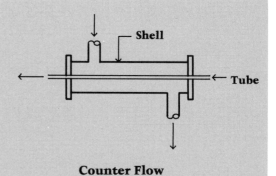

Counter Flow

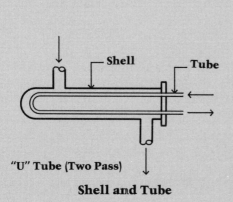

"U" Tube (Two Pass)

Shell and Tube

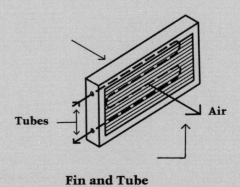

Fin and Tube

Hydronic Coil
Used in ductwork or
fan coil arrangements

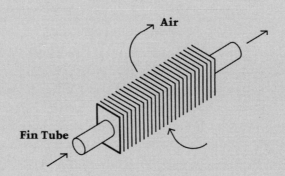

Radiator or Convector

Types of Heat Exchangers

Figure 9.1

Entering secondary fluid condition: The temperature and pressure of the fluid entering the tubes of the heat exchanger.

Leaving secondary fluid condition: The temperature and pressure of the fluid leaving the tubes of the heat exchanger. It is usually 15 – 20°F [8 – 12 °C] above the entering secondary fluid temperature for hot water systems.

Approach temperature: The entering primary temperature minus the leaving secondary fluid temperature. An approach temperature of zero is impossible. For practical considerations, an approach temperature of 5 – 20°F [3 – 12 °C] is usually selected.

Secondary fluid rate: The flow (in gpm [L/s] or pounds [kg] of steam per hour) through the secondary (tube) side of the heat exchanger.

Each of the above factors is generally determined by a trained engineer in consultation with an equipment manufacturer. The primary temperatures are made as high as possible to improve overall efficiency. The secondary temperatures are usually determined by room or terminal unit performance criteria recommendations, 130°F [55 °C] warm air supply, 200°F [93 °C] hot water supply.

To specify a heat exchanger, the required MBH [kW], primary and secondary fluid conditions, and flow rates must be determined. Reduction in the pressures of the fluids as they proceed through the shell and tube must be compensated for in calculating the total head for the primary and secondary pumps.

Heat Exchanger Performance

A perfect heat exchanger would make the temperature of the water *leaving* the tubes equal to the temperature of the water *entering* the shell. However, because this is theoretically impossible, a 5 – 20°F [3 – 12 °C] difference is usually selected between the entering temperature of the primary medium and the exiting temperature of the secondary medium. This difference is called the *approach temperature*.

Figure 9.2 illustrates a heat exchanger sized for a 15°F [8 °C] approach temperature. A lower approach temperature requires a larger, more efficient heat exchanger. The size of the heat exchanger is influenced by the following factors:

- Total Btu/hour [W] to be transferred
- Gallons per minute [L/s] flow in the primary and secondary loops
- Fluids used in each loop (i.e., water, glycol, steam, or refrigerant)
- Whether the heat exchanger is parallel, counterflow, or multipass
- Long-term cleanliness (fouling factor) of the coils

The theoretical equations for heat exchangers are complex. Manufacturers' tables and charts are typically used in the selection of shell and tube heat exchangers.

Fin and Tube Heat Exchangers

Fin and tube heat exchangers exist in many forms. The most common types used are *fin tube baseboards* for heating and *direct expansion coils* (DX) in window air conditioners. In both cases, the primary fluid is circulated in the tube and air moves around the tube (or coil), typically between parallel, closely-spaced metal fins, for increased efficiency. Fin and tube coils, in one form or another, are required somewhere in most HVAC systems.

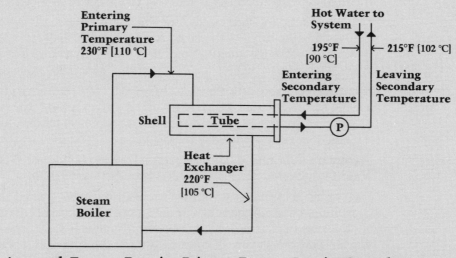

Approach Temp. = Entering Primary Temp. - Leaving Secondary Temp.
= 230°F - 215°F = 15°F Temperature Drop
[110 °C] [105 °C] [5 °C]

(Table below allows for preliminary sizing with the heat exchanger based on an inlet temperature of 200°F. This is slightly conservative compared to the 230°F entering primary temperature shown above.)

Heat Exchanger Performance

Preliminary Heat Exchanger Selection for Shell and Tube Heat Exchangers				
Shell Diameter (in.) [cm]	Length (ft.) [m]	Temperature Rise		
		40 – 140°F [4.5 – 60 °C]	40 – 180°F [4.5 – 82 °C]	140 – 180°F [60 – 82 °C]
4 [10]	3 [0.9]	100 [380]	30 [114]	100 [380]
4 [10]	4 [1.2]	150 [565]	45 [170]	150 [565]
4 [10]	5 [1.5]	215 [815]	60 [225]	205 [775]
4 [10]	6 [1.8]	275 [1040]	85 [320]	260 [985]
6 [15]	3 [0.9]	230 [870]	60 [225]	250 [945]
6 [15]	4 [1.2]	360 [1360]	95 [360]	365 [1380]
6 [15]	5 [1.5]	490 [1855]	145 [550]	475 [1800]
6 [15]	6 [1.8]	625 [2365]	200 [755]	590 [2235]
8 [20]	4 [1.2]	745 [2820]	190 [720]	720 [2725]
8 [20]	5 [1.5]	1,020 [3860]	300 [1135]	960 [3635]
8 [20]	6 [1.8]	1,270 [4805]	390 [1475]	1,200 [4540]
8 [20]	7 [2.1]	1,500 [5675]	480 [1815]	1,440 [5450]
10 [25]	4 [1.2]	1,285 [4865]	325 [1230]	1,200 [4540]
10 [25]	5 [1.5]	1,755 [6640]	480 [1815]	1,560 [5905]
10 [25]	6 [1.8]	2,100 [7950]	635 [2405]	1,920 [7265]
10 [25]	7 [2.1]	2,460 [9310]	765 [2895]	2,260 [8555]
12 [30]	4 [1.2]	1,770 [6700]	455 [1725]	1,440 [5450]
12 [30]	5 [1.5]	2,430 [9195]	645 [2440]	2,040 [7720]
12 [30]	6 [1.8]	3,000 [11 355]	840 [3180]	2,640 [9990]
12 [30]	7 [2.1]	3,510 [13 285]	1,020 [3860]	3,240 [12 260]

All numbers shown in brackets are metric equivalents.
GPH [L/s] of Water Inside Tubes (Supply Water to Shell at 200°F [93 °C])

Figure 9.2

Fin and tube exchangers are selected using a method similar to that used for shell and tube exchangers. The pressure required to drive the air through the coil must be added to the fan head.

Steam Traps

In selecting heat exchangers, proper sizing of the steam trap requires careful analysis. Steam traps are subject to large fluctuations in load and considerable thermal stresses.

Steam traps must be sized properly. An undersized steam trap will not drain the condensate and air quickly enough – the piping and terminal unit will become backed up, causing slower heat-up and lower capacities for heating terminal units. An oversized steam trap will not open, causing permanent backup, or will cause the pin in the trap to "ride" the seat, wearing the trap rapidly and shortening its life.

There are several variables in the selection of the proper type and size of steam trap for a particular application. These include:
- Volume of condensate flow
- Rate of condensate flow — steady or variable
- Operating temperature
- Air volume to be handled
- Warm-up loads
- Pressure differential between trap inlet and outlet
- Condensate temperature

Steam traps are subject to great temperature extremes and numerous repeated cycles. They wear out and must be actively maintained. Because of the number of variables that govern steam traps, there is rarely one type that can meet all trapping requirements.

The first step is to select the correct trap for the application; once this has been done, the trap can be sized.

Types of Steam Traps

There are four basic types of steam traps (see Figure 11.10 in Chapter 11 for illustrations). Their applications are as follows:

A *float trap* is normally closed and contains a ball that floats up as condensate enters the trap, allowing the condensate to drain. The ball floats down and closes to seal in the steam when the condensate has gone. It operates by moving a pin up and down and is used in low-pressure applications. Float traps are used when condensate removal does not also require air removal. Besides their use in heating systems, they are used at flash tanks and at compressed air receivers. Generally, they are the least-used trap.

An *inverted bucket trap* is a float trap with an open (inverted) bucket instead of a closed floating ball. Steam enters the inverted bucket, causing it to rise and seal the steam in the radiator. Bucket traps are rugged, but noisy, so they should be used for industrial applications and less so for hospitals, classrooms, or offices. They can operate at higher temperatures, because the bucket is stronger than the balls and thermostatic elements, but they have limited capability to control air.

A *thermostatic trap* is normally open and contains a thermal element, usually comprised of diaphragms that expand with the increasing temperature caused by steam entering the trap and pushing the drain pin closed. Thermostatic traps are primarily used for radiators in two-pipe systems.

A *float and thermostatic trap* is normally closed and combines the qualities of the float and thermostatic traps. Float and thermostatic traps are used where large quantities of air are likely to be part of the total load. They are commonly used at steam-to-water heat exchangers, heating coils, and at the end of mains and risers before a condensate receiver on low-pressure heating systems.

Sizing Steam Traps

Once the type of trap is selected, the size of the trap can be selected from a manufacturer's table. The variables that must be known to select a trap size are:

- The expected flow of condensate from the heating unit
- The safety factor, to be applied to the condensate flow – usually between 2 and 4
- The differential pressure between the inlet and outlet of the trap
- The required pressure rating of the trap

The most significant factor in sizing steam traps is the amount of condensate that will flow through the traps per hour and the safety factor used to adjust the condensate rate. The heat capacity of the heating unit is used to determine the design flow of condensate to the trap, in pounds per hour (lbs./hr.) [kg/hr.] or square feet of equivalent direct radiation (EDR). (To obtain lbs./hr. from EDR, simply divide the EDR value by four.)

A safety factor should then be applied to the design condensate flows from the heating unit to allow for the increased flow of condensate during the warm-up of the system. This safety factor should be no more than 4 and no less than 2 times the calculated condensate design flow rates; manufacturers of traps should be consulted for recommended safety factors. A note of caution: Some manufacturers rate their steam traps for condensate flows with a safety factor incorporated. Such tables give net ratings of traps and no safety factor should be applied before selecting the trap.

After determining the design condensate flow rate and the safety factor, the second significant factor to be determined is the pressure differential across the trap. The differential pressure between the inlet and outlet of the trap affects the performance of the trap. At lower differential pressures, the trap has a lower capacity; at higher differential pressures, the trap has a greater capacity. For some traps the difference between a 1/4 psi [2 kPa] and a 15 psi [103 kPa] differential could mean a difference of four times the capacity of the trap. The pressure differential is dependent on the design of the system and the configuration of the piping at the trap; in the simplest case, the pressure differential at the trap is the difference between the supply steam pressure and the return condensate system. For example, if condensate was draining from a steam system at 5 psig into a return line at 0 psig, the pressure differential would be $(5-0) = 5$ psig.

A properly selected and sized trap will maintain the correct pressure differential between the steam and condensate piping systems. It is this pressure differential that maintains the flow of steam throughout the system. If the traps in a system were all open, the pressures on the steam and condensate lines would equalize and there would be very little, if any, flow throughout the system. Consequently, poor heating would result and steam would be wasted.

Pressure-rating Steam Traps

Finally, the pressure rating of the trap should be considered. Most manufacturers rate their traps in three categories for maximum operating pressure:

Low Pressure	to 25 psi [275 kPa]
Medium Pressure	to 50 psi [450 kPa]
High Pressure	to 125 psi [965 kPa] or higher

For most residential and commercial heating applications the low-pressure traps are sufficient. For larger commercial, industrial, or district heating applications, medium- or high-pressure traps are necessary.

Example

Using Figure 9.3, select a thermostatic steam trap for a steam radiator unit of rated capacity 880 EDR [100 kg/hr.], if the steam pressure is 3 psig [122 kPa] and the return pressure is 1 psig [108.25 kPa]. The radiator is on the top floor of a residential building and will discharge into the condensate return line in the ceiling of the floor below.

Solution

Since the trap is to discharge to the floor below, an angle pattern trap is required. The pressure differential across the trap is 3 − 1 = 2 psig [122 − 108.25 = 13.75 kPa]. The steam capacity in lbs./hr. [kg/hr.] is 880 EDR/4 = 220 lbs./hr. [100 kg/hr.]. A safety factor of 2 will be applied for start-up load, so condensate load is 220 × 2 = 440 lbs./hr. [200 kg/hr.]. A 3/4" [37.5 mm] trap will discharge 465 lbs./hr. [211 kg/hr.] at 2psi [16 kPa] differential.

Capacities of Thermostatic Radiator Traps

Type	Size (in.) [cm]	Capacity—lbs./hr. [kg/hr.] steam Differential Pressure Across Traps—psi [kPa]							
		1/4 [1.7]	1/2 [3.5]	1 [7]	1-1/2 [10]	2 [14]	5 [35]	10 [69]	15 [103]
angle	1/2 [1.3]	85 [38.5]	120 [54.5]	165 [75]	200 [91]	235 [106.3]	370 [168]	530 [240]	640 [290]
swivel	1/2 [1.3]								
vertical	1/2 [1.3]								
angle	3/4 [1.9]	165 [75]	230 [104.5]	330 [150]	400 [181.5]	465 [211]	730 [331]	1050 [476]	1300 [590]
straight-away	3/4 [1.9]								
angle	1 [2.54]	290 [131.5]	410 [186]	580 [263]	700 [317.5]	810 [367]	1280 [581]	1840 [835]	2300 [1043]

All numbers shown in brackets are metric equivalents.

Figure 9.3

Equipment Data Sheets

The equipment data sheets for coils and heat exchangers shown in Figure 9.4 summarize the useful life, capacity range, typical uses, and special considerations for heat exchangers. A Means line number, where applicable, is included in order to determine costs in the current edition of *Means Mechanical Cost Data*.

Problems

9.1 How big is the heat exchanger for heating hot water from 140°F [60 °C] to 180°F [82 °C] with a flow of 1,000 gallons per hour [3780 L/s] with a supply temperature of 200°F [93 °C]?

9.2 What type of steam trap should be used for the following:
   a. Radiator in a health clinic
   b. Steam unit heater in an automotive garage
   c. Bottom of a riser in a low-pressure system

9.3 Using the steam distribution system in Figure 7.2, determine the type of each trap and label it on the drawing.

9.4 Select a thermostatic steam trap for a steam radiator unit of rated capacity 200 MBH [58.5 kW] if the steam pressure is 2 psig [115 kPa] and the return pressure is 1/2 psig [105 kPa]. The radiator is on the top floor of a residential building and will discharge into the condensate return line in the ceiling of the floor below.

Heat Exchangers Equipment Data Sheets

Heat Exchangers	Shell and Tube—Gas to Liquid	Means Number 155-601, 157-270
Useful Life: 20+ Years	Typical Use and Consideration	Capacity Range 8 – 250 GPM [0.5 – 16 L/s]
Straight Tube Heat Exchanger (Steam in, Condensate out, HW / CW)	The most common use of this heat exchanger concept is for heating water with steam in the shell. Other uses are for refrigerant gas to cool water or for water to cool and condense refrigerant gas. This type of heat exchanger is found built into a package chiller assembly. These heat exchangers may be straight tube or U tube depending on the design function. Weight consideration will mandate structural support in larger sizes.	
Heat Exchangers	Shell and Tube—Liquid to Liquid	Means Number 155-601
Useful Life: 20+ Years	Typical Use and Consideration	Capacity Range 7 – 152 GPM [0.45 – 10 L/s]
U Tube Heat Exchanger (HTHW, CW, HW)	This type of heat exchanger, when used with liquid in the shell as well as in the tubes, is used in numerous applications, such as for high temperature hot water to heat lower temperature water for domestic hot water purposes or building heating systems. When used as a below the water line heater in a steam boiler or as a tank heater for crude oil, the application and heat transfer principle is the same. On the cooling side, brine or glycol solutions can be used in these heat exchangers to chill or cool water. These heat exchangers may be U tube or straight tube depending on the design function. Weight consideration will mandate structural support in large sizes.	
Heat Exchangers	Heat Transfer Package—Shell and Tube	Means Number 155-610
Useful Life: 10 – 15 Years	Typical Use and Consideration	Capacity Range 35 – 800 GPM [2.2 – 50 L/s]
Primary Supply, Expansion Tank, HX, CW, HW Return	This hydronic package consists of a U tube heat exchanger—either steam or high-temperature hot water—as the primary source to produce hot water for heating or process use. The package contains prepiped and prewired pumps and controls. It also includes expansion tank, valves, and air eliminators, all mounted on a structural frame and base.	
Heat Exchangers	Fin and Tube	Means Number 157-201, 270
Useful Life: 20+ Years	Typical Use and Consideration	Capacity Range cfm/GPM
Duct Coil	This type of heat exchanger (coil) is used to heat or cool air. The primary source may be hot or chilled water, steam, or refrigerant. Coils may be mounted separately in ductwork or may be one of the components in a fan coil or air handling unit. The same principle is used in baseboard or wall-hung fin tube radiation, as well as in air-cooled and evaporative condensers. An air filter upstream from a duct or unit coil protects the fins and maintains heat transfer efficiency.	

All numbers shown in brackets are metric equivalents.

Figure 9.4

Chapter Ten

DISTRIBUTION AND DRIVING SYSTEMS

Distribution is the movement of heated or conditioned air to desired locations. Distribution systems include any devices that connect generating equipment to the terminal units, such as ductwork and piping, as well as the pumps, fans, and accessories that are separate from the generating equipment and conduct the steam, water, or air through the distribution system. (For information on valves and dampers, refer to Chapter 11, "Terminal Units.") This chapter addresses the basic principles involved in selecting, laying out, and sizing distribution equipment. Cost guides such as *Means Mechanical Cost Data* and manufacturers' price lists should be consulted for the specific costs of the components that make up distribution and driving systems.

The distribution equipment is part of either an *open* or *closed* loop system. In an open system, the fluid leaves the pipe or duct and is "open" to the atmosphere. In a closed system, the fluid remains in the containing pipes and pressure vessels, usually completing a recirculating loop. In general, most systems involving liquids (water) are closed, and most systems involving air are open. Open and closed systems are illustrated in Figure 10.1.

The basic types of distribution systems and their ranges are listed below.

Steam	High-pressure steam	above 15 psig [205 kPa]*
	Low-pressure steam	0 – 15 psig [101 – 205 kPa]
Water	High-temperature water	above 250°F [121 °C]
	Hot water	100 – 250°F [38 – 121 °C]
	Chilled water	35 – 100°F [1.5 – 38 °C]
Air	High velocity air	above 2,000 fpm [100 m/s] and above 2.5" wg [735 Pa]
	Low velocity air	0 – 2,000 fpm [0 – 100 m/s] and below 2.5" wg [735 Pa]
Gases	Refrigerants Natural gas Propane	varies with gas

*Throughout the book, SI metric equivalents are provided for most imperial units. The SI units appear in brackets [ ].

Distribution equipment conveys the heating or cooling medium (steam, water, air, or gas) from the generation equipment to the terminal units. When the medium used in the generating system is different from that of the distribution system, a heat exchanger and additional pumps or fans are required. The following sections describe distribution system layouts for steam, hot water, and air. Figure 10.2 is a selection guide for choosing the appropriate distribution system for a particular building.

Steam Distribution

Distribution by steam transfers the most heat per pound of water. At 212°F, one pound of steam releases 970 Btu's. [At 100 °C, one kilogram of steam releases 2250 kilojoules of energy.] Accounting for the additional heat from the condensate and using a slightly higher operating temperature, the customary value of 1,000 Btu's per pound of steam is used for most applications. The term *square foot of radiation* (SFR) is usually used to size radiation for steam (1 SFR is sometimes also expressed as equivalent direct radiation, or EDR). One square foot of radiation equals 240 Btu/hr., which is the amount of heat delivered by one square foot of radiator surface with a one-pound steam output. Water systems are rated as follows: 150/Btu/s.f./hr. equals one square foot of radiation.

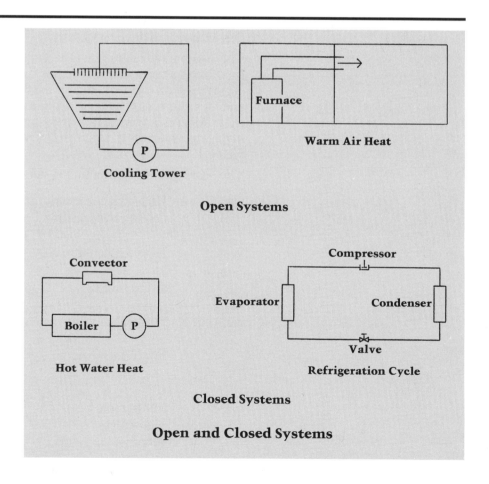

Figure 10.1

The force to drive steam is created by the boiler, which increases steam volume and pressure, and by the condensing process, which draws more steam into the system. Thus, steam systems are largely self-propelled, requiring no pumps. With proper maintenance, they are also long-lasting, because they contain no motor-driven equipment.

Selection Guide for Steam, Water, and Air Distribution Systems

Distribution System	Common Usage	Type of Piping Used
Steam 0 – 15 psig [0 – 104 kPa] – low-pressure, 212 – 250°F [100 – 120 °C] Over 15 psig [104 kPa] – high-pressure	Hospitals, industrial plants, laundries, restaurants. Supply to units for heating coils of central systems for dry cleaning, commercial buildings. Campus heating systems between buildings, district heating. Insulation required.	Carbon steel
Hot Water 30 psig [207 kPa] – low-temperature, 250°F [120 °C] 150 psig [1035 kPa] – medium-temperature, 350°F [175 °C] 300 psig – [2070 kPa] high-temperature, 450°F [230 °C]	Residences, offices, commercial, institutional. Widest use for loads under 5,000 MBH [1500 kW]. Campus type heating systems. Supply to convectors, cast iron radiators, baseboard fin tubes, VAV coils, fan coils, air handling units. Insulation required.	Carbon steel or copper tubing
Chilled Water 125 psig – [863 kPa] 40 – 55°F [4 – 12 °C]	Offices, commercial, institutional. Supplies coils, fan coil units, VAV coils, air handling units. Also used in campus type cooling systems. Vapor barrier insulation mandatory.	Carbon steel or copper tubing
Warm Air 130°F [54 °C]	Residences, offices, commercial, institutional. Supplies warm air from furnaces, air handling units, and discharges through grilles, registers or diffusers. Insulation and sound lining optional.	Aluminum, galvanized steel, or fiberglass duct
Cool Air 55°F [12 °C]	Residences, offices, commercial, institutional. Supplies cool air from cooling units such as package units and discharges through grilles, registers or diffusers. Sound lining is optional, vapor barrier insulation is mandatory in unconditioned spaces.	Aluminum, galvanized steel, or fiberglass duct
Mixed Air 55 – 130°F [12 – 54 °C]	Residences, offices, commercial, institutional. Supplies cool or heated air through common ductwork. Ductwork must be insulated for cooling.	Aluminum, galvanized steel, or fiberglass duct

Figure 10.2

A particular advantage of using steam in high-rise buildings is that it does not present the pressure problems associated with water distribution systems. Steam heating is often used for industrial applications, hospitals, and large facilities, since the steam is already necessary for cleaning, sterilization, food preparation, and generation of hot water.

Layout

The condensed steam (condensate in the form of hot water) is returned to the boiler for re-use through either a one-pipe or a two-pipe system. In a one-pipe system, the condensate returns to the boiler in the same pipe that supplies steam to the terminal unit. A two-pipe system uses the supply pipe for steam and a separate return pipe for the condensate. Two-pipe systems are more expensive, because they contain roughly twice as much piping. However, they do offer better overall control of steam systems.

Because steam is lighter than air, it displaces the heavier air and rises upward through risers to the upper floors. As steam condenses and becomes water, the water is returned to the boiler through pipes. All pipes in both one- and two-pipe steam systems are pitched toward the boiler or mechanical room to return the condensate by gravity directly to the boiler or a condensate return pump.

Most units are *upfed*, meaning that the terminal unit is located above the steam main. When steam flows down to a terminal unit below the main, the unit is said to be *downfed*. Vertical pipes that feed the system are called *risers*. Horizontal pipes are called *mains* or *branches*. Sloped pipes between the mains and risers or radiators are called *runouts*.

Types of Steam Systems

There are three types of steam heating systems, characterized by the way that the steam is propelled and/or the condensate is returned. The three types of steam heating systems are *gravity*, *vapor*, and *vacuum* systems.

Gravity Systems

Gravity systems are usually single-pipe, low-pressure systems. Steam is propelled by the initial boiler pressure, and by the fact that it is lighter than air, to radiators in upper floors. The steam and condensate are carried in the same pipe, or main. The main is pitched and gravity carries the condensate downward, sometimes in the opposite direction of steam flow, back to the boiler.

Two-pipe gravity systems are used on larger systems where a separate return pipe from the terminal units enters a return main and flows by gravity back to the boiler. Two-pipe systems generally require smaller-diameter pipe, but require twice as much pipe as a one-pipe system.

Vapor Systems

A vapor system is a two-pipe, low-pressure steam heating system utilizing thermostatic traps and combination float and thermostatic traps at terminal units and at drip connections between supply and return mains and risers. Traps let water and air pass through them; the thermostatic element prevents steam from getting by. This arrangement allows the unwanted water and air to flow through the return piping back to the boiler, while retaining the steam in the system. A condensate return unit is often utilized in this type of system.

Vacuum Systems

A vacuum added to a vapor system reduces pressure in the return pipe and improves performance and control. Because the pressure in a vacuum system is below atmospheric pressure, performance and control are improved. A vacuum return system improves steam distribution in the supply main.

Pressure

A particular advantage of steam systems is that the pressure can be adjusted to increase the heat content. By varying the pressure to meet the demand, a smooth and efficient delivery of energy can be achieved. The ranges of pressure for steam systems are shown below.

Above 15 psig [205 kPa]   High-pressure system (most efficient; strict safety controls)

0 – 15 psig [101 – 205 kPa]   Low-pressure steam (most common)

Less than 0 psig [101 kPa]   Vacuum system (vacuum pumps required to maintain pressure below atmospheric pressure)

Note that 15 psig means that the pressure is 15 pounds per square inch (psi) above atmospheric or gauge pressure. Atmospheric pressure is 14.7 psi [101 kPa] at sea level. A gauge pressure of 15 psig combined with atmospheric pressure equals 29.7 psia [205 kPa], and is known as the absolute pressure of the 15 psig reading.

The higher the pressure, the more efficient the system. Utility companies routinely supply steam at a rate of 80 – 200 psig [653 – 1480 kPa]. Large boilers generating steam above 15 psig [205 kPa] usually require a licensed attendant at all times. Power plants, hospitals, or other industrial buildings usually contain such boilers. High-pressure steam from a boiler room or utility must go through a pressure-reducing station – a system of pressure-reducing valves – to meet such standards. In commercial and residential buildings, pressures above 15 psig [205 kPa] are considered dangerous; these buildings must use low-pressure or vacuum systems.

Venting Steam Heating Systems

With steam systems, air must be removed from the pipe prior to boiler start-up and must be continually removed from the piping, the boiler, and the heated water to ensure that the steam flows evenly. Venting must be performed for gravity, vacuum, and vapor systems.

In gravity systems, air is vented throughout the piping network at the high points to allow gravity to drain the condensate back to the boiler. Air vents are located at the high point of each terminal unit, on the high point of the main, at the top of each riser, and at the boiler. Air vents are specifically designed for radiators, mains, and other pieces of equipment.

Air in vapor and vacuum systems is vented centrally at a condensate receiver, usually at the end of a system. No air vents are provided at the terminal units. In addition, steam traps are provided at each terminal unit and at the drip connections between the supply and return main. The traps allow the steam condensate to return to the boiler, but trap the steam on the supply side. Systems with traps are sometimes called "mechanical," "trapped," or "vapor" systems.

The most frequently used steam systems are illustrated in Figure 10.3.

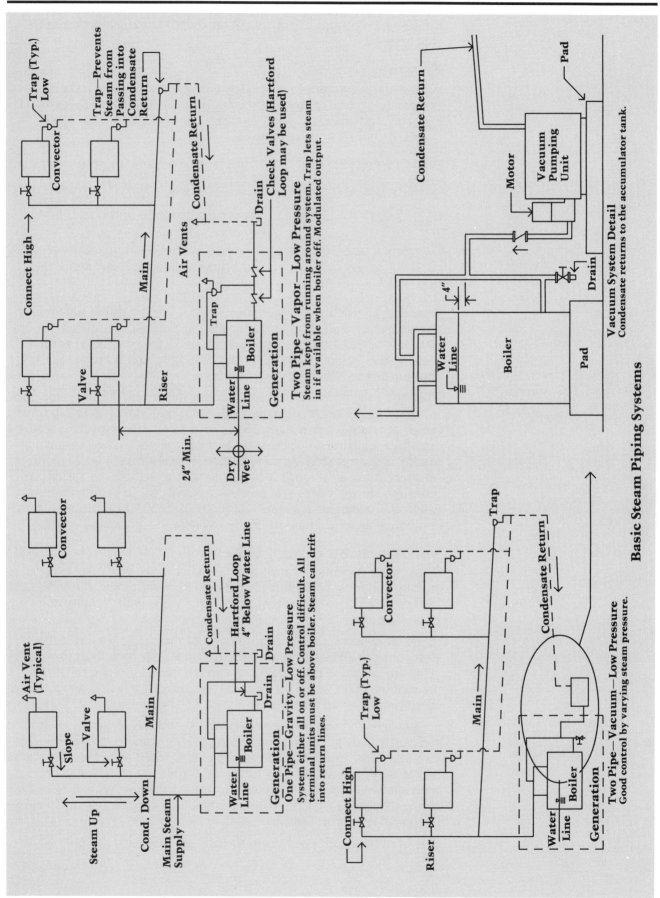

Figure 10.3

Steam Pipe Sizing

The size of steam pipe must be determined for each branch or run of pipe. Generally, the larger the pipe, the more pounds of steam per hour it will carry and the more heat will be delivered. The variables that affect pipe sizing include the pipe material, pipe length, type of system, system pressure, condensate flow, size of risers, and heat capacity.

- **Pipe material:** The surface roughness of the pipe and the actual inside pipe diameter affect flow. Schedule 40 steel pipe is the most frequently used pipe material for low-pressure steam and is the basis of most steam pipe tables.
- **Pipe length:** The total equivalent pipe length equals the length of the straight pipe plus the equivalent length of all fittings (see Figure 10.4 in Imperial units, and Figure 10.4a in SI Metric units). When fittings are not known, total equivalent length is assumed to be twice the straight length.
- **Type of system:** See the sizing data for one- and two-pipe systems for low-pressure steam in Figures 10.5 through 10.7. More piping is required for a two-pipe system.
- **Pressure of system:** Vacuum, vapor, low-pressure, and high-pressure systems deliver more heat by increasing steam pressure, which compresses the steam. The data shown in Figures 10.5 through 10.7 are for 3.5 psig [125.5 kPa] average pressure, the most common low-pressure system. (Any system under 15 psig [205 kPa] is considered a low-pressure system.) Figures 10.5a and 10.7a are in SI Metric units.
- **System pressure:** The difference in pressure between the supply and return pipe is called the *pressure drop*. The difference in pressure conveys the steam through the system. Most systems operate below 1 psi pressure drop (1 psi = 144 pounds per square foot, or a column of water 28" high). Most radiators are placed at least 28" above the boiler outlet as a safety factor to avoid condensate flooding the system. Most low-pressure and vacuum systems use a 1/16 psi [101.8 kPa] (1 ounce) pressure drop for systems under 200 feet [60 m] total equivalent length (pipe and fittings).

Assume the pressure drop to be one-third to one-half of the average operating gauge pressure. This figure is converted to pressure drop per 100 feet [per 1 meter]. Thus, for a boiler operating at 5 psig [135 kPa] on a system with 300 total equivalent feet [90 m] of length, the pressure drop would be calculated as follows:

$$1/3 \times 5 \times 100/300 = 5/9 \text{ psi}$$

$$[1/3 \times \frac{(135-101)}{1} \times \frac{1}{90} = 126 \text{ Pa}]$$

The 5/9 could be reduced, or rounded, to 1/2 psi, thus producing the figure of 1/2 psi pressure drop per 100 feet of pipe.

- **Condensate flow (with or against steam):** When the condensate flows in the same direction as the steam, the pipe has a greater capacity. If the condensate flows against the steam, the capacity is reduced and larger-diameter pipe is required.
- **Risers:** Riser pipe sizes have a greater steam capacity because of the effect of gravity, which removes the interference from the condensate flow. (In an upfed riser, the steam is rising and the condensate is flowing down.)

Equivalent Pipe Lengths for Steam Fittings

Length in Feet of Pipe to be Added to Actual Length of Run — Owing to Fittings — To Obtain Equivalent Length

Size of Pipe (Inches)	Length in Feet to be Added to Run				
	Standard Elbow	Side Outlet Tee[b]	Gate Valve[a]	Globe Valve[a]	Angle Valve[a]
1/2	1.3	3	0.3	14	7
3/4	1.8	4	0.4	18	10
1	2.2	5	0.5	23	12
1¼	3.0	6	0.6	29	15
1½	3.5	7	0.8	34	18
2	4.3	8	1.0	46	22
2½	5.0	11	1.1	54	27
3	6.5	13	1.4	66	34
3½	8	15	1.6	80	40
4	9	18	1.9	92	45
5	11	22	2.2	112	56
6	13	27	2.8	136	67
8	17	35	3.7	180	92
10	21	45	4.6	230	112
12	27	53	5.5	270	132
14	30	63	6.4	310	152

[a]Valve in full open position.
[b]Values given apply only to a tee used to divert the flow in the main to the last riser.

Example: Determine the length in feet of pipe to be added to actual length of run illustrated.

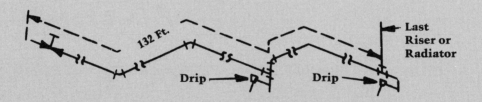

Measured Length	=	132.0 ft.
4 in. Gate Valve	=	1.9 ft.
4-4 in. Elbows	=	36.0 ft.
2-4 in. Tees	=	36.0 ft.
Equivalent	=	205.9 ft.

(Copyright 1993 by ASHRAE, from *1993 Fundamentals*. Used by permission.)

Figure 10.4

- **Capacity:** Pipe size varies with the number of Btu's per hour (W) that must be delivered. Generally, one pound of steam equals 1,000 Btu's, or 1 MBH, for both low-pressure and vacuum systems.

Adequate steam pipe size is the most critical during system start-up. Air must be removed from the piping system, which may be overloaded with additional condensate and produce excessive noise. To prevent this, it is good practice to design the horizontal steam mains larger than two inches in diameter (at least 2-1/2") [65 mm]. In addition, all pipe should be pitched for drainage. The recommended minimum pitch for mains is 1/4" per 10 feet [2 mm per meter] and 1/2" per foot [4 mm per meter] for branches (runouts) to risers and radiators. Pipe that cannot be adequately pitched must be increased in size to compensate for any hindrance to the steam and condensate flow.

Nominal Pipe Diameter, mm	Standard Elbow	Side Outlet Tee[b]	Gate Valve[a]	Globe Valve[a]	Angle Valve[a]
\multicolumn{6}{c}{Length in Meters of Pipe to Be Added to Actual Length of Run}
15	0.4	1	0.1	4	2
20	0.5	1	0.1	5	3
25	0.7	1	0.1	7	4
32	0.9	2	0.2	9	5
40	1.1	2	0.2	10	6
50	1.3	2	0.3	14	7
65	1.5	3	0.3	16	8
80	1.9	4	0.4	20	10
100	2.7	5	0.3	28	14
125	3.3	7	0.7	34	17
150	4.0	8	0.9	41	20
200	5.2	11	1.1	55	28
250	6.4	14	1.4	70	34
300	8.2	16	1.7	82	40
350	9.1	19	1.9	94	46

[a] Valve in full open position.
[b] Values apply only to a tee used to divert the flow in the main to the last riser.

Example: Determine the length in meters of pipe to be added to actual length of run illustrated.

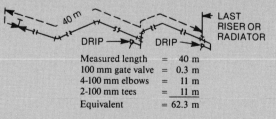

```
Measured length    =  40 m
100 mm gate valve  =  0.3 m
4-100 mm elbows    =  11 m
2-100 mm tees      =  11 m
Equivalent         =  62.3 m
```

(Copyright 1993 by ASHRAE, from *1993 Fundamentals, SI Edition.* Used by permission.)

Equivalent Pipe Lengths for Steam Fittings – In SI Metric Units

Figure 10.4a

Example

Size the main riser in an upfeed two-pipe low-pressure (4.0 psig) [130 kPa] system for 40 MBH [12 kW] where the actual length of the pipe is 175 feet [55 m]. What pressure drop should be used in sizing the runs of pipe?

In this example, the steam and condensate flow in opposite directions; therefore, the information in Figure 10.6 is used to size the main riser. A 40 MBH [12 kW] load requires 40 pounds of steam per hour [19 kg/hr]. In a two-pipe system, 1-1/2" [40 mm] vertical pipe is suitable, because it can provide 48 pounds [22 kg] of steam per hour (see Column B in Figure 10.6).

The pressure drop for this example is calculated using the tables in Figures 10.5 and 10.7. The calculations for a 4.0 psig system would be:

$$1/3 \times 4.0 = 1-1/3 \text{ psi (20 oz.)} \quad [1/3 \times (130-101) = 9.67 \text{ kPa}]$$

The pipe length with fittings is estimated initially at twice the straight length, or 350 feet [110 m] total equivalent pipe length.

Flow Rate of Steam in Schedule 40 Pipe[a] at Initial Saturation Pressure of 3.5 and 12 Psig[b]
(Flow Rate Expressed in Pounds per Hour)

Nom. Pipe Size (Inches)	1/16 Psi (1 oz.) Sat. press. psig		1/8 Psi (2 oz.) Sat. press. psig		1/4 Psi (4 oz.) Sat. press. psig		1/2 Psi (8 oz.) Sat. press. psig		3/4 Psi (12 oz.) Sat. press. psig		1 Psi Sat. press. psig		2 Psi Sat. press. psig	
	3.5	12	3.5	12	3.5	12	3.5	12	3.5	12	3.5	12	3.5	12
3/4	9	11	14	16	20	24	29	35	36	43	42	50	60	73
1	17	21	26	31	37	46	54	66	68	82	81	95	114	137
1¼	36	45	53	66	78	96	111	138	140	170	162	200	232	280
1½	56	70	84	100	120	147	174	210	218	260	246	304	360	430
2	108	134	162	194	234	285	336	410	420	510	480	590	710	850
2½	174	215	258	310	378	460	540	660	680	820	780	950	1,150	1,370
3	318	380	465	550	660	810	960	1,160	1,190	1,430	1,380	1,670	1,950	2,400
3½	462	550	670	800	990	1,218	1,410	1,700	1,740	2,100	2,000	2,420	2,950	3,450
4	640	800	950	1,160	1,410	1,690	1,980	2,400	2,450	3,000	2,880	3,460	4,200	4,900
5	1,200	1,430	1,680	2,100	2,440	3,000	3,570	4,250	4,380	5,250	5,100	6,100	7,500	8,600
6	1,920	2,300	2,820	3,350	3,960	4,850	5,700	5,700	7,000	8,600	8,400	10,000	11,900	14,200
8	3,900	4,800	5,570	7,000	8,100	10,000	11,400	14,300	14,500	17,700	16,500	20,500	24,000	29,500
10	7,200	8,800	10,200	12,600	15,000	18,200	21,000	26,000	26,200	32,000	30,000	37,000	42,700	52,000
12	11,400	13,700	16,500	19,500	23,400	28,400	33,000	40,000	41,000	49,500	48,000	57,500	67,800	81,000

(Copyright 1993 by ASHRAE, from 1993 Fundamentals. Used by permission.)

[a]Based on Moody Friction Factor, where flow of condensate does not inhibit the flow of steam.
[b]The flow rates at 3.5 psig can be used to cover sat. press. from 1 to 6 psig, and the rates at 12 psig can be used to cover sat. press. from 8 to 16 psig with an error not exceeding 8%.

Steam Supply Pipe Sizes When Condensate Flows in Same Direction as Steam

Figure 10.5

$$\frac{20 \text{ oz.} \times 100}{350} \quad \left[\frac{9.67 \text{ kPa} \times 1000}{110} \right]$$

Therefore, the pressure drop per 100 feet [per meter] is approximately 6 oz. [88 Pa].

Safety and Operational Features

There are several safety and operational features in steam systems, including provisions for monitoring both the water level in the boiler and temperature changes, properly supporting the pipe, and maintaining uniform flow and traps.

The *Hartford Loop* is a safety feature designed to prevent the water level in the boiler from dropping too low (see Figure 10.3). Whenever the water level falls below the Hartford Loop connection tie-in (approximately two to four inches below water level in the boiler), the return line is no longer isolated from the supply line by the water in the boiler. As a result, the pressures equalize and the abnormal siphon that caused the boiler water level to drop is broken.

Because steam pipes undergo a wide range of temperature changes, *expansion loops*, or joints, and *swing joints* are provided for expansion

Flow Rate of Steam in Schedule 40 Pipe

Nominal Pipe Size, mm	14 Pa/m Sat. Press., kPa		28 Pa/m Sat. Press., kPa		58 Pa/m Sat. Press., kPa		113 Pa/m Sat. Press., kPa		170 Pa/m Sat. Press., kPa		225 Pa/m Sat. Press., kPa		450 Pa/m Sat. Press., kPa	
	25	85	25	85	25	85	25	85	25	85	25	85	25	85
20	4	5	6	7	9	11	13	16	16	20	19	23	27	33
25	8	10	12	14	17	21	24	30	31	37	37	43	52	62
32	16	20	24	30	35	44	50	63	64	77	73	91	105	127
40	25	32	38	45	54	67	79	95	99	118	112	138	163	195
50	49	61	73	88	106	129	152	186	191	231	218	268	322	386
65	79	98	117	141	171	209	245	299	308	372	354	431	522	621
80	144	172	211	249	299	367	435	526	540	649	626	758	885	1090
90	210	249	304	363	449	552	640	771	789	953	907	1100	1340	1560
100	290	363	431	526	640	767	898	1090	1110	1360	1310	1570	1910	2220
125	544	649	762	953	1110	1360	1620	1930	1990	2380	2310	2770	3400	3900
150	871	1040	1280	1520	1800	2200	2590	3180	3270	3900	3810	4540	5400	6440
200	1770	2180	2530	3180	3670	4540	5170	6490	6580	8030	7480	9300	10 900	13 400
250	3270	3990	4630	5720	6800	8260	9530	11 800	11 900	14 500	13 600	16 800	19 400	23 600
300	5170	6210	7480	8850	10 600	12 900	15 000	18 100	18 600	22 500	21 800	26 100	30 800	36 700

Notes: 1. Flow rate is in kg/h at initial saturation pressures of 25 and 85 kPa (gage). Flow is based on Moody friction factor, where the flow of condensate does not inhibit the flow of steam.
2. The flow rates at 25 kPa cover saturated pressure from 7 to 41 kPa, and the rates at 85 kPa cover saturated pressure from 55 to 110 kPa with an error not exceeding 8%. All pressures are above atmospheric.
3. The steam velocities corresponding to the flow rates given in this table can be found from the basic chart and velocity multiplier chart, Figure 11.

(Copyright 1993 by ASHRAE, from 1993 Fundamentals, SI Edition. Used by permission.)

Steam Supply Pipe Sizes When Condensate Flows in Same Direction as Steam — In SI Metric Units

Figure 10.5a

and contraction throughout a system. Expansion loops and joints are used at connectors to mains at terminal units and on straight runs of more than 40 feet [12 meters].

Expansion loops and joints also require that the piping be properly anchored and guided. Steam pipes should be *insulated* and *supported* on properly designed hangers at suitable spacings.

In order to maintain continuous flow of condensate and avoid water hammer (water flowing through the pipes at too high a speed), *eccentric pipe reducers* are required to maintain uniform flow at pipe inverts whenever pipe size reduces in the direction of the condensate flow. Eccentric reducers reduce the pipe off-center, maintaining a flat invert. The small end of a *concentric reducer* is positioned centrally at the end of the fitting. Eccentric and concentric reducers are shown in Figure 10.8.

Steam systems undergo more stress than water or air systems. The pipes can accumulate dirt from corrosion, which can further reduce overall

Steam Pipe Capacities for Low-Pressure Systems
(For Use on One-Pipe Systems or Two-Pipe Systems)

Nominal Pipe Size, (Inches)	Two-Pipe Systems — Condensate Flowing Against Steam		One-Pipe Systems		
	Vertical	Horizontal	Supply Risers Up-Feed	Radiator Valves and Vertical Connections	Radiator and Riser Runouts
A	B[a]	C[c]	D[b]	E	F[c]
3/4	8	7	6	—	7
1	14	14	11	7	7
1¼	31	27	20	16	16
1½	48	42	38	23	16
2	97	93	72	42	23
2½	159	132	116	—	42
3	282	200	200	—	65
3½	387	288	286	—	119
4	511	425	380	—	186
5	1,050	788	—	—	278
6	1,800	1,400	—	—	545
8	3,750	3,000	—	—	—
10	7,000	5,700	—	—	—
12	11,500	9,500	—	—	—
16	22,000	19,000	—	—	—

Note: Steam at an average pressure of 1 psig is used as a basis of calculating capacities.

[a]Do not use Column B for pressure drops of less than 1/16 psi per 100 ft. of equivalent run.
[b]Do not use Column D for pressure drops of less than 1/24 psi per 100 ft. of equivalent run except on sizes 3 in. and over.
[c]Pitch of horizontal runouts to risers and radiators should be not less than 1/2 in. per ft. Where this pitch cannot be obtained, runouts over 8 ft. in length should be one pipe size larger than called for in this table.

(Copyright 1993 by ASHRAE, from <u>1993 Fundamentals</u>. Used by permission.)

Steam Supply Pipe Sizes When Condensate Flow is Opposite to Steam

Figure 10.6

performance. Consequently, steam systems should have *strainers* installed before traps and control valves in order to keep the systems as clean as possible. All units must be properly trapped, and dirt pockets, or drain cocks, should be provided at the base of every riser and at main drips to remove sludge build-up.

Water Distribution

Most hot and chilled water piping networks are closed-loop water distribution systems. The four basic types of water distribution systems – series, monoflow, direct return, and reverse return – are illustrated in Figure 10.9. When separate heating and cooling are provided to the same terminal unit, three- or four-pipe systems are generally used.

Steam Pipe Capacities for Low-Pressure Systems

Capacity, kg/h

Nominal Pipe Size, mm	Two-Pipe System – Condensate Flowing against Steam		One-Pipe Systems		
	Vertical	Horizontal	Supply Risers Up-feed	Radiator Valves and Vertical Connections	Radiator and Riser Runouts
A	B[a]	C[b]	D[c]	E	F[b]
20	4	3	3	—	3
25	6	6	5	3	3
32	14	12	9	7	7
40	22	19	17	10	7
50	44	42	33	19	10
65	72	60	53	—	19
80	128	91	91	—	29
90	176	131	130	—	54
100	232	193	172	—	84
125	476	357	—	—	126
150	816	635	—	—	247
200	1700	1360	—	—	—
250	3180	2590	—	—	—
300	5220	4310	—	—	—
400	9980	8620	—	—	—

Notes:
1. For one-pipe or two-pipe systems in which condensate flows against the steam flow.
2. Steam at an average pressure of 7 kPa above atmospheric is used as a basis of calculating capacities.

[a] Do not use Column *B* for pressure drops of less than 13 Pa/m of equivalent run. Use Figure 10 or Table 13 instead.
[b] Pitch of horizontal runouts to risers and radiators should be not less than 40 mm/m. Where this pitch cannot be obtained, runouts over 2.5 m in length should be one pipe size larger than called for in this table.
[c] Do not use Column *D* for pressure drops of less than 9 Pa/m of equivalent run, except on sizes 80 mm and over. Use Figure 10 or Table 13 instead.

(Copyright 1993 by ASHRAE, from *1993 Fundamentals, SI Edition*. Used by permission.)

Steam Supply Pipe Sizes When Condensate Flow is Opposite to Steam – In SI Metric Units

Figure 10.6a

Pipe Size (Inches)	1/32 Psi or 1/2 Oz. Drop per 100 Ft.			1/24 Psi or 2/3 Oz. Drop per 100 Ft.			1/16 Psi or 1 Oz. Drop per 100 Ft.			1/8 Psi or 2 Oz. Drop per 100 Ft.			1/4 Psi or 4 Oz. Drop per 100 Ft.			1/2 Psi or 8 Oz. Drop per 100 Ft.		
	Wet	Dry	Vac.	Wet	Dry	Vac.	Wet	Dry	Vac.	Wet	Dry	Vac.	Wet	Dry	Vac.	Wet	Dry	Vac.
G	H	I	J	K	L	M	N	O	P	Q	R	S	T	U	V	W	X	Y
Return Main																		
3/4	—	—	—	—	—	42	—	—	100	—	—	142	—	—	200	—	—	283
1	125	62	—	145	71	143	175	80	175	250	103	249	350	115	350	—	—	494
1¼	213	130	—	248	149	244	300	168	300	425	217	426	600	241	600	—	—	848
1½	338	206	—	393	236	388	475	265	475	675	340	674	950	378	950	—	—	1,340
2	700	470	—	810	535	815	1,000	575	1,000	1,400	740	1,420	2,000	825	2,000	—	—	2,830
2½	1,180	760	—	1,580	868	1,360	1,680	950	1,680	2,350	1,230	2,380	3,350	1,360	3,350	—	—	4,730
3	1,880	1,460	—	2,130	1,560	2,180	2,680	1,750	2,680	3,750	2,250	3,800	5,350	2,500	5,350	—	—	7,560
3½	2,750	1,970	—	3,300	2,200	3,250	4,000	2,500	4,000	5,500	3,230	5,680	8,000	3,580	8,000	—	—	11,300
4	3,880	2,930	—	4,580	3,350	4,500	5,500	3,750	5,500	7,750	4,830	7,810	11,000	5,380	11,000	—	—	15,500
5	—	—	—	—	—	7,880	—	—	9,680	—	—	13,700	—	—	19,400	—	—	27,300
6	—	—	—	—	—	12,600	—	—	15,500	—	—	22,000	—	—	31,000	—	—	43,800
Riser																		
3/4	—	48	—	—	48	143	—	48	175	—	48	249	—	48	350	—	—	494
1	—	113	—	—	113	244	—	113	300	—	113	426	—	113	600	—	—	848
1¼	—	248	—	—	248	388	—	248	475	—	248	674	—	248	950	—	—	1,340
1½	—	375	—	—	375	815	—	375	1,000	—	375	1,420	—	375	2,000	—	—	2,830
2	—	750	—	—	750	1,360	—	750	1,680	—	750	2,380	—	750	3,350	—	—	4,730
2½	—	—	—	—	—	2,180	—	—	2,680	—	—	3,800	—	—	5,350	—	—	7,560
3	—	—	—	—	—	3,250	—	—	4,000	—	—	5,680	—	—	8,000	—	—	11,300
3½	—	—	—	—	—	4,480	—	—	5,500	—	—	7,810	—	—	11,000	—	—	15,500
4	—	—	—	—	—	7,880	—	—	9,680	—	—	13,700	—	—	19,400	—	—	27,300
5	—	—	—	—	—	12,600	—	—	15,500	—	—	22,000	—	—	31,000	—	—	43,800

*Reference to this table is made by column letter G through Y.

(Copyright 1993 by ASHRAE, from 1993 Fundamentals. Used by permission.)

Steam Pipe Sizes for Condensate Returns

Figure 10.7

Series/Monoflow

The series and monoflow loops are similar configurations routinely used for systems and subsystems under 50 MBH [15 kW]. They require the least amount of piping and have the lowest cost, but offer the least control. A series circuit connects all convectors/radiators by a single pipe in one continuous loop. A monoflow circuit has an uninterrupted loop and the convectors/radiators are tapped off the loop.

The series circuit has no control; shutting off one terminal shuts off the entire circuit. Each terminal unit in a monoflow system may be individually controlled. In either system, the radiators at the end of the run should be sized based on a lower supply water temperature; the water temperature is lower as it reaches the end of the loop.

Direct/Reverse Return

In a *direct return* water distribution system, the return pipe takes the shortest, most direct route back to the pump. The distance that the water travels from the pump to the first terminal unit and back is shorter than the loop distance to the last terminal unit (see Figure 10.9). In a *reverse return* system, the return piping leaving a terminal unit follows the supply piping to the next terminal unit. The result is that the loop distance to all terminal units is equal. Some terminals in a reverse return system have short supply pipe and long return pipe runs, while others have long supply

Return Main and Riser Capacities for Low-Pressure Systems, kg/h*

	Pipe Size, mm	7 Pa/m			9 Pa/m			14 Pa/m			28 Pa/m			57 Pa/m			113 Pa/m		
		Wet	Dry	Vac.	Wet	Dry	Vac.	Wet	Dry	Vac.	Wet	Dry	Vac.	Wet	Dry	Vac.	Wet	Dry	Vac.
		G	H	I	J	K	L	M	N	O	P	Q	R	S	T	U	V	W	X	Y
Return Main	20	—	—	—	—	—	19	—	—	45	—	—	64	—	—	91	—	—	128
	25	57	28	—	66	32	65	79	36	79	113	47	113	159	52	159	—	—	224
	32	97	59	—	112	68	111	136	76	136	193	98	193	272	109	272	—	—	385
	40	153	93	—	178	107	176	215	120	215	306	154	306	431	171	431	—	—	608
	50	318	213	—	367	243	370	454	261	454	635	336	644	907	374	907	—	—	1 280
	65	535	345	—	717	394	616	762	431	762	1 070	558	1 080	1 520	617	1 520	—	—	2 150
	80	853	662	—	966	708	989	1 220	794	1 220	1 700	1 020	1 720	2 430	1 130	2 430	—	—	3 430
	90	1 250	894	—	1 500	998	1 400	1 810	1 130	1 810	2 490	1 470	2 580	3 630	1 620	3 630	—	—	5 130
	100	1 760	1 330	—	2 080	1 520	2 040	2 490	1 700	2 490	3 520	2 190	3 540	4 990	2 440	4 990	—	—	7 030
	125	—	—	—	—	—	3 570	—	—	4 390	—	—	6 210	—	—	8 800	—	—	12 400
	150	—	—	—	—	—	5 720	—	—	7 030	—	—	9 980	—	—	14 100	—	—	19 900
Riser	20	—	22	—	—	22	65	—	22	79	—	22	113	—	22	159	—	—	224
	25	—	51	—	—	51	111	—	51	136	—	51	193	—	51	272	—	—	385
	32	—	112	—	—	112	176	—	112	215	—	112	306	—	112	431	—	—	608
	40	—	170	—	—	170	370	—	170	454	—	170	644	—	170	907	—	—	1 280
	50	—	340	—	—	340	616	—	340	762	—	340	1 080	—	340	1 520	—	—	2 150
	65	—	—	—	—	—	989	—	—	1 220	—	—	1 720	—	—	2 430	—	—	3 430
	80	—	—	—	—	—	1 470	—	—	1 810	—	—	2 580	—	—	3 630	—	—	5 130
	70	—	—	—	—	—	2 030	—	—	2 490	—	—	3 540	—	—	4 990	—	—	7 030
	100	—	—	—	—	—	3 570	—	—	4 390	—	—	6 210	—	—	8 800	—	—	12 400
	125	—	—	—	—	—	5 720	—	—	7 030	—	—	9 980	—	—	14 100	—	—	19 900

(Copyright 1993 by ASHRAE, from 1993 Fundamentals, SI Edition. Used by permission.)

* Reference to this table is made by column letter G through Y.

Steam Pipe Sizes for Condensate Returns – In SI Metric Units

Figure 10.7a

and short return pipe runs, but the total loop length to all terminal units is equal. In short, in a direct return system, the first terminal unit fed is also the first one returned to the boiler, whereas in a reverse return system, the first one fed is the last one returned and the last one fed is the first returned.

Reverse return piping systems are sometimes called *three-pipe systems*, because of the apparent extra return pipe. This situation can be seen by comparing the elevations of the direct and reverse piping systems shown in Figure 10.9.

The reverse return piping layout provides the best overall control and reliability. It supplies the same temperature water to each terminal unit (minus negligible line losses) and distributes the proper quantities of water in each pipe, because all loops are the same length and no short-circuiting can occur.

The direct return system is usually used in loops or sub-loops under 50 MBH. Terminal units closest to the pump typically should have partially closed balancing valves to constrict flow on the short runs and to encourage water to flow to the more remote units.

Heat Per GPM

One gallon of water weighs 8.1 pounds at 180°F. [One liter of water weighs 1 kg at 82 °C.] One Btu is released into a room for each 1°F drop in temperature for each pound of water at a terminal unit. [4.2 kilojoules are released into a room for each 1 °C drop in temperature for each kilogram of water at a terminal unit.] For heating systems, using a 20°F temperature

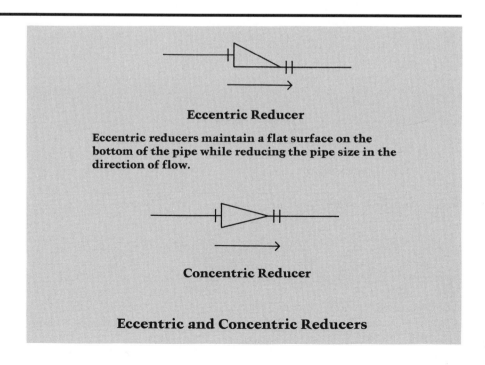

Figure 10.8

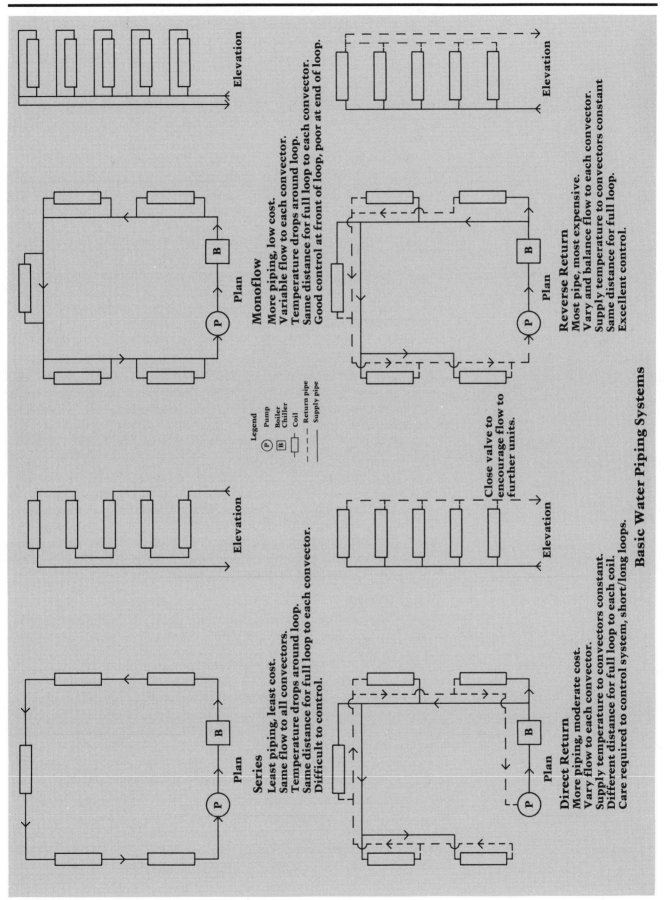

Figure 10.9

Basic Water Piping Systems

drop between the supply and return water temperatures is an industry standard. A 1 gallon per minute [1 liter per second] system equals 60 gallons per hour.

1 gpm – 60 × 8.1 = 486 lbs./hr.

[1 (L/s – 1 × 1) = 1 kg/s]

The result is rounded to 500 pounds per hour. If each pound of water drops 20°F, it loses 20 Btu/hr. Therefore, the heat given up by 1 gallon per minute equals:

500 lbs./hr. × 1 Btu/lb.°F × 20°F = 10,000 Btu/hr., or 10 MBH

If each kilogram of water drops 12 °C, it loses 50 kJ. Therefore, the heat given up by 1 L/s equals:

[1 kg/s × 4.2 kJ/kg °C × 12 °C = 50 kW]

The general rule for heating systems is: 1 gallon per minute equals 10 MBH at a 20°F temperature drop.

This calculation is easy to use because it directly converts heat loss to flow. A building with a 100 MBH heat load requires 10 gallons of water per minute with a 20°F drop in temperature between supply and return. A room in that building with a heat loss of 3,500 Btu/hr. would require 0.35 gpm. [A building with a 50 kW heat load requires 1 L/s of water with a 12 °C drop in temperature.]

Chilled water, condenser water, and high-temperature water systems use other standards for the difference between supply and return temperature. The rule is adjusted proportionally for these other systems. For example, a condenser pump to a cooling tower is likely to operate at a lower temperature drop, such as 8°F, between supply and return, and therefore requires more water flow (20/8).

To lay out a piping system, first determine the load (MBH) [kW] for the entire building, and then for each room. At least one terminal unit should be placed in each room. Next, select an appropriate combination of series, monoflow, direct, or reverse return systems. Then the pump, boiler, chiller, and other major equipment are connected to the terminal units. The load and flow requirements should be recorded next to each pipe length designation. This process is illustrated in Figure 10.10.

Water Piping

Water is the most frequently used medium in HVAC piping systems. A typical low-temperature hot water piping network is used as a base guide, but the rules used for sizing the piping for hot water apply to all water systems. As the temperature varies in water systems from 40°F [5 °C] for chilled water to 250°F [120 °C] for low-temperature hot water systems, the only significant change is in the viscosity of the water. Dual systems use the same pipes for chilled water in summer and hot water in winter. Many types of pipe material are used; the most common are Schedule 40 steel pipe, copper tubing (types K, L, and M), and Schedule 80 PVC and CPVC (plastic) pipe. Figure 10.10 illustrates an example of pipe layout and sizing. The method used is explained in the following paragraphs.

Pipe size is based on either velocity or head loss, which provide roughly equal results. Good design practice provides pipe sizes large enough to avoid the velocity noise caused by rapid water flow. Large pipe sizes also reduce the amount of friction that pumps must overcome. Velocities of three to six feet per second [one to two meters per second] are considered acceptable. Pressure head losses of two to four feet of head per 100 feet [0.2 to 0.4 kPa/m] of pipe are also acceptable. Generally, a rating of 2.5 feet of head per 100 feet [0.25 kPa/m] of pipe is used as design criteria.

To select a pipe size, determine the design flow for the particular run of pipe based on the recommended head loss per 100 feet [per meter] (see Figures 10.11 and 10.11a). For example, select a pipe for a flow of 30 gallons per minute at 2.5 feet of head per 100 feet in steel, copper type L, and Schedule 80 PVC.

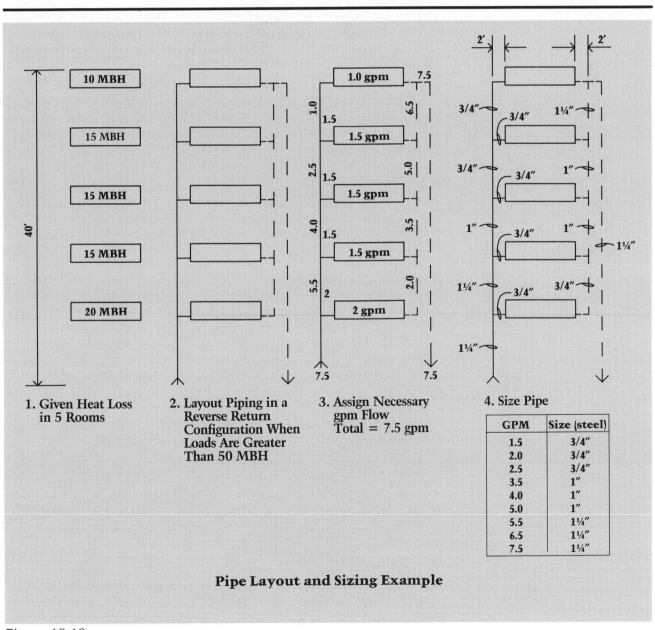

Pipe Layout and Sizing Example

Figure 10.10

Type	Interior Diameter	Nominal Size to Use	Velocity in Selected Pipe
Steel (Sched. 40)	2.067"	2"	2.9 fps
Type L Copper	1.985"	2"	3.0 fps
PVC (Sched. 80)	1.913"	2"	3.5 fps

For this example, a two-inch diameter pipe is selected, the water velocity is approximately 3 fps, and the pressure drop is 2.5 feet of head for each 100 feet of pipe.

Flow Temperature and Viscosity

The charts in Figure 10.11 and Figure 10.11a (in SI Metric units) are based on a water temperature of 60°F [15.5 °C], which means that for chilled water systems (40 – 55°F) [5 – 13 °C], they are reasonably accurate; for hot water systems (200 – 250°F) [93 – 121 °C], they are slightly conservative. However, the actual head losses will be slightly lower, because of the lower viscosity of the warmer water – the warmer the water, the less force is required to make it move. When friction losses for hot water are calculated, the pipe section sized from this data should be oversized by 15 to 20 percent. PVC pipe should not be used in systems with water temperatures greater than 110°F [44 °C].

Many hydronic systems use a mixture of glycol and water, which has a low freezing point, to prevent freezing in pipes. Water in outside air intake coils, makeup air units, solar systems, and pipes that pass through unheated spaces are sometimes filled with this mixture to minimize the problems associated with freezing. A fifty-fifty water/glycol mixture has a freezing point of -340°F $[-171$ °C], which is low enough for most applications.

The selection of pipe size is affected when glycol is used, because glycol has a higher viscosity and requires more force to pump. Correction factors for glycol mixtures are used to account for friction losses (see Figure 10.12).

Pipe Length/Pump Size

Like steam systems, hot and chilled water systems are also affected by friction losses for fittings such as valves, elbows, tees, and specialties. The pump must be sized for the total flow (in gpm) [in L/s] of the total equivalent run of pipe and the total head. The total flow is calculated from the heat load and temperature drop of the water. The total equivalent run of pipe equals the measured straight run of pipe plus the straight run equivalent for the fittings. The total head is determined by adding up the friction head loss for both the pipes and the fittings and adding (for open systems only) the lift head from one elevation to another plus losses in components (coils, radiators, boilers, and velocity head).

Figure 10.13 lists the equivalent length for elbows over a range of velocities and pipe sizes. (Figure 10.13a is the same chart in SI Metric units.) For example, a two-inch elbow in a system with water flowing at 3 feet per second has an equivalent straight run of 5.4 feet. An open gate valve for steel pipe has an elbow equivalent of 0.5. Therefore, the equivalent length of the valve is

$0.5 \times 5.4 = 2.7$ ft.

For initial trial designs, it is commonly assumed that the friction of all fittings around a loop will equal that of the straight run of pipe, meaning that the total equivalent run of pipe equals twice the measured run of

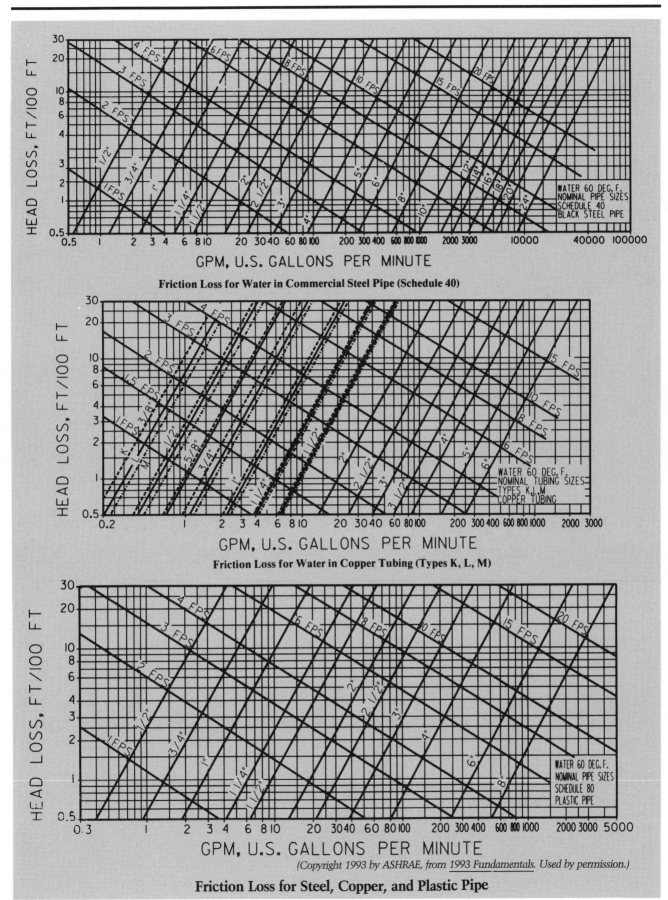

Friction Loss for Steel, Copper, and Plastic Pipe

Figure 10.11

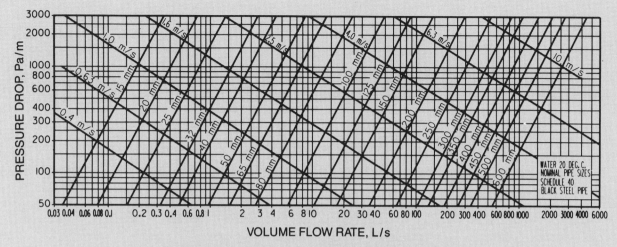

Friction Loss for Water in Commercial Steel Pipe (Schedule 40)

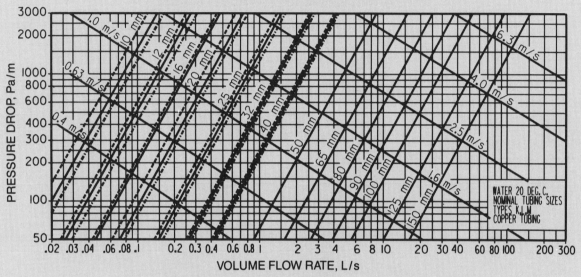

Friction Loss for Water in Copper Tubing (Types K, L, M)

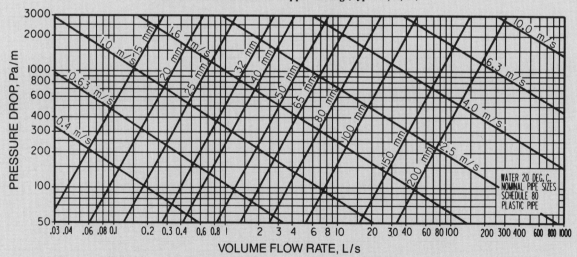

Friction Loss for Water in Plastic Pipe (Schedule 80)

(Copyright 1993 by ASHRAE, from 1993 Fundamentals, SI Edition. Used by permission.)

Friction Loss for Steel, Copper, and Plastic Pipe — In SI Metric Units

Figure 10.11a

straight pipe. This assumption should be confirmed after the design layout is completed.

For the system shown in Figure 10.10, the pump head required is based on the friction head loss around one radiator loop. A one-inch pipe in the system has water flowing at about 1.5 feet per second; the elbow equivalent is 2.3 feet (see Figure 10.13).

Correction Factors for Glycol Mixtures

Increased Flow Requirement for Same Heat Conveyance 50% Glycol as Compared with Water

Fluid Temperature (°F)	Flow Increase Needed for 50% Glycol as Compared with Water
40	1.22
100	1.16
140	1.15
180	1.14
220	1.14

Pressure Drop Correction Factors; 50% Glycol Solution Compared with Water

Fluid Temperature (°F)	Pressure Drop Correction Flow Rates Equal	Combined Pressure Drop Correction; 50% Glycol Flow Increased as Shown Above
40	1.45	2.14
100	1.1	1.49
140	1.0	1.32
180	.94	1.23
220	.9	1.18

(Courtesy ITT Fluid Handling Sales)

Figure 10.12

k Values—Screwed Pipe Fittings

Nominal Pipe Dia., in.	90° Ell Reg.	90° Ell Long	45° Ell	Return Bend	Tee-Line	Tee-Branch	Globe Valve	Gate Valve	Angle Valve	Swing Check Valve	Bell Mouth Inlet	Square Inlet	Projected Inlet
3/8	2.5	—	0.38	2.5	0.90	2.7	20	0.40	—	8.0	0.05	0.5	1.0
1/2	2.1	—	0.37	2.1	0.90	2.4	14	0.33	—	5.5	0.05	0.5	1.0
3/4	1.7	0.92	0.35	1.7	0.90	2.1	10	0.28	6.1	3.7	0.05	0.5	1.0
1	1.5	0.78	0.34	1.5	0.90	1.8	9	0.24	4.6	3.0	0.05	0.5	1.0
1-1/4	1.3	0.65	0.33	1.3	0.90	1.7	8.5	0.22	3.6	2.7	0.05	0.5	1.0
1-1/2	1.2	0.54	0.32	1.2	0.90	1.6	8	0.19	2.9	2.5	0.05	0.5	1.0
2	1.0	0.42	0.31	1.0	0.90	1.4	7	0.17	2.1	2.3	0.05	0.5	1.0
2-1/2	0.85	0.35	0.30	0.85	0.90	1.3	6.5	0.16	1.6	2.2	0.05	0.5	1.0
3	0.80	0.31	0.29	0.80	0.90	1.2	6	0.14	1.3	2.1	0.05	0.5	1.0
4	0.70	0.24	0.28	0.70	0.90	1.1	5.7	0.12	1.0	2.0	0.05	0.5	1.0

Source: *Engineering Data Book*, Hydraulic Institute, 1979.

k Values—Flanged Welded Pipe Fittings

Nominal Pipe Dia., in.	90° Ell Reg.	90° Ell Long	45° Ell Long	Return Bend Reg.	Return Bend Long	Tee-Line	Tee-Branch	Globe Valve	Gate Valve	Angle Valve	Swing Check Valve
1	0.43	0.41	0.22	0.43	0.43	0.26	1.0	13	—	4.8	2.0
1-1/4	0.41	0.37	0.22	0.41	0.38	0.25	0.95	12	—	3.7	2.0
1-1/2	0.40	0.35	0.21	0.40	0.35	0.23	0.90	10	—	3.0	2.0
2	0.38	0.30	0.20	0.38	0.30	0.20	0.84	9	0.34	2.5	2.0
2-1/2	0.35	0.28	0.19	0.35	0.27	0.18	0.79	8	0.27	2.3	2.0
3	0.34	0.25	0.18	0.34	0.25	0.17	0.76	7	0.22	2.2	2.0
4	0.31	0.22	0.18	0.31	0.22	0.15	0.70	6.5	0.16	2.1	2.0
6	0.29	0.18	0.17	0.29	0.18	0.12	0.62	6	0.10	2.1	2.0
8	0.27	0.16	0.17	0.27	0.15	0.10	0.58	5.7	0.08	2.1	2.0
10	0.25	0.14	0.16	0.25	0.14	0.09	0.53	5.7	0.06	2.1	2.0
12	0.24	0.13	0.16	0.24	0.13	0.08	0.50	5.7	0.05	2.1	2.0

Source: *Engineering Data Book*, Hydraulic Institute, 1979.

Approximate Range of Variation for k Factors

90° Elbow	Regular screwed	±20% above 2 in. ±40% below 2 in.	Tee	Screwed, line or branch Flanged, line or branch	±25% ±35%
	Long radius screwed	±25%	Globe valve	Screwed	±25%
	Regular flanged	±35%		Flanged	±25%
	Long radius flanged	±30%	Gate valve	Screwed	±25%
45° Elbow	Regular screwed	±10%		Flanged	±50%
	Long radius flanged	±10%	Angle valve	Screwed	±20%
Return bend (180°)	Regular screwed	±25%		Flanged	±50%
	Regular flanged	±35%	Check valve	Screwed	±50%
	Long radius flanged	±30%		Flanged	+200% −80%

Source: *Engineering Data Handbook*, Hydraulic Institute, 1979.

Equivalent Length in Feet of Pipe for 90° Elbows

Velocity, ft/s	Pipe Size														
	1/2	3/4	1	1-1/4	1-1/2	2	2-1/2	3	3-1/2	4	5	6	8	10	12
1	1.2	1.7	2.2	3.0	3.5	4.5	5.4	6.7	7.7	8.6	10.5	12.2	15.4	18.7	22.2
2	1.4	1.9	2.5	3.3	3.9	5.1	6.0	7.5	8.6	9.5	11.7	13.7	17.3	20.8	24.8
3	1.5	2.0	2.7	3.6	4.2	5.4	6.4	8.0	9.2	10.2	12.5	14.6	18.4	22.3	26.5
4	1.5	2.1	2.8	3.7	4.4	5.6	6.7	8.3	9.6	10.6	13.1	15.2	19.2	23.2	27.6
5	1.6	2.2	2.9	3.9	4.5	5.9	7.0	8.7	10.0	11.1	13.6	15.8	19.8	24.2	28.8
6	1.7	2.3	3.0	4.0	4.7	6.0	7.2	8.9	10.3	11.4	14.0	16.3	20.5	24.9	29.6
7	1.7	2.3	3.0	4.1	4.8	6.2	7.4	9.1	10.5	11.7	14.3	16.7	21.0	25.5	30.3
8	1.7	2.4	3.1	4.2	4.9	6.3	7.5	9.3	10.8	11.9	14.6	17.1	21.5	26.1	31.0
9	1.8	2.4	3.2	4.3	5.0	6.4	7.7	9.5	11.0	12.2	14.9	17.4	21.9	26.6	31.6
10	1.8	2.5	3.2	4.3	5.1	6.5	7.8	9.7	11.2	12.4	15.2	17.7	22.2	27.0	32.0

Equivalent Lengths of Pipe Fittings

Figure 10.13

Straight runs of pipe

3 Risers at 40 ft.		=	120 ft.
Radiator runouts	(2 × 2 ft.)	=	4
			124

Fittings on one loop

15 Elbows (estimated)	(15 × 1 elbow equiv. × 2.3)	=	34.5
1 Boiler	(1 × 3 elbow equiv. × 2.3)	=	6.9
1 Radiator	(1 × 3 elbow equiv. × 2.3)	=	6.9
2 Tees	(2 × 0.5 elbow equiv. × 2.3)	=	2.3
Total equivalent straight pipe		=	174.6 ft.

Because all pipe was sized not to exceed 2.5 feet of head per 100 feet, the pump head equals

$2.5/100 \times 174.6 = 4.4$ feet of head

Therefore, a pump that delivers 7.5 gpm with a head of 4.4 feet is required. These requirements, combined with the information from the pump curves (see Figure 8.3), result in the choice of a 1" AA pump with 3-7/16" impeller. Figure 10.14 illustrates the basic piping layouts for various low-pressure systems.

Hydronic Specialties

Hot water systems must remove accumulated air. Air vents or bleed valves at terminal units and high points allow air to be mechanically "bled." Air scoops are also used in the mechanical room to remove air.

Pipe runs over 40 feet [12 meters] long should have provisions for expansion. In addition to anchors, guides, loops, or expansion joints, additional elbows (swing joints) are normally used at branch takeoffs to allow this kind of flexibility. All piping runs should be pitched to draw off valves for drainage.

Iron and Copper Elbow Equivalents[a]		
Fitting	**Iron Pipe**	**Copper Tubing**
Elbow, 90°	1.0	1.0
Elbow, 45°	0.7	0.7
Elbow, 90° long turn	0.5	0.5
Elbow, welded, 90°	0.5	0.5
Reduced coupling	0.4	0.4
Open return bend	1.0	1.0
Angle radiator valve	2.0	3.0
Radiator or convector	3.0	4.0
Boiler or heater	3.0	4.0
Open gate valve	0.5	0.7
Open globe valve	12.0	17.0

[a] See Table for equivalent length of one elbow.
Source: Giesecke (1926) and Giesecke and Badgett (1931, 1932).

(Copyright 1993 by ASHRAE, from *1993 Fundamentals*. Used by permission.)

Equivalent Lengths of Pipe Fittings (continued)

Figure 10.13 (continued)

k Values—Screwed Pipe Fittings

Nominal Pipe Dia., mm	90° Ell Reg.	90° Ell Long	45° Ell	Return Bend	Tee-Line	Tee-Branch	Globe Valve	Gate Valve	Angle Valve	Swing Check Valve	Bell Mouth Inlet	Square Inlet	Projected Inlet
10	2.5	—	0.38	2.5	0.90	2.7	20	0.40	—	8.0	0.05	0.5	1.0
15	2.1	—	0.37	2.1	0.90	2.4	14	0.33	—	5.5	0.05	0.5	1.0
20	1.7	0.92	0.35	1.7	0.90	2.1	10	0.28	6.1	3.7	0.05	0.5	1.0
25	1.5	0.78	0.34	1.5	0.90	1.8	9	0.24	4.6	3.0	0.05	0.5	1.0
32	1.3	0.65	0.33	1.3	0.90	1.7	8.5	0.22	3.6	2.7	0.05	0.5	1.0
40	1.2	0.54	0.32	1.2	0.90	1.6	8	0.19	2.9	2.5	0.05	0.5	1.0
50	1.0	0.42	0.31	1.0	0.90	1.4	7	0.17	2.1	2.3	0.05	0.5	1.0
65	0.85	0.35	0.30	0.85	0.90	1.3	6.5	0.16	1.6	2.2	0.05	0.5	1.0
80	0.80	0.31	0.29	0.80	0.90	1.2	6	0.14	1.3	2.1	0.05	0.5	1.0
100	0.70	0.24	0.28	0.70	0.90	1.1	5.7	0.12	1.0	2.0	0.05	0.5	1.0

Source: *Engineering Data Book*, Hydraulic Institute, 1979.

k Values—Flanged Welded Pipe Fittings

Nominal Pipe Dia., mm	90° Ell Reg.	90° Ell Long	45° Ell Long	Return Bend Reg.	Return Bend Long	Tee-Line	Tee-Branch	Globe Valve	Gate Valve	Angle Valve	Swing Check Valve
25	0.43	0.41	0.22	0.43	0.43	0.26	1.0	13	—	4.8	2.0
32	0.41	0.37	0.22	0.41	0.38	0.25	0.95	12	—	3.7	2.0
40	0.40	0.35	0.21	0.40	0.35	0.23	0.90	10	—	3.0	2.0
50	0.38	0.30	0.20	0.38	0.30	0.20	0.84	9	0.34	2.5	2.0
65	0.35	0.28	0.19	0.35	0.27	0.18	0.79	8	0.27	2.3	2.0
80	0.34	0.25	0.18	0.34	0.25	0.17	0.76	7	0.22	2.2	2.0
100	0.31	0.22	0.18	0.31	0.22	0.15	0.70	6.5	0.16	2.1	2.0
150	0.29	0.18	0.17	0.29	0.18	0.12	0.62	6	0.10	2.1	2.0
200	0.27	0.16	0.17	0.27	0.15	0.10	0.58	5.7	0.08	2.1	2.0
250	0.25	0.14	0.16	0.25	0.14	0.09	0.53	5.7	0.06	2.1	2.0
300	0.24	0.13	0.16	0.24	0.13	0.08	0.50	5.7	0.05	2.1	2.0

Source: *Engineering Data Book*, Hydraulic Institute, 1979.

Approximate Range of Variation for k Factors

90° Elbow	Regular screwed	±20% above 50 mm	Tee	Screwed, line or branch	±25%
		±40% below 50 mm		Flanged, line or branch	±35%
	Long radius screwed	±25%	Globe valve	Screwed	±25%
	Regular flanged	±35%		Flanged	±25%
	Long radius flanged	±30%	Gate valve	Screwed	±25%
45° Elbow	Regular screwed	±10%		Flanged	±50%
	Long radius flanged	±10%	Angle valve	Screwed	±20%
Return bend (180°)	Regular screwed	±25%		Flanged	±50%
	Regular flanged	±35%	Check valve	Screwed	±50%
	Long radius flanged	±30%		Flanged	+200% −80%

Source: *Engineering Data Handbook*, Hydraulic Institute, 1979.

Equivalent Length in Metres of Pipe for 90° Elbows

Velocity, m/s	Pipe Size, mm													
	15	20	25	32	40	50	65	90	100	125	150	200	250	300
0.33	0.4	0.5	0.7	0.9	1.1	1.4	1.6	2.0	2.6	3.2	3.7	4.7	5.7	6.8
0.67	0.4	0.6	0.8	1.0	1.2	1.5	1.8	2.3	2.9	3.6	4.2	5.3	6.3	7.6
1.00	0.5	0.6	0.8	1.1	1.3	1.6	1.9	2.5	3.1	3.8	4.5	5.6	6.8	8.0
1.33	0.5	0.6	0.8	1.1	1.3	1.7	2.0	2.5	3.2	4.0	4.6	5.8	7.1	8.4
1.67	0.5	0.7	0.9	1.2	1.4	1.8	2.1	2.6	3.4	4.1	4.8	6.0	7.4	8.8
2.00	0.5	0.7	0.9	1.2	1.4	1.8	2.2	2.7	3.5	4.3	5.0	6.2	7.6	9.0
2.35	0.5	0.7	0.9	1.2	1.5	1.9	2.2	2.8	3.6	4.4	5.1	6.4	7.8	9.2
2.67	0.5	0.7	0.9	1.3	1.5	1.9	2.3	2.8	3.6	4.5	5.2	6.5	8.0	9.4
3.00	0.5	0.7	0.9	1.3	1.5	1.9	2.3	2.9	3.7	4.5	5.3	6.7	8.1	9.6
3.33	0.5	0.8	0.9	1.3	1.5	1.9	2.4	3.0	3.8	4.6	5.4	6.8	8.2	9.8

Equivalent Lengths of Pipe Fittings—In SI Metric Units

Figure 10.13a

Temperature/Pressure Considerations

Medium- and high-temperature water systems are designed using principles similar to those applied to chilled water and low-temperature hot water systems. An important difference is the pressure in the piping. At 212°F [100 °C], the atmospheric pressure of 14.7 psi [101 kPa] is the minimum pressure that is needed to keep water from boiling or "flashing" into the system. At 250°F [121 °C], 29.8 psi [205 kPa] is needed; at 350°F [177 °C], 134.6 psi [928 kPa]; and at 450°F [232 °C], 422.6 psi [2914 kPa] is the pressure exerted by the hot water against the piping. To avoid future problems, care must be taken in selecting safety features and materials and ensuring good construction techniques for hot water systems.

Air Distribution Systems

Air systems are widely used for both heating and cooling. Air systems offer the advantages of being free of freeze-up problems except at water or steam coils; providing full temperature control treatment of a space, including filtration, humidification, and cleaning; and being easily adapted for alterations. In addition, air systems have few mechanical or electrical devices to repair, besides the main equipment areas. Some disadvantages of air systems include large ducts, which can be difficult to conceal, and balancing air quantities for different seasons. Rapid temperature changes, duct leakage, and temperature losses in transmission of the air are also concerns.

Overall, air is the most flexible of any type of system and is used in office, commercial, and industrial buildings, as well as laboratories, universities, and hotels. Air systems are also used when needed to meet certain ventilation requirements.

Heat per Cubic Foot [per Liter]

Air transfers the least amount of heat per cubic foot. At room conditions, cool air occupies approximately 13.3 cubic feet per pound of air and one cubic foot of air equals 1/13.3, or 0.075, pounds per cubic foot [cool air

---

Iron and Copper Elbow Equivalents[a]

Fitting	Iron Pipe	Copper Tubing
Elbow, 90°	1.0	1.0
Elbow, 45°	0.7	0.7
Elbow, 90° long turn	0.5	0.5
Elbow, welded, 90°	0.5	0.5
Reduced coupling	0.4	0.4
Open return bend	1.0	1.0
Angle radiator valve	2.0	3.0
Radiator or convector	3.0	4.0
Boiler or heater	3.0	4.0
Open gate valve	0.5	0.7
Open globe valve	12.0	17.0

[a] See Table 4 for equivalent length of one elbow.
Source: Giesecke (1926) and Giesecke and Badgett (1931, 1932).

(Copyright 1993 by ASHRAE, from 1993 Fundamentals, SI Edition. Used by permission.)

Equivalent Lengths of Pipe Fittings—In SI Metric Units (continued)

Figure 10.13a (continued)

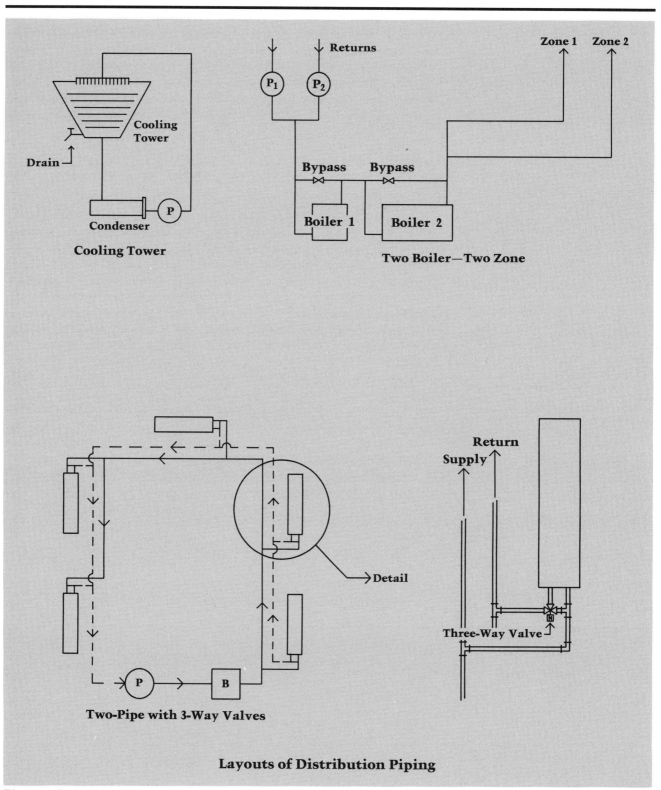

Figure 10.14

occupies approximately 0.83 cubic meters per kilogram of air and one cubic meter of air equals 1/0.83, or 1.2 kg/m^3]. 0.244 Btu's are required to heat or cool one pound of air 1°F [1.021 kilojoules are required to heat or cool one kg of air 1 °C]. To cool one cubic foot per minute of air 1°F [one liter per second of air 1 °C],

0.075×0.244 (lbs./ft.^{3}/min. $\times$ Btu/lb.) $= 0.0183$ Btu/ft.^{3}/min. or

$0.0183 \times 60 = 1.1$ Btu/ft.^{3}/hr.

$$[\frac{1.2 \times 1.021}{1000} \frac{\text{kg/m}^3/\text{s} \times \text{kJ/kg}}{\text{L/m}^3}$$
$$= 1.23 \times 10^{-3} \text{ kJ/L.s}$$
$$= 1.23 \text{ J/L.s}$$
$$= 1.23 \text{ W/L}]$$

Therefore,

H = 1.1 cfm ΔT [1.23 · L/s · ΔT] Cooling
 = 1.08 cfm ΔT Heating [1.2 · L/s · ΔT] (the air is even less dense)

The factor of 1.1 [1.23] will be used throughout the rest of the book.

Layout

Duct systems are generally classified as single- or dual-duct systems. Single-duct systems, the most common type, have one supply duct to the terminal units and are limited to either heating or cooling at a given time. Dual-duct systems have two supply air ducts, one with warm air and the second with cool air, which go to a mixing box above the space. The mixing box proportions the mixture of air to the space to achieve the design temperature. Both single- and dual-duct systems may be constant or variable volume, single or multi-zone, high or low velocity (see Figure 10.15).

High Velocity Systems

High velocity systems (in which the velocity of air in the ductwork exceeds 2,000 fpm [10 m/s] or the pressure exceeds 2.5″ [0.623 kPa]) are used for special applications. High velocity systems allow the use of smaller duct sizes, but require stronger ducts (or pipe) to transmit the high-pressure/high-velocity air. These systems also require specialized high-pressure fans and sound attenuating capabilities. Induction systems in high-rise buildings represent a common application for high velocity systems. High velocity systems are designed and sized in the same manner as low velocity systems.

VAV Systems

Variable air volume (VAV) systems are economical and energy-efficient. In a VAV system, the amount of air sent to a space is varied to meet the load. At the peak hour, the dampers are fully open and the maximum air is delivered to a space. At other times, the dampers in the VAV box close partially, and less air is delivered to the space to match the decreased load. The ducts for VAV systems are sized using the same methods as for low velocity systems. However, special care must be taken to select each part of the variable system based on a worst case scenario. In any system, a minimum amount of air must be circulated for health and comfort and to meet ventilation requirements. The main advantage of the VAV system is that the equipment is sized for the maximum simultaneous peak in each terminal, rather than the sum of all individual peaks, as for constant volume systems.

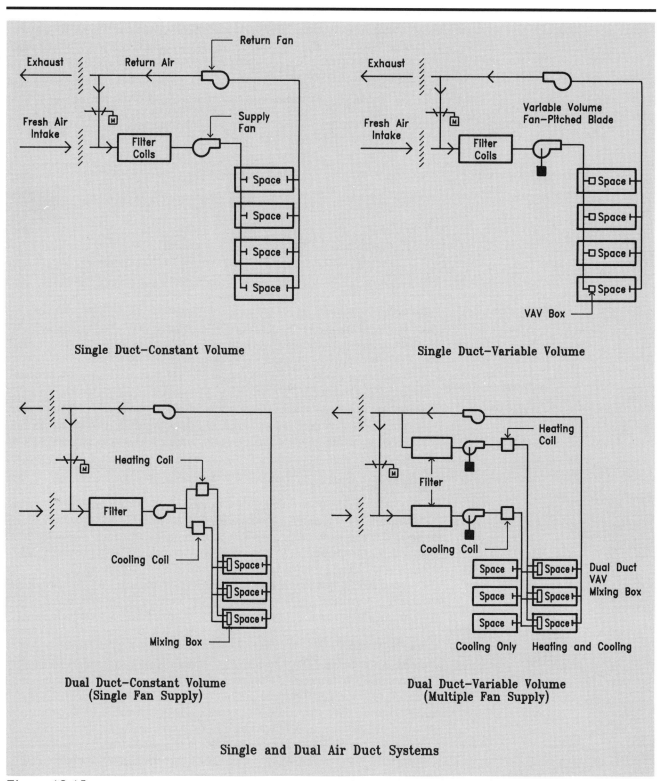

Figure 10.15

Reheat

Many single-duct systems in larger buildings have reheat coils incorporated into the terminal units. These reheat and booster coils may be hydronic or electric. Air is supplied to the single duct system at a fixed temperature of 55°F. The reheat coils raise the temperature of the 55°F air, as required, to satisfy respective space thermostats. For example, the south side of an office building, with heavy sun exposure, may need cooling and could either take the 55°F air directly or use a minimum amount of reheat to temper the air. The north side of the building would be cooler and thus would use more reheat to satisfy space thermostat requirements. Reheat systems can provide excellent comfort conditions in a building, but they are extremely inefficient from an energy standpoint because the air must be first cooled and then reheated.

Grilles, Registers, and Diffusers

In Chapter 1, Figure 1.21 illustrates the general principle for locating supply diffusers. Basically, for heating, supply low at the perimeter; for cooling, supply high near the perimeter.

A basic principle in locating air supply registers is the throw. Throw is the distance travelled by air from the diffuser to the point at which its velocity falls below 50 feet per minute [0.25 m/s]. Diffusers should be located so that the space is fully covered. The "dead" band (between diffuser throws) should not exceed 25 percent of the throw. This principle is illustrated in Figure 10.16.

Ideally, the return grilles should be diagonally opposite the supply diffusers. In laying out a system, the number and location of supply air outlets must be considered in terms of the effect on room comfort. Supply air outlets should be located so as to avoid drafts and convection currents, while providing even, overall distribution. Because most air systems are located above ceilings, they are ideally suited for cooling. Some form of perimeter heat (low at the exterior windows) must often be added to supplement the air distribution system and cover winter conditions.

Once the location of the diffusers has been determined, the distribution ductwork can be laid out and sized.

Duct Sizing

The following four methods are used to size ducts in order of increasing complexity (see Figure 10.17):

1. Velocity method
2. Constant pressure drop method
3. Balanced pressure drop method
4. Static regain method

Each method builds on the techniques of the previous method. The methods become progressively more involved and are used for more complicated layouts.

The velocity method sizes ducts by targeting an appropriate velocity. This method is basically a rough guess and is used to size simple, short-air systems. The second and third methods, constant and balanced pressure drop, are used by designers to accurately determine appropriate fan pressures. The last method, static regain, is the most difficult to design. It is also the most economical to build, although it requires advanced techniques.

The Velocity Method

The following formula is used to determine duct size using the velocity method.

Q = VA

Q = the quantity of air in a duct (cfm) [m³/s]

V = the velocity of the air in the duct (fpm) [m/s]

A = the duct cross-section area (s.f.) [m²]

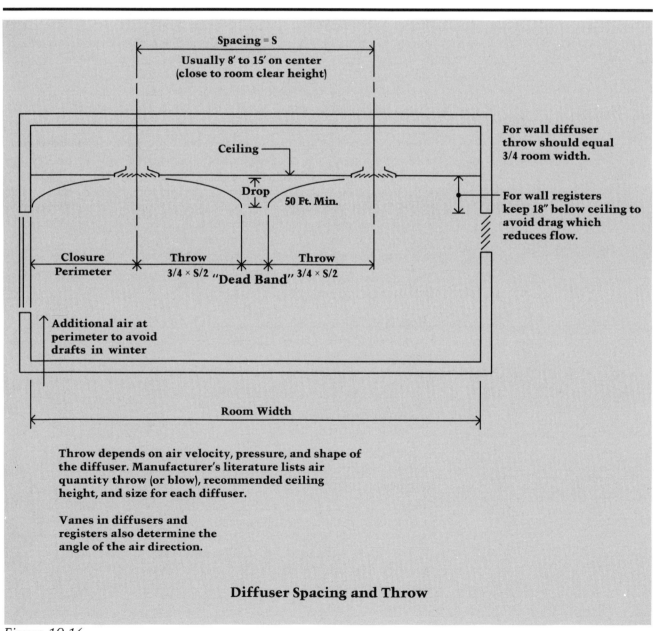

Diffuser Spacing and Throw

Figure 10.16

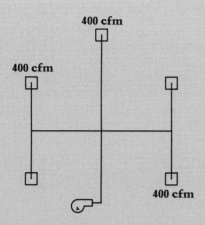

Velocity Method
Equal distance from fan to all registers.
Registers distribute same air quantity.
Easy to size ducts.
Volume dampers and field balance critical.
Small systems.

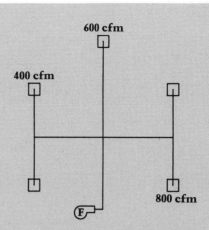

Constant Pressure Drop Method
Equal distance from fan to all registers.
Varied air quantity to each register.
Size duct from equivalent round.

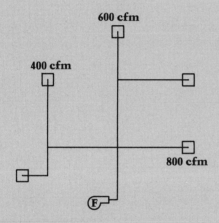

Balanced Pressure Drop Method
Varied distance from fan to registers.
Varied air quantity to each register.
Size duct by balancing pressures.

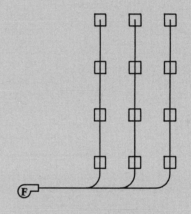

Static Regain Method
Registers take off along duct.
Varied air quantity to each register.
Most difficult to design.
Lowest fan and duct cost.

Legend

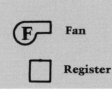

Basic Duct Sizing Methods

Figure 10.17

As the quantity of air to be moved in the duct increases, the required duct size increases. The velocity is selected utilizing half the values shown in Figure 10.18.

When using the velocity method, duct sizes are selected independent of shape, layout, or duct length; therefore, the pressures of the fan to drive the system can only be assumed. The velocity method ignores aspect ratio, length, turns, dampers, registers, and grilles. The inability to more precisely determine pressure drops in the system is the main disadvantage of the velocity method. The quantity of air required is proportional to the load. For warm air,

$H_T = 1.1\ Q\ (\Delta T)\ [1.23\ Q\ (\Delta T)]$

H_T = net gain or heat loss of the space in Btu/hr. [kW]

1.1 = constant for cool air (1.08 [1.2] for warm air)

Q = quantity of air to be delivered (cfm) [L/s] (Note: There are 1000 liters in one cubic meter)

ΔT = the difference between the room design temperature and the temperature of the air in the duct. For cooling, $\Delta T = 15 - 20°F$ [8 – 12 °C] maximum. For heating, $\Delta T = 40 - 70°F$ [22 – 40 °C].

Using these formulas, the recommended velocity, the required air quantity, and the area of the duct can be determined (see Figure 10.19).

The velocity method is best suited to calculations involving small systems, such as residential heating and cooling. The fans for these systems are

RECOMMENDED MAXIMUM DUCT VELOCITIES FOR LOW VELOCITY SYSTEMS (FPM)

APPLICATION	CONTROLLING FACTOR NOISE GENERATION Main Ducts	CONTROLLING FACTOR—DUCT FRICTION			
		Main Ducts		Branch Ducts	
		Supply	Return	Supply	Return
Residences	600	1000	800	600	600
Apartments Hotel Bedrooms Hospital Bedrooms	1000	1500	1300	1200	1000
Private Offices Directors Rooms Libraries	1200	2000	1500	1600	1200
Theatres Auditoriums	800	1300	1100	1000	800
General Offices High Class Restaurants High Class Stores Banks	1500	2000	1500	1600	1200
Average Stores Cafeterias	1800	2000	1500	1600	1200
Industrial	2500	3000	1800	2200	1500

(Courtesy Carrier Corporation, McGraw-Hill Book Company)

Note: Use one-half of these values for recommended velocity when using velocity method.

Figure 10.18

1. For each run of duct, determine the air quantity (Q).

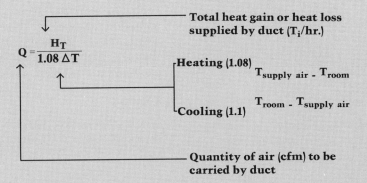

- H_T = Total heat gain or heat loss supplied by duct (T_i/hr.)
- Heating (1.08): $T_{supply\ air} - T_{room}$
- Cooling (1.1): $T_{room} - T_{supply\ air}$
- Q = Quantity of air (cfm) to be carried by duct

$$Q = \frac{H_T}{1.08\,\Delta T}$$

2. Determine the recommended velocity from Figure 10.18.

3. Determine duct area (A) required.

$$A = \frac{Q}{V}$$

- Q = Air quantity in duct (cfm) (see above)
- V = Recommended air velocity for duct (fpm) (from Figure 10.18)

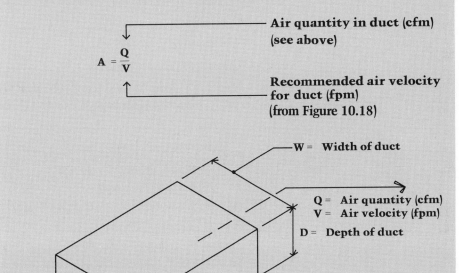

- W = Width of duct
- Q = Air quantity (cfm)
- V = Air velocity (fpm)
- D = Depth of duct
- A = Duct area = W × D

4. Determine duct width and depth.

A = W × D — Make D as deep as possible to fit construction.
   (1 : 1) Ideal
   (3 : 1) Maximum

The Velocity Method

Figure 10.19

usually slightly oversized and dampers can be used to control flow. When the velocity method is used, the duct should be as nearly square or circular as possible. The aspect ratio of 1:1 would be a square duct, with an equal proportion of width to height. The aspect ratio for the velocity method should not exceed 3:1. High aspect ratios cause additional friction, which affects overall balancing. This problem must be corrected using one of the other sizing methods. In duct design, a 4" round or 6" × 6" square are generally the smallest sizes used.

Example:

Size a cooling duct to carry 4,000 Btu/hr. [1175 W] in a main supply duct (for a store). Referring to Figures 10.18 and 10.19,

$$Q = \frac{H_T}{1.1 \times \Delta T} \left[\frac{H_T}{1.23 \times \Delta T} \right] = \frac{4,000}{1.1 \times 15} \left[\frac{1175}{1.23 \times 8} \right] = 242 \text{ cfm}$$

(rounded to 250 cfm)
[120 L/s = 0.12 m³/s]

$$A = \frac{Q}{V}$$

A = duct area

Q = required air quantity

V = 1,800 fpm [9 m/s] recommended velocity for a main supply duct in an average store (see Figure 10.18)

$$A = \frac{250 \text{ cfm}}{1800 \text{ fpm}} \left[\frac{0.12 \text{ m}^3/\text{s}}{9 \text{ m/s}} \right]$$

= 0.14 s.f. [0.013 m²]
= 0.14 s.f. × 144 sq. in. per s.f. [0.013 × 10 000 cm² per m²]

A = 20 sq. in. [130 cm²]

Therefore, use a 6" x 6" [12 cm x 12 cm] duct, because six inches is the smallest rectangular duct size recommended.

Constant Pressure Drop Method

Using the constant pressure drop method, the maximum limit for pressure drop per 100 feet of duct is set and the duct is made large enough to maintain this criteria. Familiarity with the terms *pressure drop per 100 feet* and *equivalent round* is necessary in order to understand the constant pressure drop method.

Pressure drop per 100 feet [per meter]: In a low velocity system, it is common to select 0.1 inches of water per 100 feet [1 Pa per meter] as a design pressure loss rate in supply ducts, and 0.05 [0.5 Pa] in return and exhaust ducts. This means that, when selected, the duct will require (or use up) 0.1 inches of water (0.0036 lbs. per square inch – the pressure under a column of water 0.1" high) [1 Pa] for every 100 feet [1 meter] of straight duct as it transports the required air quantity. Consider a duct that is 12 inches [30 cm] in diameter, 100 feet [30 meters] long, and open at both ends. Without a fan to push it, the air will not move. Now suppose one end has a pressure of 0.1 inches (0.0036 psi) higher than that of the other end. The added pressure will force air through the duct. Figure 10.20 illustrates the relationship between duct diameter, friction loss per 100 feet [per meter], and air quantity. For the 12-inch [30 cm] diameter duct at 0.1 inches per 100 feet [1 Pa per meter], it can be seen that 700 cfm [340 L/s] will move in the duct at 1,000 fpm [4.8 m/s] velocity.

Equivalent round: Because ducts are sized to account for friction, they can be interchanged with no substantial effect in performance, provided they have the same equivalent round diameter. Figure 10.20 can be used to determine the diameter of a circular duct. Round and spiral ductwork are gaining popularity. However, most buildings have traditionally used rectangular or square ducts. Figure 10.21 is a chart that can be used to convert a round duct to its equivalent size rectangular duct. Using Figure 10.21, a 12-inch [30 cm] diameter equivalent would be a square 11" × 11" [27 cm × 27 cm], close to the 12-inch, but less to account for the corners. An advantage to the constant pressure drop method is that it allows the designer to properly select from a wide range of ducts. Based on Figures 10.20 and 10.21 (Figures 10.20a and 10.21a in SI Metric units), the following ducts will all conduct 700 cfm [340 L/s] of air at 0.1 inches per 100 feet [1 Pa per meter]:

12" [30 cm] diameter	113 sq. in. [707 cm^2]
11" × 11" [27 × 27]	121 sq. in. [729 cm^2]
10" × 12" [25 × 29]	120 sq. in. [725 cm^2]
9" × 14" [24 × 30]	126 sq. in. [720 cm^2]
8" × 16" [21 × 35]	128 sq. in. [735 cm^2]
7" × 18" [18 × 42]	126 sq. in. [750 cm^2]
6" × 22" [15 × 50]	132 sq. in. [750 cm^2]

When using the velocity method, it is important to keep the aspect ratio close to 1:1, thereby minimizing friction errors. Using the constant pressure drop method, a wide range of duct sizes can be selected to clear obstructions and yet meet the flow requirements.

To size a system by the constant pressure drop method, select the circular duct required from Figure 10.20 for each branch according to the cfm in the duct. Then use Figure 10.21 to determine an equivalent rectangular duct. This procedure is illustrated in Figure 10.22. All duct runs, long and short, are then sized by this system, which is similar to the procedures for sizing water piping described in the water distribution section.

Fan Size: Fans for constant pressure drop (and balanced pressure drop) design methods are sized to supply the proper quantity of air and total pressure head to drive the air through the system. The total quantity of air equals the total air simultaneously supplied to all spaces in the building, plus an allowance for leakage. Leakage losses vary from a low of about 2 percent for tight, caulked and taped systems, to about 10 percent for those that have loose, plain joints. The pressure head is the total pressure necessary to drive the air from the fan to the room supply. The required total fan head is the sum of the items shown at the top of page 330.

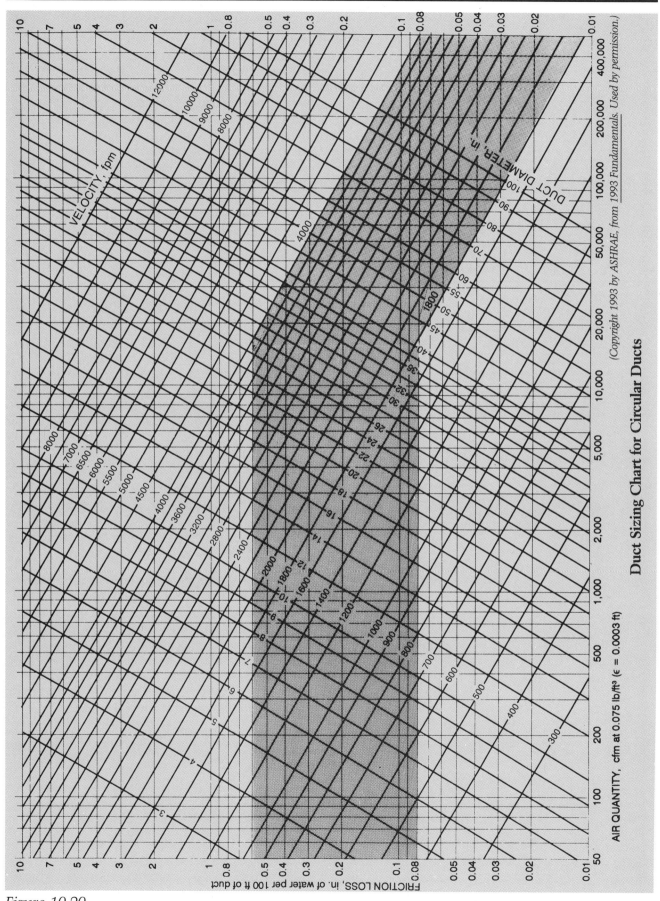

Figure 10.20

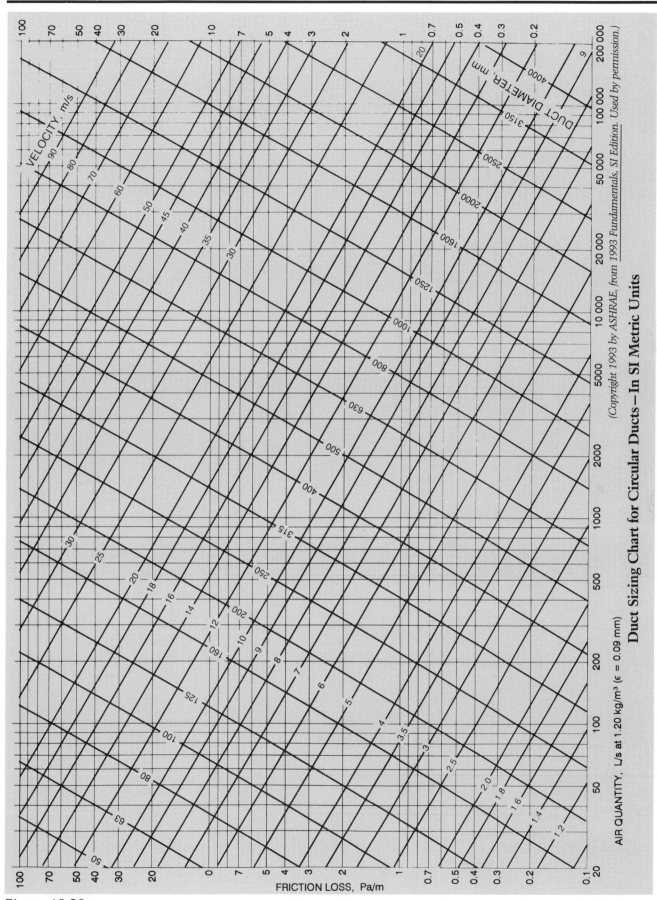

Figure 10.20a

Equivalent Rectangular Duct Dimension

Duct Diameter, in.	Rectangular Size, in.	Aspect Ratio													
		1.00	1.25	1.50	1.75	2.00	2.25	2.50	2.75	3.00	3.50	4.00	5.00	6.00	7.00	8.00
6	Width	—	6													
	Height	—	5													
7	Width	6	8													
	Height	6	6													
8	Width	7	9	9	11											
	Height	7	7	6	6											
9	Width	8	9	11	11	12	14									
	Height	8	7	7	6	6	6									
10	Width	9	10	12	12	14	14	15	17							
	Height	9	8	8	7	7	6	6	6							
11	Width	10	11	12	14	14	16	18	17	18	21					
	Height	10	9	8	8	7	7	7	6	6	6					
12	Width	11	13	14	14	16	16	18	19	21	21	24				
	Height	11	10	9	8	8	7	7	7	7	6	6				
13	Width	12	14	15	16	18	18	20	19	21	25	24	30			
	Height	12	11	10	9	9	8	8	7	7	7	6	6			
14	Width	13	14	17	18	18	20	20	22	24	25	28	30	36		
	Height	13	11	11	10	9	9	8	8	8	7	7	6	6		
15	Width	14	15	17	18	20	20	23	25	24	28	28	35	36	42	
	Height	14	12	11	10	10	9	9	9	8	8	7	7	6	6	
16	Width	15	16	18	19	20	23	23	25	27	28	32	35	42	42	48
	Height	15	13	12	11	10	10	9	9	9	8	8	7	7	6	6
17	Width	16	18	20	21	22	25	25	28	27	32	32	35	42	49	48
	Height	16	14	13	12	11	11	10	10	9	9	8	7	7	7	6
18	Width	16	19	21	23	24	25	28	28	30	32	36	40	42	49	56
	Height	16	15	14	13	12	11	11	10	10	9	9	8	7	7	7
19	Width	17	20	21	23	24	27	28	30	30	35	36	40	48	49	56
	Height	17	16	14	13	12	12	11	11	10	10	9	8	8	7	7
20	Width	18	20	23	25	26	27	30	30	33	35	40	45	48	56	56
	Height	18	16	15	14	13	12	12	11	11	10	10	9	8	8	7
21	Width	19	21	24	26	28	29	30	33	33	39	40	45	54	56	64
	Height	19	17	16	15	14	13	12	12	11	11	10	9	9	8	8
22	Width	20	23	26	26	28	32	33	36	36	39	44	50	54	56	64
	Height	20	18	17	15	14	14	13	13	12	11	11	10	9	8	8
23	Width	21	24	26	28	30	32	35	36	39	42	44	50	54	63	64
	Height	21	19	17	16	15	14	14	13	13	12	11	10	9	9	8
24	Width	22	25	27	30	32	34	35	39	39	42	48	55	60	63	72
	Height	22	20	18	17	16	15	14	14	13	12	12	11	10	9	9
25	Width	23	25	29	30	32	36	38	39	42	46	48	55	60	70	72
	Height	23	20	19	17	16	16	15	14	14	13	12	11	10	10	9
26	Width	24	26	30	32	34	36	38	41	42	46	52	55	66	70	72
	Height	24	21	20	18	17	16	15	15	14	13	13	11	11	10	9
27	Width	25	28	30	33	36	38	40	41	45	49	52	60	66	70	80
	Height	25	22	20	19	18	17	16	15	15	14	13	12	11	10	10
28	Width	26	29	32	35	36	38	43	44	45	49	56	60	66	77	80
	Height	26	23	21	20	18	17	17	16	15	14	14	12	11	11	10
29	Width	27	30	33	35	38	41	43	44	48	53	56	65	72	77	88
	Height	27	24	22	20	19	18	17	16	16	15	14	13	12	11	11
30	Width	27	31	35	37	40	43	45	47	48	53	60	65	72	77	88
	Height	27	25	23	21	20	19	18	17	16	15	15	13	12	11	11
31	Width	28	31	35	39	40	43	45	50	51	56	60	70	78	84	88
	Height	28	25	23	22	20	19	18	18	17	16	15	14	13	12	11
32	Width	29	33	36	39	42	45	48	50	54	56	60	70	78	84	96
	Height	29	26	24	22	21	20	19	18	18	16	15	14	13	12	12
33	Width	30	34	38	40	44	47	50	52	54	60	64	75	78	91	96
	Height	30	27	25	23	22	21	20	19	18	17	16	15	13	13	12
34	Width	31	35	39	42	44	47	50	52	57	60	64	75	84	91	96
	Height	31	28	26	24	22	21	20	19	19	17	16	15	14	13	12
35	Width	32	36	39	42	46	50	53	55	57	63	68	75	84	91	104
	Height	32	29	26	24	23	22	21	20	19	18	17	15	14	13	13
36	Width	33	36	41	44	48	50	53	55	60	63	68	80	90	98	104
	Height	33	29	27	25	24	22	21	20	20	18	17	16	15	14	13
38	Width	35	39	44	47	50	54	58	61	63	67	72	85	96	105	112
	Height	35	31	29	27	25	24	23	22	21	19	18	17	16	15	14

Equivalent Rectangular Duct Dimension

Figure 10.21

Equivalent Rectangular Duct Dimension (*Continued*)

Duct Diameter, in.	Rectangular Size, in.	Aspect Ratio													
		1.00	1.25	1.50	1.75	2.00	2.25	2.50	2.75	3.00	3.50	4.00	5.00	6.00	7.00	8.00
40	Width	37	41	45	49	52	56	60	63	66	70	76	90	96	105	120
	Height	37	33	30	28	26	25	24	23	22	20	19	18	16	15	15
42	Width	38	43	48	51	56	59	63	66	69	74	80	90	102	112	120
	Height	38	34	32	29	28	26	25	24	23	21	20	18	17	16	15
44	Width	40	45	50	54	58	61	65	69	72	81	84	95	108	119	128
	Height	40	36	33	31	29	27	26	25	24	23	21	19	18	17	16
46	Width	42	48	53	56	60	65	68	72	75	84	88	100	114	126	136
	Height	42	38	35	32	30	29	27	26	25	24	22	20	19	18	17
48	Width	44	49	54	60	62	68	70	74	78	88	92	105	120	126	136
	Height	44	39	36	34	31	30	28	27	26	25	23	21	20	18	17
50	Width	46	51	57	61	66	70	75	77	81	91	96	110	120	133	144
	Height	46	41	38	35	33	31	30	28	27	26	24	22	20	19	18
52	Width	48	54	59	63	68	72	78	83	84	95	100	115	126	140	152
	Height	48	43	39	36	34	32	31	30	28	27	25	23	21	20	19
54	Width	49	55	62	67	70	77	80	85	90	98	104	120	132	147	160
	Height	49	44	41	38	35	34	32	31	30	28	26	24	22	21	20
56	Width	51	58	63	68	74	79	83	88	93	102	108	125	138	147	160
	Height	51	46	42	39	37	35	33	32	31	29	27	25	23	21	20
58	Width	53	60	66	70	76	81	85	91	96	105	112	130	144	154	168
	Height	53	48	44	40	38	36	34	33	32	30	28	26	24	22	21
60	Width	55	61	68	74	78	83	90	94	99	109	116	130	144	161	
	Height	55	49	45	42	39	37	36	34	33	31	29	26	24	23	
62	Width	57	64	71	75	82	88	93	96	102	112	120	135	150	168	
	Height	57	51	47	43	41	39	37	35	34	32	30	27	25	24	
64	Width	59	65	72	79	84	90	95	99	105	116	124	140	156		
	Height	59	52	48	45	42	40	38	36	35	33	31	28	26		
66	Width	60	68	75	81	86	92	98	105	108	119	128	145	162		
	Height	60	54	50	46	43	41	39	38	36	34	32	29	27		
68	Width	62	70	77	82	90	95	100	107	111	123	132	150	168		
	Height	62	56	51	47	45	42	40	39	37	35	33	30	28		
70	Width	64	71	80	86	92	99	105	110	114	126	136	155			
	Height	64	57	53	49	46	44	42	40	38	36	34	31			
72	Width	66	74	81	88	94	101	108	113	117	130	140	160			
	Height	66	59	54	50	47	45	43	41	39	37	35	32			
74	Width	68	76	84	91	98	104	110	116	123	133	144	165			
	Height	68	61	56	52	49	46	44	42	41	38	36	33			
76	Width	70	78	86	93	100	106	113	118	126	137	148	165			
	Height	70	62	57	53	50	47	45	43	42	39	37	33			
78	Width	71	80	89	95	102	110	115	121	129	140	152				
	Height	71	64	59	54	51	49	46	44	43	40	38				
80	Width	73	83	90	98	104	113	118	124	132	144	156				
	Height	73	66	60	56	52	50	47	45	44	41	39				
82	Width	75	84	93	100	108	115	123	129	135	147	160				
	Height	75	67	62	57	54	51	49	47	45	42	40				
84	Width	77	86	95	103	110	117	125	132	138	151	164				
	Height	77	69	63	59	55	52	50	48	46	43	41				
86	Width	79	88	98	105	112	119	128	135	141	154	168				
	Height	79	70	65	60	56	53	51	49	47	44	42				
88	Width	80	90	99	107	116	124	130	138	144	158					
	Height	80	72	66	61	58	55	52	50	48	45					
90	Width	82	93	102	110	118	126	133	140	147	161					
	Height	82	74	68	63	59	56	53	51	49	46					
92	Width	84	94	104	112	120	128	138	143	150	165					
	Height	84	75	69	64	60	57	55	52	50	47					
94	Width	86	96	107	116	124	131	140	146	153	168					
	Height	86	77	71	66	62	58	56	53	51	48					
96	Width	88	99	108	117	126	135	143	151	159						
	Height	88	79	72	67	63	60	57	55	53						
98	Width	90	100	111	119	128	137	145	154	162						
	Height	90	80	74	68	64	61	58	56	54						
100	Width	91	103	113	123	132	140	148	157	165						
	Height	91	82	75	70	66	62	59	57	55						
102	Width	93	105	116	124	134	142	153	160	168						
	Height	93	84	77	71	67	63	61	58	56						
104	Width	95	106	117	128	136	146	155	162							
	Height	95	85	78	73	68	65	62	59							

Equivalent Rectangular Duct Dimension (continued)

Figure 10.21 (continued)

Equivalent Rectangular Duct Dimension (*Concluded*)

Duct Diameter, in.	Rectangular Size, in.	Aspect Ratio													
		1.00	1.25	1.50	1.75	2.00	2.25	2.50	2.75	3.00	3.50	4.00	5.00	6.00	7.00	8.00
106	Width	97	109	120	130	140	149	158	165							
	Height	97	87	80	74	70	66	63	60							
108	Width	99	110	122	131	142	151	160	168							
	Height	99	88	81	75	71	67	64	61							
110	Width	101	113	125	135	144	153	163								
	Height	101	90	83	77	72	68	65								
112	Width	102	115	126	137	146	158	165								
	Height	102	92	84	78	73	70	66								
114	Width	104	116	129	140	150	160									
	Height	104	93	86	80	75	71									
116	Width	106	119	131	142	152	162									
	Height	106	95	87	81	76	72									
118	Width	108	121	134	144	154	164									
	Height	108	97	89	82	77	73									
120	Width	110	123	135	147	158										
	Height	110	98	90	84	79										

*Shaded area not recommended.

Equivalent Spiral Flat Oval Duct Dimensions

Duct Diameter, in.	Major Axis (a), in. / Minor Axis (b), in.														
	3	4	5	6	7	8	9	10	11	12	14	16	18	20	22	24
5	8															
5.5	9	7														
6	11	9														
6.5	12	10	8													
7	15	12	10	8												
7.5	19	13	—	9												
8	22	15	11	—												
8.5		18	13	11	10											
9		20	14	12	—	10										
9.5		21	18	14	12	—										
10			19	15	13	11										
10.5			21	17	15	13	12									
11				19	16	14	—	12								
11.5				20	18	16	14	—								
12				23	20	17	15	13								
12.5				25	21	—	—	15	14							
13				28	23	19	17	16	—	14						
13.5				30	—	21	18	—	16	—						
14				33	—	22	20	18	17	15						
14.5				36	—	24	22	19	—	17						
15				39	—	27	23	21	19	18						
16				45	—	30	—	24	22	20	17					
17				52	—	35	—	27	24	21	19					
18				59	—	39	—	30	—	25	22	19				
19						46	—	34	—	28	23	21				
20						50	—	38	—	31	27	24	21			
21						58	—	43	—	34	28	25	23			
22						65	—	48	—	37	31	29	26			
23						71	—	52	—	42	34	30	27			
24						77	—	57	—	45	38	33	29	26		
25								63	—	50	41	36	32	29		
26								70	—	56	45	38	34	31		
27								76	—	59	49	41	37	34		
28										65	52	46	40	36		
29										72	58	49	43	39	35	
30										78	61	54	46	40	38	
31										81	67	57	49	44	39	37
32											71	60	53	47	42	40
33											77	66	56	51	46	41
34												69	59	55	47	44
35												76	65	58	50	46
36												79	68	61	53	49
37													71	64	57	52
38													78	67	60	55
40														77	69	62
42															75	68
44															82	74

(Copyright 1993 by ASHRAE, from *1993 Fundamentals*. Used by permission.)

Equivalent Rectangular Duct Dimension (continued)

Figure 10.21 (continued)

Circular Equivalents of Rectangular Duct for Equal Friction and Capacity[a]

Lgth Adj.[b]	\multicolumn{20}{c}{Length of One Side of Rectangular Duct (a), mm}
	100	125	150	175	200	225	250	275	300	350	400	450	500	550	600	650	700	750	800	900
100	109																			
125	122	137																		
150	133	150	164																	
175	143	161	177	191																
200	152	172	189	204	219															
225	161	181	200	216	232	246														
250	169	190	210	228	244	259	273													
275	176	199	220	238	256	272	287	301												
300	183	207	229	248	266	283	299	314	328											
350	195	222	245	267	286	305	322	339	354	383										
400	207	235	260	283	305	325	343	361	378	409	437									
450	217	247	274	299	321	343	363	382	400	433	464	492								
500	227	258	287	313	337	360	381	401	420	455	488	518	547							
550	236	269	299	326	352	375	398	419	439	477	511	543	573	601						
600	245	279	310	339	365	390	414	436	457	496	533	567	598	628	656					
650	253	289	321	351	378	404	429	452	474	515	553	589	622	653	683	711				
700	261	298	331	362	391	418	443	467	490	533	573	610	644	677	708	737	765			
750	268	306	341	373	402	430	457	482	506	550	592	630	666	700	732	763	792	820		
800	275	314	350	383	414	442	470	496	520	567	609	649	687	722	755	787	818	847	875	
900	289	330	367	402	435	465	494	522	548	597	643	686	726	763	799	833	866	897	927	984
1000	301	344	384	420	454	486	517	546	574	626	674	719	762	802	840	876	911	944	976	1037
1100	313	358	399	437	473	506	538	569	598	652	703	751	795	838	878	916	953	988	1022	1086
1200	324	370	413	453	490	525	558	590	620	677	731	780	827	872	914	954	993	1030	1066	1133
1300	334	382	426	468	506	543	577	610	642	701	757	808	857	904	948	990	1031	1069	1107	1177
1400	344	394	439	482	522	559	595	629	662	724	781	835	886	934	980	1024	1066	1107	1146	1220
1500	353	404	452	495	536	575	612	648	681	745	805	860	913	963	1011	1057	1100	1143	1183	1260
1600	362	415	463	508	551	591	629	665	700	766	827	885	939	991	1041	1088	1133	1177	1219	1298
1700	371	425	475	521	564	605	644	682	718	785	849	908	964	1018	1069	1118	1164	1209	1253	1335
1800	379	434	485	533	577	619	660	698	735	804	869	930	988	1043	1096	1146	1195	1241	1286	1371
1900	387	444	496	544	590	663	674	713	751	823	889	952	1012	1068	1122	1174	1224	1271	1318	1405
2000	395	453	506	555	602	646	688	728	767	840	908	973	1034	1092	1147	1200	1252	1301	1348	1438
2100	402	461	516	566	614	659	702	743	782	857	927	993	1055	1115	1172	1226	1279	1329	1378	1470
2200	410	470	525	577	625	671	715	757	797	874	945	1013	1076	1137	1195	1251	1305	1356	1406	1501
2300	417	478	534	587	636	683	728	771	812	890	963	1031	1097	1159	1218	1275	1330	1383	1434	1532
2400	424	486	543	597	647	695	740	784	826	905	980	1050	1116	1180	1241	1299	1355	1409	1461	1561
2500	430	494	552	606	658	706	753	797	840	920	996	1068	1136	1200	1262	1322	1379	1434	1488	1589
2600	437	501	560	616	668	717	764	810	853	935	1012	1085	1154	1220	1283	1344	1402	1459	1513	1617
2700	443	509	569	625	678	728	776	822	866	950	1028	1102	1173	1240	1304	1366	1425	1483	1538	1644
2800	450	516	577	634	688	738	787	834	879	964	1043	1119	1190	1259	1324	1387	1447	1506	1562	1670
2900	456	523	585	643	697	749	798	845	891	977	1058	1135	1208	1277	1344	1408	1469	1529	1586	1696

Lgth Adj.[b]	\multicolumn{20}{c}{Length of One Side of Rectangular Duct (a), mm}
	1000	1100	1200	1300	1400	1500	1600	1700	1800	1900	2000	2100	2200	2300	2400	2500	2600	2700	2800	2900
1000	1093																			
1100	1146	1202																		
1200	1196	1256	1312																	
1300	1244	1306	1365	1421																
1400	1289	1354	1416	1475	1530															
1500	1332	1400	1464	1526	1584	1640														
1600	1373	1444	1511	1574	1635	1693	1749													
1700	1413	1486	1555	1621	1684	1745	1803	1858												
1800	1451	1527	1598	1667	1732	1794	1854	1912	1968											
1900	1488	1566	1640	1710	1778	1842	1904	1964	2021	2077										
2000	1523	1604	1680	1753	1822	1889	1952	2014	2073	2131	2186									
2100	1558	1640	1719	1793	1865	1933	1999	2063	2124	2183	2240	2296								
2200	1591	1676	1756	1833	1906	1977	2044	2110	2173	2233	2292	2350	2405							
2300	1623	1710	1793	1871	1947	2019	2088	2155	2220	2283	2343	2402	2459	2514						
2400	1655	1744	1828	1909	1986	2060	2131	2200	2266	2330	2393	2453	2511	2568	2624					
2500	1685	1776	1862	1945	2024	2100	2173	2243	2311	2377	2441	2502	2562	2621	2678	2733				
2600	1715	1808	1896	1980	2061	2139	2213	2285	2355	2422	2487	2551	2612	2672	2730	2787	2842			
2700	1744	1839	1929	2015	2097	2177	2253	2327	2398	2466	2533	2598	2661	2722	2782	2840	2896	2952		
2800	1772	1869	1961	2048	2133	2214	2292	2367	2439	2510	2578	2644	2708	2771	2832	2891	2949	3006	3061	
2900	1800	1898	1992	2081	2167	2250	2329	2406	2480	2552	2621	2689	2755	2819	2881	2941	3001	3058	3115	3170

[a] Table based on $D_e = 1.30(ab)^{0.625}/(a + b)^{0.25}$
[b] Length of adjacent side of rectangular duct (b), mm.

Equivalent Rectangular Duct Dimension – In SI Metric Units

Figure 10.21a

Item	Total Pressure in Water Gauge [Pa]
Change the velocity of air from 0 to 1,000 feet per minute [0 to 5 m/s] at the fan	0.07 [20.6]

Internal equivalent losses
    coils 0.15 [44.1]
    filters 0.10 [29.4]
	0.25 [73.5]

Duct (say 200 equivalent feet at 0.1″/100′ for longest run)	0.20 [59] (main system variable)
Pressure to drive across registers to room	0.15 [44.13]
Additional reserve for dampers and balancing (ten percent of above)	0.07 [20.6]
Required total fan head =	0.74, use 3/4″ [220 Pa]
Required external fan pressure =	0.20 + 0.15 + 0.07 [59 + 44.1 + 20.6]
=	0.42, use 1/2″ [125 Pa]

The procedure for sizing fans is similar to that used to size water pumps. The external fan pressure is often listed by the manufacturer. This figure represents the fan pressure still remaining after overcoming the resistance

Equivalent Spiral Flat Oval Duct Dimensions

Duct Diameter, mm	Major Axis (a), mm / Minor Axis (b), mm																
	70	100	125	150	175	200	250	275	300	325	350	375	400	450	500	550	600
125	205																
140	265	180															
160	360	235	190														
180	475	300	235	200													
200		380	290	245	215												
224		490	375	305	—	240											
250			475	385	325	290											
280				485	410	360	—	285									
315				635	525	—	—	345	325								
355				840	—	580	460	425	395	375							
400				1115	—	760	—	530	490	460	435						
450				1490	—	995	—	675	—	570	535	505					
500						1275	—	845	—	700	655	615	580				
560						1680	—	1085	—	890	820	765	720				
630								1425	—	1150	1050	970	905	810			
710										1505	1370	1260	1165	1025			
800											1800	1645	1515	1315	1170	1065	
900												2165	1985	1705	1500	1350	
1000													2170	1895	1690		
1120														2455	2170	1950	
1250															2795	2495	

(Copyright 1993 by ASHRAE, from 1993 Fundamentals, SI Edition. Used by permission.)

Equivalent Rectangular Duct Dimension – In SI Metric Units (continued)

Figure 10.21a (continued)

encountered in the fan equipment. This is the pressure available to drive air through the distribution system. Most systems in small buildings require between 3/8″ [110 Pa] and 1/2″ [150 Pa] fan pressure.

Equivalent straight run of duct: The pressure drop through grilles, fittings, and dampers must be determined for the longest run. Each fitting equivalent is shown in Figure 10.23.

Using these tables, the equivalent length of a fitting can be determined. The equivalent length of a fitting is added to the straight length to arrive at the overall straight run of duct. This is similar to the calculation for equivalent pipe lengths.

To size the fan, select the longest total equivalent run. Then add up all of the friction losses from elbows, fittings, filters, and registers that the fan must overcome.

Elbow equivalents can be obtained using Figure 10.23. For example, for a duct carrying 2,000 cubic feet of air per minute [945 L/s], based on an average velocity of 1,000 feet per minute [5 m/s], the approximate duct area

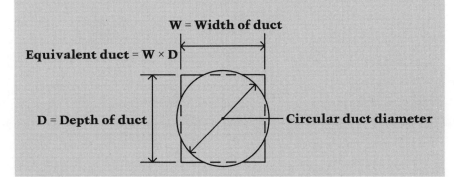

1. **For each run of duct determine the air quantity (Q).**

$$Q = \frac{H_T}{1.08 \Delta T}$$

(Refer to velocity method for additional information.)

2. **Determine the recommended pressure drop per 100′.**
   Low-pressure systems—0.1″ supply, 0.07″ return
   High-pressure systems—1.7″ supply, 1.4″ return

3. **Determine the circular duct diameter (see Figure 9.20)** using the air quantity (Q) and the recommended pressure drop.

   W = Width of duct
   Equivalent duct = W × D
   D = Depth of duct
   Circular duct diameter

4. **Determine the duct width and depth using Figure 10.20.**

Constant Pressure Drop Method

Figure 10.22

will be two square feet [0.18 m²]. The elbow equivalent length depends on the duct depth, which must be approximated in order to size the duct. A 1' × 2' [30 cm × 60 cm] duct would have an area of two square feet [0.18 m²]. Referring to Figure 10.23, the duct has an L/D ratio of 9 on a 90-degree smooth elbow. Therefore, the equivalent length of the duct elbow would be 9' × 2', or 18 feet.

Balanced Pressure Drop Method

The balanced pressure drop method is based on the constant pressure drop method, adapted for all duct runs regardless of length or air quantity. The longest duct run is sized using a constant pressure drop. All other takeoffs from the longest duct are sized in such a way that their friction loss equals that of the longest run. This procedure is illustrated in Figure 10.24.

The balanced pressure drop method enables the designer to size a system with a wide variety of duct runs and "balances" the system so that the fan will provide the proper quantities of air in each run.

Static Regain Method

Static regain refers to the increase (or regain) of static pressure in ductwork when the air velocity decreases. A main trunk duct, for example, may have diffusers at constant intervals for air takeoff (see the static regain example in Figure 10.17). If the main duct size remains constant, the air velocity in the duct will decrease after each takeoff. As the air slows down, it "bunches up" and its pressure is increased; the energy of the fast-moving air before the takeoff is converted to higher pressure/lower velocity air after the takeoff. By carefully sizing the ductwork, a designer uses the increased pressure from one takeoff to overcome pressure friction losses on the run to the next takeoff. If the air starts at about 1,600 feet per minute [8 m/s] and loses velocity with each succeeding takeoff, it is possible to put about five takeoffs on a run before the supply velocity reaches the low cut-off value of about 800 feet per minute [4 m/s].

This method of duct sizing reduces the velocity in the ductwork after each branch takeoff to an amount sufficient to produce enough static regain to compensate for the friction losses of the succeeding section. Static regain can occur only when the velocity is reduced. In any duct system the total energy for velocity and pressure always decreases in the direction of the flow.

A duct system sized by the static regain method will generally have a high initial velocity in beginning sections; this velocity will reduce for each downstream section, resulting in a system with slightly larger downstream duct sizes than systems sized by other methods. A system sized by static regain, however, will have lower noise levels and smaller fan size and energy consumption.

The following formula is used to determine the increase in static pressure:

$$\Delta P = K((V_1/4005)^2 - (V_2/4005)^2)$$
$$[\Delta P = K((0.602 \, V_1^2) - (0.602 \, V_2^2))]$$

ΔP = increase in static pressure after the takeoff in inches of water [Pa]

K = between 0.5 and 0.75 for decrease in velocity, -0.7 and -1.1 for pressure loss

V_1 = velocity of air (fpm) [m/s] before the takeoff

V_2 = velocity of air (fpm) [m/s] after the takeoff

FRICTION OF ROUND DUCT SYSTEM ELEMENTS

ELEMENT	CONDITION	L/D RATIO*
90° Smooth Elbow	R/D = 1.5	9
90° 3-Piece Elbow	R/D = 1.5	24
90° 5-Piece Elbow	R/D = 1.5	12
45° 3-Piece Elbow	R/D = 1.5	6
45° Smooth Elbow	R/D = 1.5	4.5
90° Miter Elbow	Vaned Not Vaned	22 65

ELEMENT	CONDITION	VALUE OF n†
90° Tee‡ and 90°, 135° & 180° Cross‡ Pressure Loss Thru Branch = nhv₂	$\frac{V_2}{V_1} = \begin{cases} 0.2 \\ 0.5 \\ 1.0 \\ 5.0 \end{cases}$	4.0 2.0 1.75 1.6
45° Tee‡ Pressure Loss Thru Branch = nhv₂	$\frac{V_2}{V_1} = \begin{cases} 0.8 \\ 1.0 \\ 2.0 \\ 3.0 \end{cases}$	.10 .44 1.21 1.47
90° Conical Tee and 180° Conical Cross Pressure Loss Thru Branch = nhv₂	$\frac{V_2}{V_1} = \begin{cases} 0.5 \\ 1.0 \\ 2.0 \\ 5.0 \end{cases}$	0.2 0.5 1.0 1.2

(Courtesy Carrier Corporation, McGraw-Hill Book Company)

Equivalent Length Factors for Duct Fittings

Figure 10.23

FRICTION OF RECTANGULAR DUCT SYSTEM ELEMENTS

Rectangular Radius Elbow

W/D	R/D = .5	R/D = .75	R/D = 1.00	R/D = 1.25*	R/D = 1.50
	L/D Ratio				
.5	33	14	9	5	4
1	45	18	11	7	4
3	80	30	14	8	5
6	125	40	18	12	7

Rectangular Vaned Radius Elbow

Number of Vanes	R/D = .50	R/D = .75	R/D = 1.00	R/D = 1.50
	L/D Ratio			
1	18	10	8	7
2	12	8	7	7
3	10	7	7	6

X° Elbow

Element	Conditions	L/D Ratio
X° Elbow	Vaned or Unvaned Radius Elbow	X/90 times value for similar 90° elbow
Rectangular Square Elbow — No Vanes		60
Rectangular Square Elbow — Single Thickness Turning Vanes		15
Rectangular Square Elbow — Double Thickness Turning Vanes		10
Double Elbow (W/D = 1, R/D = 1.25*)	S = O	15
Double Elbow (W/D = 1, R/D = 1.25*)	S = D	10
Double Elbow (W/D = 1, R/D = 1.25*)	S = O	20
Double Elbow (W/D = 1, R/D = 1.25*)	S = D	22
Double Elbow (W/D = 1, R/D = 1.25* For Both)	S = O	15
Double Elbow (W/D = 1, R/D = 1.25* For Both)	S = D	16
Double Elbow (W/D = 2, R₁/D = 1.25*, R₂/D = .5)	Direction of Arrow	45
Double Elbow (W/D = 2, R₁/D = 1.25*, R₂/D = .5)	Reverse Direction	40
Double Elbow (W/D = 4, R/D = 1.25* for both elbows)	Direction of Arrow	17
Double Elbow (W/D = 4, R/D = 1.25* for both elbows)	Reverse Direction	18

(Courtesy Carrier Corporation, McGraw–Hill Book Company)

Equivalent Length Factors for Duct Fittings (continued)

Figure 10.23 (continued)

FRICTION OF RECTANGULAR DUCT SYSTEM ELEMENTS (Contd)

ELEMENT	CONDITIONS	VALUE OF n‡
Transformer	$V_2 = V_1$ S.P. Loss = nhv_1	.15

Expansion

"n"

	Angle "a"					
v_2/v_1	5°	10°	15°	20°	30°	40°
.20	.83	.74	.68	.62	.52	.45
.40	.89	.83	.78	.74	.68	.64
.60	.93	.87	.84	.82	.79	.77

S.P. Regain = $n(hv_1 - hv_2)$

Contraction

a	30°	45°	60°
n	1.02††	1.04	1.07

S.P. Loss = $n(hv_2 - hv_1)$ ††Slope 1″ in 4″

Abrupt Entrance	S.P. Loss = nhv_1	.35
Bellmouth Entrance		.03
Abrupt Exit	S.P. Loss or Regain Considered Zero	
Bellmouth Exit		
Re-Entrant Entrance	S.P. Loss = nhv_1	.85

Sharp Edge Round Orifice

A_2/A_1	0	.25	.50	.75	1.00
n	2.5	2.3	1.9	1.1	0

S.P. Loss = nhv_2

Abrupt Contraction

V_1/V_2	0	.25	.50	.75
n	1.34	1.24	.96	.52

S.P. Loss = nhv_2

Abrupt Expansion

V_2/V_1	.20	.40	.60	.80
n	.32	.48	.48	.32

S.P. Regain = nhv_1

Pipe Running Thru Duct

E/D	.10	.25	.50
n	.20	.55	2.00

S.P. Loss = nhv_1

Bar Running Thru Duct

E/D	.10	.25	.50
n	.7	1.4	4.00

S.P. Loss = nhv_1

Easement Over Obstruction

E/D	.10	.25	.50
n	.07	.23	.90

S.P. Loss = nhv_1

(Courtesy Carrier Corporation, McGraw–Hill Book Company)

Equivalent Length Factors for Duct Fittings (continued)

Figure 10.23 (continued)

The pressure head is measured in feet of air. As the movement of the air slows down, the pressure head increases as follows:

$$P = 1/2 \times V_s^2/g$$

P = pressure in feet of air

V_s = velocity of air (ft./sec.)

g = gravitational constant 32.2 ft./sec.^{2}

The pressure in feet of air must be converted to head in inches of water. A cubic foot of 68°F air weighs 0.075 pounds, which is equivalent to a column of water one foot square ((0.075/62.3) = 0.0012 feet) and approximately 0.01445 inches high. Velocity in feet per second must be multiplied by 60 to convert it to the standard feet per minute.

$$P \text{ (feet of air)} = P \text{ (in. wg)} \frac{1}{.01445}$$

$$P \text{ (in. wg)} = \frac{1}{2} \frac{(V)^2}{(60)} \frac{(1)}{(32.2)} (.01445)$$

$$= \frac{.01445 \, V^2}{3{,}600 \times 32.2}$$

$$= \frac{(V^2)}{(16{,}044{,}000)} = \frac{(V)^2}{(4{,}005)}$$

P = velocity pressure (in. wg)

V = fluid mean velocity (fpm)

Air = 0.075 lbs./ft.^{3} at standard conditions

NOTES FOR TABLE 9

*L and D are in feet. D is the elbow diameter. L is the additional equivalent length of duct added to the measured length. The equivalent length L equals D in feet times the ratio listed.

†The value of n is the loss in velocity heads and may be converted to additional equivalent length of duct by the following equation.

$$L = n \times \frac{h_v \times 100}{h_f}$$

where: L = additional equivalent length, ft

h_v = velocity pressure at V_2, in. wg (conversion line on Chart 7 or Table 8).

h_f = friction loss/100 ft, duct diameter at V_2, in. wg (Chart 7).

n = value for tee or cross

‡Tee or cross may be either reduced or the same size in the straight thru portion

NOTES FOR TABLE 10:

*1.25 is standard for an unvaned full radius elbow.

†L and D are in feet. D is the duct dimension illustrated in the drawing. L is the additional equivalent length of duct added to measured duct. The equivalent length L equals D in feet times the ratio listed.

‡The value of n is the number of velocity heads or differences in velocity heads lost or gained at a fitting, and may be converted to additional equivalent length of duct by the following equation:

$$L = n \times \frac{h_v \times 100}{h_f}$$

where: L = additional equivalent length, ft.

h_v = velocity pressure for V_1 or V_2, in. wg (conversion line on Chart 7 or Table 8).

h_f = friction loss/100 ft, duct cross section at h_v, in. wg (Chart 7).

n = value for particular fitting.

(Courtesy Carrier Corporation, McGraw-Hill Book Company)

Footnote: Cross-references in table footnotes pertain to the source from which tables were taken; they do not refer to chapters or pages in this text.

Equivalent Length Factors for Duct Fittings (continued)

Figure 10.23 (continued)

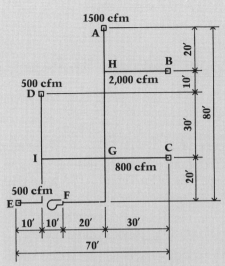

Given—Duct layout shown with dimensions and air quantities shown. Keep duct depth below 12".

Find—Size ducts and fan by balanced pressure drop method.

For 90° round elbows L/D = 9.
HB 2,000 cfm = 12 x 24 duct approx. at 1,000 fpm.
2' wide x L/D of 9' = 18' for elbows.

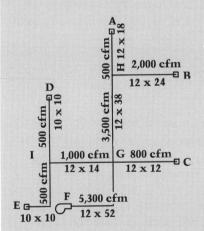

Solution

1. **Select "longest run" from fan to furthest register.**
   F-A at 20' + 80' = 100' is the longest.

2. **Determine air quantities in each run of duct. These are shown in the lower diagram.**

3. **Size each length of duct on longest run by the constant pressure drop method. (0.1"/100').**

Run	Air Quantity	(Fig. 10.20) Equivalent Round	(Fig. 10.21) Rect. Size
FG	5,300 cfm	26"	12 x 52"
GH	3,500 cfm	22"	12 x 38"
HA	1,500 cfm	16"	12 x 18"

4. **For each take-off from longest run, determine the design pressure loss per 100' by proportion to match losses on longest run.**

Run	Lengths Straight	Elbows	Total
HA	20'	0	20'
HB	30'	18	48'
GA	60'	0	60'
GC	30'	9	39'
GID	60'	9	69'
IE	30'	5	35'

Run	Ratio	Pressure	cfm	Eq. Round	Rect.
HA	$\frac{20}{48}$ x .1	—	—	—	—
HB		0.04	2,000	21½"	12 x 24
GA	$\frac{60}{39}$ x .1	—	—	—	—
GC		0.15	800	12"	12 x 12
GID	$\frac{60}{69}$ x .1	0.09	1,000	14"	12 x 14
		.10	500	11"	10 x 10
IE	$\frac{35}{30}$ x .09	1.1	500	11"	10 x 10

Fan—Longest Run F-A × 100' $9 \times \frac{52}{12} = 140'$

Fan Pressure Required:
1. Velocity (0 to 1,000 fpm)   0.07
2. Internal coils   0.10
3. Filters   0.10
4. Duct 140' at 0.1/100   0.14
5. Residual pressure grilles   0.15
6. Reserve for balancing   0.05

                                0.61 use 0.75"

Use 3/4" pressure at 5,300 cfm fan

Balanced Pressure Drop Duct Sizing Example

Figure 10.24

Either the diameter in inches [meters], velocity in fpm [m/s], or volume flow in cfm [m³/s] may be calculated, if two of the three values are known, by the following formula:

$$\text{fpm} = \frac{\text{cfm}}{(\pi/4)(d^2/144)} \qquad \left[\text{m/s} = \frac{\text{m}^3/\text{s}}{(\pi/4)d^2}\right]$$

The velocity pressure (Vp) in a section can be calculated when the velocity is known by the following:

$$V_p = \frac{(\text{fpm})^2}{(4005)^2} \qquad [V_p = 0.602\,(\text{m/s})^2]$$

When the length (L) of a straight run of duct is known, and the pressure drop per 100 feet [per meter] of duct is read from Figure 10.26 for a given flow (cfm) [L/s] and duct size, the static pressure loss (Ps) in the section is given by

$$P_s = (L/100) \times \text{pressure drop}/100\text{ ft.} \quad [L \times \text{pressure drop/meter}]$$

Example: Size the duct system for the conditions shown in Figure 10.25. Utilize the static regain sizing method where possible. The outlet pressures at the branches should average 0.25 inches water, and the base initial velocity should be approximately 2500 fpm. The ductwork should be round spiral ducts.

Solution: The first section is sized according to the parameters set out in the problem. This results in a duct diameter of 14". The pressure loss for section AB is calculated at 0.078 inches of water. For an average branch pressure of 0.25 inches, the first outlet should have a pressure of 0.3 inches and the final outlet should have a pressure of 0.2 inches. The required pressure at the fan, therefore, is the sum of the pressure required at point B and the pressure loss in section AB (0.3 + 0.078 = 0.378 inches). The calculations show that the pressure at each branch outlet varies as will happen in the design of any duct system; further variations will occur when the system is installed. Therefore, dampers will be required at all branch outlets regardless of how well the system is designed. This will result in additional pressure in the system, which we have not included so as to simplify the problem and illustrate more clearly the static regain method. The calculations also show that the average pressure at branch outlets is 0.253 inches.

The table in Figure 10.25a summarizes the calculations of the problem.

Actual details of duct construction must be taken into account. For example, smooth connections result in ease of flow, whereas abrupt connections cause turbulence in the air stream, thereby diminishing the maximum theoretical pressure head. Thus, for the decrease in velocity after a takeoff (where V_1 is greater than V_2 and pressure is regained), a factor between 0.5 and 0.75 is generally used. When conditions cause the post-takeoff velocity to increase, V_2 is greater than V_1 and a factor between -0.7 and -1.1 is used. In this situation, there is a theoretical decrease in pressure, with K values between -0.7 and -1.1. (The $(-)$ sign means that the pressure drops, which is the opposite of regain).

Using the static regain method requires more care and experienced judgment than the other duct sizing methods. Solutions may sometimes be based on "trial and error" and may be sensitive to any changes in air balancing and variable air flow. Figure 10.26 contains charts used to determine static regain.

Miscellaneous Fluids

In addition to steam, water, and air, other fluids are transported in an HVAC system, such as the fuels and refrigerants listed below.

- **Fuels**
  oil
  propane
  natural gas
- **Refrigerants**
  ammonia
  brine
  fluorocarbons
  other refrigerants

The piping design for these systems is regulated by a wide range of codes. The fuel piping in particular generally follows local fire, plumbing, and gas fitting codes. Requirements for refrigerants are often established by equipment manufacturers. Basic characteristics of typical refrigerants are shown in Figure 10.27. (For further information, refer to Chapter 2, "Codes, Regulations, and Standards.")

Distribution Equipment Data Sheets

The data sheets in Figure 10.28 summarize the efficiency, useful life, typical uses, capacity range, advantages and disadvantages, special considerations, accessories, and additional equipment needed for distribution system components. A Means line number, where applicable, is included in order to determine costs in the current edition of *Means Mechanical Cost Data*.

Problems

10.1 For the building shown in Figure 10.29, design a hydronic system:
  a. Size the boiler(s).
  b. Lay out a reverse return system.
  c. Size the piping.
  d. Give the pump flow and head characteristics.

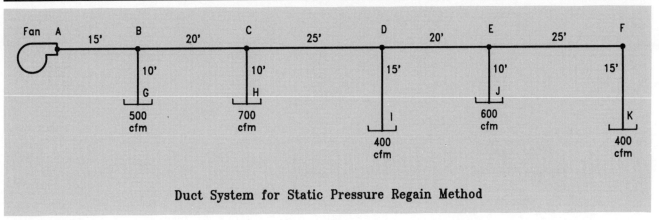

Duct System for Static Pressure Regain Method

Figure 10.25

Static Pressure Regain Method of Duct Sizing

Section	Entity	Value	Units	Remarks
Section AB				
	Section volume flow =	2600	cfm	Given
	Section length =	15	Feet	Given
	Required section velocity =	2500	Fpm	Given
	Calculated duct diameter =	13.81	Inches	
	Actual duct diameter =	14	Inches	Calculated rounded up
	Actual velocity of section =	2433	Fpm	
	Velocity pressure (Vp) of section =	0.369	Inches	
	Pressure loss of duct =	0.52	Ins./100 Ft.	
	Loss in section =	0.078	Inches	
	Net loss in section =	0.078	Inches	Sum of losses
	Resultant pressure required at Point B =	0.3	Inches	Given
	Static pressure required at fan =	0.378	Inches	Sum of net loss and branch pressure requirement
Section BC				
	Branch section volume flow =	500	cfm	Given
	Previous section volume flow =	2600	cfm	From previous
	Section volume flow =	2100	cfm	Previous flow — Branch flow
	Section length =	20	Feet	Given
	Static pressure required at C =	0.275	Inches	Given
	Vp of previous section =	0.369	Inches	From previous
	Trial duct size =	14	Inches	Trial and error
	Velocity of this section =	1965	Fpm	
	Vp of this section =	0.241	Inches	
	Sp/100 Ft. =	0.4	Inches	
	Loss in section =	0.080	Inches	
	Static pressure regain (SPR) for section =	0.064	Inches	
	Net loss in section =	0.016	Inches	Loss in section — SPR
	Resultant pressure at Point C =	0.284	Inches	(Pressure @ B-Section net loss) Close enough to required
Section CD				
	Branch section volume flow =	700	Cfm	Given
	Previous section volume flow =	2100	Cfm	From previous
	Section volume flow =	1400	Cfm	Previous flow — Branch flow
	Section length =	25	Feet	Given
	Static pressure required at D =	0.25	Inches	Given
	Vp of previous section =	0.241	Inches	From previous
	Trial duct size =	13	Inches	Trial and error
	Velocity of this section =	1520	Fpm	
	Vp of this section =	0.144	Inches	
	Sp/100 Ft. =	0.35	Inches	
	Loss in section =	0.088	Inches	
	Static pressure regain (SPR) for section =	0.048	Inches	
	Net loss in section =	0.039	Inches	Loss in section — SPR
	Resultant pressure at Point D =	0.245	Inches	(Pressure @ C-Section net loss) Close enough to required

Figure 10.25a

10.2 A 200,000 square foot [18,600 square meter] office tower is to be heated using steam delivered to main units located in the penthouse. The design heat load averages 25 Btu/hr. per square foot [80 watts per square meter]. If the system is a vapor system, what are the sizes

Static Pressure Regain Method of Duct Sizing (continued)

Section	Entity	Value	Units	Remarks
Section DE				
	Branch section volume flow =	400	Cfm	Given
	Previous section volume flow =	1400	Cfm	From previous
	Section volume flow =	1000	Cfm	Previous flow − Branch flow
	Section length =	20	Feet	Given
	Static pressure required at E =	0.225	Inches	Given
	Vp of previous section =	0.144	Inches	From previous
	Trial duct size =	12	Inches	Trial and error
	Velocity of this section =	1274	Fpm	
	Vp of this section =	0.101	Inches	
	Sp/100 Ft. =	0.21	Inches	
	Loss in section =	0.042	Inches	
	Static pressure regain (SPR) for section =	0.021	Inches	
	Net loss in section =	0.021	Inches	Loss in section − SPR
	Resultant pressure at Point D =	0.224	Inches	(Pressure @ D-Section net loss) Close enough to required
Section EF				
	Branch section volume flow =	600	cfm	Given
	Previous section volume flow =	1000	cfm	From previous
	Section volume flow =	400	cfm	Previous flow − Branch flow
	Section length =	25	Feet	Given
	Static pressure required at E =	0.2	Inches	Given
	Vp of previous section =	0.101	Inches	From previous
	Trial duct size =	9	Inches	Trial and error
	Velocity of this section =	906	Fpm	
	Vp of this section =	0.051	Inches	
	Sp/100 Ft. =	0.16	Inches	
	Loss in section =	0.040	Inches	
	Static pressure regain (SPR) for section =	0.025	Inches	
	Net loss in section =	0.015	Inches	Loss in section − SPR
	Resultant pressure at Point E =	0.210	Inches	(Pressure @ E-Section net loss) Close enough to required
Average Static Pressure				
	Resultant average static pressure at branch takeoffs =	0.253	Inches	

Figure 10.25a (continued)

of the steam main and the main condensate return from the service to the penthouse? What is the pressure drop per 100 feet [per meter]?

10.3 A 50-50 water glycol solution is to be used in a hydronic system to prevent freezing. Normally it delivers 100 MBH at 220°F [30 kW at 105 °C].
   a. How many gpm [L/s] of the mixture are required to deliver adequate heat?
   b. Assuming 1500 feet [460 m] total equivalent straight run of pipe, what pump head is required for the mixture to drive the required quantity through the system?

10.4 A 100′ × 100′ [30 m × 30 m] 10-story building with a basement is to be heated with a 4-pipe fan coil system. The convectors are roughly 20′ [6 m] on center around the perimeter and the floor-to-floor heights are 10′ [3 m]. There are two possible distribution systems:
   a. A main perimeter loop in the basement with risers at 20′ [6 m] to the convectors above.
   b. A main vertical riser with perimeter distribution on each floor.

   Calculate the length of pipe for each system and give the advantages and disadvantages of each. State which building types would favor each system.

10.5 Determine the pump characteristics for a conventional hot water system serving a building having a heat loss of 2,500,000 Btu/hr. [735 000 W] and a total straight run of pipe of 2,000 feet [600 m]. The system is a reverse return and all the pipes were sized to not exceed a head loss of 2.5 feet [0.25 kPa per meter] of head per 100 feet of pipe. Allow for other line losses.

10.6 For the same problem as #5 above, assume the system has a 50-50 glycol solution. Find the pump characteristics.

10.7 Determine the pump capacity and head required for an air-cooled condenser whose capacity is 3,000 MBH [900 kW] but whose

Calculations Summary Table

Section	Flow (cfm)	Velocity Used (FPM)	Diameter of Duct (Ins.)	Pressure Loss (Ins. Wat)	Static Pressure Regain (Ins. Wat)	Net Pressure Loss (Ins. Wat)	Static (Ins. Wat)	Pressure at Point
A	—	2433	14	—	—	—	0.378	A
AB	2600	2433	14	0.078	—	0.078	0.300	B
BC	2100	1965	14	0.080	0.064	0.016	0.284	C
CD	1400	1520	13	0.088	0.048	0.039	0.245	D
DE	1000	1274	12	0.042	0.021	0.021	0.224	E
EF	400	906	9	0.040	0.025	0.015	0.210	F
FK	400	906	9	0.024	—	0.024	0.186	K

Figure 10.25a (continued)

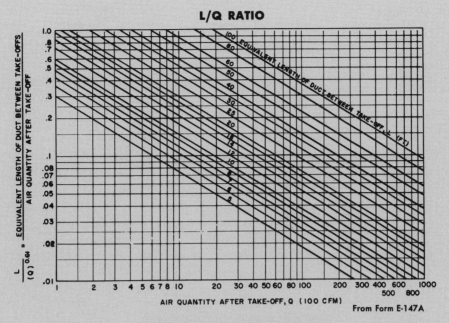

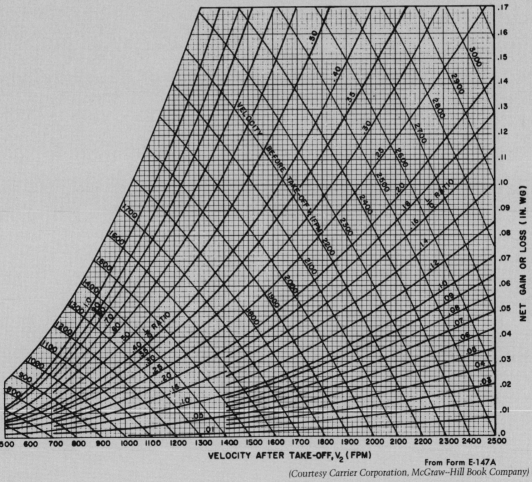

Static Regain Charts

Figure 10.26

REFRIGERANT NUMBER (ARI DESIGNATION)	11	12	22	113	114	500
Chemical Name	Trichloromono-fluoromethane	Dichlorodi-fluoromethane	Monochlorodi-fluoromethane	Trichlorotri-fluoroethane	Dichlorotetra-fluoroethane	Azeotrope of Dichlorodi-fluoromethane and Difluoroethane
Chemical Formula	CCl_3F	CCl_2F_2	$CHClF_2$	$CCl_2F\text{-}CClF_2$	$C_2Cl_2F_4$	73.8% CCl_2F_2 26.2% CH_3CHF_2
Molecular Wt	137.38	120.93	86.48	187.39	170.93	99.29
Gas Constant, R (ft-lb/lb-R)	11.25	12.78	17.87	8.25	9.04	15.57
Boiling Point at 1 atm (F)	74.7	−21.62	−41.4	117.6	38.4	−28.0
Freezing Point at 1 atm (F)	−168	−252	−256	−31	−137	−254
Critical Temperature (F)	388.0	233.6	204.8	417.4	294.3	221.1
Critical Pressure (psia)	635.0	597.0	716.0	495.0	474.0	631.0
Specific Heat of Liquid, 86 F	.220	.235	.335	.218	.238	.300
Specific Heat of Vapor, C_p 60 F at 1 atm	*	.146	.149	*	.156	.171
Specific Heat of Vapor, C_v 60 F at 1 atm	*	.130	.127	*	.145	.151
Ratio $\frac{C_p}{C_v} = K$ (86 F at 1 atm)	1.11	1.14	1.18	1.12	1.09	1.13
Ratio of Specific Heats Liquid, 105 F / Vapor, C_p, 40 F sat. press.	2.04	1.55	2.14	1.47	1.59	1.77
Liquid Head (ft), 1 psi at 105 F	1.61	1.84	2.04	1.51	1.65	2.10
Saturation Pressure (psia) at: −50 F	0.52	7.12	11.74	*	1.35	*
0 F	2.55	23.85	38.79	0.84	5.96	27.96
40 F	7.03	51.67	83.72	2.66	15.22	60.94
105 F	25.7	141.25	227.65	11.58	50.29	167.85
Net Refrigerating Effect (Btu/lb) 40 F-105 F (no subcooling)	67.56	49.13	66.44	54.54	43.46	59.82
Cycle Efficiency (% Carnot Cycle) 40 F-105 F	90.5	83.2	81.8	87.5	84.9	82.0
Solubility of Water in Refrigerant / Miscibility with Oil / Toxic Concentration (% by vol)	Negligible Miscible Above 10%	Negligible Miscible Above 20%	Negligible Limited *	Negligible Miscible *	Negligible Miscible Above 20%	Negligible Miscible Above 20%
Odor	Ethereal, odorless when mixed with air	Same as R 11	Same as R 11	Same as R 11	Same as R 11	Same as R 11
Warning Properties	None	None	None	None	None	None
Explosive Range (% by vol)	None	None	None	None	None	None
Safety Group, U.L.	5	6	5A	4-5	6	5A
Safety Group, ASA B9.1	1	1	1	1	1	1
Toxic Decomposition Products	Yes	Yes	Yes	Yes	Yes	Yes
Viscosity (centipoises) Saturated Liquid 95 F	.3893	.2463	.2253	.5845	.3420	.2150
105 F	.3723	.2395	.2207	.5472	.3272	.2100
Vapor at 1 atm 30 F	.0101	.0118	.0120	.0097	.0108	*
40 F	.0103	.0119	.0122	.0098	.0109	*
50 F	.0105	.0121	.0124	.0100	.0111	*
Thermal Conductivity (k) Saturated Liquid 95 F	.0596	.0481	.0573	.0512	.0435	*
105 F	.0581	.0469	.0553	.0500	.0421	*
Vapor at 1 atm 30 F	.0045	.0047	.0060	.0037	.0056	*
40 F	.0046	.0049	.0061	.0039	.0057	*
50 F	.0046	.0051	.0063	.0040	.0059	*
Liquid Circulated, 40 F-105 F (lb/min/ton)	2.96	4.07	3.02	3.66	4.62	3.35
Theoretical Displacement, 40 F-105 F (cu ft/min/ton)	16.1	3.14	1.98	39.5	9.16	2.69
Theoretical Horsepower Per Ton 40 F-105 F	0.676	0.736	0.75	0.70	0.722	0.747
Coefficient of Performance 40 F-105 F (4.71/hp per ton)	6.95	6.39	6.29	6.74	6.52	6.31
Cost Compared With R 11	1.00	1.57	2.77	2.15	2.97	2.00

*Data not available or not applicable.

(Courtesy Carrier Corporation, McGraw-Hill Book Company)

Characteristics of Typical Refrigerants

Figure 10.27

Distribution Equipment Data Sheets

Designation	Carbon Steel Pipe	Means Number 151-701
Piping	**Typical Use and Consideration**	**Size Range (Diameter)**
Black Steel Schedule 40	Schedule 40 indicates the wall thickness and is also known as the standard weight. Used for low- and medium-pressure hydronic service as well as chilled water, condenser water, and gas service and distribution.	1/4" through 12"
Galvanized Steel Schedule 40	Galvanized pipe is used in corrosive atmospheres or where exposed to the elements. This piping is usually found in drainage, waste, vent, and condenser water applications. Galvanized (zinc-coated) pipe should not be used for glycol solutions nor should it be welded after galvanizing.	1/4" through 12"
Black Steel Schedule 80	Schedule 80, or extra heavy steel pipe, is used for high-pressure service and for underground installations.	1/4" through 12"
Black Steel Schedule 10	This thin wall lightweight pipe is used in many HVAC applications. Special groove-type fittings and valves are used to join the piping in low-pressure systems.	**151-801** 2" through 10"

Steel pipe, regardless of wall thickness, retains its designated outside diameter. For example, 3" pipe, regardless of wall thickness or schedule number, is always 3-1/2" in diameter. This is necessary for threading, socket welding, and other fitting and mating purposes. The inside diameter decreases or increases depending on the wall thickness specified. The larger the schedule number, the heavier the wall.

Piping	Copper Tubing	Means Number 151-401
Type K	This heavy wall tubing, available in hard drawn straight lengths or in soft-drawn coils, is found in severe applications such as underground water piping and higher pressures. K is also used in some refrigeration work.	1/4" through 8"
Type L	This medium wall tubing, available in hard-drawn straight lengths or in soft-drawn coils, is used predominantly in indoor hot or cold water piping throughout HVAC systems.	1/4" through 8"
Type M	This thin wall tubing, available in straight lengths only, is used in non-critical (low-pressure) hydronic heating and cooling systems.	1/4" through 8"
Type ACR	This tubing, available in hard-drawn straight lengths or soft-drawn coils, is actually Type L tubing that has been cleaned and capped prior to shipping. As the ACR designation implies, it is used for air conditioning and refrigeration work. This tubing is called by its actual outside diameter rather than the usual nominal sizing.	3/8" OD through 4-1/8" OD
Type DWV	This tubing, available in straight lengths only, is produced for drainage, waste, and vent applications.	1-1/4" through 8"

Copper tubing, regardless of type, has the same outside diameter per size. Only one class of fittings for copper tubing is made, whether they are soldered, brazed, flared, or compression type.

Figure 10.28

Distribution Equipment Data Sheets (continued)

Designation	Plastic Pipe and Tubing	Means Number 151-551
Piping	**Typical Use and Consideration**	**Size Range (Diameter)**
Type PVC (Polyvinyl Chloride)	This rigid pipe is used in many HVAC applications, both above and below ground, such as chilled water, drains, wastes and vents, oil and gas. Available in Schedule 40 and 80 with the same outside diameters as steel pipe, it may be cut and threaded using standard steel pipe dies. Flanged, victaulic, and socket weld fittings and valves are available. Hot gas or solvent welds are used with the socket fitings.	1/2" through 16"
Type CPVC (Chlorinated Polyvinyl Chloride)	CPVC piping has all the features and dimensions of PVC but has the added feature of higher operating temperatures in the 120°F to 180°F range. This piping is very difficult to weld by the hot gas method; only highly qualified welders should attempt this procedure. This piping can be used in low-temperature hot water and heat pump applications.	1/2" through 8"
Polybutylene SDR-11 (Standard Dimension Ratio)	This tubing is available in both straight lengths and coils. Having an operating temperature of 180°F under pressure, and being available in 1,000 foot coils, makes it a popular selection for low-temperature hot water radiant panels and earth-coupled heat pump systems. Available in both copper tubing size (CTS) and iron pipe size (IPS), it is joined by insert type fittings and bands.	1/2" through 3"
Ductwork	**Rigid Rectangular**	**Means Number 151-250**
Aluminum Alloy	Used in low-pressure, low-velocity air conditioning systems, especially in residential work. Aluminum expands and contracts to a greater extent than steel and allowances must be made to control this movement. Additional bracing and support is required to make up for its lack of rigidity compared to steel.	Varies (priced by weight)
Galvanized Steel	Used in both low- and high-pressure velocity systems, particularly in commercial and institutional HVAC applications. This is the most popular material used for sheet metal ductwork in the HVAC industry.	Varies (priced by weight)
Stainless Steel Type 304	Used in lieu of aluminum or galvanized ductwork in corrosion-resistant applications such as autoclave, dishwashers, or showers. Pitch ductwork to drain.	Varies (priced by weight)
Fiberglass Rigid Board	Used extensively in residential and light commercial applications where thermal insulation and sound isolation are considerations. Restricted to areas where physical abuse is not a factor.	Varies (priced by the square foot)

Figure 10.28 (continued)

temperature difference is only 5°F [3 °C]. All pipes are designed not to exceed 2.5 feet of head per 100 feet [0.25 KPa of static pressure per meter] of pipe, and the head loss through the condenser is 60 feet [180 kPa]. Allow for other line losses. The straight run of pipe is 100 feet [30 m].

10.8 Given the air distribution system shown in Figure 10.30:
   a. Determine the cfm [L/s] in each run of duct.
   b. Determine the longest run.
   c. Size all ducts by the constant pressure drop method.
   d. Size the fan – allow for elbows and fittings.

10.9 You have a duct system that is designed for 20 tons [70 kW] of cooling. State your assumptions and size the duct by the velocity method and the constant pressure drop method. Choose a rectangular duct with an aspect ratio of approximately 1:3.

10.10 For the air system shown in Figure 10.31, all registers deliver 1,000 cfm [475 L/s] at 55°F [13 °C] to a space set at 70°F [21 °C].
   a. What is the total sensible cooling that the system can handle?
   b. Size all the ducts by the constant pressure drop method.
   c. Size the fan.

Distribution Equipment Data Sheets (continued)

Designation	Rigid Round	Means Number 151-250
Ductwork	**Typical Use and Consideration**	**Size Range (Diameter)**
Spiral	This type of duct is available in galvanized steel, aluminum, and stainless steel. It is found in low and high velocity systems. Its ease of installation and fabrication is an economic consideration compared to shop-fabricated rectangular duct.	3" through 48"
PVC (Polyvinyl Chloride)	Used in special corrosive exhaust applications such as laboratory hoods.	6" through 48"
Ductwork	**Flexible Round**	**Means Number 151-250**
Various Materials	This form of ductwork is available in many forms and materials, from uninsulated flexible metal and uninsulated helical wound wire with aluminum or poly foil wrap, to insulated adaptations of the same. Flexible round PVC duct is also available. Flexible duct is a labor-saving device in HVAC systems in limited space areas where many special fabricated fittings can be eliminated. The use of flexible duct in lengths over fifteen feet may be restricted by local labor agreements. General takeoffs from sheet metal are often less than eight feet in length.	3" through 20"

Figure 10.28(continued)

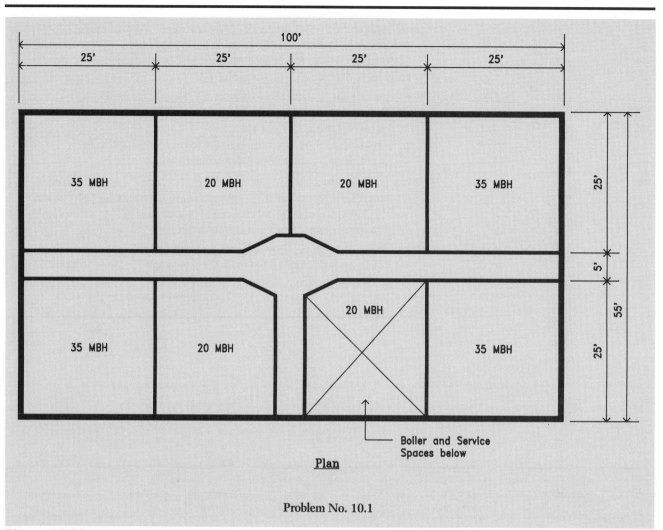

Problem No. 10.1

Figure 10.29

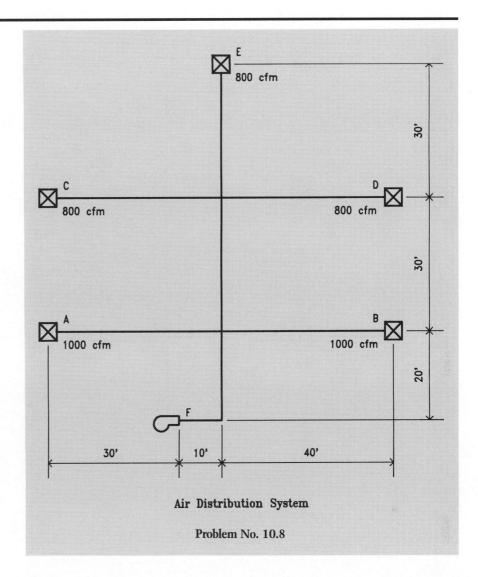

Figure 10.30

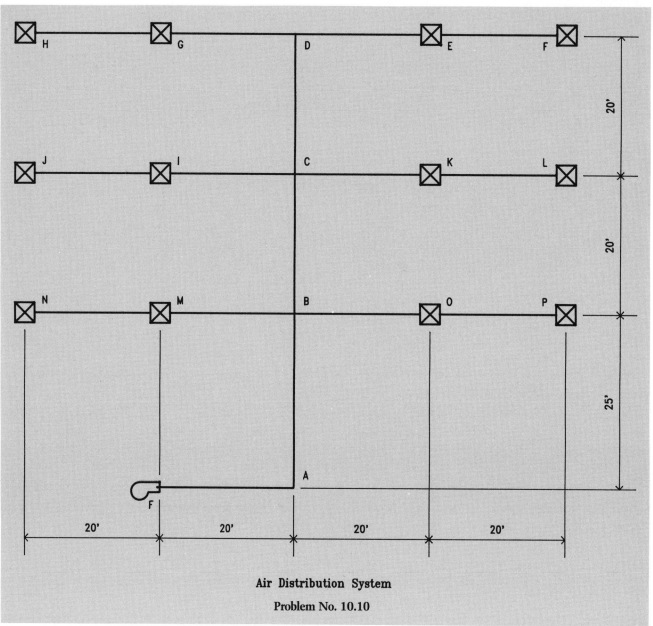

Figure 10.31

Chapter Eleven

TERMINAL UNITS

A *terminal unit* is a device at the end of a duct or pipe that transfers heating or cooling from the distribution system to the conditioned space. The end product of a terminal unit is conditioned (warm or cool) air. With the exception of air distribution systems, all terminal units are localized heat exchangers; they take in heat (in the form of steam or hot water) or cooling (from the direct expansion or chilled water system) and transfer it to the air in the conditioned space. Hydronic systems use two-, three-, and four-pipe systems with two- or three-way control valves. Air systems are connected directly to terminal units (grilles, registers, and diffusers) or through mixing boxes or VAV (variable air volume) units.

Steam Systems

The terminal units used for steam and hot water (hydronic) systems are outlined below.

Terminal Unit	Advantages	Disadvantages	Common Use
Cast iron convector or radiator	Rugged	Most expensive and cumbersome	Residential
Fin tube commercial	Inexpensive; blankets exterior wall	Lose wall space	Commercial
Fin tube baseboard	Inexpensive	Covers come loose; interference by drapes, etc.	Residential
Fan coils	Heating and cooling uses	Space considerations	Office and institutional
Unit heaters (hydronic coils with fans)	Inexpensive	Noisy	Garages and factories

The size of the unit depends on the heating or cooling load and the temperature of the steam or hot water. Figure 11.1 lists the sizes of typical hydronic terminal units. Accessories for steam terminal units are described in the following sections.

Valves

Steam radiator valves control the flow of steam to the heating element. Radiator valves are furnished in straight or angle patterns and have a union tailpiece connection. Automatic radiator valves, either self-contained or controlled by a remote thermostat, are popular in most commercial installations.

Air Vents

Air vents eliminate air from steam systems by using a float mechanism that rises when buoyed by the condensate to prevent the discharge of steam. Air vents are used on individual radiators and at the high points of risers and mains in gravity systems.

Traps

Traps are used to keep steam in the terminal units and to pass only condensate and air to the return system. Two types of steam traps that are commonly used are thermostatic (radiator) and float and thermostatic (combination) traps. Thermostatic traps are activated by temperature, and expand an internal bellows. Float and thermostatic traps contain a float that rises to let condensate pass and closes when there is no condensate, as well as a thermostatic element to pass the air. Float and thermostatic traps are used for main and riser drips and for heavy equipment condensate loads. The accessories for steam systems are illustrated in Figure 11.2.

Water Systems

Hydronic systems are among the most frequently used heating systems. Hydronic equipment should be sized properly to match the loads required to heat or cool the space. Common terminal units for water systems include the following:

- Cast iron radiators/convectors (heating only)
- Fin tube baseboard (heating only)
- Unit heaters (heating only)
- Fan coil units (heating and cooling)
- Induction units (heating and cooling)
- Heat pumps – water source (heating and cooling)

There are many manufacturers of terminal units, which differ in appearance, color, enclosure, mounting, and controls. Typical performance characteristics for these units are shown in Figures 11.1 and 11.3. To select the proper unit, determine the load that the unit must deliver to the space (refer to Chapter 3 for heating and cooling loads); determine the entering water temperature, which is usually done by the designer; and read the size of the unit required (from manufacturer's data).

Valves

There are many types of valves used in hydronic systems. Valves are usually sized to match the piping in the system. Automatic temperature control valves and pressure regulators may be one or two sizes smaller than line size because of precise body sizing for actual capacity. Figure 11.4 illustrates typical valves and fittings utilized with hydronic terminal units.

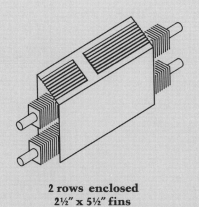

2 rows enclosed
2½" x 5½" fins

1 row enclosed
2" x 3" fins

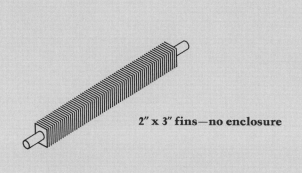

2" x 3" fins—no enclosure

Commercial Fin Tube Radiators

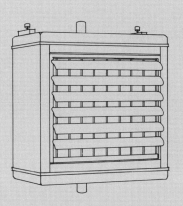

Hydronic Unit Heater

Sizes for Typical Hydronic Terminal Heating Units

Fin Tube Radiation (Output in Btu/hr./l.f.) [W/m]						
Temp. (°F) [°C]	170° [77°]	180° [82°]	190° [88°]	200° [93°]	210° [99°]	220° [104°]
2 rows	1,000 [960]	1,150 [1105]	1,300 [1250]	1,425 [1370]	1,575 [1515]	1,725 [1660]
1 row	700 [675]	800 [770]	875 [840]	975 [940]	1,075 [1035]	1,175 [1130]
No enclosure	575 [550]	650 [625]	750 [720]	825 [795]	900 [865]	1,000 [960]

Unit Heaters (Output in MBH) [kW]						
Temp. (°F) [°C]	180° [82°]	200° [93°]	220° [104°]	240° [115°]	260° [127°]	300° [147°]
cfm = 500 [250 L/s]	14 [4]	19 [6]	24 [7]	28 [9]	33 [10]	43 [13]
cfm = 1,000 [500 L/s]	31 [9]	40 [12]	48 [14]	56 [17]	65 [20]	80 [24]
cfm = 1,500 [750 L/s]	50 [15]	62 [18]	74 [22]	85 [25]		
cfm = 2,000 [1000 L/s]	85 [25]	100 [30]	118 [36]	135 [40]		

All numbers shown in brackets are metric equivalents.

Figure 11.1

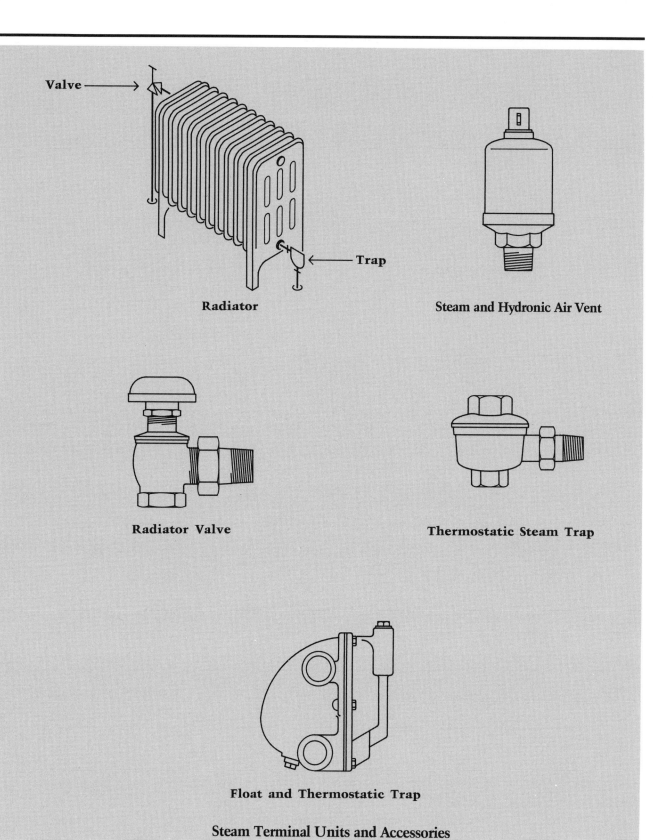

Figure 11.2

Air Vents

Manual air valves and automatic air vents are used to purge air from the system. Located at or near the high point of each radiator riser and main, they permit air to escape the system. Even though hydronic systems are typically closed, fresh makeup water and minor reactants contain air bubbles, which must be released from the piping and coils via air vents.

Performance Characteristics of Hydronic Unit Heating and Cooling Coils

Size	32 x 25 x 9 [800 x 635 x 230]	40 x 25 x 9 [1015 x 635 x 230]	44 x 25 x 9 [1115 x 635 x 230]	56 x 25 x 9 [1420 x 635 x 230]	60 x 28 x 11 [1525 x 710 x 280]	72 x 28 x 11 [1830 x 710 x 280]
cfm [L/s]	200 [100]	300 [150]	400 [200]	600 [300]	800 [400]	1,000 [500]
MBH [kW] Heating	20 [6]	25 [7.5]	32 [9.5]	40 [12]	58 [17]	68 [20]

Fan Coils—Rated at 180°F [81°C] Water

Size	25 x 44 x 11 [635 x 1115 x 280]			25 x 55 x 12 [635 x 1397 x 305]	
cfm [L/s]	230 [115]	270 [135]	325 [160]	380 [190]	530 [260]
MBH [kW] Cooling	7,000 [2050]	8,700 [2550]	11,500 [3370]	15,000 [4400]	18,000 [5300]

All numbers shown in brackets are metric equivalents.   Heat Pumps

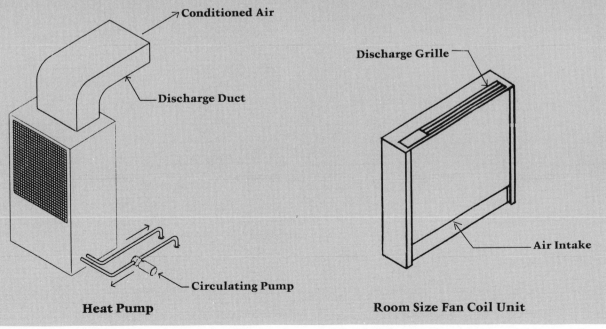

Figure 11.3

The flow of fluids in a piping system is controlled or regulated by the use of valves. Valves are used to start, stop, divert, relieve, or regulate the flow, pressure, or temperature in a piping system. Valves are manufactured in several configurations according to use. Some of the types include: gate, globe, angle, check, ball, butterfly, and plug. Valves are further classified by their piping connections, stem position, pressure and temperature limits, as well as by the materials from which they are made.

Stainless steel, or steel alloy, valves can be used effectively in most instances for corrosion protection.

Bronze is one of the oldest materials used to make valves. It is most commonly used in steam, hot- and cold-water systems, and other non-corrosive services. Bronze is often used as a seating surface in larger iron-body valves to ensure tight closure. Pressure ratings of 300 psi [2070 kPa] and temperatures up to 150°F [66° C] are typical.

Iron valves are normally used in medium to large pipe lines to control non-corrosive fluids and gases. Pressures for these valves should not exceed 250 psi [1725 kPa] at 450°F [232° C], or 500 psi [3450 kPa] cold working pressures for water, oil, or gas.

Carbon steel is a high-strength material, and the valves made from this metal are therefore used in higher-pressure services, such as steam lines up to 600 psi [4150 kPa] at 850°F [454° C]. Many steel valves are available with butt-weld ends for economy and are generally used in high-pressure steam service, as well as other higher-pressure, non-corrosive services.

Forged steel valves are made of tough carbon steel. They are used at pressures up to 2,000 psi [13 800 kPa] and temperatures up to 1,000°F [540° C].

Plastic is used for a great variety of valves, generally in high-corrosive service, at low temperatures and low pressures. Plastic lining of metal valves for corrosive service and high-purity applications and temperatures are also available.

Gate valves provide full flow, minute pressure drop, and minimum turbulance. They are normally used where operation is infrequent, such as for equipment isolation.

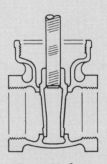

Gate Valve

Globe valves are designed for throttling and/or frequent operation with positive shutoff. Particular attention must be paid to the several types of seating material available to avoid unnecessary wear. The seats must be compatible with the fluid in service and may be composition or metal in construction. The configuration of the globe valve opening causes turbulence, which results in increased flow resistance.

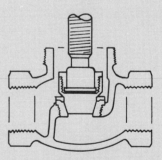

Globe Valve

Check valves are one-way valves and are designed to prevent backflow by automatically seating when the direction of fluid is reversed. Swing check valves are generally installed with gate valves, as they provide comparable full flow, and are usually recommended for lines where flow velocities are low. They should not be used on lines with pulsating flow. They are also recommended for horizontal installation, or in vertical lines where flow is only upward.

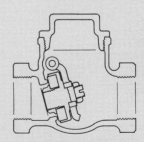

Check Valve

Hydronic Terminal Units—Valves and Fittings

Figure 11.4

Expansion Tank

Because water expands and contracts as it is heated or cooled, an air cushion must be provided in all hydronic systems to maintain system pressure. In small residential systems, a 30-gallon [115-liter]* tank is usually sufficient, but actual size depends on the gallons of water contained

Throughout the book, SI metric equivalents are provided for most imperial units. The SI units appear in brackets [ ].

Ball valves are light and easily installed, yet because of modern elastomeric seats, they provide tight closure. Flow is controlled by rotating 90° a drilled ball that fits tightly against resilient seals. This ball seals with flow in either direction, and the valve handle indicates the degree of opening. This type of valve is recommended for frequent operation, such as for tank filling, and is readily adaptable to automation. Ball valves are ideal for installation where space is limited.

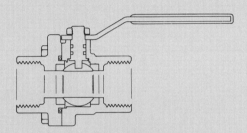

Ball Valve

Butterfly valves provide bubble-tight closure with excellent throttling characteristics. They can be used for full-open, closed, and for throttling applications. The butterfly valve consists of a disc, controlled by a shaft, within the valve body. In its closed position, the valve disc seals against a resilient seat. The disc position throughout the full 90° rotation is visually indicated by the position of the operator. Butterfly valves are only a fraction of the weight of a gate valve and require no gaskets between flanges in most cases. They are recommended for frequent operation and are adaptable to automation where space is limited. Wafer- and lug-type bodies, when installed between two pipe flanges, can be easily removed from the line. The pressure of the bolted flanges holds the valve in place. Locating lugs makes installation easier.

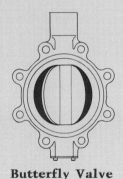

Butterfly Valve

Backflow preventers are mechanical devices installed to stop contaminated fluids from entering the potable water system. This reversal of flow may be caused by back pressure or back syphonage. The assembly incorporates double check valves, a relief valve, and vacuum breaker all contained in one housing.

Backflow Preventer

Hydronic Terminal Units—Valves and Fittings (cont.)

Figure 11.4 *(continued)*

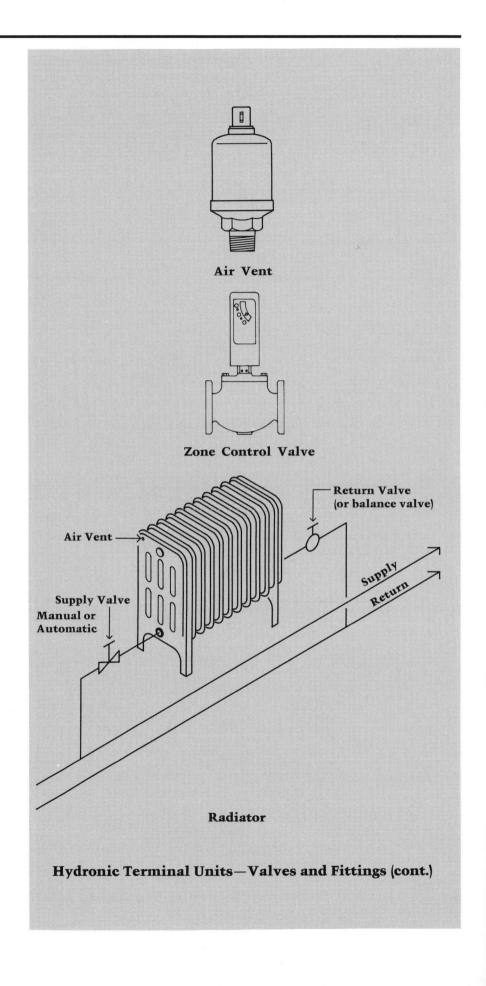

Figure 11.4 (continued)

in the entire system. Manufacturers usually provide sizing data based on overall system capacity. Tanks are also available with a diaphragm that separates the water from the air cushion.

Automatic Control

Two-way valves, the most frequently used type of valve, normally have two positions, open or closed. When open, two-way valves allow flow (typically water) to a coil or convector. When closed, no flow occurs on the loop. As a result, the overall system balancing and the distribution and flow rates throughout the system are affected. These valves may be controlled from a built-in (self-contained) thermostatic element or a remote thermostat.

Three-way valves act as bypass valves. On call, they send water to the unit. On no call, they direct the flow past the unit, but maintain overall flow conditions. Three-way valves are also used as mixing or tempering valves. They may be self-contained, but usually are remote-controlled. Two- and three-way valves are available as modulating valves. Figure 11.5 illustrates typical uses of two- and three-way valves.

Two-, Three-, and Four-Pipe Systems

The number of pipes feeding a piece of equipment varies according to the type of system. If the equipment provides heating only or cooling only, it is fed by a two-pipe system, one supply and one return. This is the most common arrangement. Equipment that can heat and cool independently of each other are fed by a two-, three-, or four-pipe system (see Figure 11.5). The three- and four-pipe systems permit either heating or cooling to occur as required in a particular space.

Air Systems

Air is distributed through ductwork in a heating or cooling system using single-duct, dual-duct, constant-volume, and variable-volume systems (see Figure 11.6).

Single-duct systems are the most effective and most often used system for common applications.

Dual-duct systems (one heating and one cooling) with mixing boxes are used for hospitals, laboratories, and special control environments. Dual-duct systems are more expensive to build and operate than single-duct systems.

Constant-volume systems contain ducts with a constant velocity at all times. They may be single- or dual-duct systems. Constant-volume systems work best for spaces with relatively even loads or where the temperature of the air must be adjusted to meet varying loads.

Variable-volume systems use variable air volume (VAV) boxes, which vary the amount of air to a space to meet the load. VAV systems are efficient because they supply the precise amount of air needed. A VAV box has a variable-position damper that controls flow of air to a space. The VAV box may also contain a reverse-acting thermostatic control for cooling, a reheat coil, and a plenum recirculating feature. Air system terminal units consist of grilles, registers, diffusers, and louvers, which supply or return air from spaces.

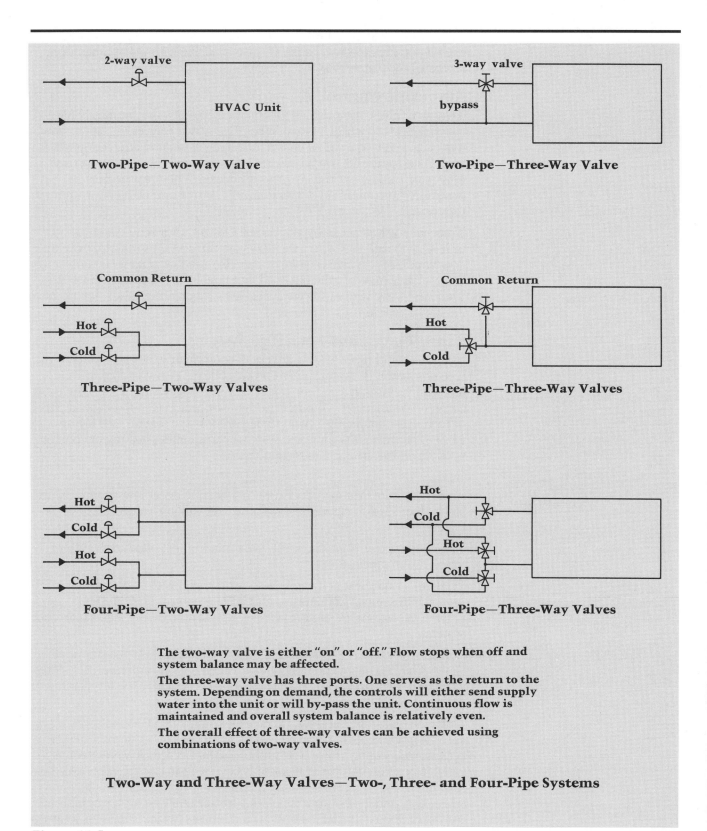

Figure 11.5

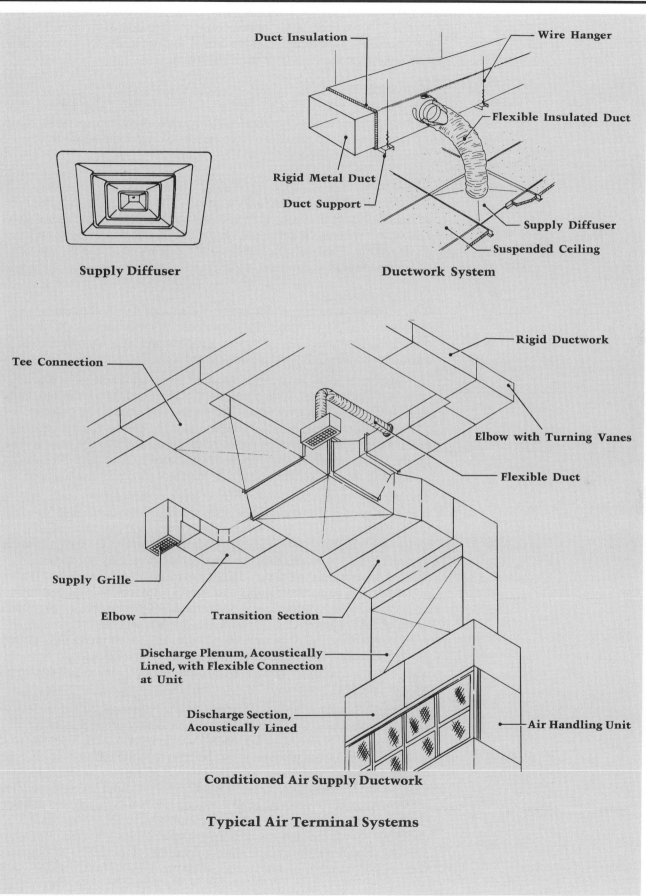

Figure 11.6

Grilles

A grille is a screened or perforated connection to ductwork, usually used for return air. Grilles are nondirectional and typically act as a cover to visually screen the ductwork from the room.

Registers

A register is a grille with a manually adjusted damper behind it, used to control the flow of the conditioned air into a conditioned space. Registers are usually wall-mounted and have either no directional capability or limited ability to direct air up to the ceiling.

Diffusers

Diffusers direct the conditioned air stream into a space through fixed or adjustable blades and may also have an integral damper to control flow. Diffusers are normally mounted on the ceiling and can disperse air in all directions. The most frequently used types are ceiling, wall, and slot or linear diffusers.

Sizing Diffusers

The selection, placement, and sizing of diffusers is critical to the proper distribution of the air. Once they have been properly located and the assigned quantities of air (at a given temperature) have been determined for the space, the diffusers can be selected.

Diffusers, as well as most other terminal units, are available in a wide variety of shapes, sizes, configurations, throws (the distance the air travels), and spreads (the area over which the air is distributed). The designer should first determine where the terminal units can be placed, avoiding any lights as well as existing pieces of equipment above the ceiling. The height and area of the space should be evaluated to determine the layout that would most effectively eliminate stagnant air zones in the space.

The velocity, quantity, and direction of the air coming from each terminal unit is important. Air should be designed to leave the diffuser at 50 – 150 feet per minute [0.25 – 0.75 m/s], although it is substantially reduced within 4 – 6 inches [100 – 150 mm]. Many terminal units have the ability to direct the air at angles ranging from vertical to nearly horizontal. In spaces where the ceiling is very high, too low an air velocity or too shallow an angle can lead to stratification. In spaces with low ceilings, too high an air velocity or an angle that is too vertical will be uncomfortable in both heating and cooling cycles. A disadvantage associated with high exit velocities is noise. Most manufacturers provide noise criteria (NC rating) for each terminal unit, given the amount of air passing through that unit. For most occupied spaces, an NC rating of 20 – 30 is considered a good design goal.

Suppose that a space with a 10' [3 m] ceiling requires 1,200 cfm [560 L/s]. Assume that four diffusers with an air deflection of 30° (from horizontal) at 300 cfm [140 L/s] each should perform well. Referring to standard manufacturers' data, it can be seen that the diffusers in this cfm range have a throw of about 9 – 11' [2.75 – 3.35 m] (at 100 fpm) [0.5 m/s], so the diffusers themselves should be placed about 25' apart to allow some overlap (see Figure 10.16 in Chapter 10). If this places one diffuser closer to one of the interior walls, it should throw the air in three directions (in other words, a three-way diffuser), with 100 cfm [50 L/s] in each direction. Looking at typical manufacturers' data, a 9" × 9" diffuser with an NC rating of 21 and a static pressure of 0.117" WC [35 Pa] could be selected. The 0.117" [35 Pa] would be considered in sizing the fan.

There are other considerations in selecting diffusers. Opposed blade dampers are options in most units, allowing more flexibility for the user. Throw-reducing vanes may be added to provide a more comfortable flow of air. These options add comfort and convenience, but they also increase the static pressure necessary to drive the air into the space. After selections are made, the system should be evaluated, and if the static buildup is found to be too high, then larger diffusers, with more free area, should be considered.

Louvers

Louvers are openings in exterior walls to intake or exhaust air. Louvers typically have blades to shed rain, insect or bird screens, and many have dampers. Some are covered by rain hoods or cowls.

Dampers

Dampers are devices within a terminal unit that provide manual or automatic control of air flow. Dampers operate in a manner similar to the valves in hydronic systems. Figure 11.7 illustrates typical air system terminal units.

Electric Systems

Several types of electric terminal units are used to provide electric heat:
- Baseboard radiators
- Unit heaters
- Wall heaters
- Radiant heaters
- In-line duct heaters

Electric terminal units are used for air heating (see Figure 11.8). (Many of the previously mentioned pieces of hydronic equipment use electric power, but are not considered electric terminal units.)

Spot coolers are used for cooling individual rooms or pieces of equipment on a temporary or emergency basis. Spot coolers run on the same principal as window or through-the-wall residential air conditioners, but have special features that should be considered.

A window air conditioner is contained in a box and mounted in a window – the condenser coil and fan are outdoors and the evaporative coil and room fan are indoors. The condenser and evaporator sections are isolated from each other by the window or by a divider in place of the window. The air conditioner uses outside air as a condensing medium and recirculates and cools the room air. A spot cooler is similar in that it is a self-contained unit, but it is portable and mounted on wheels. The condensing and evaporating sections are not isolated from each other by a window, wall, or divider. In fact, the condensing and evaporating sections are in the same room. In order to avoid short-circuiting of the evaporator air to the condenser and overheating of the area being cooled by the warm condenser air, flexible ducts are connected to the two sections to direct the airstreams away from each other (see Figure 11.9).

One disadvantage of spot coolers is that the heat taken from one area of the room is discharged in another portion of the same room. Spot coolers rely on a building's main system to absorb the heat rejected locally. Spot coolers are generally used to provide emergency cooling of equipment, laboratories, or patient areas, where localized cooling can be provided while heat is rejected to the rest of the space. It is possible to reduce the spot cooler's heat load by:
- Extending the discharge-air flexible duct from the condensing section

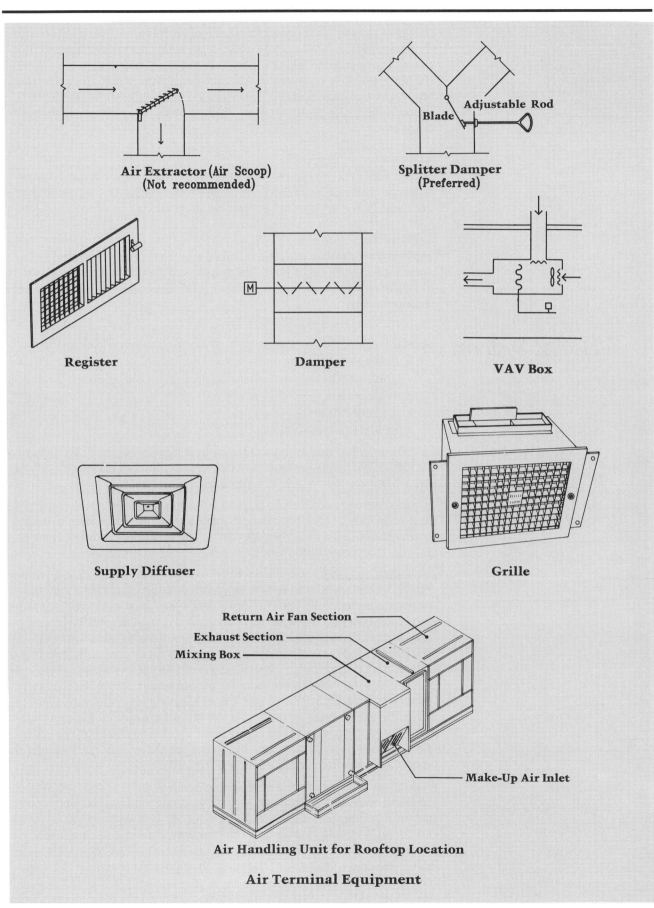

Figure 11.7

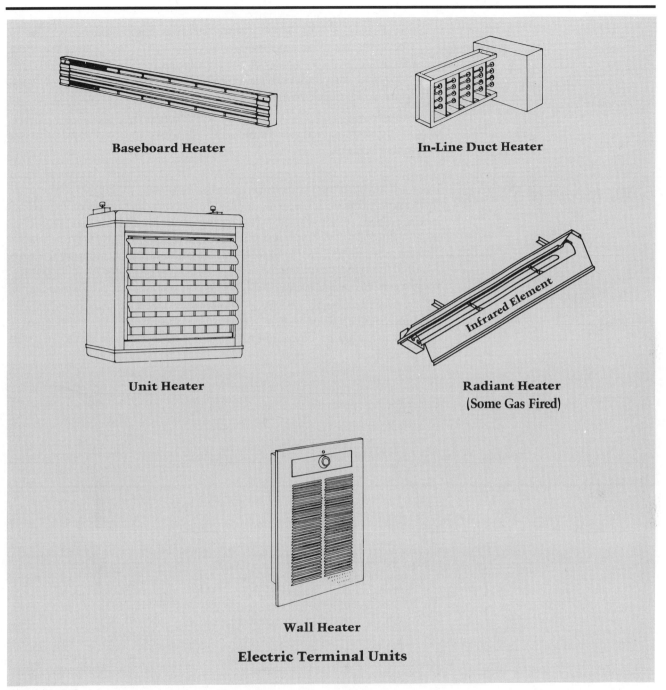

Figure 11.8

to a window, door, exhaust grille, or other opening so the warm air is discharged outside the room. This may be possible only if the condenser fan has adequate static pressure.
- Locating the spot cooler at the area or equipment to be cooled and directing the condenser air into an airstream of a temporary exhaust fan in the door or window.

Care should also be taken with spot coolers when they are used in areas with a high latent load, as condensate will form and may leak from the evaporator section.

Floor Heating System

Recent developments in technology utilizing plastic piping have made hydronic heating with under-floor piping attractive. Previously called radiant heating, these systems used metal piping or tubing that in time would leak at the joints. With plastic piping, the leaking problem is minimized and the material is capable of distributing water at moderate temperatures. An in-slab piping system delivers approximately 20 Btu/hr./s.f. [65 W/m^2] with water temperatures at 100°F [38 °C].

The piping can be installed in or below most floors. Hot water travels through small (5/8" diameter) [16 mm] plastic tubes. These tubes are typically arranged in a loosely-knit pattern across the floor of the space to

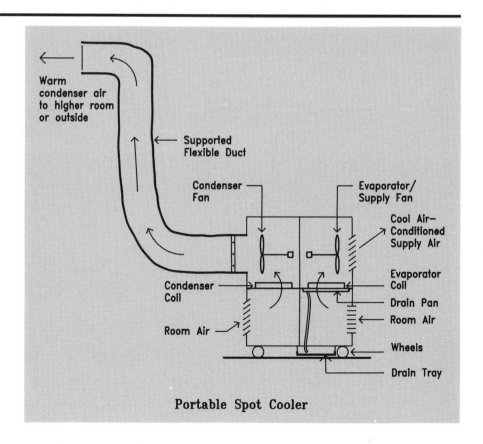

Figure 11.9

be heated, leaving 7 – 12 inches [175 – 300 mm] between tubes depending on the heating output desired.

In-slab systems have the double advantage of not taking up any floor space with terminal units and of providing a comfortable, warm floor in winter. Heated slabs are particularly desirable in bathrooms and healthcare and child-care spaces, where contact with the floor is common. These systems also utilize lower water temperatures than conventional hydronic systems (90 – 150°F) [32 – 66 °C], which makes them very useful in energy-scavenger, cogeneration, and solar applications.

A particular concern of plastic piping, however, is that it is susceptible to oxygen penetration through the walls of the piping. Special precautions should be taken to remove the oxygen or use cross-linked plastic piping chemistry to limit oxygen in the water.

Equipment Data Sheets

The equipment data sheets for terminal units (Figure 11.10) can be used in the selection of each piece of equipment. These sheets provide the efficiency, useful life, typical uses, capacity range, advantages and disadvantages, accessories, and additional equipment for each component listed. A Means line number, where applicable, is provided. Costs can be determined by looking up the line number in the current edition of *Means Mechanical Cost Data*.

Problems

11.1 For a space that has a design heat loss of 35,000 Btu/hr. [10 250 W] and a supply water temperature of 200°F [94 °C], answer the following:
  a. How many linear feet of single-row baseboard are needed?
  b. How many linear feet of 2-row baseboard are needed?
  c. What size fan coil will handle the space if the water temperature is 180°F [82 °C]?

11.2 An office space of 30' × 50' [9 m × 15 m] with a 10' [3 m] ceiling is to have an air handling system capable of 6 air changes per hour. Calculate the air required, lay out the ductwork, and select a typical ceiling diffuser.

11.3 Suppose you have a 4-pipe fan coil unit to be installed on a floor. Draw a schematic diagram of the unit, showing the piping to the supply and return mains below the floor as well as the control valves and thermostats that will control the unit.

11.4 A spot cooler of 5 tons [17.5 kW] is brought into a space. It has a C.O.P. of 3. What is the net total load that the building system must provide to meet the additional loads of the spot cooler operating in the building?

Terminal Units Equipment Data Sheets

Terminal Units	Hydronic Automatic Air Vents	Means No. 156-201
Useful Life: 10-20 years	**Typical Use and Consideration**	**Size Range**
Float-Type Air Vent	Air vents are used on radiators, unit heaters, and high points of mains and risers in all gravity steam systems. Air vents are used on radiators, coils, and high points of mains and risers in hot water systems. Air vents are used on coils and high points of mains and risers in chilled water systems. Air must be vented from the piping systems, as it may impede or even stop the flow of the heating/cooling medium. Manual air vents are sometimes used. Air vents in hard to reach places often have the discharge piped to an accessible location.	1/8"–2" [4–50 mm]
Useful Life: 7-15 years	**Steam Specialties**	**Means No. 156-272, 156-240**
Angle Radiator Valve	Manual radiator valves, both angle and straightway, are used to control the steam or hot water supply to radiators and convectors. In many instances these manual valves are replaced by thermostatically controlled valves, either self-contained or with remote thermostats.	1/2"–2" [15–50 mm]
Angle Thermostatic (Radiator) Trap	Radiator traps are installed on the return end of the two-pipe steam radiators and convectors. They do not have sufficient capacity to remove the condensate from unit heaters or coils. A thermostatic bellows stops the flow of steam and reopens upon temperature drop to allow the flow of condensate. In rare instances under light loads, a thermostatic trap can be used as a drip trap.	1/2"–1" [15–25 mm]
Combination Float and Thermostatic Trap	Combination float and thermostatic (F & T) traps are designed to handle large volumes of air and condensate. A built-in float mechanism controls the flow of condensate and a thermostatic bellows vents the accumulated air. It is good practice to precede an F & T trap with a sediment strainer and to connect it with a three-valve by-pass. F & T traps handle the large condensate loads generated by piping drips, heating coils and other low-pressure steam points of condensation. For high-pressure steam and industrial applications, bucket traps are used rather than F & T.	3/4"–2" [20–50 mm]

Figure 11.10

Terminal Units Equipment Data Sheets (continued)		
Terminal Units	**Hydronic Cast Iron Convectors and Radiators**	**Means No. 155-630**
Useful Life: 20-40 years	**Typical Use and Consideration**	**Capacity Range**
Cast Iron Tubular Radiator	Cast iron has an advantage not only in durability but in its capability of transferring heat evenly due to its mass. This heat is given up both by radiation and convection. The type of radiator pictured here can be free standing or wall hung. It consists of sectional elements which may be replaced or even added to in the field in the event of leakage or undersizing. A smaller cast iron element enclosed in a steel compartment with inlet and outlet grilles allowing air to pass over the element increases the capacity by convection. Cast iron, like other hydronic convectors or radiators, can be efficiently used for steam or hot water.	Infinite — priced by the section
Useful Life: 15-25 years	**Hydronic Baseboard Radiation Cast Iron and Fin Tube**	**Means No. 155-630**
Baseboard Radiation	Baseboard radiation is extensively used in residential applications, and in some light commercial. Due to cost considerations, fin tube is more popular than the more solid and substantial cast iron. Baseboard radiation is almost exclusively used in forced hot water heating systems, as the ports are too small for gravity flow systems.	Infinite — priced by the linear foot
Useful Life: 10-25 years	**Hydronic Fan Coil Units**	**Means No. 157-150**
Concealed Fan Coil Element	Most commonly used in commercial applications where space consideration precludes the use of direct radiation. With the addition of a drain pan, a forced hot water unit may be used for cooling applications. Steam coils are a heating option and direct expansion coils are a cooling option. Built-in fan speed control is a popular option also. As fan coil capacities increase in size, separate coils for heating, cooling, preheat and/or reheat modules may be added.	1/2–50 tons [2–175 kW]
Useful Life: 7-20 years	**Hydronic Unit Heaters**	**Means No. 157-630**
Unit Heater	A unit heater is simply a hydronic coil within the same enclosure as an air moving fan. These units may discharge horizontally (as pictured) or may project vertically. Unit heaters are primarily used in industrial or commercial applications where noise or air movement is not objectionable. Steam or hot water coils are the available hydronic choice. Electric resistance unit heaters are also available.	14.7 MBH to 520 MBH [4–155 kW]

Figure 11.10 (continued)

Terminal Units Equipment Data Sheets

Terminal Units	Grilles, Registers and Diffusers	Means No. 157-460
Useful Life: 20-40 years	**Typical Use and Consideration**	**Size Range**
Floor Grille	Grilles, registers and diffusers are the supply and return faces used in ductwork systems. Grilles are the least expensive and are generally used for exhaust and return air inlets requiring deflection but having no throttling capability. They are often used in non-critical supply applications as well.	6" x 6" to 48" x 48" [150 mm × 150 mm to 1200 mm × 1200 mm]
Ceiling Register		**Means No. 157-470**
	Registers are used on both supply and return duct openings. A register differs from a grille in that it has a volume control, whereas the grill does not have this throttling feature. Supply registers also have deflection capabilities.	4" x 6" to 48" x 24" [100 mm × 150 mm to 1200 mm × 600 mm]
Supply Diffuser		**Means No. 157-450**
	Supply diffusers distribute the conditioned air in accordance with the specified requirements for volume velocity and direction. A major concern with diffusers is noise criteria. Diffusers have the capability of various flow patterns. Diffusers are available as fixed or adjustable flow. Some diffusers include both supply and return air connections. Refer to manufacturer's data for the many shapes, materials, and finishes that are available for grilles, register, and diffusers.	6" x 6" to 36" x 36" [150 mm × 150 mm to 925 mm × 925 mm]
Useful Life: 20-40 years	**Air Mixing Boxes**	**Means No. 157-425, 480**
Air Mixing Box	These mixing boxes are used to blend warm and cool air from dual duct supply systems to a final preset discharge temperature. This system has the capability of supplying heating or cooling simultaneously for special environmental requirements. Reheat coils and humidification capabilities can also be accommodated. Mixing boxes are used in constant volume systems as well as in variable air volume systems. Damper control is critical for this application.	150–1900 cfm [70 – 900 L/s]
Useful Life: 20-40 years	**Variable Air Volume Boxes**	**Means No. 157-425, 480**
Variable Air Volume Box	The VAV box detailed here incorporates a heating coil, a variable or modulating damper, plus has the capability of utilizing the warm air trapped in the ceiling plenum. This economizer feature is supplemented by the reheat coil when plenum temperatures are not sufficient. The control station air handling unit provides the capacity to vary the primary air flow.	150–1900 cfm [70 – 900 L/s]

Figure 11.10 (continued)

Terminal Units Equipment Data Sheets (continued)		
Terminal Units	**Electric Baseboard Heaters**	**Means No. 168-130**
Useful Life: 10-20 years	**Typical Use and Consideration**	**Capacity Range**
Electric Baseboard Heater	Electric baseboard heat, like most electric heat, is the least expensive type of mechanical heat to install. Operating costs of electric resistance heaters are high.	375-1875 watts
	Major features of this form of electric heat are built-in individual thermostatic controls, individual room thermostats, and the capability of individual metering.	
	This type of heat is used in residential and light commercial applications.	
Useful Life: 10-20 years	**Electric Unit Heaters**	**Means No. 168-130**
Electric Unit Heater	Electric unit heaters consisting of an electric resistance element and a fan within the same enclosure are for residential, commercial and industrial applications.	.75 to 50 kilowatts
	Electric unit heaters can be floor or wall mounted and the heavy duty commercial and industrial models are commonly suspended from overhead.	
	Not susceptible to freezing, these units can be located in extreme areas for localized or spot heating.	
Useful Life: 10-20 years	**Computer Room Units**	**Means No. 157-130**
Computer Room Unit	These dedicated-use heating and cooling units provide conditioned air with temperature and humidity control for the close tolerances required in computer rooms.	
	They may be completely self-contained or may obtain heating or cooling from a remote source.	
	Air distribution may be free blow or ducted. Air may be distributed from above a hung ceiling or from beneath a raised floor.	
	Computer room units are built to much higher standards than conventional air conditioners to give trouble-free operation for their critical function.	

Figure 11.10 (continued)

Chapter Twelve
CONTROLS

Controls are manual or automatic devices used to regulate HVAC systems. A bathroom light switch that also activates an exhaust fan is an example of a manual control; a time clock on an office air conditioner is an automatic control. The simplest controls turn equipment on or off. Advanced controls may allow the equipment to modulate in response to demand.

Types of Operation

The control systems most frequently used today are electric, electronic, pneumatic, direct digital control (DDC), and self-contained control systems. The choice of system often depends on the designer's preference, but also hinges on the owner's requirements, budget, and the technical level of the owner's operating personnel.

Electric Controls

Electric controls are used in residential and small commercial buildings where there are few pieces of equipment to control. Electric controls operate on line or low voltage (see Figure 12.1). Line voltage can be single or three-phase, 60 Hz, 120, 208, 240, 277, or 480 volts ac. Low voltage operates at below 30 volts; usually 24 volts. Electric systems provide control by starting and stopping the supply of electric current and by varying the voltage or current by use of a rheostat or wheatstone bridge circuit. Electric control circuits use mechanical means such as bimetallic strips and bellows to actuate a switch or position a potentiometer; some circuits also incorporate relays and switches.

Electronic Controls

Electronic controls are used in small and medium-sized buildings. Electronic systems operate on low voltages and currents and use solid-state components to amplify signals to perform control tasks such as positioning an actuator or switching a relay (see Figure 12.1). Electronic circuits provide fixed control sequences based on the logic of the solid-state components. Typical characteristics of an electronic system include a low voltage solid-state controller, inputs of 0 – 1V dc, 0 – 10V dc, and 4 – 20 mA, resistance-temperature sensors, thermistor, thermocouples, and outputs of 2 – 10V dc or 4 – 20 mA. Electronic systems use four control modes: two-position, proportional, proportional-integral, and step.

The sensors and output devices used in an electronic system are usually the same type used in a DDC system, which is described below.

Pneumatic Controls

Pneumatic controls are used in medium-sized and large buildings and utilize low pressure (0 – 15 psi) [100 – 200 kPa]* compressed air (see Figure 12.1). The sensor causes a change in air pressure, causing an actuator

Throughout the book, SI metric equivalents are provided for most imperial units. The SI units appear in brackets [ ].

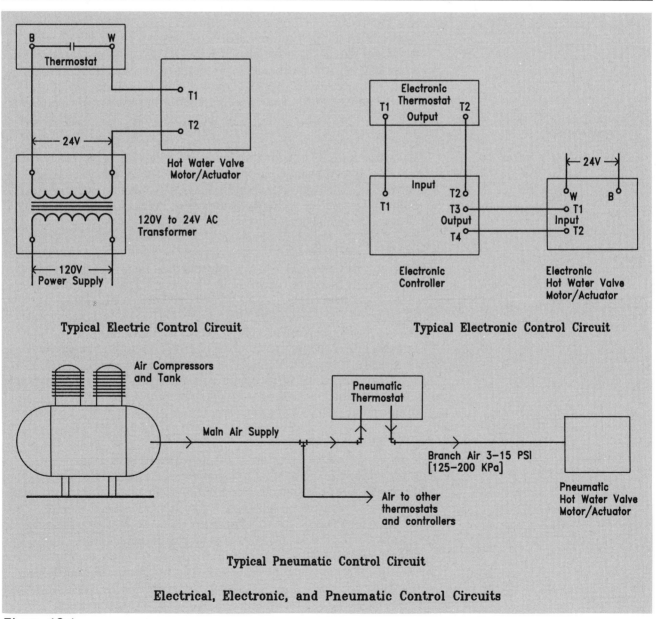

Electrical, Electronic, and Pneumatic Control Circuits

Figure 12.1

or motor to react or move. Air pressure acts against spring-loaded pistons or mechanisms to cause motion in the actuator.

Direct Digital Controls

Direct Digital Controls (DDC) are used in medium-sized to large buildings and in facilities comprised of a number of buildings. DDC systems are microprocessor-based – a signal from a sensor is measured, an algorithmic control routine is performed in a software program within a microprocessor-based controller, and a corrective signal is sent to an actuator or controlled device (see Figure 12.2).

DDC controllers consist of digital and analog inputs and outputs, convertors, a power supply, and software to perform preprogrammed HVAC and energy-management routines. Microprocessor-based controllers can be used as stand-alone controllers that perform a few local tasks. They can also be incorporated in an automated building management system utilizing a personal computer to provide additional enhancement.

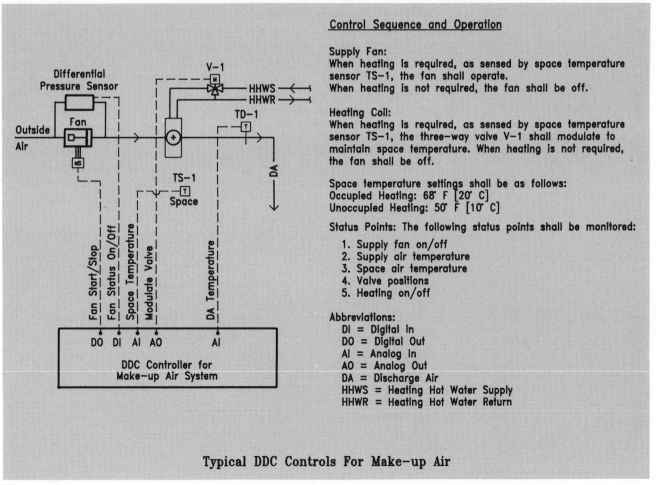

Figure 12.2

A *control loop* consists of a process wherein a *sensor* reads a condition and sends a signal to a controller; the *controller* positions an *actuator* or device to accommodate the sensor's reading. Smaller controllers generally have 8 – 10 control loops; however, some controllers are large enough for 30 – 40 loops.

Controllers in a DDC system are networked together and can share functions and readings. For example, only one outdoor air temperature sensor is required to control the economizer sections of two or more air handling units; alternatively, the occupied/unoccupied times may be programmed in the PC and read by all controllers.

DDC control systems generally utilize electronic sensors and electronic, electric, or pneumatic actuators. The electronic input signals must be converted to digital or analog inputs for the controller to understand; the digital and analog outputs from the controller must be converted to electric or pneumatic signals for the actuators to operate. Digital signals are two-position signals: for example, on/off, start/stop, freezing/not freezing, and open/closed. Analog signals are varying or continuously-changing signals. These signals are used for measuring space temperature or humidity or for positioning a valve at a certain point – for example, the valve is 54% closed, the temperature is 74°F [22 °C], the humidity is 63%.

DDC systems are more expensive than other systems but give very accurate control, can easily incorporate energy-management programs and graphic system displays, and can be tied into most building automation systems such as fire alarms, security, and lighting controls.

Self-contained Controls

Self-contained controls are control systems in which the sensor controller and actuator are part of a single package, such as thermostatic valves. These self-contained controls incorporate sensor, controller, and controlled device in one package. No external source is required. An example of a self-contained control is one in which a gas or fluid in a bulb expands and retracts to operate the controlled device.

Sensors, Controllers, and Actuators

Control systems can be divided into three categories: *sensors* sense the controlled medium; *controllers* read the sensor's signal and convert it to a response signal based on predetermined settings; *actuators* receive the controller's signal and actuate a piece of equipment to cause a change in the controlled medium.

Each process that senses, reads, and responds (sensor – controller – actuator) is also called a control loop. A time clock detects the "on" or "off" time and a thermostat detects temperature – these events activate a motor or a valve. Examples of other control loops are shown in Figure 12.3.

Internal and External Controls

Most pieces of HVAC equipment have *internal* controls. For example, chillers, oil burners, heat pumps, window air conditioners, and cooling towers all come from the factory with prewired controls, which activate the internal parts of the equipment in the proper sequence. Certain safety features are also provided with the equipment; for example, a low water cut-off or stack temperature time delay. Both of these features are designed to ensure safety; the first monitors the water level in the boiler to ensure that it is not too low and the other knows when the fuel being injected has not ignited.

In addition to internal safety controls, manual or automatic *external* controls are added to start and stop the equipment or to regulate its output. A familiar example of an external control is a thermostat in a house. When it measures that the space temperature is below the set point, it automatically sends a signal to the circulating pump to turn on. Meanwhile, the internal controls of the boiler maintain the water temperature.

Zoning

For greater design flexibility and energy economics, an approach known as zoning is commonly used in HVAC systems. By subdividing a building into a series of zones, the set conditions can be maintained properly for a specific space and the equipment need service only the areas that require heating or cooling. Combining zoning with night temperature set-back controls and a time clock is an effective means of achieving occupied and unoccupied building controls.

Although it is common to have many zones within one building, the number of occupied and unoccupied modes allowable in one day is normally limited to two each in any one zone; in other words, the system can switch from occupied to unoccupied mode and back to occupied twice daily.

Figure 12.4 (a & b) illustrates simple two-zone occupied/unoccupied heating systems and operations common to residences, commercial buildings, and institutions. Commercial buildings and institutions may have many more than two zones.

Figure 12.4 (c) illustrates an air-handling system that utilizes an outside air temperature economizer cycle and an occupied/unoccupied time clock to determine the position of the dampers.

Simple Control Loops		
Sensors	**Controllers**	**Controlled Devices**
Thermostat	⟶	VAV damper/motorized valve/zone pump
Pressure sensor	⟶	Fan motor, space heater, bypass valve
High temperature limit	⟶	Oil/gas burner
Duct smoke sensor	⟶	Equipment shutdown, damper motor
Duct freeze thermostat	⟶	Equipment shutdown
Outdoor air sensor	⟶	Boiler start-up
Time clock	⟶	Pump motor, package unit
Humidistat	⟶	Steam valves/humidifiers
Aquastat	⟶	Oil/gas burner, fan motor
Enthalpy sensor	⟶	Return, exhaust, and intake dampers

Figure 12.3

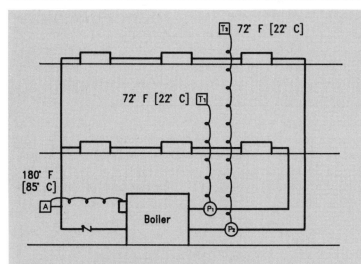

(A) Two-Zone Heating System

T_1 activates Pump P and maintains 72° F [22° C].
T_2 activates Pump P_1 and maintains 72° F [22° C].
A activates burner when return water falls below 180° F [85° C].

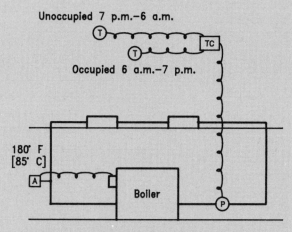

(B) Occupied/Unoccupied (Night Setback) Heating System

72° F [22° C] during the day (6 a.m.–7 p.m.).
60° F [10° C] during the night (7 p.m.–6 a.m.).
A activates burner.

KEY

T = Thermostat   ⋀⋁⋀ = Motorized Damper
A = Aquastat   M = Damper Motor
TC = Time Clock

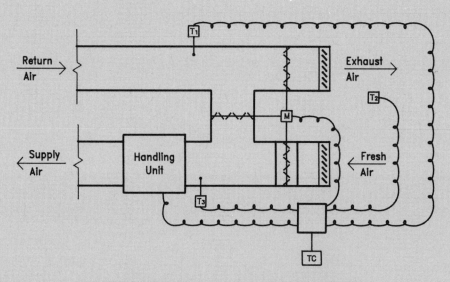

(C) Modulating Damper and Air Handling Unit

T_1 senses return air temperature and determines if heating or cooling is required.
T_2 senses outdoor temperature to determine optimum position of dampers, operated by damper motor (M).
T_3 senses mixed air temperature to determine if air handling unit heating or cooling is required.
The time clock determines the occupied or unoccupied mode. When the system is in the unoccupied mode, the damper motor closes the fresh air and exhaust dampers.

Basic Control Diagrams

Figure 12.4

Control Diagrams and Sequences

Creating control diagrams is an essential and valuable task in designing and laying out a control system. There are four types of control diagrams: oneline schematic diagrams, wiring diagrams, ladder diagrams, and algorithms.

Oneline Schematic Diagrams

In a *oneline schematic diagram*, the HVAC system is laid out in its simplest format, including major pieces of equipment, control components, wiring, important pipes, control valves, control dampers, and controller. Figure 12.5 illustrates a control oneline schematic diagram for a typical air handling unit. Figure 12.6 illustrates a control oneline schematic diagram for a steam-to-water heat exchanger and associated pumps.

Oneline schematic and wiring diagrams should be generated for all types of control systems because they give the designer, installer, and operator a clear picture of the intent of the control system. These diagrams should clearly show the different types of control components, DDC wiring, pneumatic tubing and actuators, electric motors, and so on, as well as the number and size of wires required for each portion of the system.

Schematic diagrams are normally laid out for different functions or small portions of a larger system, such as for the air handling unit or the heat exchanger in Figures 12.5 and 12.6. These systems may be just part of a complex system in a large building. Once all necessary controllers have been determined, an overall schematic of how these controllers interconnect can be generated. Figure 12.7 illustrates how different stand-alone controllers in a DDC system are interconnected via a personal computer.

Wiring Diagrams

Wiring diagrams outline all wires and connection points (see Figure 12.8). Wiring diagrams are normally found in manufacturers' literature and installation drawings and are most useful to the installer of the system. These diagrams should be generated for electric, electronic, and DDC systems. Wiring diagrams for pneumatic systems should include all control components, tubing connections, and required air pressures. Wiring diagrams should be first generated by the manufacturer with a description of operation and then analyzed by the designer for correctness and intent.

Ladder Diagrams

Ladder diagrams show the logic and connection configuration of electric control systems. Ladder diagrams are essential for designers and installers as they outline the control components required for a system and the interrelationship of the control components.

Figure 12.9 illustrates a typical electric ladder diagram for an economizer cycle.

Algorithms

Algorithms are logic diagrams that determine the correct sequence of events necessary for the system to operate. Algorithms are necessary for the correct programming of software in microprocessor-based systems. Figure 12.10 illustrates a simple algorithm for a DDC controller for a simple single zone heating system.

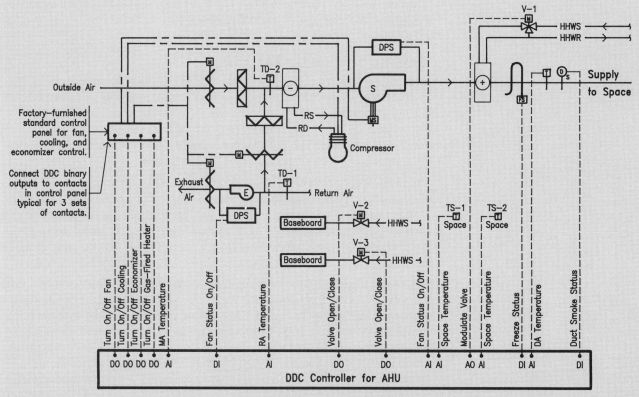

Oneline Schematic (DDC) for Air Handling Unit

Control Sequence and Operation

Supply and Exhaust Fans:
During the occupied mode the supply and exhaust fans shall operate continuously in both heating and cooling conditions. The supply fan shall supply a constant cfm. The exhaust fan shall exhaust a minimum cfm of air during normal occupied times, and shall exhaust a maximum cfm during the economizer cycle as outlined by the economizer paragraph below. The exhaust fan shall be off during the occupied mode.

Supply, return, and exhaust air dampers:
During the occupied mode the OA and EA dampers shall be at minimum cfm open positions and the RA damper shall be fully open. During the unoccupied mode the OA and EA dampers shall be closed and the RA damper shall be fully open. During the economizer cycle the OA and EA dampers shall be fully open and the RA damper shall be fully closed.

Cooling coil and compressor:
When cooling is required, as sensed by RA duct temperature sensor TD-1, the compressor and the condenser fan shall operate to activate the cooling coil and maintain required space temperature in the occupied mode. When cooling is not required, when heating is required, or during the unoccupied mode, the compressor and the condenser fan shall be off.

Heating coil:
The heating coil shall be used to temper air for ventilation. The ventilation air temperature shall be 68°F [20 °C]. Ventilation shall be required when cooling is not required, or when heating is required. When the RA temperature is below 68°F [20 °C], as sensed by duct temperature sensor TD-2, and the unit is not in the cooling mode, the three-way valve V-1 shall modulate to maintain the OA temperature at 68°F [20 °C], as sensed by duct temperature sensor TD-3.

Notes:
1. Occupied temperature settings shall be as follows:
   Heating = 68°F [20 °C]
   Cooling = 78°F [26 °C]
2. Unoccupied temperature settings shall be as follows:
   Heating = 50°F [10 °C]
   Cooling = No Setting

Baseboard heating for offices and pub:
There are two separate baseboard systems in each space. When heating is required, as sensed by space temperature sensor TS-1 (TS-2), the two-way valve V-2 (V-3) shall open fully to maintain required space temperature. When required space temperature has been achieved, valve V-2 (V-3) shall close. The baseboard heating shall operate in the unoccupied mode to maintain the unoccupied setback temperature of 55°F [13 °C].

Economizer cycle:
When cooling is required and the OA temperature is below 65°F [18 °C], as sensed by the outdoor air temperature sensor at the controller, the economizer cycle shall be activated. The OA and EA dampers shall open fully, the OA damper shall close, and the exhaust fan shall exhaust a maximum cfm as outlined in the schedules. When the OA temperature rises above 65°F [18 °C], the economizer cycle shall end and the dampers and fans shall return to normal operating positions.

Safeties, cutouts, and diagnostics:
Freeze protection: When the OA temperature, as sensed by freeze stat F3-4, drops below 35°F [2 °C] the supply and exhaust fans shall shut down. The OA and EA dampers shall close fully, the RA damper shall open fully, valve V-1 shall open fully to the coil, and the heating circulating pumps shall be on. An emergency signal shall be sent to the central control computer and the printer shall print out an emergency message.

Smoke detection: When the duct smoke detector senses smoke in the supply air, supply and exhaust fans shall shut down immediately and an emergency signal shall be sent to the central control computer and the printer shall print out an emergency message. The fire alarm system shall be activated.

Cooling system status:
When cooling is required and TD-3 senses temperatures above 80°F [26 °C], indicating that the cooling system is not working, the unit shall be shut down and a signal shall be sent to the central computer indicating that the compressor or condensor is not operating and that the unit has been shut down.

Figure 12.5

Control Sequence and Operation

Pumps:
Pumps P-1, P-1A, P-2, and P-2A shall be on when the outside temperature, as sensed by outside air temperature sensor TD-3, is below 65°F [18 °C] from October through May 15th.

Pumps P-1 and P-1A shall serve the heating coils and shall run in a lead/lag alternating fashion.

Pumps P-2 and P-2A shall serve the baseboard heat and the unit heaters and shall run in a lead/lag alternating fashion.

Hot water temperature:
When the leaving hot water supply requires heating, as determined by the outdoor air temperature and the leaving hot water return temperature, the steam valve V-1 shall modulate to maintain the leaving hot water supply temperature at the correct setting as follows.

When the outdoor air temperature is 64°F [18 °C] the supply water temperature shall be 100°F [37 °C]. When the outdoor air temperature is 6°F [–14 °C] the supply water temperature shall be 190F [88 °C]. The supply water temperature shall be set proportionately to the outdoor air temperature in relation to the above maximum and minimum values; for example, at 35°F [2 °C] outdoor air temperature the supply water temperature shall be 145°F [63 °C].

Pressure relief bypass valve:
When the differential pressure sensor DPS-ER, across Pumps P-2 and P-2A, senses an increase in system pressure, the pressure relief bypass valve V-2 (V-3) shall open proportionately to relieve the increase in pressure. The bypass valve shall be sufficient to allow 100% flow to bypass the system.

Exhaust fan EF-8:
When the steam room temperature rises above 95°F [35 °C], as sensed by space temperature sensor TS-4, the exhaust fan EF-2 shall be on. When the room temperature is below 85°F [30 °C] the fan shall be off.

System status and diagnostics:
The following status points shall be monitored:
1. Pump P-1 On/Off
2. Pump P-1A On/Off
3. Pump P-2 On/Off
4. Pump P-2A On/Off
5. Pump P-1 Start/Stop
6. Pump P-1A Start/Stop
7. Pump P-2 Start/Stop
8. Pump P-2A Start/Stop
9. Supply water temperature
10. Return water temperature
11. Outside air temperature
12. Room air temperature
13. Exhaust fan status On/Off
14. Pressure relief bypass valve V-2 position
15. Pressure relief bypass valve V-3 position
16. Differential pressure across Pumps P-1/P-1A
17. Differential pressure across pumps P-2/P-2A

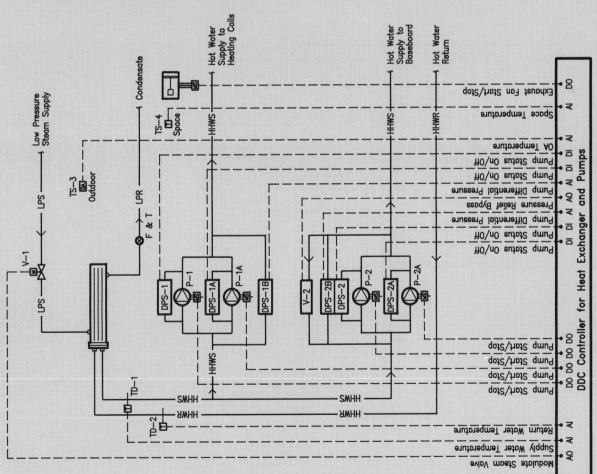

Oneline Schematic (DDC) for Heat Exchanger and Pumps

Figure 12.6

Control Sequences

Control sequences describe the intent and required operation of the control system. The sequence expresses, in words, how the system should operate to achieve the desired results. Control sequences should be established for all types of control systems no matter how small the system is. They are a valuable reference when complications occur in control diagrams.

The *settings chart* works in conjunction with the control sequences. The settings chart summarizes, in tabular form, the settings of each piece of equipment in occupied and unoccupied, summer and winter modes, and when these modes should be in effect. Figure 12.11 depicts a simple settings chart.

Energy Management Systems

Energy management systems (EMS) are used in commercial and institutional buildings and facilities. They are usually operated with a computer and can generate impressive financial paybacks that will offset the installation cost. EMS will monitor and control a building's HVAC and electrical systems so they use the least amount of energy possible. Chapter 4 discusses the overall benefits of EMS as well as the methods to use in calculating paybacks for the use of such systems.

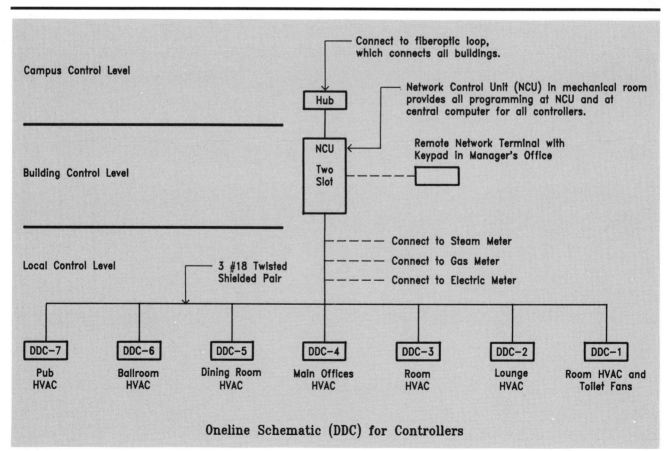

Figure 12.7

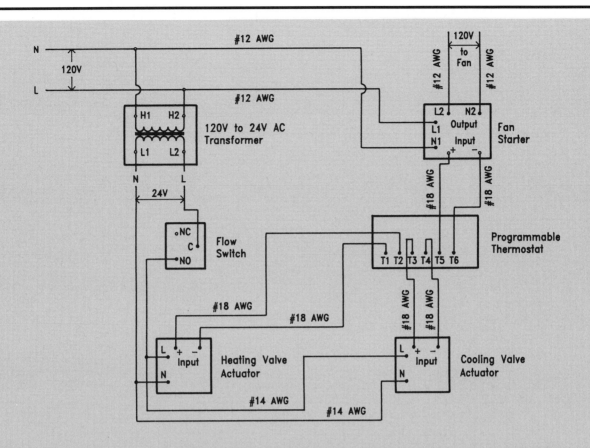

Simple Fan Coil Unit Wiring Diagram

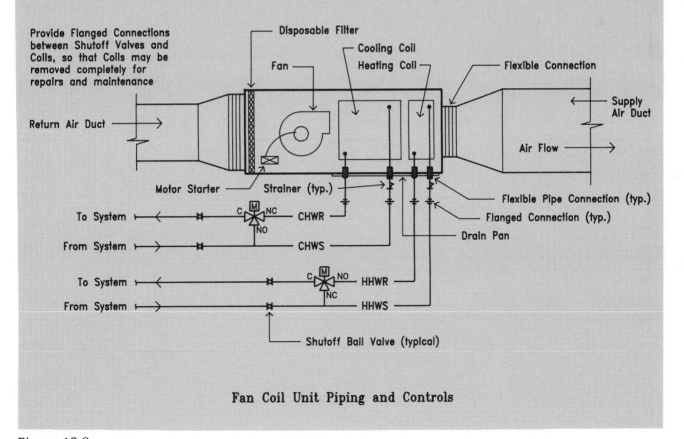

Fan Coil Unit Piping and Controls

Figure 12.8

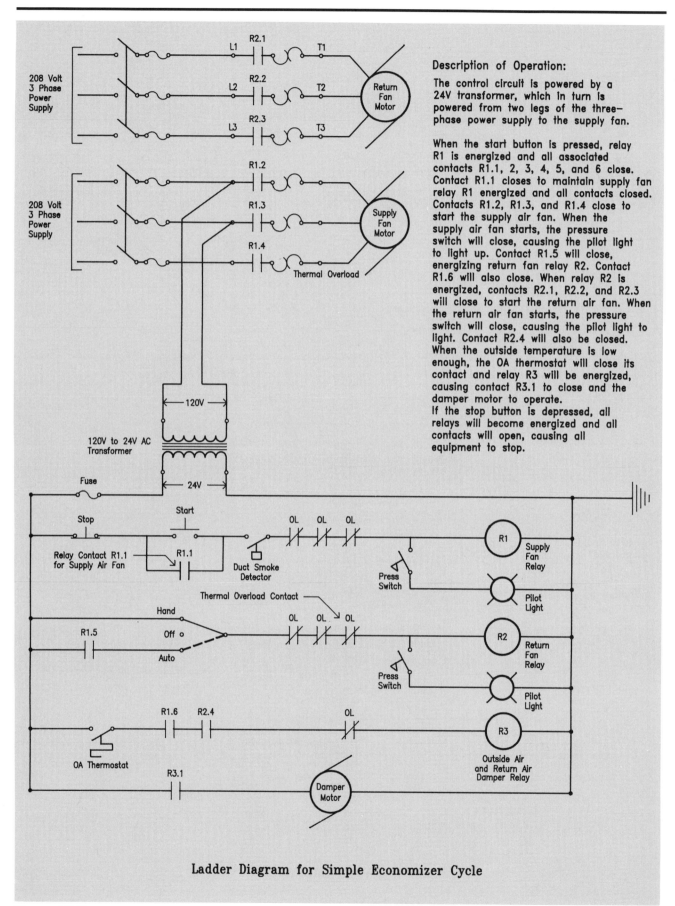

Ladder Diagram for Simple Economizer Cycle

Figure 12.9

The energy consumed in a typical office building can be divided as follows:
- Electricity, 67%:
  - Cooling systems (27%)
  - Lighting systems (22%)
  - Heating (8%)
  - Other (10%)
- Oil, gas, and purchased steam or hot water (33%)

Energy management systems incorporate programs that address all of these areas of energy consumption.

Careful control of HVAC systems can result in significant energy savings. For example, electricity can be saved by reducing pump or fan operation time and intensity. Reduction of outside air intake will save on the energy used by cooling and heating equipment. Economizer sections on air handling units can use outside air for free cooling, which reduces the load

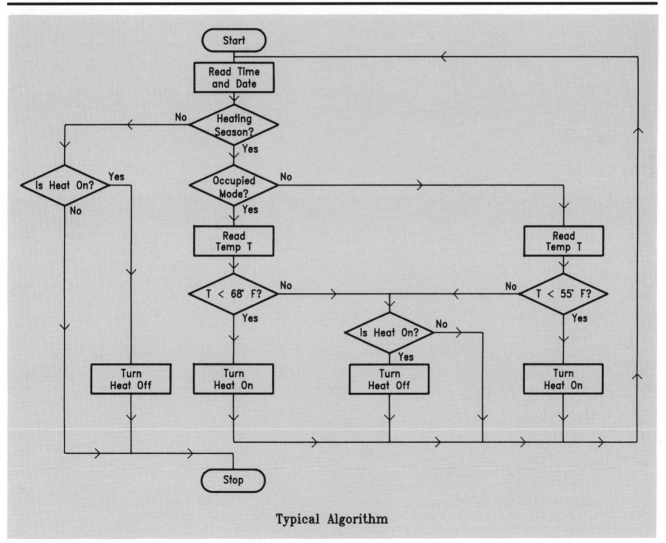

Figure 12.10

on the cooling equipment. Outdoor temperature reset programs for heating and cooling systems will lower or raise supply temperatures of heating and chilled water, reducing the energy required by boilers and chillers. This avoids overcooling and overcycling of the equipment, leading to longer life and fewer equipment repairs.

Most electrical supply companies promote rebate incentive programs to reduce the demand for electrical power in buildings and facilities. Replacing inefficient light fixtures with new efficient ones, reducing lighting levels,

Temperature and Control Setting Chart

Unit Time Shall Be Adjustable	Winter (Oct 1-May 15)		Summer (May 15-Oct 1)	
	Occupied 6am-12pm**	Unoccupied 6am-12pm**	Occupied 6am-12pm**	Unoccupied 6am-12pm**
AHU Lobby	68° [20 °C]	Off	75° [24 °C]	Off
AHU Dining Room	68° [20 °C]	50° [10 °C]	75° [24 °C]	Off
AHU Kitchen	68° [20 °C]	Off	75° [24 °C]	Off
AHU Bar Room	68° [20 °C]	Off	75° [24 °C]	Off
AHU Dining Room	68° [20 °C]	Off	75° [24 °C]	Off
AHU Ballroom 1	68° [20 °C]	50° [10 °C]	75° [24 °C]	Off
AHU Ballroom 2	68° [20 °C]	50° [10 °C]	75° [24 °C]	Off
Baseboard Units	68° [20 °C]	50° [10 °C]	Off	Off
UH-Storage 1	* 50° [10 °C]	50° [10 °C]	Off	Off
UH-Storage 2	* 50° [10 °C]	50° [10 °C]	Off	Off
UH-Mechanical	* 50° [10 °C]	50° [10 °C]	Off	Off
UH-Pantry	68° [20 °C]	50° [10 °C]	Off	Off
UH-Garbage	* 50° [10 °C]	50° [10 °C]	Off	Off
UH-Electric	* 50° [10 °C]	50° [10 °C]	Off	Off
EF-Toilet Rooms	On	Off	On	Off
EF-Storage 1	* 85° [30 °C] On			
EF-Storage 2	* 85° [30 °C] On			
EF-Mechanical	* 85° [30 °C] On			
EF-Dish Room	ETR—Tied to Dishwasher			
EF-Electrical	* 85° [30 °C] On			
EF-Bar Room	On	Off	On	Off
Dining Room	On	Off	On	Off
Dining Room	Wall Switch			
EF-Kitchen	* 85° [30 °C] On			
Heat Exch.	On	On	Off	Off
P1.1A	Alternate On	Alternate On	Off	Off
P2.2A	Alternate On	Alternate On	Off	Off

All numbers shown in brackets are metric equivalents.
*Provide local manual override.
** Manager shall have override of occupied/unoccupied settings at the NCU terminal in the manager's office.
Note: Each controller or thermostat shall have a deadband of 4°F [2 °C] minimum between heating and cooling.

Figure 12.11

and turning lights off during unoccupied times are all examples of reducing electrical energy consumption and demand.

Despite the effectiveness of energy management programs, great care must be taken to apply them correctly – a poorly-applied program could have adverse affects and even damage equipment. The following sections describe some of the energy management programs in use today.

Deadband Programs

Deadband programs set the heating and cooling design temperatures so that they are separated by a "deadband" in which neither heating nor cooling is required. For example, in an office space set at 65°F [18 °C] for heating and 78°F [25 °C] for cooling, there is a 13°F [7 °C] (78−65) deadband. The temperature in the deadband is allowed to drift and no energy, except fan power for fresh air, is consumed by the HVAC system.

Duty-cycling Programs

Duty-cycling is a method of shutting down equipment that might otherwise run continuously for a portion of each hour. Fans and fan coil units can often be shut down for 15 minutes per hour with little detrimental effect and only a modest swing in space temperature. By alternating the equipment, the peak electrical load is lowered and less energy is used.

Some precautions are recommended when applying duty cycling. The fan should have a minimum off time, as constantly switching the fan on and off will overheat the motor and cause damage. Duty cycling should not be applied to packaged air conditioners, because the packaged controls are not set up for this program and damage to related controls or portions of the air conditioner could result. In the spring and fall, adequate fresh air must be supplied to the building during long periods of shutdown; this may be compensated by supplying larger volumes of fresh air when the fan is running.

Demand-limiting Programs

Electric bills are divided into three categories:
- A charge for each *kilowatt-hour* consumed during the billing period. This charge is normally on a declining block rate system.
- A *fuel charge*, which is also based on the kilowatt-hour consumption and is a constant charge for each kilowatt-hour.
- A *demand charge*, which is based on the building's maximum draw (demand) in kilowatts on the power supply during the billing period. The more demand each consumer has on the system, the more strain it causes on the utility company's generating equipment.

The demand charge may be further increased by a "ratchet" clause. Many utilities have a ratchet clause wherein the utility charges the consumer for the maximum demand drawn for the preceeding eleven months on every bill, regardless of the lower demand drawn in a subsequent billing period. For this reason, it is in the consumer's best interest to lower demand as much as possible.

A *demand-limiting program* monitors a building's current draw. When that draw begins to reach a predetermined maximum, which usually occurs during the day, the system will begin to switch off or "shed" loads (see Figure 12.12). Loads should be shed in order of importance – for example, the temperature of the water coolers should be increased before the temperature of the executive offices is increased. The loads that have been shut down or reduced should be restored in reverse order; that is, last off should be first on.

Unoccupied-period Programs

A typical office building has set occupied and unoccupied periods; for instance, occupied from 8 am to 6 pm and unoccupied from 6 pm to 8 am. Office buildings are occupied by the majority of the workers for only about 48 hours per week. For the rest of the time, the building should be set to an unoccupied set-back mode – turning down the heat to 50°F [10 °C] or 55°F [13 °C] and turning off the air conditioning. Employees can request to have their area heated or cooled in the occupied mode if they wish to work late or on weekends.

Unoccupied-period programs make the application of zoning and override capabilities of a control system worthwhile. Most override modes of operation last for three hours; the system then returns to the unoccupied mode or returns to override mode for another three hours.

Optimum Start/Stop Programs

HVAC systems in commercial buildings typically need a warm-up time in the winter and a cool-down time in the summer before the occupants arrive. On a design heating day a warm-up time of two hours is not unusual; however, a design day occurs less than 5% of the time. The

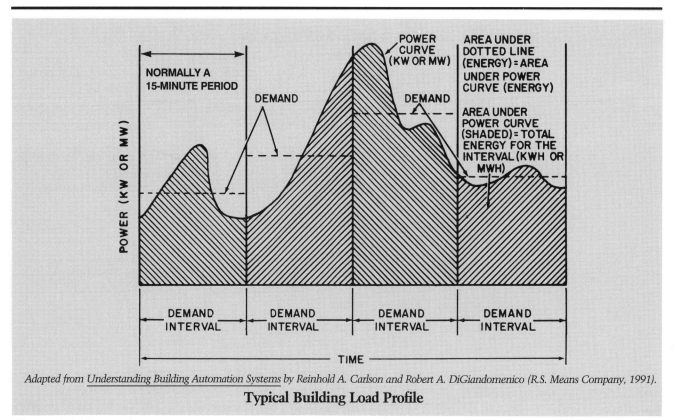

Adapted from Understanding Building Automation Systems by Reinhold A. Carlson and Robert A. DiGiandomenico (R.S. Means Company, 1991).

Typical Building Load Profile

Figure 12.12

reverse is true in the summer. On intermediate days in the winter and summer, when the temperatures are 10°F [−12 °C] or more above or below the design temperature and the indoor temperatures are also more favorable, warm-up and cool-down periods can be reduced. Thus the energy required to run the equipment will also be reduced.

An *optimum start/stop program* monitors outdoor and indoor temperatures and turns the equipment on at the optimum time to save energy. Optimum stop time is often incorporated to shut down the equipment before the occupants leave but not so early that the building becomes uncomfortable before the unoccupied mode of operation.

Night-purge Programs

The temperature of the air often drops quickly one or two hours before sunrise, and during many summer nights the outdoor temperature drops when the sun sets. The temperature of the outdoor air is often low enough that the air handling units can purge the building of all the stored heat that accumulated during the working day. The *night-purge program* monitors the indoor and outdoor temperature and, if conditions are favorable, opens the fresh-air and exhaust-air dampers fully and purges the building of all excess heat. This program often works so well that mechanical air conditioning may not be required until halfway through the working day.

Enthalpy Programs

The *enthalpy program* works in conjunction with an economizer control. In an economizer section, temperature sensors monitor inside and outside air and allow the lower-temperature air to pass over the cooling coil. This works well most of the time; however, in cases when the outside air is very humid, it is better to pass the return air over the cooling coil.

The enthalpy program requires temperature and humidity sensors in both the return air duct and outdoors. The program monitors all sensors, calculates the enthalpy (or total energy) of each airstream, and positions the dampers accordingly.

Control of Building Systems

Building automation systems (BAS) go one step further than DDC and EMS systems because they can control and monitor all of the building's automated systems from a single computer. Control systems for HVAC, fire alarms, security, card readers, lighting, and energy management can become one integrated building system. A system like this requires only one or two operators to monitor the building's systems, rather than one or two operators per system.

A typical BAS is illustrated in Figure 12.13. Note the additional printer for emergency printouts. Both BAS and EMS systems can be accessed by remote computers or touch-tone telephones via a modem.

Buildings with complete control and communication systems, such as infrared detectors to turn lights on and off when rooms are occupied or unoccupied, and cable, computer, satellite, and telecommunication hookups with fully automated operations, are referred to as "smart" buildings. A selection of typical smart controls for automated buildings is outlined in Figure 12.14.

Equipment Data Sheets

The data sheets for controls (Figure 12.15) summarize the useful life, typical uses, capacity range, advantages, disadvantages, and special considerations for each device. These sheets can be used in the selection of each piece of equipment. A Means line number, where applicable, is provided. Costs can be determined by looking up the line number in the current edition of *Means Mechanical Cost Data*.

Summary

Controls are the most cost-effective way to manage energy. They can represent 3–15% of the cost of an HVAC system but, when properly designed, can save at least 30% of operations costs. In small residential systems, a simple occupied/unoccupied heating thermostat will be effective to control the heating boiler. In larger buildings with many pieces of equipment, control systems must be designed with the use of logic diagrams and a hierarchy of control devices. Control systems are the "brains" of an HVAC system and have an important effect on the overall success of any system.

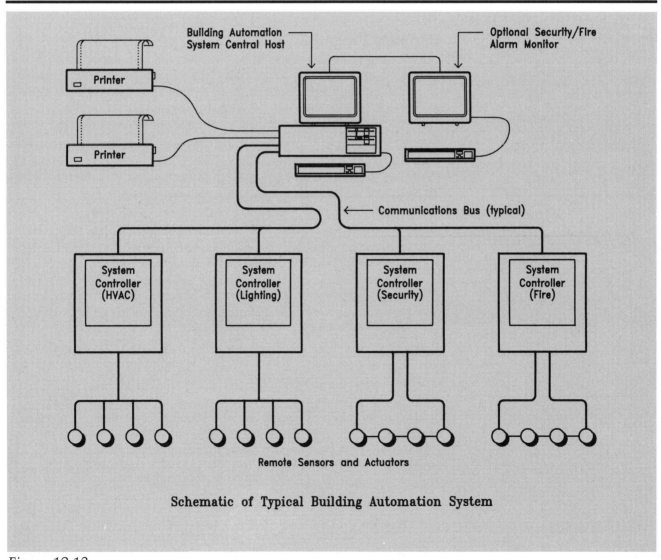

Figure 12.13

Typical Controls in "Smart" Buildings

Control System
  Time shift load response (alternates equipment operation to avoid simultaneous peaks)
  Automatic temperature response for individual spaces
  Energy management control system for entire system

Heating, Ventilating, and Air Conditioning
  Zoning
  Heat reclamation
  Off-hour energy storage

Electrical
  Lighting
    High voltage
    Photocell actuators
    Infrared switches
    Delayed time ballast
  Power
    Flexible grid for tenants
  Emergency systems
    Fire alarm
    Emergency lighting
    Smoke evacuation
    Security
    Fire detection

Communications
  Telephone
    Re-routing call
    Intercom/conference
    Modem/data bank/EMCS access
    Automatic billing
  Computer access for tenants
    Data processing
    Word processing
    Data base
  Fiber optics
  Satellite communications
  Pneumatic delivery
  Cable communications

Transportation
  Elevators
    Voice synthesizers
    Variable mode control

Building Services
  Trash removal
    Vacuum system
    Compactors
  Cleaning
    Window system equipped for wash down

Figure 12.14

Typical Controls in "Smart" Buildings (continued)

Technical Support
- Computer programmers
- Secretarial pool
- Central library and information center
- Mail delivery
- Copy services
- Computer software security
- Banking services

Executive Support
- Dining room
- Meeting rooms
- Transportation services
- Messenger services
- Valet services
- On-call meal and office services

Building Organization
- Scheduled spaces
  - Meetings
  - Off-hours use
- Orientation
  - Sun
  - Traffic

Figure 12.14 (continued)

Control Equipment Data Sheets

Controls	Outdoor Reset	Means Number 157-420
Useful Life 15-25 years	**Typical Use and Consideration**	**Capacity Range** 0-70°F/210-100°F
	The outdoor sensor (bulb) for this control is mounted on the building wall out of direct sunlight or is provided with a sun shield. When there is a drop in outdoor temperature, this control will raise (reset) the boiler water operating temperature. Conversely, if there is a rise in outdoor temperature, the boiler operating temperature will fall.	
Controls	**Thermostatic Heating Control Valves (Non-Electric)**	**Means Number** 156-240
Useful Life 10-20 years	**Typical Use and Consideration**	**Size Range** 1/2" to 1-1/4"
	These non-electric, self-contained valves can be controlled by a thermostat via a capillary tube, or may have a built-in thermostat. A dial on the valve handle allows for temperature adjustment.	
Controls	**Controllers**	**Means Number** 157-420
Useful Life 10-20 years	**Typical Use and Consideration**	**Capacity Range** Various
	Compares measured quantity (temperature, humidity, pressure) with set point and sends or halts an activating signal. Controllers act like switches. When "on call" they make or break contact, depending on the controlled device requirements. Many controllers are integral with the sensing device. Most thermostats, for example, contain the controller as part of the thermostat unit. Coordinate controllers with control manufacturer and control system.	
Controls	**Gauges and Thermometers**	**Means Number** 157-240
Useful Life 10-20 years	**Typical Use and Consideration**	**Capacity Range** 40-240°F 30"—0-30 lbs.
	Thermometers are installed in equipment, piping, and ductwork to read water and air temperatures. Thermometers are provided as dial type, in diameters of 2", 4", or 6", or as stem type, 5", 7", or 12" long. Thermometers and gauges can be supplied as straight or angle configuration. For piping installation, thermometers are threaded into a separable socket or well so that the instrument may be removed without draining the pipe line.	

Figure 12.15

Control Equipment Data Sheets (continued)		
Controls	Gauges and Thermometers	**Means Number 157-240**
Useful Life 10-20 years	Typical Use and Consideration	**Capacity Range 40-240°F 30"—0-30 lbs.**
	Gauges are installed in pipe lines and equipment to read pressure, vacuum, or both (compound gauge). A gauge cock is installed at the inlet to allow removal of the gauge without losing line pressure. Steam gauges are normally provided with a pigtail or siphon to protect the working parts from prolonged contact with the steam. Gauges in lines subject to pulsation or vibration should be installed with pressure snubbers.	
Controls	Temperature Control Devices	**Means Number 157-420**
Thermostat	The most common temperature control instrument is the space thermostat. Thermostats sense temperature, and at a variation from their set point, transmit a signal or command to a piece of mechanical equipment to turn on or off. Thermostats operate burners (gas or oil), pumps, unit heaters, fan coils, valves, dampers, etc. Some of the options available when selecting a thermostat are: line or low voltage; pneumatic; indicating or non-indicating; occupied/unoccupied; night set-back locking enclosures; or built-in clock.	
Aquastat	Aquastats may be surface-mounted (strap-on), remote bulb, or inserted into the piping or equipment. They sense water temperature and, at a variation from the set point, send a signal to the controlled equipment. Aquastats are used primarily as limit or regulating controls on boilers, condensate return lines, pumps, converters, etc.	
Humidistat	Humidistats control the humidity in a conditioned space by sensing the moisture in the space or in the air supply duct. When the moisture content varies from the set point, a signal is sent to operate the humidifier or dehumidifier.	
	Time clocks control the hours of operation for designated periods. They are available in 24-hour or seven day arrangements. Holiday and weekend skips are an optional feature of time clocks, as are battery backups.	
	Other control devices such as sensors, which read certain conditions, and relays, which transfer a control signal from a controller of one voltage to an operator of another voltage, are used throughout temperature control systems. Switches are also available to convert electrical/pneumatic signals.	

Figure 12.15 (continued)

Control Equipment Data Sheets (continued)		
Controls	**Automatic Control Valves**	**Means Number 157-420**
Useful Life 10-20 years	**Typical Use and Consideration**	**Capacity Range 1/2" - 6"**
Motorized Zone Valve 3 Way Valve Globe Valve	Automatic control valves control the flow in hydronic circuits. These valves may have electric, pneumatic, or magnetic (solenoid) operators. Valve body determination is similar to manual valves, i.e., gate type for open and closed functions, and globe type for modulating or throttling. Motor-operated valves may be used to control individual components in a system or to separate various piping loops into zones.	

Figure 12.15 (continued)

Problems

12.1 For each of the following determine if it is a sensor, controller or actuator.
   a. Room temperature sensor
   b. Thermostat
   c. Bellows on a pneumatic damper
   d. Humidistat
   e. DDC controller
   f. Time clock
   g. Automatic damper
   h. Pressure sensor

12.2 For the fan coil shown in Figure 12.8 write a control sequence. Include occupied and unoccupied periods, summer and winter.

12.3 For the hot water distribution system shown in Figure 7.3, show the location of temperature sensors and write the control sequence.

12.4 Given a system such as that shown in Figure 7.7, how would the control system achieve each of the following? Indicate where control points would be required and what equipment they would be programmed to control.
   a. Deadband programs
   b. Duty cycle
   c. Demand limiting
   d. Optimum start/stop

12.5 Make a list of ten "smart" building devices that are currently available.

Chapter Thirteen

ACCESSORIES

Accessories are necessary adjuncts to HVAC equipment operation and performance. Flues, chimneys, motor starters, draft inducers, hangers, supports, insulation, noise and vibration reducers, and air cleaning and filtration equipment are some of the important accessories used in HVAC systems.

Flues and Chimneys

All combustion systems produce hot gases, which are vented to the outside above the roof of a building through a system of flues and chimneys remotely located from air intake louvers and enclosed in fire-rated shafts, typically rated for two hours of fire resistance. The inside chamber that conducts the gases upward is called the *flue*. The entire structural assembly is called the *chimney*.

Chimneys are made from many materials – brick, clay tile, concrete, single- or double-wall steel, or aluminum – and come in standard sizes from 4" to 48" [10.2 – 122 cm].* Local codes determine the materials permitted for a given use. Since combustion gases move at relatively low velocities (20 to 40 feet per second) [6.1 – 12.2 m/s], the surface roughness of the chimney does not significantly affect the flow of gas. Flue diameters are relatively unaffected by the materials used for the flue.

The main principle of a chimney is that the hot combusted gases rise relative to colder outside air. The main factors that affect chimney size are the temperature, pressure, and velocity of the gas being conveyed.

Gas Temperature

As the temperature of the flue gas rises, its density relative to ambient air decreases. The gas is lighter and more buoyant, producing a higher pressure, or draft, to drive the gases up the chimney. Hotter flue gases produce better draft and require smaller chimneys.

Pressure

The following formula is used to calculate the pressure that acts to push air up a chimney:

---

*Throughout the book, SI metric equivalents are provided for most imperial units. The SI units appear in brackets [ ].

$$P_d = 0.255 P_{(atm)} L \left(\frac{1}{T_a} - \frac{1}{T_f}\right) \left[0.044 P_{atm} L \left(\frac{1}{T_a} - \frac{1}{T_f}\right)\right]$$

P_d = pressure in inches of water [kPa] at the base of the chimney, which is theoretically available to drive the flue gas

$P_{(atm)}$ = atmospheric pressure expressed in inches H_g (typically 29.92 inches H_g) [101.3 kPa]

L = height of chimney (feet) [meters]

T_a = outside air temperature (boiler room temperature if conservative) ($°F_{(abs)}$) [K]

T_f = average flue gas temperature ($°F_{(abs)}$) [K]

$F_{(abs)}$ = temperature °F + 460 [°C + 273]

Common flue gas properties for chimneys are shown below.

Fuel	Average Temperature Rise (°F) [°C]	Approximate Value ($T_f - °F_{(abs)}$) [$°C_{(abs)} = K$]	Mass lbs. Gas per 1,000 Btu) [kg/kWh]
Gas (no draft hood)	300–400 [147–202]	800 [442]	1.1 (1.6 w/hood) [1.7 (2.5)]
Oil	450–550 [230–285]	950 [525]	1.2 [1.8]
Incinerators	1,300–1,400 [697–752]	1,800 [992]	1.6 [2.5]

Velocity

As the velocity of the flue gas increases, the required diameter of the flue decreases. This is expressed as follows:

$$V = \frac{HMT_f}{26 P_{(atm)} D^2} \left[\frac{HMT_F}{.95 P_{(atm)} D^2}\right]$$

$$D^2 = \frac{HMT_f}{26 P_{(atm)} V} \left[\frac{HMT_F}{.95 P_{(atm)} V}\right]$$

V = velocity of flue gas (ft./sec.) [m/s]
H = heat input (MBH) [kW]
M = lbs. mass gas in flue per 1,000 Btu/hr. [kg/kWh]
T_f = average flue gas temperature $°F_{(abs)}$ [$°C_{(abs)} = K$]
$P_{(atm)}$ = atmospheric pressure (in H_g) [kPa]
D = flue inside diameter (in.) [cm]

A velocity of 20 feet per second [6.1 m/s] is a conservative estimate used for typical installations.

Always slightly increase chimney size to account for start-up flue temperatures, which are colder than during actual operation. Use the equivalent diameter charts in duct design to select a rectangular flue size of equivalent circular diameter.

Example

Find the diameter of a flue for a boiler with an input of 1,500,000 Btu/hour [440 kW] burning oil.

H = 1,500 MBH [440 kW]
M = 1.2 [1.8 kg/kWh]
T_g = 950°$F_{(abs)}$ [525 °C]
P = 29.92 in H_g [101.3 kPa]

$$V = 20 \text{ ft./sec (minimum/conservative) [6.1 m/s]}$$

$$D_2 = \frac{1{,}500 \times 1.2 \times 950}{26 \times 29.92 \times 20} = 109.9 \left[\frac{440 \times 1.8 \times 525}{.95 \times 101.3 \times 6.1} = 708 \right]$$

$$D = 10.5" \text{ [26.6 cm]}$$

Therefore, use a 12" [30 cm] diameter flue.

Lining

Masonry chimneys should be lined with tile or metal to prevent deterioration of the mortar by acids; sulfuric acid, for example, forms with water vapor from combustion waste gases such as sulphur dioxide. For this reason, special care must be used in selecting chimneys for fuels with a high sulphur content.

Ventilation

Metal flues for gas systems typically require a type B vent, which is a double-wall metal vent with insulation. The metal should be corrosion-resistant (e.g., stainless steel, aluminum, or galvanized steel), because gas exit temperatures are low – under 550°F [285 °C] – and an insulating wall is necessary to prevent the gas from cooling too much before it leaves the flue. Other flues include type BW, type L, and masonry chimneys. BW and L are also reserved for low-temperature flue gases. Local codes vary and should be checked for permitted flue types.

Examples of flues and chimneys are illustrated in Figure 13.1.

Draft Inducers

Draft inducers are high-temperature and chemical-resistant fans that aid in the removal of hot gases and products of combustion from boilers. Draft inducers are used when there is a restriction on the height or cross-sectional area of a flue.

As we know from the previous discussion, the capacity of a chimney to convey all the products of combustion from the boiler to the outside safely and at an adequate velocity depends, among other variables, on the height and size of the flue. The shorter the flue, the smaller the draft; the smaller the cross-sectional area of the flue, the greater the friction loss and, consequently, the lower the capacity of the chimney. If the capacity of the chimney is too low for the boiler, the draft inducer can be connected to the boiler so that it will shut off as a result of a buildup of gases in the firing chamber; otherwise, the mechanical room will fill with smoke.

To overcome the problem of a chimney that is too small, a draft inducer can be installed in the chimney system. The draft inducer creates the additional draft required in the flue for the gases to be discharged into the atmosphere safely and with adequate velocity.

The draft inducer is controlled by the burner on the boiler, so it operates whenever the burner operates. It is very important to size a draft inducer carefully, with the help of the boiler manufacturer and the draft-inducer manufacturer. A draft inducer that is too small will not solve the problem; a draft inducer that is too big may create adverse negative pressures in the combustion chamber and may even cause the pilot flame to be extinguished.

New boilers that replace existing boilers may be oversized for the existing chimneys for a number of reasons, including the following:

- The existing boiler may have been replaced with a new, smaller, more efficient boiler.

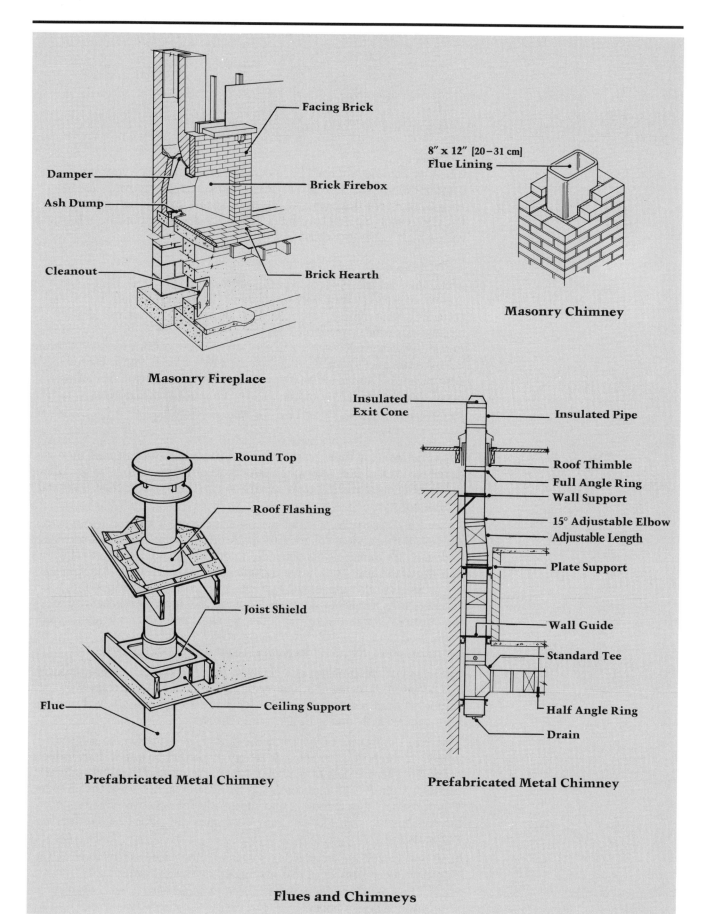

Figure 13.1

- The flue height may have been reduced because of structural problems, appearance, or space constraints.
- The flue may be too small because of space constraints.

Draft inducers should be avoided whenever possible, as they are not 100 percent reliable. They can break down and consume more energy in the system because of their electrical load. A chimney system of adequate height and size will always work, and there are no controls or moving parts to break down.

Motor Starters

Electric motors for HVAC equipment are turned on and off by motor starters. Motor starters are activated by either an *automatic controller*, such as a thermostat, or by a *manual switch*, which closes contacts in the starter. This, in turn, permits a current to flow in the motor, causing the motor to run.

The starter, a separate piece of equipment, may be mounted adjacent to the motor it serves or located remotely, as in a motor control center. Motor controllers typically are designed to:

- start and stop the motor as directed by control device;
- protect against short circuits or grounds;
- ignore the high initial motor inrush current, but protect the motor from excessive current draw, or *overloading*, which damages insulation and shortens motor life. The overload relays that protect the motor may be magnetic, thermal (melting alloy), or bi-metallic. The relays monitor the current or temperature of the motor and shut it down during overload.

Functions in operational controls, such as reversing, changing speed, and other rapid sequences are more common in industrial plants.

Manual Starters

Manual starters (on/off switches) are often located so that they can be operated remotely to draw the operator away from the machine, to allow the machine a few minutes to cool during an overload, or to prevent an overload.

Motors less than one horsepower (H.P.) [.75 kW] single phase under 230 volts use *fractional horsepower motor starters*. Single-phase motors up to 5 H.P. [3.75 kW] and three-phase motors up to 10 H.P. [7.5 kW] commonly use *integral horsepower manual starters*. These operate similar to fractional horsepower starters, but are rated for the heavier daily loads and multiple phases.

Magnetic (Automatic) Starters

When motors are to be activated automatically or by a remote device, *magnetic starters* are used. Magnetic starters are the most common type of starters used for HVAC equipment. Electromagnets, powered by the controllers, are used to make contacts in relays, which remotely activate the motor. These starters must be selected according to the National Electric Manufacturers Association (NEMA) standards for voltage, horsepower, phase, current rating, and power.

Most major pieces of equipment have motor starters tied to the automatic control system. They also have an automatic required reset on power failure to prevent an equipment operator from being injured inadvertently when power is resumed.

Enclosures

Motor starters, as well as other electrical equipment, must be encased in properly specified enclosures. The following are the main NEMA-approved enclosure types:

NEMA 1	General purpose	Normal indoor use
NEMA 3	Dust-tight, rain-tight	Ordinary outdoor use
NEMA 3R	Rainproof, select-resistant*	Severe outdoor use
NEMA 4	Water-tight	Shipyards, dairies, hose test stations, etc.

*Applies to outdoor use in extreme conditions.

Other NEMA classifications include 4X, water-tight, and corrosion-resistant enclosures; class I or class II, hazardous locations; 12, industrial use; and 13, oil-tight and dust-tight.

Magnetic starters are wired into the control circuit, usually incorporated into the control wiring sequence. Some examples of motor starters are shown in Figure 13.2.

Noise and Vibration

All mechanical equipment vibrates and makes noise, which must be reduced to minimize discomfort for those who must work near the equipment. It is not possible to eliminate all noise; some installations actually call for a small amount of background noise to mask conversations and incidental sounds. However, noise and vibrations must be modified when they are objectionable.

In rare situations it is possible to place equipment in remote locations, avoiding the problem of noise entirely. In other situations, the equipment is either placed on isolated footings or encased to control sound.

Vibration Isolation

The most common solution for excessive noise or vibration is to examine the mass and frequency of the equipment, then select appropriate noise and vibration control devices. For example, vibration absorbers absorb, or dampen, the noise or vibrations. Another common technique is to absorb sound with insulation.

A machine rigidly bolted to a structure will transmit all of its vibration to the building. If a machine is mounted on springs or absorbent cushions (pads), its movements can be dampened considerably, effectively controlling sound.

A critical relationship exists between the frequency of the disturbing vibration (from a fan or pump) and the natural frequency of the mounting (vibration isolator). When properly selected, the natural frequency of the vibration isolator should be three to five times lower than that of the machine. If it is less than three times lower than the force of the machine, very large forces may be transmitted to the structure; more than five has a negligible increase in overall effectiveness. Roughly ten percent of the forces on an isolator with one-third of the frequency of the machine and four percent of the forces on an isolator of one-fifth the frequency of the machine will be transmitted to the structure.

Example

Select a vibration isolator for a 1,000-pound [460 kg] piece of equipment operating at 600 rpm (the disturbing frequency) that develops a 50-pound [222.5 newtons] force. 600 rpm = 600/60, or 10 hertz (Hz). The isolator that is selected should be at least 1/3 10 Hz, or 3.3 Hz. The isolator will transmit a maximum of 10 percent of the 50 pounds [222.3 newtons], or 5 pounds [22.3 newtons], to the structure, a negligible increase over 1,000 pounds [460 kg]. To determine the necessary vibration isolator for the selected 3.3 Hz, the spring constant for the isolator must be determined. The spring constant, k, is based on the mass of the equipment and is expressed in pounds per inch.

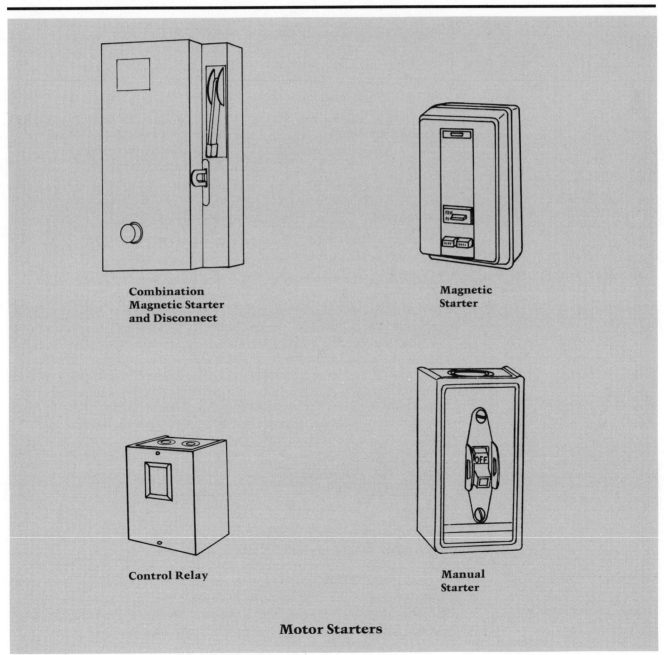

Figure 13.2

$$k = \frac{(\text{natural freq. of isolator (Hz)})^2 \times \text{equip. (lbs)}}{9.8}$$

$$\left[k = \frac{Hz^2 \times \text{equip. (kg)}}{25}\right]$$

where the natural frequency of the isolator equals one-third to one-fifth of the disturbing frequency.

Therefore, for the above example,

$$k = \frac{(3.3)^2 \times (1,000)}{9.8} \text{ or } \frac{1,111 \text{ lbs/in.}}{\text{max.}} \left[k = \frac{3.3^2 \times 460}{25} = 200 \text{ kg/cm}\right]$$

This is illustrated in Figure 13.3.

An important factor in vibration isolation is the need for *flexible connections*, because the equipment moves as it vibrates on the isolator pads or springs. The piping, ductwork, and even electrical connections must be isolated from these vibrations so that they will not be transmitted throughout the building or shake the pipes.

The natural frequency of the floor system should differ from the vibration isolator's natural frequency to avoid catastrophic resonant reinforcement, which could initiate uncontrolled amplified vibrations. For this reason, the frequency should be verified with the structural designer.

Sound transmission in ductwork can be reduced by the use of insulation and elbows. An effective technique is to insulate the interior of ductwork and/or provide an elbow within ten feet [3 m] of air handling equipment.

Vibration and noise control devices are shown in Figure 13.4.

Air Cleaning and Filtration

Besides being the proper temperature and humidity, the air supplied to a space must also be clean and free of odors and pollutants. There are several factors that have recently magnified the need for more attention to odor and pollution control. As modern buildings became tighter as a result of better insulating and building techniques, odors were more likely to be trapped inside. Current energy codes and practice encourage proper amounts of outdoor air for ventilation. Fifteen cfm [7.1 L/s] per person is common today, whereas 5 cfm [2.4 L/sec] per person was used ten years ago, which means that more air is introduced into a building today to dilute unpleasant conditions. The outdoor air quality generally worsens each year; therefore, the introduction of outdoor air deserves careful attention. Finally, many buildings use 100 percent outdoor air in the spring and fall, dramatically increasing the need for air treatment, particularly in urban and industrial areas.

Odors

Most odors are generated within a building. These odors must be prevented from accumulating or being recirculated in the building's HVAC system. The odors controlled are primarily from tobacco smoke and people. Others include cooking fumes, paint spray, animals, bacteria, pollen, and cleaning chemicals. Buildings that have locker rooms, smoking rooms, bars, commercial kitchens, toilets, painting booths, animal shelters, dry cleaners, and laundries must have special provisions for odor control.

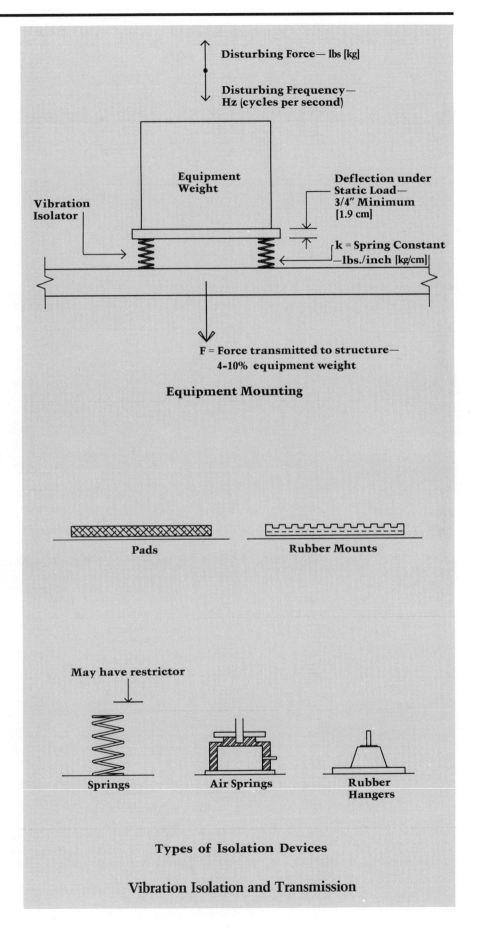

Figure 13.3

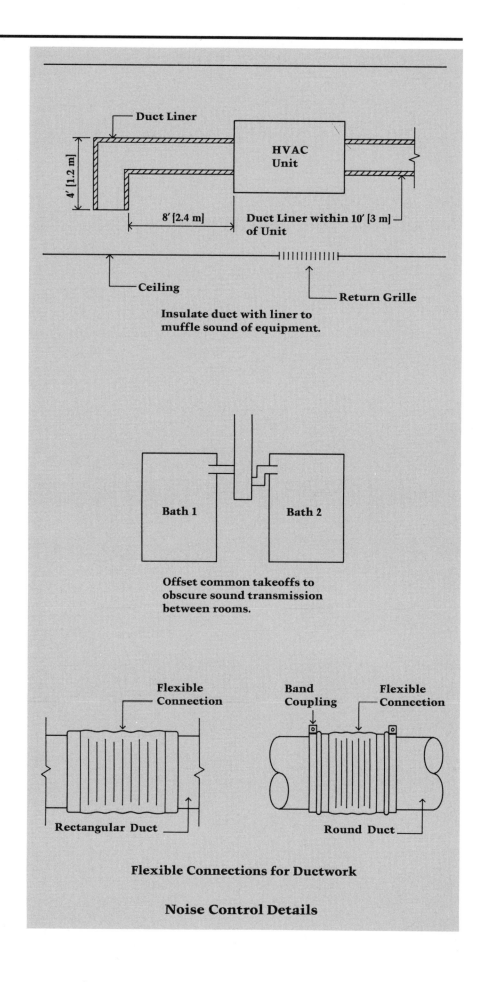

Figure 13.4

Methods of Odor Control

The most frequently used methods for controlling odor and air quality include:

- Ventilation and exhaust (dilution or removal)
- Air washing (air cleaning, scrubbing)
- Absorption (activated charcoal)
- Odor masking (perfume)

Common methods of odor control are shown in Figure 13.5 and described in the following paragraphs.

Ventilation

For odor control, an exhaust system removes contaminated air from a building. Fresh outdoor air, when introduced into a supply air system, reduces odors by diluting inside air with fresh air. The American Society of Heating, Refrigeration, and Air Conditioning Engineers (ASHRAE) standards provide recommended air quantities for ventilation.

Air Washing

Water is the most common air-washing fluid. When a fine mist spray is introduced into the air, odor-causing particles are absorbed by the liquid, collected in a pan, and removed. In process industries, chemicals other than water are selected to neutralize elements in the air. Another air-washing method utilizes scrubbers, which are highly developed air-washing machines.

Absorption

Activated charcoal beds are the most common type of absorption systems used to eliminate odors. The air (or gas) is passed over the absorbing material. Because of the porosity and vast surface area of this granular material, odor-causing particles adhere to the surface. The beds of material, however, should be changed or regenerated periodically in order to work properly.

Odor Masking

The least desirable, but often the only available, cure for eliminating odors is the use of "perfume." Introducing a more pleasant odor into the air to overcome an unpleasant one is often a manageable way to control odors.

Pollutants

Automobile fumes, industrial pollutants, dust, and grime all exist to some extent in otherwise "fresh" outdoor air. These pollutants must be minimized, as they may contain sulfur oxide, carbon monoxide, hydrocarbons, nitrogen oxides, and exhausts from nearby buildings.

Methods of Pollution Control

Pollution control is often accomplished in conjunction with odor control. With pollution control, however, as the size of the particle to be removed decreases, the cost of the cleaning system increases. In addition to air washing, other common air-cleaning devices are *filters* and *electronic cleaners*.

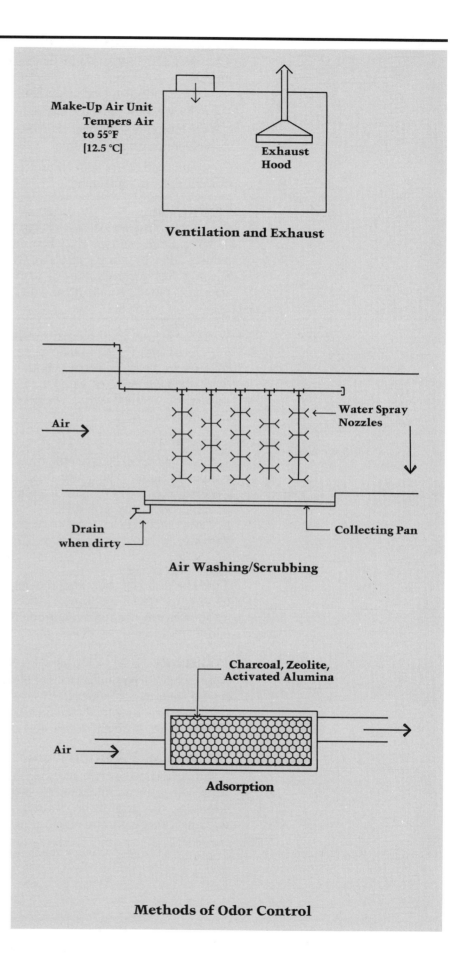

Figure 13.5

When selecting filters, it is necessary to know what needs to be removed from the air. Bird screens filter birds and rodents; insect screens filter insects and smaller animals; and smaller screens control dust, grease, and specific particles. The smallest type of filter is an electronic filter, which removes molecules of an identified pollutant by ionization. Grease and tobacco smoke are usually removed by such filters. Examples of air filters are shown in Figure 13.6.

Filters are generally rated for their *efficiency* and *pressure drop*. The efficiency rating is the percent of dust arrested by the filter. The pressure drop is the decrease in air static pressure that occurs as it passes through the filter. The ratings are stated for a particular air quantity and velocity and must be converted for other design conditions. In addition, the performance of a filter varies dramatically over its life, decreasing as it becomes dirty.

Old filters should be periodically inspected and cleaned, as they may become clogged. If blockage impedes air flow, fans may become overloaded and a system failure could result.

Most filters are dry and fibrous. Some are available on continuous rolls, which continually feed new filter media to replace the old. *Dry filters* are commonly made from cellulose, glass fiber, or metal. Electronic filters (precipitators) are most effective on small particles, which are ionized and then collected on plates.

Hangers and Supports

Piping for mechanical installations must be supported by a wide variety of hangers and anchoring devices. Many different building components may be used for anchoring pipe hangers. The location of piping supports is also an important consideration. Some common locations for anchoring devices include the roof slab or floor slab, structural members, side walls, another pipe line, machinery, or building equipment (see Figure 13.7).

Pipe hangers are usually made out of black or galvanized steel. For appearance or in corrosive atmospheres, chrome, copper-plated steel, cast iron, or a variety of plastics may also be used.

Anchoring to Concrete

The method selected for anchoring the hanger depends on the type of material to which it is being secured: generally concrete, steel, or wood. If the roof or floor slab is constructed of concrete, formed, and placed at the site, concrete inserts may be nailed into the forms at the required locations prior to the placing of concrete. These inserts may be manufactured from steel or malleable iron, and either are tapped to receive the hanger rod machine thread or contain a slot to receive an insert nut. Because the slotted type of insert requires separate insertable nuts for the various rod diameters, only one size of insert need be warehoused. For multiple side-by-side runs, long slotted insert channels in up to 10' [3 m] increments are available with several types of adjustable insert nuts.

When precast slabs are used, the inserts may be installed on site in the joints between slabs or drilled or shot into the slab itself. Care must be taken to verify that no critical strands will be impaired. Electric or pneumatic drills and hammers are available to drill holes for anchors, shields, or expansion bolts. Another method of installing anchors on site utilizes a gunpowder-actuated stud driver, which partially embeds a threaded stud into the concrete.

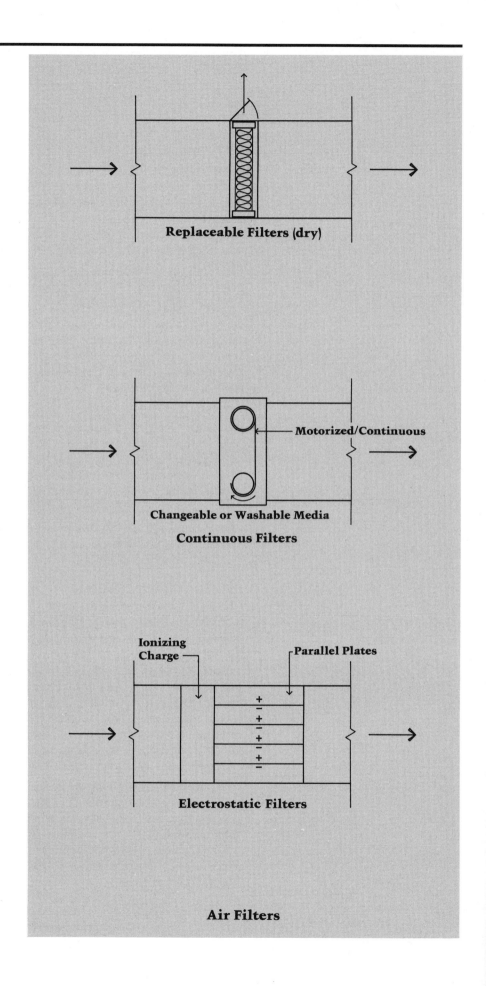

Figure 13.6

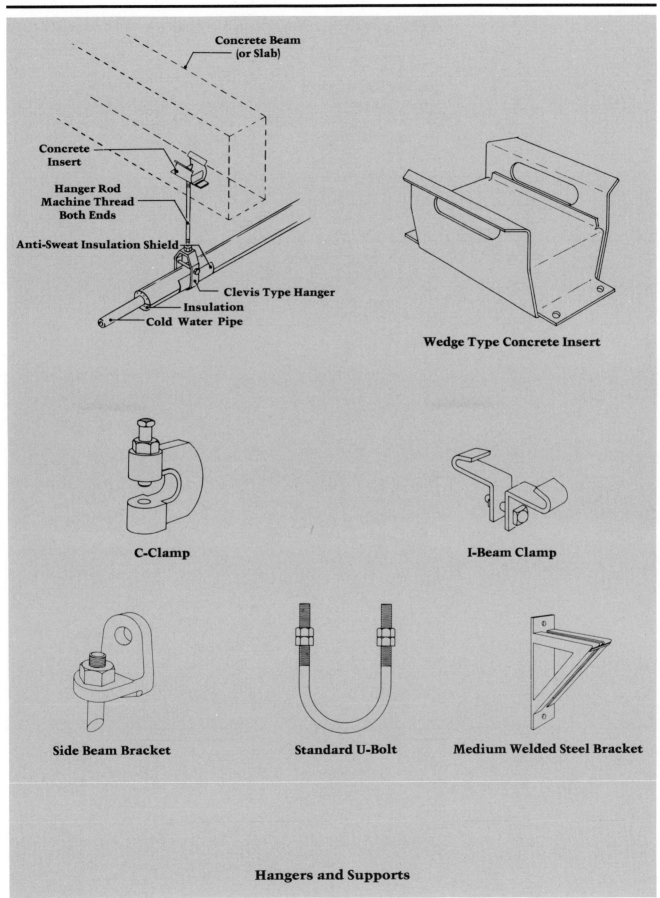

Figure 13.7

If the piping is to be supported from the sidewalls, similar methods of drilling, driving, or anchoring are used for concrete walls. Where hollow-core masonry walls are to be fitted, holes may be drilled for toggle or expansion-type bolts or anchors.

Anchoring to Steel

When piping is to be supported from the building's structural steel members, a wide variety of beam clamps, fish plates (rectangular steel

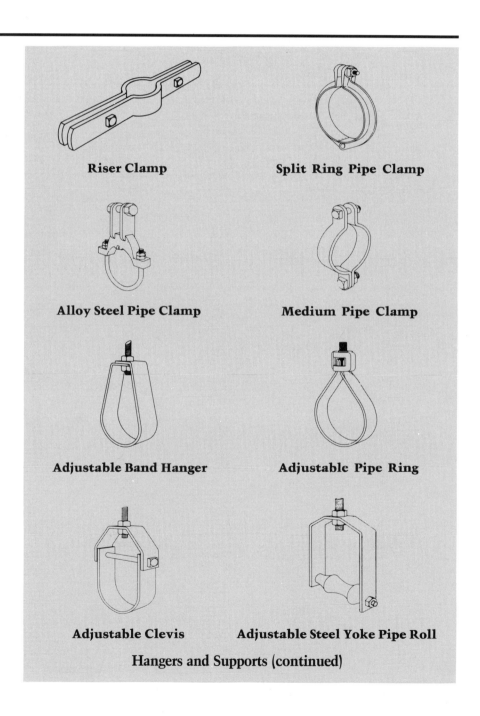

Figure 13.7 (continued)

washers), and welded attachments can be employed. If the piping is being run in areas where the structural steel is not located directly overhead, then intermediate steel is used to bridge the gap. This intermediate steel is usually erected at the piping contractor's expense. If the building is constructed of wood, then lag screws, drive screws, or nails are used to secure the support assembly.

From the anchoring device, a steel hanger rod, threaded on both ends, extends to receive the pipe hanger. One end of this rod is threaded into the anchoring device, and the other is fastened to the hanger itself by a washer and a nut. For cost-effectiveness and convenience, continuous thread rod may be used. The pipe hanger itself may be a ring band, roll, or clamp, depending on the function and size of the piping being supported. Spring-type hangers are also used when necessary to cushion or isolate vibration.

Accommodating Expansion

Piping systems subject to thermal expansion must often absorb this expansion with the use of piping loops, bends, or manufactured expansion joints. The piping must be anchored to force the expansion movement back to the joint. In order to prevent distortion of the piping joint itself, alignment guides are installed. Figure 13.8 shows alignment guides and roll hangers used in these types of installations.

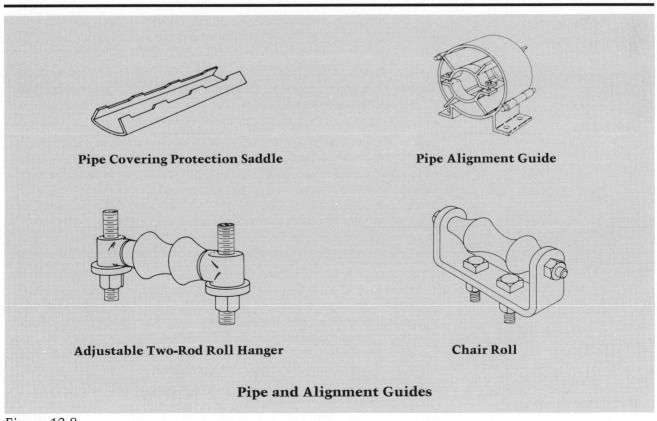

Figure 13.8

Specifically designed pipe hangers are used for fire protection piping with underwriters and factory mutual approvals.

Wood-frame Construction

In wood-frame residential construction, holes drilled through joists or studs are often the sole support for certain pipes. These pipes may occasionally be reinforced with wedges cut from a two-by-four. A piece of wood, cut and nailed between two studs, is often used to support the plumbing stubouts to a fixture. Prepunched metal brackets are available to span studs for both 16" and 24" [41 and 61 cm] centers. These brackets can be used with both plastic and copper pipe, and maintain supply stub spacing 4", 6," or 8" [10, 15, or 20 cm] on centers, giving perfect alignment and a secure time-saving support. Plastic support and alignment systems are also available for any plumbing fixture rough-in.

Insulation

Insulation is used in mechanical systems to prevent heat loss or gain and to provide a vapor barrier for piping and ductwork systems, boilers, tanks, chillers, heat exchangers, and air handling equipment casings. Most boilers, water heaters, chillers, and air handling units are provided with insulated metal jackets and require little, if any, field insulation.

Insulation is manufactured from fiberglass, cellular glass, rock wool, polyurethane foam, closed cell polyurethane, flexible elastometric, rigid calcium silicate, phenolic foam, or rigid urethane, and is available in rigid or flexible form.

Insulation is produced in a variety of wall thicknesses and may be applied in layers for extreme temperatures. Rigid board or blocks, or flexible blanket insulation, are used for ductwork and equipment. Perforated sections are available for use with pipe.

The standard length for pipe insulation has been three feet. This is still true for fiberglass and calcium silicate, but foam insulations (e.g., polyurethane, polyethylene, and urethane) are produced in four-foot lengths, which means that fewer butt joints will have to be made in straight runs of pipe. Flexible elastometric insulation is shipped in six-foot lengths.

The simplest pipe-covering installation method is elastometric insulation slipped over straight sections of pipe or tubing prior to pipe installation. This insulation can be formed around bends and elbows.

When the piping is already installed, the insulation must be slit lengthwise, placed, and both the butt joints and seams (at the slit) must be joined with a contact adhesive. This process requires more labor than the simple slipping-on method described above. Fittings, in this case, are covered by mitering or cutting a hole for a tee or a valve bonnet. Fiberglass insulation is already slit for placement around the pipe, and fittings are mitered. From straight sections, holes are cut for tees and valve bonnets using a knife. Elastometric insulation usually does not require any additional finish.

A variety of jackets and fittings are available for all types of insulation, including roofing felt wired in place and preformed metal jackets used as closures where exposed to weather. Premolded fitting covers are available to give fittings and valves a finished appearance. Factory-applied self-sealing jackets are an advantage of fiberglass insulation. Flanges, which are removed frequently, often require unique insulated metal jackets or boxes, which are fabricated by the insulation contractor.

Calcium silicate insulation is often specified for higher temperatures (above 850°F) [450 °C]. Installation of calcium silicate is more labor-intensive than installation of fiberglass or elastometric insulation. This rigid insulation is made in half sections for pipe and in three-foot lengths. Multiple segments must be used in layers to achieve larger outside diameters. The segments are wired in place nine inches on center, normally requiring two workers. Fittings for calcium silicate insulation are made by mitering sections using a saw. Flanges and valves require oversized sections cut to fit. These valve and fitting covers are wired in place and finished with a troweled coat of insulating cement. This type of insulation has a high waste factor because of crumbling and breaking during cutting and ordinary use.

Special thickness of any insulation is obtained by adding multiple layers of oversized insulation, and sealing and securing using the same method used for the base layer. This additional work should be included in the labor estimate.

Hangers and supports require special treatment for insulated piping systems. The pipe hanger or support must be placed outside the insulation to allow for expansion and contraction in steam or hot water systems and to maintain the vapor barrier in cold water systems. As a result, oversized supports should be used to fit the outer diameter of the insulation, rather than the pipe.

For heated piping, preformed steel segments are welded to the bottom of the pipe at each point of support. These saddles are sized according to the insulation thickness in addition to the pipe's outer diameter. For cold-water (anti-sweat) systems, where no metal-to-metal contact between pipe and support can be tolerated, the hanger is oversized to allow for placement of the insulation. The insulation is then protected by a sheet metal shield that matches the outer radius of the insulation.

For flexible insulation that cannot support the weight of the pipe, a rigid insert is substituted for the lower section of insulation at each point of support. The insert may be formed from calcium silicate insulation, cork, or even wood. The insulation jacket must enclose this insert within the vapor-proof envelope. The labor for insulation around hangers should be estimated carefully. The insert, for example, might be installed by the pipe coverer, while the metal shield or protector is furnished by the pipefitter or plumber.

Ducts and equipment may be insulated with a wrap-around blanket insulation, wired in place and sealed with adhesive or tape. Rigid board is secured by the use of pins, fixed to the duct exterior with mastic, or spot-welded into place. The insulation sheets are pressed onto the pins and secured with self-locking washers.

The butt joints and seams are sealed with adhesive and tape. Block and segment insulation, when installed on round or irregular shapes, is wired on, covered with a chicken-wire mesh, and coated with a troweled application of insulation cement.

Procedures for installing and finishing insulation will vary. Examples of insulation are illustrated in Figure 13.9.

Equipment Data Sheets

The equipment data sheets in Figure 13.10 can be used in the selection of accessories. A Means line number, where applicable, is provided for the component listed. Costs can be determined by looking up the line number in the current edition of *Means Mechanical Cost Data*.

Problems

13.1 True or false?

a. A NEMA 3 enclosure refers to a motor starter for ordinary outdoor use.

b. Expansion joints in steam pipes are necessary to control temperature expansion and contraction.

c. Most building operators prefer two-pipe fan coil systems over four-pipe systems for heating and cooling.

d. Most cooling systems run on electricity.

e. A series of small modular boilers is more efficient than one large boiler.

f. A 2" [5.1 cm] pipe requires approximately 9" [23 cm] of space when insulation and hangers are considered.

g. Pumps in series boost pressure more than they increase flow.

h. A 12" × 18" [31 × 46 cm] duct has an equivalent round of 16" [41 cm].

i. A 16" [41 cm] diameter duct with 2500 cfm [1180 L/s] loses more than 0.01" [0.03 cm] per 100 feet [30.5 m] in pressure.

j. Tall chimneys have less draft than short ones.

13.2 What size flue is required for a gas boiler with an input of 2 million Btu/hr. [586 kW] and no draft hood?

13.3 What type of vibration system should be used for each of the following:

a. hanging piping

b. a 1,000-ton [3517 kW] cooling tower

c. a 10-horsepower [7.5 kW] pump in a basement

d. a heat pump on a concrete floor

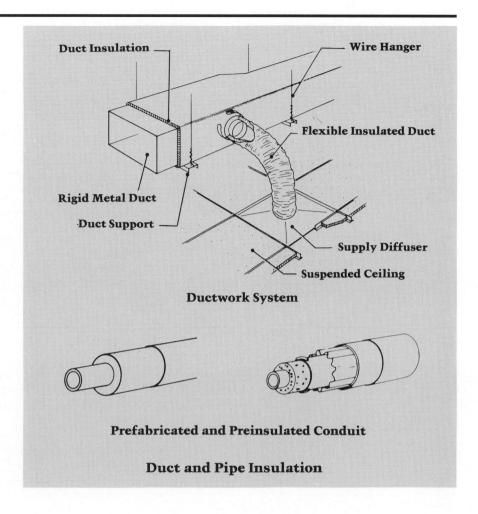

Figure 13.9

Accessory Equipment Data Sheets

Accessory Equipment	Chimneys	Means No. 155-680
Useful Life: 10-20 years	**Typical Use and Consideration**	**Size Range (Diameter) 3″–48″ [7.6-122 cm]**
Prefabricated Metal Chimney — Round Top, Roof Flashing, Joist Shield, Ceiling Support, Flue	Metal vent chimneys are available in single wall and double wall type B construction and in galvanized steel, type 304 stainless steel liner with aluminized steel outer jacket (all fuels), and with 316 stainless steel liner (more durable) and 11 gauge galvanized jackets (2,000°F [1082 °C] fuels—acid-resisting with refractory lining).	
Masonry Chimney — 8″ × 12″ [20 × 31 cm] Flue Lining	Masonry flues and chimneys have the durability and adaptability for most fuels. The weight of such chimneys requires a substantial foundation.	
	Select proper flue material to match fuel.	
$D^2 = \dfrac{HMT_f}{.95(P_{atm})V}$	D = Flue inside diameter—inches [cm] H = Heat input (MBH) [kW] M = lbs. mass gas in flue per 1,000 Btu/hr. [kg/kWh] T_f = Average flue gas temperature (degree F_{abs}) [$°C_{abs}$] $P_{(atm)}$ = Atmospheric pressure (in./H_s)—approx. 29.92 [101.3 kPa] V = Velocity of flue gas (feet/sec.)—assume 20 [6.1 m/sec]	

All numbers shown in brackets are metric equivalents.

Figure 13.10

Accessory Equipment Data Sheets (continued)

Accessory Equipment		Vibration Isolators	Means No. 157-485
Useful Life: 30+ Years	**Type**	**Typical Uses**	**Capacity Range: 10 – 30,000 lbs.**
Glass Fiber Pad	1	Advantages: Durable, provides damping under pads and slabs. Disadvantages: Concrete construction on top of pad.	1 – 4" thick, 500 psi [2.5 – 10 cm, 3448 kPa]
Rubber Mounts and Hangers	2	Put under pipe supports or for suspended equipment. Disadvantages: Limited load capability.	0.5" maximum [1.3 cm] deflection .25" [.6 cm] recommended
Spring Hangers and Isolators	3	Suspended equipment and pad- or base-mounted-durable. Most common type. Disadvantages: Provide corrosion protection if outside.	Wide range capacities
Restrained Spring Isolator	4	Advantages: Hold-down bolts or clips limit deflection when load of water or refrigerant is removed. Disadvantages: Most expensive. Horizontal restraint also important.	Use for larger boilers, chillers, cooling towers.
Thrust Restraint	5	Used with HV duct systems to dampen vibrations and thrusts of air pressure, similar to spring hanger.	
Air Springs and Bellows		Large deflections and filtering of high frequency noise. Special applications. Disadvantages: Constant dry air supply required to maintain preset pressure.	6" [15 cm] deflections

Consult manufacturer for recommendations of type of isolator and general guidelines. Check equipment weight and structural span for additional criteria. Recommended deflection varies from 1/4" to 2.5" [.6 – 6.4 cm].

Type	Common Use
1	Under pads
2	Hung or mounted small equipment, pipes, terminal units
3	Compressors, pumps, fans, air handling equipment, package units
4	Chillers, condensers, cooling towers
5	Variable high velocity ductwork

Provide proper spacing and locations

$$\text{Isolator Frequency Hz} = \frac{\text{RPM of Machine}}{6 \times (3 \text{ to } 5)} \qquad \text{Spring Constant k (lb./in.)} = \frac{(\text{isolator frequency Hz}) \times \text{Equipment/(lbs.)}}{9.8}$$

$$\left[K(kg/cm) = \frac{Hz \times \text{Equip. kg}}{25} \right]$$

All numbers shown in brackets are metric equivalents.

Figure 13.10 (continued)

Accessory Equipment Data Sheets (continued)

Accessory Equipment		Vibration Bases	Means No. 157-485
Useful Life:	Type	Typical Uses – Under Vibration Isolators Connecting to Structure	Common Use
Direct Isolation Equipment Pad	A	Advantages: Rigid equipment with no additional support.	Chillers, tank compressors, cooling towers, air handlers, fans, package units.
Structural Bases	B	Advantages: Transfers equipment concentrated loads to structural load points of building.	Fans greater than 300 rpm and 0 – 50 H.P. [0 – 38 kW]
Concrete Bases on Pad	C	Advantages: Provides needed mass to dampen impact on structure. Disadvantages: Expensive. Reinforced concrete.	Reciprocating compressors, fans greater than 300 rpm and greater than 50 H.P. [38 kW]
Curb	D	Advantages: Limits deflection to 1" and provides narrow water-tight seal for equipment with curbs.	Special applications

Accessory Equipment	Insulation	Means No. 155-651
Useful Life: 20+ years	Typical Uses – All Generation and Distribution Equipment	Size Range (Diameter) 1/2" – 24" [1.3 – 61 cm]
Equipment	Calcium silicate 1" – 3" [2.5 – 7.6 cm]	

Ductwork – Blanket Type
  Vinyl FRK wrap
  Rigid board
	Fiberglass liner 1/2" – 2" thick (1-1/2 – 2 lb. density) [1.3 – 5.1 cm (.7 – 9 kg)]
	1" – 3" thick [2.5 – 7.6 cm]
	1" – 2" thick – fire-resistant or FRK barrier [2.5 – 5.1 cm]	

Pipe Covering
  Calcium silicate
  Rubber tubing
  Urethane – ASJ cover
	1" – 3" thick – aluminum jacket available [2.5 – 7.6 cm]
	Closed cell foam 3/8" – 3/4" thick [1 – 1.9 cm]
	1" – 1-1/2" [2.5 – 3.8 cm] thick – most common – widest temperature range
	(−60° – 225°F) [−51° – 106° C]	

Special considerations: Insulation is most effective if continuous. Make hangers large enough to accommodate insulation. Insulation protection saddles must be provided to prevent crushing the insulation.		

All numbers shown in brackets are metric equivalents.

Figure 13.10 (continued)

Accessory Equipment Data Sheets (continued)

Accessory Equipment	Air Filters		Means No. 157-401
Useful Life: 1-3 months	**Typical Uses — Odor and Dust Control**		**Capacity Range:** .001 – 1,000 microns
	Advantages	Disadvantages	Remarks
Activated Charcoal	Excellent for overall cleaning.	Expensive to install and recharge this system.	
Washers/Scrubbers	Most effective overall.	Grease and oil, requires additional filter.	Sludge must be removed.
Electronic Air Cleaners	Very effective for smaller particles. Low pressure drop.	Expensive. Additional filtration for large particles may be required.	Laboratories
Permanent Washable Air Filters	System can be designed for specific filtration requirements.	Maintenance costs required. Reliability may be a problem.	
Renewable Roll Air Filters	Simple, maintains low pressure.	Maintenance costs required.	
Disposable Fiberglass Air Filters	Simple, light.	Air resistance increases with use.	Most common for small to medium sized systems.
Special considerations: Particle size to be removed should be known. Most systems should be able to handle particle sizes from 1 to 5 microns (1 micron = 0.001 millimeters) plus filter out larger particles such as smoke and dust.		The type of filter selected is heavily dependent on the maintenance regimen for the building. Sludge removal, cleaning of electronic filters, vacuuming of bag filters, and recharging charcoal beds requires an active, knowledgeable crew.	

Accessory Equipment	Motor Starters	Means No. 163-320
Useful Life: 15-25 years	**Typical Uses — Control Electrical Devices**	**Capacity Range:** Nema 1 through Nema 7
Control Stations	Used to remotely control operation of HVAC equipment from the work space.	
Motor Control Center Pilot Lights Starters Push Buttons	HVAC equipment that is controlled by automatic controls will commonly use automatic magnetic starters. Motor starters are specified by voltage, horsepower, phase, and current rating. Starters should ideally be in the view of the equipment operator.	Means No. 163-130 **Capacity Range:** 1/3 – 400 H.P. [.25 – 300 kW] Size: 00 through 6

All numbers shown in brackets are metric equivalents.

Figure 13.10 (continued)

Accessory Equipment Data Sheets (continued)

Accessory Equipment	Pipe Hangers and Supports	Means No. 151-900
Useful Life: 25-40 years	**Typical Uses — Support Pipes**	**Size Range (Diameter)** 1/4" – 8" [.6 – 20 cm]
Brackets	For wall mounting	
Clamps	For attachment to steel beams	
Rings	Size to pipe and insulation. Provide plate for insulation	
Rods	1/4" – 7/8" diameter [.6 – 2.2 cm]	
Rolls	Provide where expansion required — large temperature swings for pipe	
U hooks	Direct attachment — no rods	

Support Spacing for Metal Pipe or Tubing

Nominal Pipe Size (Inches) [cm]	Span Water (Feet) [m]	Steam, Gas, Air (Feet) [m]	Rod Size
1 [2.5]	7 [2.1]	9 [2.7]	
1-1/2 [3.8]	9 [2.7]	12 [3.7]	3/8" [1 cm]
2 [5.1]	10 [3]	13 [3.9]	
2-1/2 [6.4]	11 [3.5]	14 [4.3]	
3 [7.6]	12 [3.7]	15 [4.6]	1/2" [1.3 cm]
3-1/2 [8.9]	13 [3.9]	16 [4.9]	
4 [10.2]	14 [4.3]	17 [5.2]	5/8" [1.6 cm]
5 [12.7]	16 [4.9]	19 [5.8]	
6 [15.2]	17 [5.2]	21 [6.4]	3/4" [1.9 cm]
8 [20.3]	19 [5.8]	24 [7.3]	
10 [25.4]	20 [6.1]	26 [7.9]	7/8" [2.2 cm]
12 [31]	23 [7]	30 [9.2]	
14 [36]	25 [7.6]	32 [9.8]	
16 [41]	27 [8.2]	35 [10.7]	
18 [46]	28 [8.5]	37 [11.3]	1" [2.5 cm]
20 [51]	30 [9.2]	39 [11.9]	
24 [61]	32 [9.8]	42 [12.8]	1-1/4" [3.2 cm]
30 [76]	33 [10.1]	44 [13.4]	

All numbers shown in brackets are metric equivalents.

Figure 13.10 (continued)

Chapter Fourteen
SPECIAL SYSTEMS

In previous chapters, the design of HVAC systems has been reviewed with emphasis on determining heating and cooling loads, selecting appropriate systems, and sizing components. This chapter focuses on the following list of specific HVAC topics and the special considerations they require.

- Fire and smoke control
- Special building types: high-rise buildings, atriums, and covered malls
- Hospitals
- Commercial kitchens
- Specifications
- Coordination between the trades
- Costs: operations, maintenance, and capital improvements

Each of these topics is a specialty area in building design. An understanding of the particular concerns of these areas is useful for all designers, since the principles they utilize exist to some degree in all HVAC systems.

Fire and Smoke Control

While many aspects of building design focus on egress and compartmentation of fire zones, smoke control has become a particularly important issue in the design of life safety systems. Provisions for smoke vents at the tops of stair and elevator shafts and requirements for smoke partitions in hospitals are well recognized design elements. Compartmentation focuses on the principle that once generated, smoke should be contained by smoke barriers within compartments throughout a building. High-rise buildings, atriums, and covered malls all require some form of mechanical smoke control system.

Smoke control systems have become more widely used since the late 1960s, when it became evident that smoke inhalation is the major cause of deaths in fires. Prior to 1960, accepted engineering practice recommended that HVAC systems only had to shut down in the event of fire or alarm. At that time, it was believed that the spread of fire would acceptably decrease if air handling units shut down and building compartmentation was sufficient to contain the smoke and fire. Since then, studies of fire smoke movement have revealed that smoke tends to migrate through an HVAC system even when the equipment has been turned off. Today, designs that actively utilize HVAC systems for smoke control are encouraged.

Successful applications of smoke control in buildings should include one or more of the following:

- Control of air pressure at door openings
- Proper venting of smoke
- Adequate pressurization of all floors
- Functional dampers
- Properly designed controls
- Adequate emergency power

Smoke control systems utilize supply and exhaust fans, ducts and fire dampers to pressurize parts of a building and control smoke movement during fires. Operable, smoke-tight dampers are also used to facilitate smoke control. The systems are dynamic, able to respond automatically to changes in the building during an alarm and to be manually overridden by the fire department and fire emergency response teams. Fans may be specially dedicated to the smoke control system, or they may be part of the building's normal air distribution system. In addition, a series of supply and exhaust fans may be linked to control air pressure in a building, limiting smoke migration.

In the particular case of high-rise buildings, but generally in all buildings with an air handler larger than 2,000 cfm [950 L/s],* some smoke control is required. Chapter 2 (Figure 2.1 in particular) illustrates the basic health and safety requirements for HVAC systems that apply to all buildings. In air systems that handle more than 2,000 cfm [950 L/s], smoke detectors must be installed to shut down the air handler and prevent continued circulation of smoke throughout the building's duct system.

While national codes such as BOCA and the UBC do not specifically require zoned smoke-control systems, they do require stair pressurization and smoke evacuation systems for atriums and parking structures. In addition, many large metropolitan areas such as Boston, New York, Atlanta, Philadelphia, and Washington, D.C. do require that high-rise buildings have smoke control systems. As such, the ability to design an effective smoke control system is becoming a regulatory necessity.

General Principles of Smoke Movement

To control smoke produced during a fire, there are two important goals of life safety. First, smoke should be directed away both from the occupants who are fleeing and from firefighters who need access to the building. Second, smoke should be vented to the outside, usually at the roof. In general, smoke control systems use air flow and pressure differences to achieve these goals. Significant air flow, such as wind blowing through an open garage door, can control smoke movement if the average air flow movement is of sufficient magnitude. Velocities of 200 feet per minute (1 m/s) have been effective in preventing smoke from passing through an open stair door into the stairwell and velocities of 400 fps (2 m/s) are used by some designers in smoke control systems.

Air pressure differences are a special case of air flow. If air is forced across door cracks and similar barriers, it can inhibit smoke movement by creating pressure differences across the barrier. Figure 14.1 illustrates how air pressure in a stair tower inhibits the flow of smoke from the fire area into the stair tower, with and without the door open.

*Throughout the book, SI metric equivalents are provided for most imperial units. The SI units appear in brackets [ ].

The air flow across a narrow opening such as a door crack, gap or other flow path is approximately proportional to the square root of the pressure difference across the gap. In other words,

$$Q = K \sqrt{p} A$$

where
- Q = flow across the crack in cfm or L/s
- K = constant depending on units selected
- p = pressure difference across the crack
- A = area of crack in square feet or square meters

For example, a door crack area of 0.11 square feet with a pressure difference of .01 inches H_2O [3 Pa] would have an air leakage rate of approximately 30 cfm [14 L/s], while a pressure difference of 0.3 inches H_2O [90 Pa] would have an air leakage rate of approximately 160 cfm [75 L/s]. While the above formulas work in a laboratory, during a fire buildings can be affected by many factors. For example, the pressure across any particular gap can vary from minute to minute. During field tests of smoke control systems, fluctuations of as much as 0.02 inches H_2O [6 Pa] have been observed at cracks. Causes for the fluctuation include wind pressures, stack effects, buoyancy, door opening/closing, and the dynamic effect of the fire itself.

- *Wind force*, resulting from broken windows, has been shown to produce significant pressure fluctuations and promote air flow through buildings.
- *Stack effect* refers to the natural movement of air resulting from temperature differentials between interior and exterior environments. If outdoor temperatures are significantly below indoor temperatures, the denser outside air infiltrates through cracks in the building fabric and the warmer, lighter interior air tends to rise. In high-rise

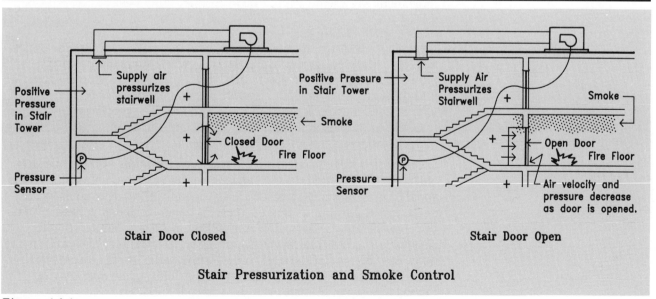

Figure 14.1

buildings, this can have a significant effect on air flow through stairwells and main HVAC risers.
- *Buoyancy* refers to high-temperature smoke that tends to rise above the denser ambient air. This effect causes smoke to rise through cracks in ceiling tiles, penetrate into ductwork, and spread throughout the building when HVAC systems are shut down.
- The simple act of *opening a door* can have significant effects on stair pressurization. As each door opens and closes, the local pressure across the door decreases. This dynamic shift in pressures is extremely difficult to model and poses a difficult challenge for the designer.
- As a fire *grows and spreads*, it consumes oxygen and the heat produced expands the gaseous combustion products. This chemical reaction in itself can dramatically affect pressure fluctuations within a space. If contained, the heat will build up until windows burst from the heat and pressure. The heat pressure pushes the created combustible gases forward into new spaces where they ignite, further accelerating the fire and the pressures.

Design and Layout of Smoke Control Systems

To combat stack effect, buoyancy, and wind, systems have been developed to control the flow of smoke during a fire and to maintain a pressure differential between the fire zone and means of egress. The first applications of smoke control involved the pressurization of stairwells. Conceptually, the idea was to use pressurization to reduce the flow of smoke into the exit stairs. This was generally accomplished by placing a fresh air intake fan in the stairwell in a single injection stair pressurization system. By supplying a greater amount of fresh air than is exhausted, a positive pressure is induced in the stairwell and the amount of smoke entering the stair from any floor is reduced (see Figure 14.2).

Current design of smoke control systems subdivides a building into zones that are interactive with the building's fire alarm and HVAC components. These zones, which resemble a multi-layered sandwich (in some cases, more than one per floor), are separated by physical barriers such as floors, walls, ceilings, and doors. In the event of fire, pressure differences and air flows, in conjunction with physical barriers, are used to reduce smoke movement (see Figure 14.3).

Stair Pressurization

As stated previously, stair pressurization is one of the primary means of smoke control. As shown in Figure 14.1, when stairs are positively pressurized relative to the floors they serve, they prevent smoke backflow into the stair. This provides a smoke-free escape route for occupants as well as a staging area for firefighting operations. The proper design of these systems is complex and requires a careful consideration of the number of open doors and design weather conditions.

Two primary systems are used for stair pressurization – single injection and multiple injection. In the *single injection* system (Figure 14.2), air is supplied into the stair at one location. Generally, the supply fan is located at the top of the stair so that open egress doors located at ground level do not short-circuit the incoming air and lead to system failure. In taller structures, *multiple injection* systems are typically employed because many doors may be open at the same time. As the name implies, these systems provide supply air at several different levels. The supply duct may be located in a separate shaft or in the stairwell itself.

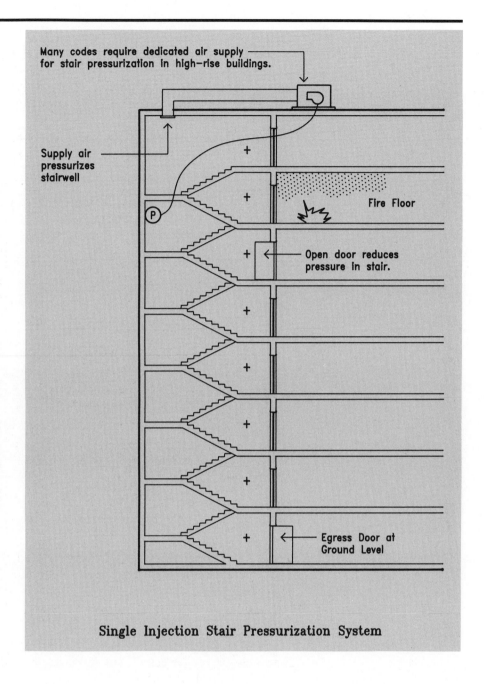

Figure 14.2

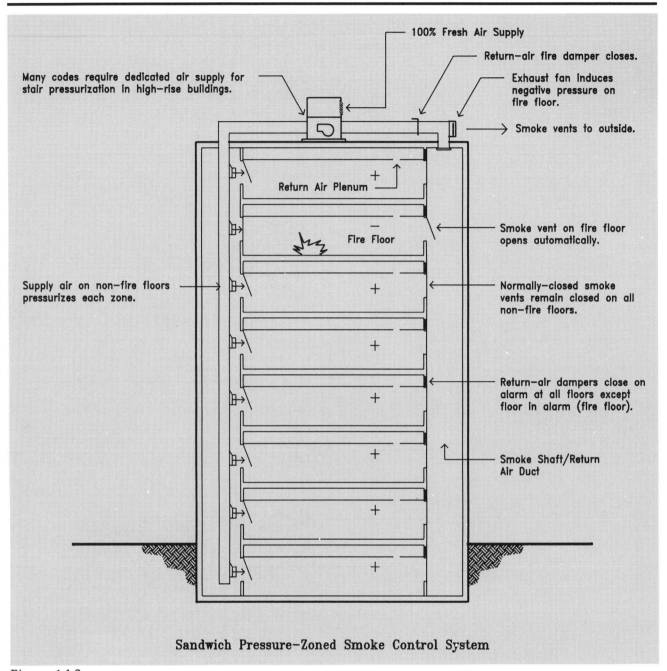

Figure 14.3

Stairwell Compartmentation

Compartmentation provides an alternative to multiple injection systems. This method divides a stair tower into two or more compartments that are separated by doors and partitions. Unfortunately, when the door between compartments is opened, the effect of compartmentation is lost. This method can be applied most effectively to high-rise buildings where refuge areas or evacuation plans exist. In buildings with high densities and no evacuation plans, compartmentation is less effective.

Venting Smoke

The key to smoke control systems is the proper venting or exhausting of the fire floors. If zones are not properly vented during a fire, occupants will be prevented from evacuating and firefighters will have great difficulty extinguishing the fire. In addition, the fire will create high temperatures with rapid gas expansion. This can lead to significant overpressures, which may result in explosions. Frequently, the building's HVAC system, in conjunction with operable fire dampers, provides the means for pressurization and venting. Other methods of ventilation used in conjunction with the operable dampers include smoke shafts and exterior wall vents (see Figure 14.3).

Smoke Shafts

A smoke shaft consists of a shaft that extends from the bottom to the top of a building. Normally-closed dampers are located at each floor and a vent to the exterior is located at the roof. These shafts act in conjunction with stair or zone pressurization to decrease smoke buildup on the fire floor and promote air movement into the shaft where it is vented to the outside. Exterior wall vents allow smoke to vent directly to the exterior from individual smoke zones. Typically, exterior vents consist of windows or wall panels that automatically open in the event of fire. Positive pressurization of the fire zone promotes the movement of smoke to the vents. These systems must be designed so that zone pressures can overcome any wind forces acting on the vent.

Specific Design Considerations

Because the design of smoke control systems is relatively recent, the design guidelines are still under development. Local codes, the National Fire Protection Association (NFPA), and ASHRAE afford some guidelines; however, many criteria still require engineering evaluation to determine whether the system will perform effectively. In large installations, computer modeling is often used to design the system and smoke-bomb testing is often employed to test the system after installation.

When designing a smoke control system, the five most important areas of consideration are as follows:

1. *Leakage Areas*. Leakage areas refer to the gaps, cracks and openings in building components. An adequate pressure difference across these surfaces acts as the primary means of smoke control. For buildings higher than 100 feet [30.5 meters] or with complex floor plans, the analysis of air movement can be extremely complex and a computer analysis is highly recommended.

2. *Weather Data*. While temperature design data has been well-established for heating and cooling, the same cannot be said for smoke control systems. In heating and cooling systems, short-term temperature fluctuations above and below the design temperatures do not generally affect the performance of the system. This is generally because of the time lag associated with thermal

movement through building envelopes. In smoke control systems, however, temperature fluctuations above or below the design temperature can have significant effects because smoke control systems have to act at a particular point in time. If the outdoor temperature is above or below the design temperature, stack effect can have a significant influence.

3. *Pressure Differences.* Both minimum and maximum pressure differences should be carefully determined for a smoke control system. If pressure is too high, doors that open into stairwells may be too difficult to open. If pressure is too low, wind, stack and buoyancy effects may overpower the smoke control system. The low pressure difference may become extremely important if a window is broken during the fire.

4. *Air Flow.* Each open door in a stairwell will leak large quantities of air and affect the overall performance of the stair pressurization system. On average, each open door will require the stair fan to pump approximately 3,000 cfm [1400 L/s] additional. Since the doors can open randomly, the building stack effect will vary pressure vertically in the stairwell. Fans have a limited capacity to respond quickly, and the pressure on the doors needs to be controlled so that occupants will be able to enter the stair in order to evacuate. The design of such systems is an extremely complex problem that is critical to the design of smoke control systems. One criterion for the design of air flow is that smoke backflow must be prevented during building evacuation. While the analysis of air flow is beyond the scope of this book, factors such as evacuation time, rate of fire growth, building configuration, and the presence of a fire suppression system are critical to proper design.

5. *Number of Open Doors.* Given the large air requirements for open doors (velocities ranging from 50 to 800 fpm [0.25 to 4 m/s]), the number of open doors can significantly affect the size, complexity and cost of smoke control systems. The design for door openings is based primarily on occupancy. In buildings with high densities, all doors may be open during evacuation. However, if the building evacuation is staged, or areas of refuge are provided, the number of open doors may significantly decrease.

Smoke Control Actions in a Fire

When a fire alarm is initiated, the following control sequence applies:
1. In the fire zone, 100 percent of return air is exhausted to the outside.
2. The supply air to the fire zone is shut off.
3. In non-fire zones, supply air should consist of 100 percent outside air.
4. The exhaust air from each non-fire zone should be shut down.
5. Stair pressurization may be used to further protect escape routes.

Figure 14.3 illustrates some of the mechanisms in smoke control systems.

During a fire, air pressures in the building will vary widely. Tests on buildings have shown pressure differences across stair doors to vary from .2 to .5 inches of water (60–150 Pa), so that a force of 25 pounds (110 newtons) is necessary to open the doors unless pressure relief devices are provided.

Fire and Smoke Dampers

It is a requirement of life safety codes that openings through fire-rated assemblies (e.g., floor, ceiling, wall) be protected. Thus when a duct passes through a fire-rated wall, it must have a fire damper that will close in the event of a fire and maintain the fire rating integrity of the wall. Fire dampers are rated in hours and for the temperature that triggers the closing damper. Corridor walls, stairwell walls, and walls between different tenants are normally fire-rated. There are four basic types of dampers:

- Fire dampers
  1.5 or 3 hours
  165°F [74 °C]
- Ceiling radiation dampers (ceiling dampers)
  1, 2, 3 or 4 hours
  165°F [74 °C]
- Smoke dampers

Leakage Classification	Maximum Leakage (cfm per ft^2 at 1 in. wg)
0	0
I	4
II	10
III	40
IV	60

- Fire and smoke dampers (Combination fire/smoke dampers are available in 1-1/2- or 3-hour ratings and for different leakage classifications; combinations will vary from manufacturer to manufacturer.)

In buildings that are sprinklered, some walls may be built unrated. Fire dampers may be triggered by a fusible link that will melt at 165°F [74 °C] and close the damper, or by an electric control that will close the damper from a central fire alarm location. It is good practice to provide access panels, access doors and spare links so that resetting the dampers is as simple as possible. Figure 14.4 shows some typical fire and smoke damper details.

High-rise Buildings

High-rise buildings require life safety provisions beyond those of basic building codes. Because of their height (generally above 75 feet [23 meters]), removal of occupants becomes more critical – upper floors are beyond the reach of fire department ladder trucks. Many codes require specific design features to decrease the danger in building evacuation during a fire. Figure 14.5 illustrates the basic principles of high-rise building systems.

The basic provisions of most codes for high-rise buildings include:

- A complete automatic sprinkler system and a standpipe system tied to the fire department.
- A complete fire detection system, including smoke detectors in all utility closets and rooms. Smoke detectors are required in mechanical and electrical rooms, on the supply side of air handlers over 2,000 cfm [950 L/s], on both the supply and return side of air handlers over 15,000 cfm [1700 L/s], corridors, storage areas, tops of shafts, and related spaces tied to central alarm and smoke evacuation.
- Smokeproof enclosures in at least one stair. Re-entry on every fourth floor maximum is usually provided to avoid descending unnecessary floors. Vestibule required (44" × 72" [1.12 × 1.82 m] minimum between tenant space and stair) with natural or mechanical ventilation. Other stairs must be pressurized.

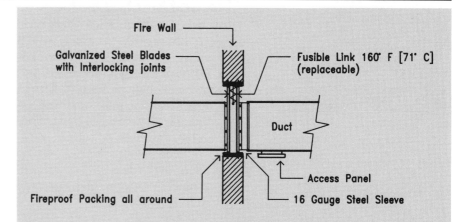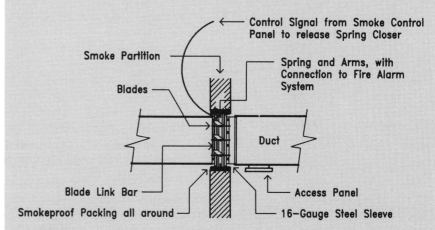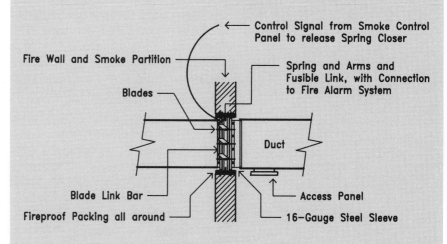

Figure 14.4

- Voice communication between a central control station and various parts of the building, including stairs, elevators, and elevator lobbies.
- A fire department control station (command center) with annunciator panels and controls for various building systems, including emergency power, AHU's, and communications. The fire department control station contains a voice alarm with PA system speakers in elevators, elevator lobbies, corridors, exit stairways, rooms, guest rooms, all tenant spaces exceeding 1000 s.f. [93 m^2], dwelling units, and hotel guest rooms (which broadcast a pre-recorded message); two-way fire department communication panel; fire detection annunciator; elevator status indicators with override; smoke evacuation indicators with override on air handlers; stair unlocking systems; sprinkler and waterflow detection and display; and telephone for the fire department.
- A smoke control system: remote control of shutters, dampers, automatic exhaust of smoke, pressurization of floors and exit stairs.
- Elevator recall: automatic delivery of all elevators to the main lobby in alarm, with one car for use by the fire department. Minimum elevator: 54" × 78" [1.4 m × 2 m].
- A fire safety plan: pre-organized management plan that utilizes building occupants trained by the fire department for use during emergencies and drilled in building evacuation techniques.
- Emergency power and systems: fire alarm, smoke control system, exit lights, elevator system, and emergency lighting all must be capable of running on emergency power, usually an emergency generator.

Stairs in high-rise buildings provide the primary means of egress during a fire. However, because they generally form continuous, unobstructed openings from the base of the building to the roof, the danger from smoke movement into these shafts is severe. To address this issue, most codes have adopted the concept of "smokeproof enclosures" as a means for providing safe and relatively smoke-free egress from buildings. Generally, a smokeproof enclosure consists of a fire-rated stair tower connected to each floor of the building by a vestibule. The towers are pressurized as described previously, using fans to minimize smoke infiltration into the stair towers.

The vestibule in a smokeproof enclosure has several key components, which are tied either directly or indirectly to the HVAC system. These include: 2-hour fire separation assemblies, self-closing doors, minimum size requirements (generally 44" × 72" [1.12 m × 1.82 m]), ceilings located 20" [0.5 m] above door openings to trap smoke, and mechanical ventilation. In terms of ventilation, NFPA suggests that one air change per minute be provided in the vestibule. In addition, NFPA recommends that the exhaust air be 150% of the supply air. This creates a negative pressure chamber in the vestibule, which moves smoke away from the positively pressurized stair towers. The supply vents should be placed a maximum of 6" [150 mm] above the floor and the exhaust vents should be placed a minimum of 6" [150 mm] below the smoke trap. A typical vestibule is shown in Figure 14.5, *Typical High-Rise Requirements*.

Other High-rise Issues

There are several other issues that may or may not be addressed directly by the model codes but which are a part of many systems for high-rise construction. These include such items as elevator recall and fire safety plans. In general, *elevator recall* refers to the automatic delivery of all

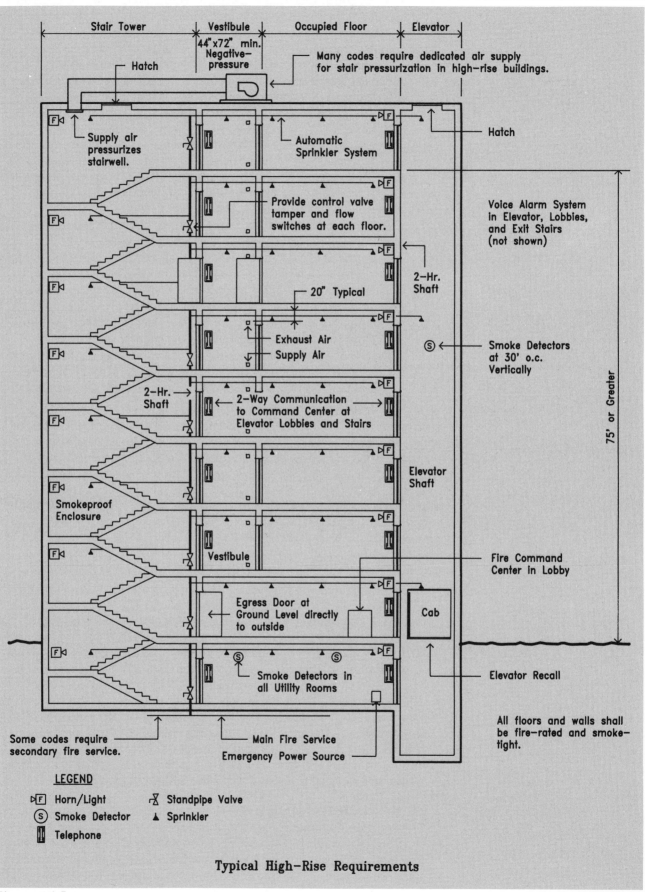

Figure 14.5

elevators to the main lobby in an alarm, with one car for use by the fire department. The minimum size for these elevators is 54" × 78" [1.4 m × 2 m], which is large enough for a gurney so that injured people can be evacuated.

Fire safety plans refer to the instruction and preparation of building staff members or tenants for use during emergencies. These individuals would be drilled by the fire department in such matters as staged evacuations, first aid procedures, organizing persons in an emergency, and how to minimize panic situations. Fire wardens are designated for each floor and are in contact with the fire department by intercom during an emergency. They direct people during an emergency in the event that fire personnel are not available. For example, the fire warden on a particular floor would direct people to remain on the floor or to evacuate, and to which floor.

Figure 14.6 summarizes the special high-rise requirements for five different building codes, including: BOCA (Building Officials of America), UBC (Uniform Building Code), SBCCI (Standard Building Code published by the Southern Building Code Conference International), MASS (Commonwealth of Massachusetts Building Code), and NFC (National Fire Codes published by the National Fire Prevention Association (NFPA)). The chart lists the similarities as well as the significant differences between these codes. Designers should be aware that different codes may conflict, which can add considerable difficulty to the design of a building. The designer should take extreme care to conform to the codes applicable to a particular project; for example, many large cities implement their own codes, which are more stringent than the state codes in which they reside. This is particularly true with regard to high-rise buildings.

Atriums and Covered Malls

Other types of buildings or parts of buildings also require special life safety requirements. Two common types are atriums and covered malls. An *atrium* is generally defined as an occupied space that includes a floor opening or several floor openings that connect two or more stories. Atriums require a smoke control system, usually triggered by the fire alarm system and designed to evacuate smoke and bring in large quantities of fresh outdoor air.

A comparison of atrium code requirements is shown in Figure 14.7. An atrium smoke control system generally consists of separate supply and exhaust fans capable of bringing in and exhausting six air changes an hour or a minimum of 60,000 cfm [28 500 L/s]. The volume of air to be changed includes the atrium and all spaces opening into it that are not separated from the atrium by a fire wall. This could be an entire floor area if, for example, the hotel rooms or shops that open onto the balcony that overlooks the atrium are of glass, have windows, or are not rated.

A *covered mall* is generally a roofed-over common pedestrian area serving more than one tenant. A covered mall is most often a building enclosing a number of occupancies such as retail stores, drinking and dining establishments, entertainment and amusement facilities, transportation terminals, offices, or other related businesses wherein two or more tenants have a main entrance into one or more malls. Anchor stores are not generally considered part of a covered mall building.

Figure 14.8 illustrates the requirements that different model codes have established for covered malls.

Design of HVAC Systems for Health Care Facilities

Of all the building types, none is more directly concerned with life safety than hospitals, or more complicated to design. In addition to providing all the systems and controls for life safety that are necessary in other buildings, a hospital must also deal with control of microorganisms; improve odor control; individualize room comfort; provide for a large range of spaces such as kitchens, bedrooms, laboratories, offices, computer

High-Rise Code Comparison				
Item	BOCA	UBC	SBCCI	MASS
Definition: Occupied floors > 75 feet [23 m] above lowest vehicle access	X	X	X	
Occupied floors > 70 feet [21.5 m] above lowest vehicle access				X
Automatic fire suppression	X	X		X
Automatic sprinkler system or areas of refuge for buildings less than 12 stories			X	
Group R buildings and Group B buildings greater than 12 stories or 150 feet [46 m] shall have an automatic sprinkler system			X	
Secondary water supply may be required	X	X		
Smoke control on all floors			X	
Smokeproof enclosure for all exit stairs	X	X		
One smokeproof enclosure; other stairs to be pressurized				X
Group B buildings of greater than 15,000 s.f. [1400 m^2] per floor are exempt from smokeproof enclosure requirements when areas of refuge are provided			X	
Smoke detectors in utility rooms	X	X		X
Smoke detectors in shafts more than 2 stories high		X	X	X
Smoke detectors in main return and exhaust air plenums for each air conditioning system		X		
Smoke detectors in main return and exhaust air plenums for each air conditioning system serving more than one story			X	
Provide smoke detectors in ceilings of elevator lobbies		X	X	
Voice alarm systems	X	X	X	X
Public address systems	X	X	X	X
Fire department communication system	X	X	X	X
Fire department central control station	X	X	X	X
Emergency power	X	X		X
Stairway communication system	X	X	X	X
Elevators shall open into lobbies that are separated from other floor areas		X	X	
Elevators cannot be vented through machine rooms		X		
All rooms and enclosures shall have at least one means of exit that does not pass through an elevator lobby			X	
In seismic zones 2, 3 and 4, anchorage of mechanical/electrical systems for life safety shall be anchored to resist earthquake loads.		X		

All numbers shown in brackets are metric equivalents.

Figure 14.6

Atrium Code Comparison

Item	BOCA	UBC	SBCCI	MASS	NFPA
Definition[1]: An occupied space that includes one or more floor openings, which connect two or more stories	X				
An automatic sprinkler system shall be installed in atrium and all areas connected to atrium	X	X		X	X
Sprinkler not required in areas separated from atrium by a 1-hour fire separation wall				X	
Sprinkler not required in areas separated from atrium by a 2-hour fire separation wall			X		
Where ceiling of atrium is greater than 55 feet [17 m], AHJ[2] may waive requirement for ceiling sprinklers			X		X
Atrium shall not be occupied for other than low hazard occupancies	X		X	X	X
A smoke control system shall not be located in the atrium	X	X	X	X	X
Smoke control system not required for atriums that connect only two stories				X	
Atrium space shall be separated from adjoining spaces by 1-hour wall or by fire windows protected by a sprinkler system	X	X	X	X	X
The adjacent spaces of any three floors do not need 1-hour separation, if these spaces are included in smoke control calculations		X	X	X	
Provide smoke control in all areas not directly connected to the atrium			X		
Fixed glazing in atriums shall be equipped with fire windows having a rating of 3/4 hour. Window area shall not exceed 25% of wall area between room and atrium.		X			
An automatic fire detection system shall be required in buildings with an atrium	X	X	X	X	X
Provide smoke detectors at ceiling and under floors extending into atrium area		X	X		
Provide smoke detectors above highest floor level and at return air intakes in atrium					X
No horizontal dimension between opposite edges of the floor opening is less than twenty feet, and the opening is a minimum of 1,000 s.f.					X
Minimum size is as follows: Stories / Minimum Dimension / Minimum Area 3–4 / 20 / 400 5–7 / 30 / 900 >8 / 40 / 1,600		X			
The exits are separately enclosed from the atrium; access to exits may be in the atrium					X

All numbers shown in brackets are metric equivalents.

[1] Definitions for atrium vary slightly from code to code.
[2] AHJ: Authorities having jurisdiction.

Figure 14.7

Covered Malls Code Comparison

Item	BOCA	UBC	SBCCI	MASS	NFPA
Definition[1]: A roofed-over common pedestrian area serving more than one tenant located within a covered mall building	X				
Tenant spaces shall be separated by 1-hour walls	X		X	X	
Walls separating tenants shall extend from the floor to the underside of the floor or roof deck above			X		X
Maximum travel distance for egress: 200 ft. [60 m]	X	X	X	X	
Anchor stores shall provide required exits directly to the exterior	X		X	X	X
Mall dead ends shall not exceed twice their width	X		X	X	
Design occupant load shall be based on following formula: OLF = (.00007) × (GLA) + 25; OLF = occupant load factor; GLA = gross leasable area	X				
Design load based on gross leasable area: < 150,000 s.f. [45 000 m^2] 30 s.f./person [2.75 m^2/person]; 150 000 to 350 000 [45 000 m^2 to 32 500 m^2] 40 s.f./person [3.75 m^2/person]; over 350,000 s.f. [32 500 m^2] 50 s.f./person [4.6 m^2/person]		X	X	X	
Distribute exits and widths equally throughout mall	X		X	X	
Every floor of the mall should have at least two remotely located exits					X
Storage is prohibited in exit corridors	X		X	X	
Minimum mall width is 20' [6 m]. Provide 10' [3 m] unobstructed exit access on either side of mall. Minimum width of exit 66" [1.7 m].	X	X	X	X	X
Covered malls less than 3 stories are exempt from area limitations; must be type 1, 2, or 4 construction	X			X	
Floor/ceiling assemblies in multilevel malls shall be of 1-hour construction				X	
Anchor stores 3 stories or less shall be exempt from area limitations if a smoke control system is installed	X			X	
Roof coverings for covered malls shall be Class A, B, or C	X			X	
Use group A-1 and A-2 shall have not more than 1/2 required exits opening into mall	X				
No less than 1/2 the required exit widths for Class A or B stores shall lead directly outside					X
Provide automatic sprinkler system throughout building	X	X	X	X	X
Mall sprinkler system shall be independent of tenant system unless each tenant space is independently zoned		X			
Provide standpipe connections at each entrance to an exit passageway; supply 250 GPM [16 L/s]	X			X	
Provide standpipe connections at each entrance to an exit passageway; at each floor landing in a stairwell, and at principal exterior entrances; supply 250 GPM [16 L/s].		X	X		
Provide smoke control system for mall and adjoining tenant spaces	X	X	X	X	X
Provide fire detection system	X	X	X	X	X
Fire protection systems and HVAC controls shall be identified and shall be accessible to fire department	X	X		X	
Meet plastic sign limitations for size and location	X	X	X	X	

All numbers shown in brackets are metric equivalents.

[1] Definitions for covered malls vary slightly from code to code.

Figure 14.8

rooms, treatment rooms and operating rooms; and accommodate a full range of technological systems such as medical gases, nurse call systems, and telecommunications. Hospitals must have special accommodations for patients in all manner of distress, special toilet facilities, isolation rooms to protect patients from getting diseases from the public, and other rooms to protect the medical staff and the public from communicable diseases of patients. Hospital buildings are awesome in the success with which they provide health care as well as the enormous efforts needed to design and maintain them.

Currently there is a shift in health care from hospitals to ambulatory care. While the definitions vary slightly from state to state, a *hospital* is generally a place where a patient may stay for more than 24 hours or be treated while incapacitated (e.g., under general anesthesia or incapable of self-preservation). Hospitals are designed for "defense in place." Hospital staff are trained to move patients to an adjoining area of refuge or evacuate if necessary during a crisis.

Ambulatory care implies that a patient walks in, through, and out of the facility. Minor surgery using local anesthesia; patient examinations; diagnosis; limited invasive treatment; endoscopy; and radiation treatments are all permitted in an ambulatory facility. In an ambulatory-care emergency, the occupants (including patients) should be able to evacuate and or be moved safely. Ambulatory-care facilities are generally designed as office buildings, so they do not need to conform to hospital standards and are not certified as hospital facilities.

Codes Pertaining to Health Care Facilities

Health care facilities are highly regulated. Building codes and standards are extensive for hospital design. In addition, all hospitals are governed by the physicians and medical staff. In the design of hospitals, the medical staff play the crucial role in determining particular requirements. Because there are often several options available in the design of a system, and a physician may have developed a successful protocol for treatment in a particular area, there is much individuality in hospital design despite the presence of a large regulatory framework.

The following is a list of codes that apply to most hospital facilities.
- State building codes cover general building issues. State codes categorize the building use group and may refer to and incorporate

Covered Mall Code Comparison (continued)						
Item	BOCA	UBC	SBCCI	MASS	NFPA
Meet special requirements for kiosks (i.e., construction, fire suppression, horizontal separation, and maximum area)	X			X	
Parking structures are generally considered separate buildings and shall be separated from mall by 2-hour fire wall	X			X	

Figure 14.8 (continued)

parts of other codes. Many state codes refer to the health care chapters of the National Fire Protection Association's *Life Safety Code 101*.
- *Guidelines for the Construction and Equipment for Hospitals and Health Care Facilities*, published by the AIA press, covers space use and mechanical standards for health care facilities. It also refers to NFPA 101.
- NFPA 101 is more specific than most state codes with regard to requirements for specific use groups such as health care facilities. This code does require that institutional sleeping floors be divided by a one-hour-rated "smoke partition."
- Many hospitals seek accreditation by the Joint Commission for Accreditation of Healthcare Organizations. To receive accreditation, conformance to NFPA 101 or the FSES (Fire Safety Evaluation System) is required. A program of preventative maintenance of equipment, and maintenance record-keeping is also required.
- The local fire department and Department of Health, as well as Building Officials, will typically have additional requirements.

The above codes and standards focus mainly on architectural design and management issues for hospitals. They also, however, contain cross-references to the provisions of model building and energy codes that are applicable. There are particular ventilation requirements for hospitals, which are discussed later in this chapter.

Infection Control: Asepsis

Within hospitals, infection control is a major concern. The single most important focus of health management in a hospital is to control the spread of disease and maintain an aseptic environment. In a hospital, which by its nature is inhabited by people who are either susceptible to disease or already infected, it is very important to control the spread of organisms. Much of medical protocol and procedure is concerned with patient care and protection of both patients and healthcare workers from disease. Infection and disease can be transmitted by direct personal contact, by airborne transfer between people, and by airborne mold transfers (e.g., via ductwork).

Personal Contact

Although research has shown that pathogenic organisms can be transferred over considerable distances via air flow, the transfer occurs mainly by personal contact. Hospitals emphasize hand-washing in order to cut down on disease transfer by personal contact. Hand-washing sinks are generally required to be located throughout a hospital, convenient to healthcare workers as they move from patient to patient.

Airborne human organisms

After direct person-to-person contact, the next most common transmission of communicable disease is via air indoors. These diseases include viral respiratory infections, viral childhood diseases (mumps, measles, and chicken pox), and tuberculosis. Methods for controlling these infections are discussed below and include special ventilation standards and use of pressurization in HVAC room control.

Molds: Aspergillus and Legionnaires Disease

Infection by airborne organisms such as mold may also occur from non-human sources in buildings. These are usually fungus diseases of the respiratory tract.

Aspergillus is a mold that can grow in ductwork, particularly if it is moist. When it exists it can become airborne, and may be released during a building renovation. It is important to make sure that ductwork is clean and dry, that there are no leaks of coils or condensate in ductwork, that humidification is carefully controlled using dry steam, and that care is used in the use of insulation.

Legionnaires Disease, like aspergillus, is not transmitted person to person, but can become airborne and has been traced to contaminated water in cooling towers and evaporative condensers of air conditioning systems. These diseases are usually acquired indoors, where the concentration of the organisms is greater and where they are transmitted into the air handling system. In air-tight buildings with a large quantity of recirculated air, these airborne transfers of organisms are of great concern.

Airborne Particle Control

Person-to-person contact is generally controlled by medical protocol – hand-washing, gowning, and other sterilization procedures. The proper design of HVAC systems helps to control the transmission of airborne human organisms and mold. The four basic methods of controlling airborne transmission of disease or infection with HVAC systems are dilution/ventilation, filtration, precipitation, and radiation.

Dilution/Ventilation

Hospitals generally require a great amount of ventilation. As the old saying goes, "the simplest solution to pollution is dilution." This still applies with regard to airborne organisms. The major obstacle to bringing in large quantities of fresh outdoor air is that it is an expensive and energy-intensive way to heat, cool, filter, and humidify/dehumidify the resulting large quantities of air.

Ventilation standards have been developed in hospitals not only for asepsis but also for comfort and odor control. This is because life in a hospital is more fragile than in other types of buildings. Odors have a powerful psychological effect on patients' comfort and stress level.

Odor control is best accomplished at the source of the odor. Hospitals are full of odor sources, such as toilets, soiled linen, laboratories, kitchens, germicidal cleaning fluids, and medicine itself. A hospital should have an exhaust system to remove the odors from most sources. Laboratory hoods and infectious disease room exhausts should be vertical directly to the outdoors without being combined with any other ducts.

Figure 14.9 shows the AIA ventilation requirements for areas affecting patient care in hospital and outpatient facilities. These tables are in addition to the ventilation requirements detailed in the BOCA and ASHRAE tables in Figures 2.2 and 2.3. The AIA chart values are designed to move sufficient air for comfort as well as to control odor and asepsis. As can be noted from the tables, most hospital spaces require 2 to 20 air changes per hour, depending on the room (from patient rooms to operating rooms). 100% fresh air is required for contamination and odor control in areas such as isolation, toilet, soiled laundry, and autopsy rooms. Recirculating-type units (fan coils) are prohibited in certain rooms (such as isolation rooms) to limit the growth of molds.

Ventilation Requirements for Areas Affecting Patient Care in Hospitals and Outpatient Facilities[1]

Area designation	Air movement relationship to adjacent area[2]	Minimum air changes of outdoor air per hour[3]	Minimum total air changes per hour[4]	All air exhausted directly to outdoors[5]	Recirculated by means of room units[6]	Relative humidity[7] (%)	Design temperature[8] (degrees F)
SURGERY AND CRITICAL CARE							
Operating/surgical cystoscopic rooms[9]	Out	3	15	—	No	50-60	70-75
Delivery room[9]	Out	3	15	—	No	45-60	70-75
Recovery room[9]	—	2	6	—	No	30-60	70
Critical and intensive care	—	2	6	—	No	30-60	70-75
Treatment room[10]	—	—	6	—	—	—	75
Trauma room[10]	Out	3	15	—	No	45-60	70-75
Anesthesia gas storage	—	—	8	Yes	—	—	—
NURSING							
Patient room	—	1	2	—	—	—	70-75
Toilet room	In	—	10	Yes	—	—	—
Newborn nursery suite	—	2	6	—	No	30-60	75
Protective isolation[11]	Out	1	6	—	No	—	70-75
Infectious isolation[12]	In	1	6	Yes	No	—	70-75
Isolation alcove or anteroom[11,12]	In/Out	—	10	Yes	No	—	—
Labor/delivery/recovery	—	—	2	—	—	—	70-75
Labor/delivery/recovery/postpartum	—	—	2	—	—	—	70-75
Patient corridor	—	—	2	—	—	—	—
ANCILLARY							
Radiology[13]							
X-ray (surgical/critical care and catheterization)	Out	3	15	—	No	45-60	70-75
X-ray (diagnostic & treat.)	—	—	6	—	—	—	75
Darkroom	In	—	10	Yes	No	—	—
Laboratory							
General[14]	—	—	6	—	—	—	—
Biochemistry[14]	Out	—	6	—	No	—	—
Cytology	In	—	6	Yes	No	—	—
Glass washing	In	—	10	Yes	—	—	—
Histology	In	—	6	Yes	No	—	—
Microbiology[14]	In	—	6	Yes	No	—	—
Nuclear medicine[13]	In	—	6	Yes	No	—	—

Notes

1. The ventilation rates in this table cover ventilation for comfort, as well as for asepsis and odor control in areas of acute care hospitals that directly affect patient care and are determined based on health care facilities being predominantly "no smoking" facilities. Where smoking may be allowed, ventilation rates will need adjustments. Refer to ASHRAE Standard 62-1989, *Ventilation for Acceptable Indoor Air Quality,* and *ASHRAE Handbook of Fundamentals,* latest edition. Areas where specific ventilation rates are not given in the table shall be ventilated in accordance with these ASHRAE publications. Specialized patient care areas, including organ transplant units, burn units, specialty procedure rooms, etc., shall have additional ventilation provisions for air quality control as may be appropriate. OSHA standards and/or NIOSH criteria require special ventilation requirements for employee health and safety within health care facilities.

2. Design of the ventilation system shall, insofar as possible, provide that air movement is from "clean to less clean" areas. However, continuous compliance may be impractical with full utilization of some forms of variable air volume and load shedding systems that may be used for energy conservation. Areas that do require positive and continuous control are noted with "out" or "in" to indicate the required direction of air movement in relation to the space named. Rate of air movement may, of course, be varied as needed within the limits required for positive control. Where indication of air movement direction is enclosed in parentheses, continuous directional control is required only when the specialized equipment or device is in use or where room use may otherwise compromise the intent of movement from clean to less clean. Air movement for rooms indicated in the table with dashes and nonpatient areas may vary as necessary to satisfy the requirements of those spaces. Additional adjustments may be needed when space is unused or unoccupied and air systems are deenergized or reduced.

3. To satisfy exhaust needs, replacement air from outside is necessary. Table 2 does not attempt to describe specific amounts of outside air to be supplied to individual spaces except for certain areas such as those listed. Distribution of the outside air, added to the system to balance required exhaust, shall be as required by good engineering practice.

4. Number of air changes may be reduced when the room is unoccupied if provisions are made to ensure that the number of air changes indicated is reestablished any time the space is being utilized. Adjustments shall include provisions so that the

Figure 14.9

Ventilation Requirements for Areas Affecting Patient Care in Hospitals and Outpatient Facilities[1]

Area designation	Air movement relationship to adjacent area[2]	Minimum air changes of outdoor air per hour[3]	Minimum total air changes per hour[4]	All air exhausted directly to outdoors[5]	Recirculated by means of room units[6]	Relative humidity[7] (%)	Design temperature[8] (degrees F)
Laboratory							
Pathology	In	—	6	Yes	No	—	—
Serology	Out	—	6	—	No	—	—
Sterilizing	In	—	10	Yes	—	—	—
Autopsy room	In	—	12	Yes	No	—	—
Nonrefrigerated body-holding room[15]	In	—	10	Yes	Yes	—	70
Pharmacy	—	—	4	—	—	—	—
DIAGNOSTIC AND TREATMENT							
Examination room	—	—	6	—	—	—	75
Medication room	—	—	4	—	—	—	—
Treatment room	—	—	6	—	—	—	75
Physical therapy and hydrotherapy	In	—	6	—	—	—	75
Soiled workroom or soiled holding	In	—	10	Yes	No	—	—
Clean workroom or clean holding	—	—	4	—	—	—	—
STERILIZING AND SUPPLY							
ETO-sterilizer room[16]	In	—	10	Yes	No	—	75
Sterilizer equipment room[16]	In	—	10	Yes	—	—	—
Central medical and surgical supply							
Soiled or decontamination room	In	—	6	Yes	No	—	—
Clean workroom and sterile storage	Out	—	4	—	No	(Max) 70	75
SERVICE							
Food preparation center[17]	—	—	10	—	No	—	—
Warewashing	In	—	10	Yes	No	—	—
Dietary day storage	In	—	2	—	—	—	—
Laundry, general	—	—	10	Yes	—	—	—
Soiled linen (sorting and storage)	In	—	10	Yes	No	—	—
Clean linen storage	—	—	2	—	—	—	—
Soiled linen and trash chute room	In	—	10	Yes	No	—	—
Bedpan room	In	—	10	Yes	Yes	—	—
Bathroom	—	—	10	—	—	—	75
Janitor's closet	In	—	10	Yes	No	—	—

direction of air movement shall remain the same when the number of air changes is reduced. Areas not indicated as having continuous directional control may have ventilation systems shut down when space is unoccupied and ventilation is not otherwise needed.

5 Air from areas with contamination and/or odor problems shall be exhausted to the outside and not recirculated to other areas. Note that individual circumstances may require special consideration for air exhaust to outside, e.g., an intensive care unit in which patients with pulmonary infection are treated, and rooms for burn patients.

6 Because of cleaning difficulty and potential for buildup of contamination, recirculating room units shall not be used in areas marked "No." Isolation and intensive care unit rooms may be ventilated by reheat induction units in which only the primary air supplied from a central system passes through the reheat unit. Gravity-type heating or cooling units such as radiators or convectors shall not be used in operating rooms and other special care areas.

7 The ranges listed are the minimum and maximum limits where control is specifically needed.

8 Dual temperature indications (such as 70-75) are for an upper and lower variable range at which the room temperature must be controlled. A single figure indicates a heating or cooling capacity of at least the indicated temperature. This is usually applicable when patients may be undressed and require a warmer environment. Nothing in these guidelines shall be construed as precluding the use of temperatures lower than those noted when the patients' comfort and medical conditions make lower temperatures desirable. Unoccupied areas such as storage rooms shall have temperatures appropriate for the function intended.

9 National Institute of Occupational Safety and Health (NIOSH) Criteria Documents regarding Occupational Exposure to Waste Anesthetic Gases and Vapors, and Control of Occupational Exposure to Nitrous Oxide indicate a need for both local exhaust (scavenging) systems and general ventilation of the areas in which the respective gases are utilized.

10 The term *trauma room* as used here is the operating room space in the emergency department or other trauma reception area that is used for emergency surgery. The first aid room and/or "emergency room" used for initial treatment of accident victims may be ventilated as noted for the "treatment room."

Figure 14.9 (continued)

11 The protective isolation rooms described in these guidelines are those that might be utilized for patients with a high susceptibility to infection from leukemia, burns, bone marrow transplant, or acquired immunodeficiency syndrome and that require special consideration for which air movement relationship to adjacent areas would be positive rather than negative. For protective isolation the patient room shall be positive to both anteroom and toilet. Anteroom shall be negative to corridor. HEPA filters should be used on air supply. Where requirements for both infectious and protective isolation are reflected in the anticipated patient load, ventilation shall be modified as necessary. Variable supply air and exhaust systems that allow maximum isolation room space flexibility with reversible air movement direction would be acceptable only if appropriate adjustments can be ensured for different types of isolation occupancies. Control of the adjustments shall be under the supervision of the medical staff.

12 The infectious isolation rooms described in these guidelines are those that might be utilized in the average community hospital. The assumption is made that most isolation procedures will be for infectious patients and that the room should also be suitable for normal private patient use when not needed for isolation. This compromise obviously does not provide for ideal isolation. The design should consider types and numbers of patients who might need this separation within the facility. When need is indicated by the program, it may be desirable to provide more complete control with a separate anteroom as an air lock to minimize potential for airborne particulates from the patients' area reaching adjacent areas. Isolation room shall be negative to anteroom and positive to toilet. Anteroom shall be negative to corridor.

13 Large hospitals may have separate departments for diagnostic and therapeutic radiology and nuclear medicine. For specific information on radiation precautions and handling of nuclear materials, refer to appropriate publications of the National Radiation Safety Council and Nuclear Regulatory Commission. Special requirements are imposed by the U.S. Nuclear Regulatory Commission (Regulatory Guide 10.8-1980) regarding use of Xenon-133 gas.

14 When required, appropriate hoods and exhaust devices for the removal of noxious gases shall be provided (see Section 7.31.D1o and NFPA 99).

15 A nonrefrigerated body-holding room would be applicable only for health care facilities in which autopsies are not performed on-site, or the space is used only for holding bodies for short periods prior to transferring.

16 Specific OSHA regulations regarding ethylene oxide (ETO) use have been promulgated. 29 CRF Part 1910.1047 includes specific ventilation requirements including local exhaust of the ETO sterilizer area. Also see Section 7.31.D1r.

17 Food preparation centers shall have ventilation systems that have an excess air supply for "out" air movements when hoods are not in operation. The number of air changes may be reduced or varied to any extent required for odor control when the space is not in use. See Section 7.31.D1 for designation of hoods.

(Copyright 1993, AIA Committee on Architecture for Health, with assistance from the U.S. Department of Health and Human Services. From Guidelines for Construction and Equipment of Hospitals and Medical Facilities. Used by permission.)

Footnote: Cross-references in table footnotes pertain to the source from which tables were taken; they do not refer to chapters or pages in this text.

Figure 14.9 (continued)

Note: Fresh air inlet louvers should be as high as reasonable. The grille bottom should be a minimum of 6 feet [1.8 m] above grade or 3 feet [.9 m] above the roof. The inlets should not be within 25 feet [7.6 m] of any sources of poor quality air such as exhaust air outlets, hood exhausts, combustion exhaust, medical vacuum exhaust, and so on.

Filtration

Because of the minute size of infectious organisms, a high efficiency particle air (HEPA) filter is required in health care facilities. These filters are 99.97% efficient in removing particles of 0.3 microns, effectively maintaining good air quality. They are used in small rooms of one to four patients.

All central ventilation and air conditioning systems should be filtered according to Figure 14.10, *Filter Efficiencies for Central Ventilation and Air Conditioning Systems in General Hospitals and Psychiatric Facilities*, from the AIA guidelines. Filter number one should be installed upstream of the air handler if two filters are listed. Where the filter efficiency listed exceeds 75, a manometer should be installed across the filter (including HEPA filters in lab hoods).

Electrostatic Precipitation

Air purification can be achieved with electrostatic precipitation. This technique is commonly used with air scrubbers on fuel-burning facilities and for removing tobacco smoke and other irritants. It is less common in hospital environments.

Ultraviolet (UV) Light Radiation

UV radiation is extremely effective in disinfecting most airborne bacteria and viruses. UV lamps are low-pressure mercury vapor lamps that emit radiation of 254 nanometers, which is effective in destroying several microbes including those responsible for infectious tuberculosis. There are some problems in the practicality of using UV light radiation. The lamps do produce some visible violet light. They are difficult to place in patient rooms because they should not be turned off at night. Although the lights should be on 24 hours a day, prolonged direct contact to the eyes is harmful to the retina. In patient waiting rooms, emergency rooms, or similar spaces that are temporarily occupied, they can be effective. If there is a high ceiling, up-lighting can be used to kill the organisms while it is shielded from the eyes by a soffit. When using UV lighting the relative humidity should be below 70, which is not difficult to obtain in most conditioned spaces.

The most effective method of providing UV protection is in ductwork that contains a UV lighting fixture, with access panels for lamp cleaning and changing, view windows, and alarm stations. These can be installed as inserts in ductwork as needed in patient rooms, and room air may be circulated over the lights for effectiveness.

HVAC Design of Hospital Spaces

There are certain rooms where HVAC systems are integrally connected to the health care system. Isolation rooms are among the most significant. There are two types of isolation rooms: negatively pressured rooms for patients with communicable diseases, such as tuberculosis, and positively pressured rooms for patients who are immune-suppressed or vulnerable to disease from others, including patients with organ transplants, burns, or AIDS. Other types of spaces where HVAC design provides assistance in the health care network include waiting rooms, operating rooms, trauma rooms, and delivery rooms. Waiting areas for emergency rooms should consider air systems to help minimize the risks of having many undiagnosed patients together. All of these spaces have unique requirements, which are discussed below.

Infectious Disease

Negative Pressure Isolation Rooms: Rooms for patients with communicable diseases should be designed to contain the disease organisms in the room or exhaust them to the outside. A room negatively pressured with respect to the rest of the hospital will draw in air (which will be exhausted) rather than leak contaminated air to the rest of the hospital.

Among communicable diseases are new highly-resistant strains of tuberculosis, which are contagious and airborne. To protect other patients and staff from infectious diseases, it is necessary to isolate the patient with an infectious disease in a private, negatively-pressured room. As noted

Filter Efficiencies for Central Ventilation and Air Conditioning Systems in General Hospitals and Psychiatric Facilities

Area designation	No. filter beds	Filter bed no. 1	Filter bed no. 2
All areas for inpatient care, treatment, and diagnosis, and those areas providing direct service or clean supplies such as sterile and clean processing, etc.	2	25	90
Protective isolation room	2	25	90
Laboratories	1	80	—
Administrative, bulk storage, soiled holding areas, food preparation areas, and laundries	1	25	—

Note: Additional roughing or prefilters should be considered to reduce maintenance required for main filters. Ratings shall be based on ASHRAE 52-76.

(Copyright 1993, AIA Committee on Architecture for Health, with assistance from the U.S. Department of Health and Human Services. From Guidelines for Construction and Equipment of Hospitals and Medical Facilities. Used by permission.)

Figure 14.10

below, negative pressure is achieved by exhausting 25% more air than is being supplied. The following features should be part of an infectious disease isolation room.

- Supply a minimum of six air changes per hour (ACH) through a 90% filter in a diffuser in the ceiling.
- Exhaust air directly to the outside; no recirculation of the air is permitted. The exhaust volume should be 1.25 × the supply air volume directly to the outside.
- The bottom of the exhaust grille should be located low in the wall, approximately 4" above the floor across the room from the ceiling diffuser. Locate the air devices so that the clean air will pass the staff and the patient and be exhausted.
- The exhaust fan should operate continuously and be located on the roof, at least 25 feet from any air intake or operable window.
- A vestibule with 2 doors should be included in the room design to provide an air lock between the corridor and the patient room. This room should be supplied with 10 air changes per hour. The patient room door should be kept closed.
- No recirculation of air should be permitted by fan coils or by gravity through concealed spaces, which are unlikely to be cleaned; bacteria could multiply. Air should be recirculated in this room only through a HEPA filter unit.
- No radiators, baseboards, or fin tubes should be installed because they are difficult to clean and are a potential source of microorganism growth.

Positive-pressure Immune Suppressed Rooms: Many patients are very susceptible to infections because their immune systems are weak. These immune-suppressed patients are usually bone-marrow-transplant, organ-transplant, AIDS, or burn patients. For these rooms, the air is positively pressured relative to the rest of the hospital, which discourages air from the hospital from leaking into the room and reaching the patient. Depending on the nature of the disease, staff may wear sterile garments when entering the room, or the patient may be kept under a sterile canopy. The overall air requirements for these rooms are similar to negative pressure isolation rooms except the exhaust air volume is lowered to maintain positive pressure and it is not as critical (although it is preferable) to have a vestibule. This type of room protects the patient from airborne organisms coming from outside the room. This patient is not a threat to the staff or other patients.

The following criteria should be followed in designing immune-suppressed rooms:

- Exhaust at least 6 ACH from low-wall return air grilles. This air can be recirculated in this room or elsewhere.
- Supply 125% of the exhaust air volume from the ceiling above the patient. This air should be 100% filtered outside air with the final filter being a HEPA-type filter. Some component of the air volume could be recirculated from within this room only if it is HEPA-filtered.
- No recirculation through a concealed space is permitted, either by fan or by gravity.
- If a vestibule is provided, the air pressure should be negative and the volume equal to 10 ACH. The patient room door should be kept closed.

Figure 14.11 shows the layout of a typical isolation room designed to switch from positive to negative pressure. It contains a vestibule, a control system that permits rebalancing of the supply and exhaust air quantities, low-air exhaust to help ensure the patient is "washed" with clean air, and the sensors and dampers that operate the system. Such rooms may also contain an alarm to notify staff in the event of failure of the air to flow and connection of the equipment to an emergency generator.

Waiting Areas

In areas where many undiagnosed people may be together, precautions should be taken against transmission of organisms. An infected patient may transmit a disease to another in a susceptible condition. This situation calls for the use of either air disinfection with UV radiation or dilution of the organisms by supplying large quantities of air with sufficient exhaust to create negative pressure. Most spaces do not have the high ceiling required for the UV system, and the violet light may be objectionable. To use the high volume air ventilation approach, the following measures are recommended:

- Supply 6–20 ACH from the center of the ceiling through 90% filters. This filtered air can be recirculated.
- Exhaust air from two or more low-wall return-air grilles at the perimeter of the space. The exhaust volume should be 125% of the supply air volume.
- The exhaust air should be exhausted directly outside to the roof by a fan and discharged 25 feet [7.6 m] from any intake louvers or windows.

Psychiatric Units

In psychiatric units, special equipment must be used in spaces where patients are housed. These patients may not be capable of self-rescue, and they may attempt to harm themselves or demonstrate destructive behavior. In these areas safety and order is preserved by staff surveillance. Areas where patients may be alone or are difficult to observe are the most hazardous. In these areas, such as patient rooms or bathrooms, it is important to use tamperproof equipment (e.g., diffusers and grilles). They should have tamperproof screws, and any openings for air should be of thick gauge material with openings too small to insert a finger. Small perforations also prevent the insertion of any implements.

Accessible or suspended ceilings should not be used in unsupervised areas, so access panels should be provided for equipment above ceilings. No mechanical devices should be used in unsupervised rooms that would require maintenance. Concealed spaces, such as fin tube radiation, should be avoided. Heating and cooling is best handled with inaccessible grilles and diffusers or by radiant panels.

Laboratories

In general, laboratories are places where carefully regulated air-flow design is quite important. Laboratories can be very specialized, varying from clinical labs designed around automated diagnostic machines to generic basic research labs. Flexibility is important to allow for future changes in equipment for different studies. It is important to provide excess capacity for air quantity and cooling capacity. Laboratories are usually packed with heat-producing equipment and odor-producing chemicals. Laboratories require a minimum of six air changes per hour. All odors and contaminants should be confined within the lab by closed doors and exhausted outside without any recirculation.

Some areas in laboratories deserve special air treatment, such as hoods and benches for cutting and staining tissue. The air flow here should be

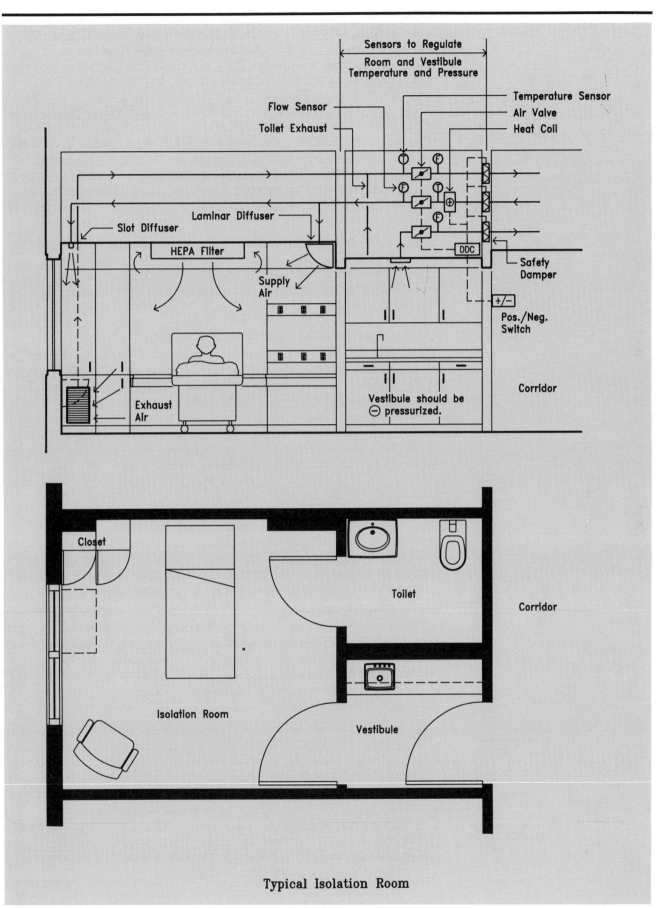
Figure 14.11

undisturbed laminar flow away from the technician. These areas should be designed in an alcove away from doorways, which would interrupt the flow when opened. The amount of exhaust is often regulated so that when hood exhaust is activated, the room exhaust is lowered. This keeps the overall pressure in the space properly adjusted and maintains the proper quantity of supply air.

Special Procedure Rooms
Hospitals contain many specialized rooms. Different types of operations and procedures are performed on patients who are then open to the possibility of infection. These rooms include operating rooms (OR), delivery rooms, catheterization rooms, and trauma rooms. The air handling in these spaces should be similar to the positive-pressure immune suppressed room.

Many studies have been done concerning the air flow in operating rooms. Generally, flexibility is important; air flow should be handled in the micro-environment with specialized equipment. Exhaust air from anesthesiologists, for example, can be taken from the ceiling directly above the area where administered.

HVAC Layout and Equipment for Hospitals
The ductwork, insulation, exhaust systems, and other HVAC devices have particular design guidelines in hospitals. The overall layout is affected by the location of smoke partitions and fire walls. NFPA 101 requires that each patient sleeping floor be divided by a "smoke partition," which should be a one-hour rated wall. In addition, corridor walls are required to be one-hour rated unless a fire protection (sprinkler) system is provided. A fire protection system can have a significant effect on floor layout as shown in Figures 14.12a and 14.12b.

Distribution Ducts
NFPA 101 requires that each patient sleeping floor be divided by a "smoke partition," which should be a one-hour rated wall. Ducts penetrating this wall are required to include smoke dampers. In addition, corridor walls are required to be one-hour unless fire protection is provided. Only the major distribution ducts should penetrate this barrier. Branch ducts should be located to serve only the spaces within each compartment.

Smoke dampers should be operated by the fire alarm system. When smoke is detected they close, limiting the recirculation of smoke on the floor. In addition, the air handling units go into a smoke control mode and the fire alarm system is activated. The staff also follow emergency procedures for the duration of an emergency. The duct systems are required to conform to NFPA 101, which among other issues provides for the following:

> *Duct lining* inside ductwork in hospitals is a sensitive issue. Acoustical separation between adjacent rooms is generally desirable, yet providing a place where bacteria could grow is a much more serious problem. Duct linings are permitted to be used in less sensitive areas, provided the lining is covered by an impermeable membrane. Duct linings should not be used in any ducts serving procedure rooms (areas where any invasive procedures are performed) or rooms where immune-suppressed patients may be housed. During renovation any existing internal duct lining should be removed. Many hospitals prohibit duct linings.

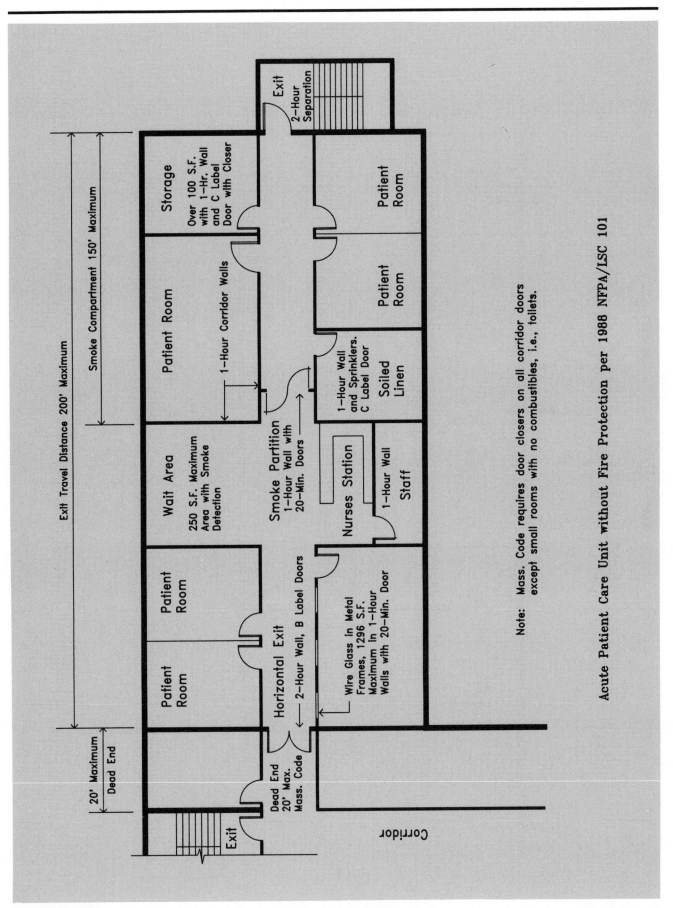

Figure 14.12a

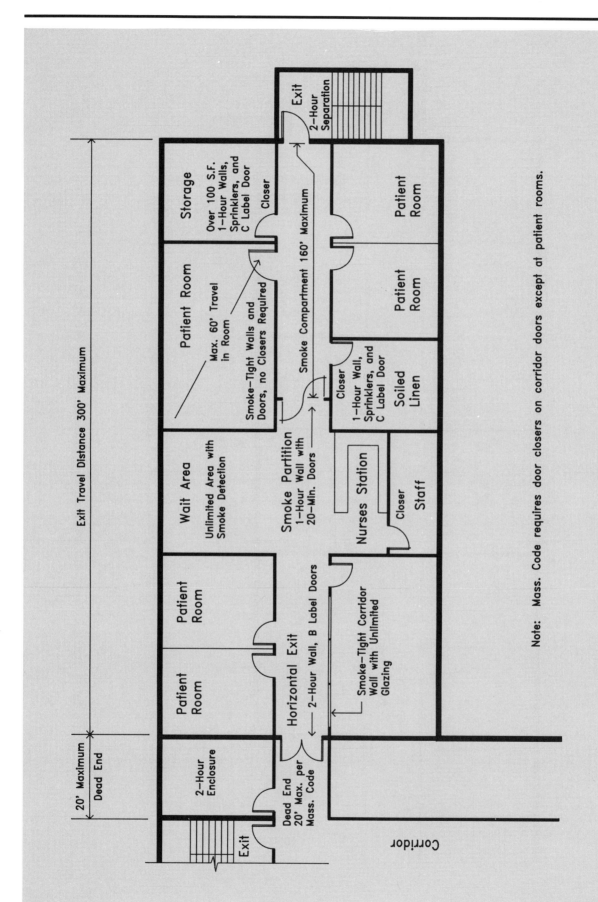

Figure 14.12b

Ducts penetrating walls with *radiation shielding* cannot compromise this protection. In X-ray rooms and radiation therapy areas, the ducts will need to be shielded with lead or make a convoluted pathway through a solid concrete radiation-absorbing wall. A physician must be consulted to determine the shielding required depending on the location with respect to the strength of the source of radiation.

Exhaust Systems

As previously discussed, areas of the hospital should be ventilated as described in Figure 14.9. Exhaust air systems should be installed to provide the required quantities of exhaust air and pressure differentials. When possible, exhausts are collected together both to provide a more economical fan layout and for energy reclamation. Caution must be used to prevent any recirculation of this exhaust air within the building. Exhaust fans should be located at the exterior of the building so that all the interior exhaust ductwork is under negative pressure. The exhaust air outlets must be 25 feet away from any air intakes or operable windows, and should be above the roof level.

Laboratory Hoods: Hoods are used to protect technicians from hazardous materials within the hood by means of a glass shield and air flow away from the opening and into the hood. The air then flows through a filter and out the duct to the exhaust fan. Manufactured hoods generally provide two-speed fans. The low speed is always on so that when the hood is not in use and the glass shield is closed, the hood space is still being evacuated to prevent the leakage of toxic fumes or organisms into the building's air system.

Each hood requires its own separate duct and exhaust fan. The fans must be outside the building, usually at the roof, and must operate continuously. The exhaust ducts cannot be combined with each other or with the building exhaust system. The exhaust ducts should be of a non-combustible and non-corrosive material such as stainless steel, or as required to meet the standards for the intended use of the hood.

Hoods should not be used for the normal exhaust of a room: care should be taken to keep the room slightly positively pressurized relative to the hood. If sufficient air is not available for the hood when it is in use, then additional make-up air should be provided in the hood area. Depending on the nature of the laboratory or uses of the hood, it may be connected to standby power. Some other considerations for laboratory hoods are as follows:

- Chemical fume hoods shall have a minimum face velocity of at least 75 fpm [0.4 m/s] and a maximum of 150 fpm [0.8 m/s].
- Hoods intended for use with biologically hazardous and radioactive (hot) material shall have a face velocity of 150 fpm [0.8 m/s]. A HEPA filter should be in the exhaust air flow. This type of hood is required to include static-pressure-operated dampers to trap hazardous material in case of a fan failure or loss of power. It should also include an alarm to alert the staff of a fan failure. The alarms should be operated by static air pressure and powered by emergency power.

Special Applications

In addition to medical gases such as oxygen and suction lines, gases are used in hospitals for anesthesia and for sterilization. Once injected into a space, these gases must be controlled and effectively exhausted.

Anesthesia Exhaust and Gas Storage Areas: Special procedure rooms generally have low-air-return grilles. However, grilles may also be installed in the ceiling over the area where anesthesia equipment is located. The purpose of this is to immediately exhaust escaped anesthesia gas before it can be inhaled by people. Rooms used for the storage of anesthesia gas cylinders shall be ventilated per NFPA 99 – natural convection or mechanical ventilation to the outside air is permitted.

ETO Sterilizers: Reusable plastic items can be sterilized by the use of low-temperature ethylene oxide (ETO) sterilizers. ETO is a toxic gas that will kill the organisms in the sterilizer; it is lethal if not properly controlled. The sterilizer is usually enclosed within a room. This area is treated as a plenum with vents to the loading side of the sterilizer above the sterilizer door. The whole plenum space then is exhausted with a dedicated welded duct to the roof-mounted exhaust fan and discharged 25 feet [7.6 m] from any intake. The capture velocity above the sterilizer door should be at least 200 feet per minute [1 m/s]. The air flow in the operator's space should be toward the sterilizer. A gas alarm system is recommended to warn of any accumulation of the gas.

Humidifiers

While humidification of air is recommended for general comfort, it is required in some areas of hospitals to prevent the drying of tissue. In particular, humidification is required in rooms where invasive procedures are performed, such as operating rooms. The humidifier installation should follow these criteria.

- Locate the duct humidifier at least 15′ [4.5 m] before the final filter.
- Power the humidifier through an in-duct air-flow switch and provide a humidistat.
- Allow duct length downstream of the humidifier for complete moisture mixing before branch duct take-offs.
- Reservoir-type humidifiers should not be used, as they provide a bacterial breeding area in stagnant water.
- Use steam-type humidifiers where possible.

Commercial Kitchens

Commercial kitchens have particular design requirements, which are meant to vent combusted gases and cooking odors from cooking stoves, and most importantly, to exhaust the grease-laden fumes safely. Because of the flammability of grease and its close proximity to combustion in commercial kitchens, kitchen hoods are an important type of exhaust system. Kitchen exhaust systems typically include the following components:

- Hood
- Grease removal device
- Exhaust duct
- Fan
- Fire extinguishing equipment

The recommended practice for these items is covered in NFPA 96 – *Standard for the Installation of Equipment for the Removal of Smoke and Grease Laden Vapors from Commercial Cooking Equipment*.

Exhaust Hoods

There are two basic types of kitchen exhaust hood systems: exhaust-only systems and exhaust-plus-makeup-air systems. The *exhaust-only system* typically consists of a hood, grease trap, filter, exhaust duct, roof curb, and fan (see Figure 14.13). These systems are generally used for low-intensity applications, such as delis and snack counters. They exhaust room air directly to the outside. These systems are generally less energy-efficient than make-up units because the exhaust fan removes 100% tempered air from the occupied space.

Exhaust-plus-make-up-air systems include all the features of exhaust-only systems as well as a supply air slot, supply duct, supply fan, intake extension and air-intake filter hood. See Figure 14.14.

In make-up air systems, fresh air is supplied to the perimeter of the exhaust hood and replenishes all or part of the air exhausted from the space. This reduces the amount of tempered air removed from the space and provides for greater energy efficiency. Further enhancements can be accomplished by supplying air around the perimeter of the exhaust duct. As the hot exhaust gases rise, heat is transferred through the walls to the exhaust duct and into the supply air stream. This warms the supply air and reduces the strain on the heating coil in the make-up unit. There are limits on how cold the incoming air from the make-up unit can be, and in winter it is normal in cold climates to temper the incoming air somewhat.

NFPA recommends that hoods be installed in all applications that produce smoke or grease-laden vapors. It is also recommended that all solid-fuel-burning equipment be served by hoods that are separate from other kitchen ventilation equipment. Recommended clearances for hoods are as follows:

- 18 inches to combustible materials
- 3 inches to limited combustible materials
- 0 inches to non-combustible materials

The hood should be designed to capture a minimum amount of air. Therefore, the hood should extend at least 6 inches (152 mm) beyond the edge of the cooking surface on all sides and exhaust 100 cfm per square foot (508 L/s per square meter) for wall canopy hoods and 150 cfm per square foot (762 L/s per square meter) for island canopy hoods. Non-canopy applications must have a hood set back no more than one foot from the edge of the cooking surface and exhaust 300 cfm per linear foot [142 L/s] of cooking surface.

The gauge of metal for the hood (canopy), exhaust duct, and supply air duct is also specifically controlled. The canopy or hood for kitchen exhausts should be constructed of either 18-gauge steel or 20-gauge stainless steel. All seams, joints, and penetrations of the hood should have a liquid-tight continuous weld. Interior joints and seams do not have to be continuously welded, but they should be completely sealed. Exhaust air volume for hoods should be sufficient to remove all grease-laden vapors. Since these factors are highly dependent on the size and type of equipment being used, designers should contact equipment manufacturers to determine specific requirements.

Exhaust ducts should be designed to transport the hot flue gases outside as directly as possible. Formed of 16-gauge steel or 18-gauge stainless steel, they should be installed without dips or traps that might collect grease. Ducts should not be connected with any other duct systems; all duct joints should be completely welded to provide a continuous liquid-tight seal.

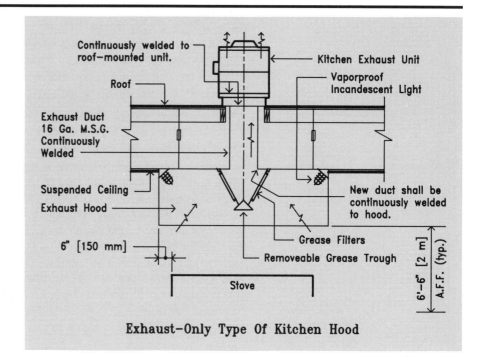

Figure 14.13

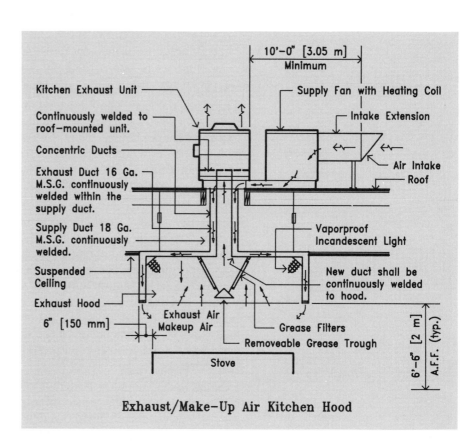

Figure 14.14

Grease Removal Devices

Grease removal devices (grease traps) should be an integral part of any kitchen exhaust system. These devices generally consist of filters, baffles, or other approved devices. NFPA recommends that mesh filters not be used.

Grease filters should have a minimum area of one square inch [6.5 cm^2] for every two cfm [1 L/s] of exhaust air from the hood.

The distance between the grease removal system and the cooking surface should be as great as possible. For charcoal or charcoal-type broilers, the minimum distance should be four feet [1.2 m]. Grease removal devices should also be protected from combustion gas outlet or from direct flame because items like deep fat fryers produce high flue gas temperatures. In cases where the minimum distance between the grease removal device and the combustion gas outlet is 18 inches [46 cm] or less, a steel baffle should be installed.

Cleaning and Inspection of Exhaust Systems

Access to ducts for cleaning purposes is essential to the proper design of a kitchen ventilation system. They should be accessible for cleaning and inspection purposes. There should be a sign placed on all duct access panels that reads: ACCESS PANEL – DO NOT OBSTRUCT. On *horizontal duct runs*, an opening 20″ × 20″ [50 cm × 50 cm] should be provided. If this is not possible, an opening large enough to allow thorough cleaning should be installed at 12 feet [3.65 m] on center. On *vertical duct runs*, an opening for personnel entry should be located at the top of the duct. If this is not feasible, access panels should be located at each floor. In *buildings greater than one story* in height, ducts should be enclosed in a continuous enclosure extending from the ceiling above the hood to the roof. In *buildings less than four stories*, the enclosure should have a one-hour fire rating (if it qualifies as a shaft, a two-hour enclosure is common). In *buildings greater than four stories*, the rating should be increased to two hours. Clearance between the duct and non-rated shafts should be 18 inches [46 cm]. Clearance between the duct and rated shafts may be reduced to 6 inches [15 cm].

Air Volumes in Exhaust Systems

Kitchen exhaust fans should provide a minimum air velocity of 1,500 feet per minute [7.6 m/s]. The air volumes should be adequate to capture and remove all grease-laden cooking vapors. In the event of a fire, exhaust fans should continue to operate unless a listed component of the exhaust system requires that the fan be shut down. It is not necessary to automatically restart the exhaust fan after the extinguishing system is activated if all the cooking equipment has been shut down. Approved up-discharge fans with motors surrounded by the air flow should be hinged and supplied with flexible wiring to allow periodic cleaning.

Fire Systems for Exhaust Hoods

Approved fire extinguishing systems should be provided for all grease removal items and hoods. These extinguishing systems should encompass all duct systems and cooking equipment that could be a source of grease ignition. NFPA recommends the following methods of fire suppression for use in kitchen exhaust systems:

Dry chemical extinguishing system	NFPA 17
Wet chemical extinguishing system	NFPA 17A

Carbon dioxide extinguishing system	NFPA 12
Sprinkler system	NFPA 13
Deluge foam water sprinkler and foam spray systems	NFPA 16

In the event that a fire suppression system activates, all sources of fuel and heat to the equipment should shut down. In addition, all gas appliances not requiring ventilation but located under the same hood should be shut down. All of these systems should be tied to the central fire alarm system in the building.

Halon is currently an acceptable system for fire suppression; however, it is considered a Class I ozone-depleting substance and is scheduled to be phased out of production at the end of 1993. For more information, see Chapter 15.

Specifications

Specifications and drawings together make "contract documents", which delineate the work to be done by a contractor. The contract documents are, in fact, the actual contract between the owner and contractor. The drawings generally show things that are best shown pictorially, such as shape, location, and interrelationship of parts. Plans, elevations, sections, and details are all examples of drawings used to delineate aspects of the work pictorially. Specifications are used to describe aspects of the work that are best illustrated in words. How payment will be made, temporary conditions, how the work will be phased, and insurance requirements are common general specification items.

MasterFormat Divisions

The Construction Specifications Institute (CSI) and the American Institute of Architects (AIA) have developed specifications formats. For each division, CSI and AIA have each classified the commonly known items in construction into a basic format. All the work to be done in construction is classified by trade, and there are 16 MasterFormat divisions. They are as follows:

Division 1   General Requirements
Division 2   Site Work
Division 3   Concrete
Division 4   Masonry
Division 5   Metals
Division 6   Wood and Plastics
Division 7   Thermal and Moisture Protection
Division 8   Doors and Windows
Division 9   Finishes
Division 10 Specialties
Division 11 Equipment
Division 12 Furnishings
Division 13 Special Construction
Division 14 Conveying Systems
Division 15 Mechanical
Division 16 Electrical

For each particular division, specifications are used to describe the items of work, the products to be used and how they are to be installed.

Each division of the specifications is subdivided. For example, Division 15, Mechanical, is divided into:

15A – Plumbing
15B – HVAC
15C – Fire Protection

The HVAC designer writes the Division 15B – HVAC Specifications in conjunction with preparing the HVAC drawings to complete the contract documents for the HVAC trade.

Proprietary and Performance Specifications

Specifications are written in one of two methods: proprietary or performance. In *proprietary specifications* the designer indicates the exact product or equipment by specifying the manufacturer and model together with the desired options and accessories by make and model. This is a straightforward method of specification and leaves little room for confusion. For example, a particular fan from a particular manufacturer would qualify as a proprietary specification. The contractor agrees under the terms of the contract to supply the equipment and products specified.

In *performance specifications* the designer describes the *result* to be achieved by the products, rather than the products themselves. In performance specifications many more items need to be determined than make and model number. For example, a fan might be described by its electrical characteristics, static pressure, cfm's delivered, discharge opening dimensions, and fan speed. In this case the contractor is allowed to provide any fan that meets all of the requirements of the specification. Care must be taken to specify maximums, minimums, or often both for certain qualities. This method allows for a wider range of choice for the contractor but can often result in items of equipment which meet the performance standards but that generally are not as rugged as may be desired. For this reason it is often wise to specify some qualities of manufacture as well. These may include equipment size, warranty periods, gauge of jacket material, types and spacing of fasteners, or similar features.

In HVAC design, performance specifications are often used to specify the automatic control systems. It is critical to thoroughly and concisely describe all of the control sequences that are to be executed by the system.

Substitutions: "or Equal"

By law, most public work requires that the products specified be available from at least three manufacturers in order to foster competitive bidding. Private work has no such mandate, but selected manufacturers do keep prices lower when they have to compete for the use of their product.

On rare occasions an exception can be made when a designer feels that a proprietary product is necessary for the satisfactory completion of the design. In the case of private work where it can be specified that no substitutions will be accepted, this sometimes prevents the contractor from "shopping" the work to his regular supplier. However, manufacturers usually increase prices or decrease discounts when they know they make the only acceptable product for a project. For private work it is often best to allow the contractor to submit substitutions but reserve the right to reject substitution at the discretion of the designer.

The consequence of substitution is that in the specification of HVAC equipment and systems, as well as other building components, designers often permit products to be "or equal" or "or equal as approved by the architect." When this is done, substitutions are likely. This can result in a contractor submitting products that may be adequate but that do not meet the intent of the designer. Even with a great deal of effort in preparing the specifications, it can be very difficult to demonstrate that a substituted product is not "equal," and a definition of "equal" is an important part of the specifications.

The need to provide competition by specifying at least three manufacturers inherently tends to lower the quality of the products to the least common denominator of the three. In addition, each product interfaces with others; so, for example, electrical characteristics, weight, and duct sizes may all be different for otherwise equal products. This variety can lead to an endless coordination dilemma.

Specification Organization

The organization of the specifications is important to their function of clearly and accurately conveying the information required. The method of dividing the system into smaller components is an effective way to accomplish this. For the HVAC system, the subcategories of products are as follows:

- Generation components
- Distribution systems
- Terminal units
- Controls and accessories

The categories listed above are also the way in which R.S. Means Co. has classified its costs; when specifying a system, related costs can be estimated.

Coordination Between the Trades

Coordination of the HVAC systems with plumbing, electrical, mechanical, fire protection, and communication systems, and the codes that govern their design and operation, has become increasingly complex. Close coordination is vital to avoiding problems during the actual construction of a project as well as problems with the continuing operation of the building by the owner.

Some common coordination problems are listed below:

- Early civil drawings may not agree with building and landscaping drawings, especially as to building footprint, floor elevations, and proper roads, paths, and dropoffs.
- Openings for ducts, conduits, and chases must be shown on architectural and structural drawings.
- Reflected ceiling plans must properly locate lights, grilles, diffusers, heat and smoke detectors, intercoms, sprinklers, soffits, box outs, and ceiling grids.
- Drawings must be checked for ductwork clearances, especially at beams where ceilings might otherwise be lowered.
- Voltage for mechanical and other equipment is often incorrectly specified.
- Motors and motor starters for plumbing, HVAC, and fire protection systems should be clarified. The trade usually supplies its own motors to the electrician.
- Controls and control wiring for plumbing, HVAC, and fire protection systems are entirely provided by the trade, not the electrician. Motor starters are generally placed between the disconnect and the equipment and are considered part of the control system.
- Disconnects are commonly missed. The electrician should generally provide power to and include a disconnect of the right NEMA class for each piece of equipment. An exception is when the disconnect comes as an integral, factory-installed part of the equipment. Clarification is required when it is a factory option.

- Construction phasing needs to be shown on the drawings. Where to cap lines, provide temporary egress, which doors and areas are to be used for workers, and which for egress should be shown.

Scope of Work

Coordination between the different specification sections and with the drawings is of key importance. For coordination between the trades, "scope of work" is written. The scope of work defines the extent of the trade's responsibility for providing or installing various aspects of the system. All of the items of work that connect various systems must be described and assigned to the appropriate trade or subcontractor. Items of work being assigned to a trade should be those items traditionally performed by that trade. When there is a choice it should be clarified. For example, unless specified clearly, control work utilizing line voltage wiring may be assumed to be the electrical contractor's work because control work wiring is usually low voltage. Unless the work is described and assigned clearly, each subcontractor may assume it is the other's responsibility, resulting in potential conflicts, delays, and extra costs. In addition, certain states may require that public projects have subcontractor trades bid work independently of the general contractor. In these cases it is critical that the scope of each subcontractor's work be defined to enable the work to be bid accurately and to avoid items of work being lost between two trades.

The scope of work section is also coupled with a "related work" section, designed to clarify the work of other trades that is closely related to the work being specified. By using the "scope of work" to define what work is to be performed, and the "related work" to establish the work being done by others, the work to be done and who does it is defined. However, the potential exists for an item of work to be listed in the scope of two trades or in the scope of neither trade. While no job is ever perfect, a careful review of all the divisions of the work will minimize these conflicts.

Coordination of Responsibilities

It is important to define who is responsible for items of work, as plumbing, mechanical, and electrical systems are routed in a building. Hangers, fasteners, and supports for a particular trade are normally provided by that trade. There may be instances, however, where runs by a number of trades would be best contained in a common support, such as for long parallel runs of different trades in a rack above a corridor ceiling. In that case it is important that the responsibility for the rack itself be clearly defined to all trades who will use the rack for support. Occasionally supports for one trade are furnished or provided by another, as in the case of rooftop equipment supports for HVAC equipment that are fabricated by the miscellaneous metals contractor. Again, a clear definition of responsibility will save time and money and avoid disputes.

Coordination of the specifications with the drawings is more straightforward. Information required for the performance of the work at the construction site should be placed on the drawings whenever feasible, as the drawings are the most frequently consulted portion of the construction documents at the job site. Specifications should contain information regarding the qualities and standards that apply to the materials, equipment, and execution of the work. There should be no duplication of information on specifications and drawings; the presence of the same information in both places provides no real advantage and usually opens the door to conflicts.

Clear and standard language is helpful in communicating the areas of responsibility. The words "furnish", "install," and "provide" have very specific and accepted meanings. *Furnish* indicates that the trade is to purchase the item and deliver it to the job site so that it will be available for mounting and connection. *Install* indicates that an item is to be physically incorporated into the project by the trade and does not include purchase and delivery. *Provide* indicates that the item will be both furnished and installed.

Mechanical & Electrical Coordination: The need for close coordination between the trades is strongest between mechanical (HVAC, plumbing, and fire protection) and electrical (power, lighting, and fire alarms). Careful and methodical coordination is required to ensure that the electrical drawings indicate the supply of power to every piece of equipment that requires power, including motors, motorized dampers, control panels, fans, chillers, AC units, fan coils, duct heaters, burners, pumps, and miscellaneous equipment. Each piece of equipment must be correctly wired for voltage; amperage (both starting and running); circuit protection (breaker, fuse, time delay); type of power (normal, emergency, stand-by); disconnects with service lockouts; phasing (3-phase, single-phase, 3-wire, 4-wire); explosion-proof wiring; and provision of multiple circuits for those pieces of equipment that require them. Care should be taken to provide receptacles on or near equipment and a source of light to facilitate repair and service. Receptacles in exterior, wet, or potentially wet areas should be GFCI type.

The installation of complete wiring for electrical motors can be complicated. Electric motors require various configurations of external accessories depending on the size of the motor, factory-installed accessories on the equipment, method of controlling the motor, and the location of the controls. The general principle is that the motor is to be provided by the contractor who provides the equipment. In most cases the equipment comes with the motor factory-installed so there is little confusion. In addition, the equipment provider should furnish accessories (motor starters, overload protection, motor controllers) that are specific to the motor. This ensures that the motor and its accessories are compatible. When the accessories are installed by the factory, they are provided by the HVAC contractor. When the accessories are to be installed separately from the equipment, they are furnished by the HVAC contractor and installed by the electrical contractor.

Motors also require one or more disconnects. Unless they are factory-installed, the disconnects are furnished and installed by the electrical contractor and should be indicated on the electrical drawings as part of the power supply for the motor. Larger motors or equipment motors that will be automatically controlled or controlled from a remote location will require either line or low-voltage control wiring. In either case the control wiring and devices are furnished and installed by the HVAC contractor.

The specifications for all trades providing motors should clearly state individual responsibilities. Schematic diagrams that indicate the various configurations of motor wiring, as well as which trade furnishes and/or installs each component, are extremely helpful in defining each trade's obligations. Coordination of electrical disconnects is shown in Figure 14.15.

Figure 14.16 illustrates an example that requires coordination of the trades: the installation of a smoke control system for an air handler that is also tied to a fire alarm system for shutdown. HVAC supplies the unit, motor starter, and controls, and does all power and control wiring from the disconnect. HVAC also installs the duct smoke detector furnished by the electrician and wires it to the control system for shutdown. The

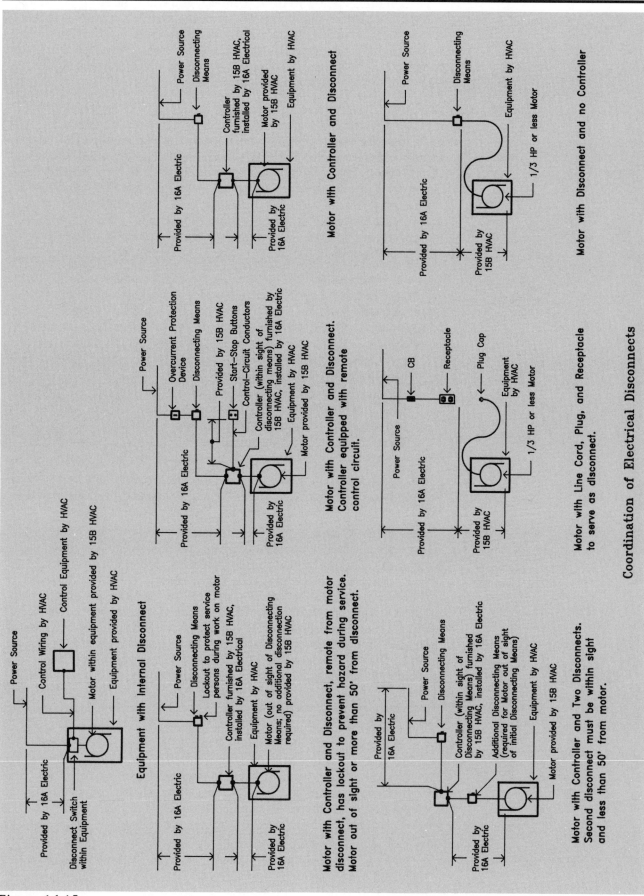

Figure 14.15 Coordination of Electrical Disconnects

electrician brings power to the disconnect, and wires the smoke detector and the fire alarm system. The design of the system is critical – it must operate correctly in order to handle smoke under a variety of fire conditions and locations. The designer must specify the operating sequence clearly and completely so that the control panels (many of which are solid-state electronics) can be programmed correctly.

Warranties: Warranties for equipment are usually the responsibility of the trade that furnishes or provides the item. Often additional connections to an item are required beyond the scope of "furnish and install"; these should be clearly defined. For example, a rooftop air conditioning unit would be provided by the HVAC contractor together with its curb. The HVAC contractor is responsible for the purchase, delivery, rigging, mounting, duct connections, control connections, testing, and warranty for that unit. The roofing contractor is responsible for the flashing of the curb into the roofing system, and the electrical contractor is responsible for the power wiring connections to the unit.

Space Coordination: The most basic of coordination issues is allocation of physical space. All systems occupy space which cannot be simultaneously occupied by other systems; the flexibility of the configuration of that space varies from system to system. For example:

- Structural systems are often quite set in their configuration and are not easily altered, while electrical wiring is often small, flexible and easily rerouted around obstacles.
- Pressurized piping and ductwork can endure moderate rerouting, while pitched drains and diffusers are often set in their location by practical or architectural constraints.
- Distribution systems are subject to potential friction losses when too many changes of direction are introduced.

It is important to know and understand the degree of flexibility that systems or components have with regard to the space they occupy. This includes building systems within and related to each designer's specialty.

Coordination of physical space includes not only the space actually occupied by the equipment itself but also the space around that equipment required for inspection, maintenance, repair, insulation, access to neighboring systems, air circulation, and similar requirements. These requirements include the swing area of access doors, area for the removal and insertion of filters, hand and visual access to valves and dampers, space above lay-in tile ceilings to allow for tile installation, air circulation space around air-cooled units, separation between air exhausts and intakes to prevent short-circuiting, and allowance for clearance between pipes and ducts for insulation installation. Remember that equipment has to get to its final location and to be removed when obsolete. An eight-foot-diameter tank will not fit through a six-foot double door.

In typical construction projects, a great deal of work occupies the space above the ceilings. For coordination purposes the reflected ceiling plans become extremely important. Once the location on the ceiling is determined for grilles, diffusers, sensors, lights, sprinkler heads, smoke detectors, security devices, access panels soffits, and so on, the routing of all the systems must be reviewed for conflicts.

CAD can be a useful tool for the examination of preliminary space conflicts. The work of different trades is typically shown on different "layers," which can be turned on or off when drawings are printed. Printing all the layers – in other words, showing all the work of all the trades on one drawing –

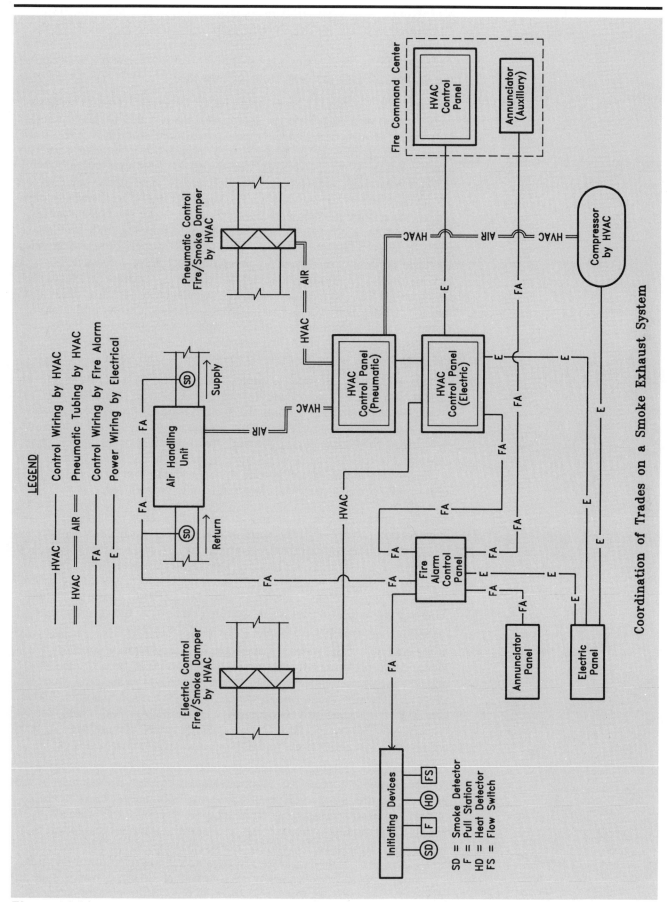

Figure 14.16

is too confusing to be practical for construction drawings. However, printing all of the trades at once can indicate graphically which areas contain the greatest overlap of trades; these areas should be examined carefully for spatial conflicts. This is not a substitute for careful review of the design but it offers an easy first screening for conflicts.

Cutting and Coring for Pipes and Ducts: When distribution systems pass through building elements, coordination is required to determine the boundaries of work between the mechanical trades and the architectural trades. These points should be delineated on both the architectural and the mechanical drawings. New pipes, ducts, and conduits that will pass through new walls or decks often employ sleeves. The sleeves are large enough for the pipe to pass through and slightly longer than the wall or deck thickness, and are furnished by the trade whose work is to pass through. The sleeves are incorporated into the wall or deck by the trade building the wall or deck. The trade whose work is to pass through the sleeve provides both the sleeve and the layout for the exact location of the sleeve. For new work to pass through existing walls or decks, the trade whose work penetrates the walls or deck typically is responsible for cutting, drilling, or coring the hole. Sometimes this involves specialized cutting and could be performed by another trade. An example is saw-cutting a stone facade for a new louver. The HVAC contractor would lay out the hole, but the cutting work would be done by the masonry contractor.

When systems pass through fire-rated assemblies (walls, decks, or shafts) code requires that the space around the penetration be fire- and smoke-sealed. The trade passing the system through the assembly also provides the fire and smoke stopping. As the actual layout is made in the field during the course of construction, the number of penetrations often changes. For this reason it would be difficult for the caulking contractor to accurately bid the work. By making the penetration and firestopping the responsibility of one party the chance of conflict is reduced.

Other Considerations: It is important to note that on a typical construction project all of the work of one trade is not necessarily done by one subcontractor. For example, an electrical subcontractor will usually furnish and install all of the building's permanent power wiring, but separate electricians may install temporary lights and power for the general contractor or line voltage controls for plumbing, HVAC, or fire protection. These separate electricians would be hired directly by the contractor or subcontractor responsible for that portion of the work.

Some portions of the mechanical system are visible and become architectural features; it is important that these items are coordinated with the architect. Diffusers, grilles, registers, fan coil units, and heat pump units are all visible and come in assorted shapes, sizes, and colors. These should be considered and coordinated. Exterior louvers and screens are usually provided by specialties or miscellaneous metals and need to be coordinated for connections to ductwork and architectural features.

Exposed piping and ductwork should be coordinated with architectural requirements for appearance, durability, and corrosion. If work is to be left exposed, an additional level of workmanship is often required for neatness and attention to accessories such as hangers, supports, transitions and fittings. The quality, workmanship and jacket material of insulation may be upgraded for exposed work. Often exposed pipes, ducts and conduits will be painted; care should be given to specifying paintable materials. Layout of exposed work is sometimes a consideration if the work is to be an architectural feature, in which case the drawings and specifications should reflect the appropriate level of detail. Controls (valves, dampers, actuators) on exposed work should be located or designed so as to avoid tampering by occupants.

Figure 14.17 shows areas of responsibility for work that is commonly coordinated between the trades.

Building Management Costs

Designers are often asked to balance the first costs of a building against the consequential operating costs. For example, ball valves are more expensive than globe valves but are more reliable. During the operation of a building, ball valves do not break or lock as globe valves are prone to do, and ball valves do not need any repacking. The cost of removing and replacing one defective globe valve during the life of a building may pay for having used several ball valves in the first place.

The economic analysis of first costs versus operating costs using present worth analysis is beyond the scope of this book. However, the overall principle that careful initial design can affect the costs of operating a building during its lifetime is readily understood.

Annual Operation and Maintenance Costs

Once a building has been commissioned, the owner will incur expenses for operations and maintenance as well as capital expenses. Operations and maintenance are generally those annual costs that can be expensed each year for tax purposes. Salaries of staff, utility bills, and disposable materials all fall under the general category of operations and maintenance costs. Figure 14.18 lists typical categories for annual operations and maintenance expenses for a commercial building.

The costs for operations and maintenance are a considerable portion of the overall annual costs of a facility; the costs associated with HVAC represent approximately 50% of those expenses. As discussed in the hospital and college facilities section of Chapter 4, there are many energy-saving items that can be incorporated into a facility to save energy and the cost of operations. Much of maintenance staff time in buildings is spent answering local calls for lack of heat or cooling. Many of these problems can be solved by activating equipment or resetting central controls. Others require repairs to equipment, usually valves or other parts of the control system. These problems may be solved by considering the initial cost of an energy management system (EMS).

Capital Improvement Costs

Those expenses that improve a building but that have a useful life of more than one year cannot be expensed in the year they were spent because they are not considered maintenance expenses. They must be depreciated over time. Such expenditures are termed *capital improvement costs*. Figure 14.19 lists common capital improvement costs.

For a given building, monies are generally expended for capital improvements every year. These costs can be budgeted each year and funded on an as-needed (or how-badly-needed) basis or funded from a capital reserve fund that averages out the expected expenditures and allows for rational budgeting.

The graph in Figure 14.20 is a rough average of anticipated annual capital improvement costs based on initial building value.

Coordination Between the Trades

Item	Responsibility		Comments
Sleeves			
Framed slots and openings in walls, decks, and slabs	3A Concrete 5B Masonry 5A Steel 6A Rough Carpentry	P P P P	Framed openings by trade responsible for structural framing.
Sleeves through slabs, decks, and walls	3A Concrete 5B Masonry 5A Steel 6A Rough Carpentry 15A Plumbing 15B Fire Protection 15C HVAC 16A Electrical	I I I I F F F F	Sleeve furnished by trade whose work will pass through and installed by trade building material penetrated. Coordination drawings are required from mechanical and electrical trades.
Waterproofing/Dampproofing/Roofing			
Sealing of pipes, ducts, conduits, etc., passing through sleeves	7A Caulking	P	Includes sealing around cables in sleeves provided for telephone, data, control and other low voltage wiring not contained in conduit.
Fire stopping of pipes, ducts, conduits, etc., passing through fire-rated slabs, decks, and walls	7A Caulking	P	
Sealing pipes, ducts, conduits, etc., passing through existing slabs, decks, and walls	7A Caulking	P	Hole cut or cored by trade whose work will pass through material penetrated.
Shower pan floor drain flashing	15A Plumbing	P	Coordinate with ceramic tile.
Roof drains	15A Plumbing	P	Flashed into roofing system by 7D Roofing.
Flashing for pipes, ducts, conduits, etc., which penetrate roof	7D Roofing	P	
Flashing of rooftop equipment, supports, and screens	7D Roofing	P	Supports provided by 15C HVAC or 5A miscellaneous Metals.
Flashing of rooftop equipment curbs	7D Roofing	P	Curbs furnished and set by trade providing equipment.
Temporary roofing	7D Roofing	P	
Division 1E Temporary Conditions			
Temporary heat	1E General	P	
Temporary water	1E General	P	Maintain existing services.
Temporary light and power	1E General	P	
Temporary toilets	1E General	P	
Temporary fire protection	1E General	P	Includes coordination of shutdowns and maintenance of system in project spaces during and installation of bypass during shutdowns.
Temporary staging above 8' high	1E General	P	7D Roofing may provide staging on sloped roofs from eaves up.
Removal of scrap and debris from all construction trades	1E General	P	Removal of existing materials by 2A Demolition.
Division 2B Earthwork			
Excavation, trenching, and backfill	2B Earthwork	P	Coordinate layout with trades requiring underground installation.
Shoring and dewatering excavations	2B Earthwork	P	

P = Provide
F = Furnish
I = Install

Figure 14.17

Coordination Between the Trades (continued)

Item	Responsibility		Comments
Division 3A Concrete			
Concrete encasement of underground utilities	3A Concrete	P	16A Electric to provide above-buried power lines.
Concrete foundations, footings, and pads	3A Concrete	P	Trade providing equipment to furnish anchors, inserts, vibration mounts, and mounting templates.
Concrete thrust blocks	3A Concrete	P	
Division 5A Miscellaneous Metals			
Catwalks to equipment	5A Miscellaneous Metals	P	Trade providing equipment to supply list of locations where required.
Ladders to equipment	5A Miscellaneous Metals	P	Trade providing equipment to supply list of locations where required.
Ornamental grilles	5A Miscellaneous Metals	P	Duct connections by 15C HVAC.
Division 9E Painting			
Prime and finish painting of exposed piping, ductwork, and conduit	9E Painting	P	Rust-proofing by trade making field cuts.
Field touch-up factory applied finishes	Trade that provided equipment	P	If known, may be backcharged to party causing damage.
Division 10A Specialties			
Toilet room accessories	10A Specialties 6B Finish Carpentry	F I	6A Rough Carpentry to install blocking.
Louvers	10A Specialties 6B Finish Carpentry	F I	15B HVAC to make duct connections.
Fire extinguishers and cabinets	10A Specialties 6B Finish Carpentry	F I	
Items Provided by Each Trade for Itself			
Hoisting, rigging, and setting of equipment	Trade providing equipment	P	
Starting up to 8'	Trade performing work	P	
Coring, cutting, drilling holes for the passage of pipes, ducts, or conduits	Trade installing pipes, ducts, or conduit	P	
Fasteners, hangers, braces, and supports required for the installation of the system	Trade installing the system	P	
Special tools for equipment maintenance	Trade providing equipment	P	
Rubbish removal to a central location on site for removal from site by GC	Trade performing work	P	
Division 15A Plumbing			
Make up water supply for HVAC including backflow preventor	15A Plumbing	P	15B HVAC provides connection from backflow preventor to equipment.
Corian integral countertops and sinks	6B Finish Carpentry	P	15A Plumbing to layout holes for faucets, holes by 6B Finish Carpentry.
Faucets for Corian integral countertops and sinks	15A Plumbing	P	6B Finish Carpentry to drill holes for faucets.
Precast, stone or molded shower bases and mop receptors	15A Plumbing	P	
Copper shower pans	15A Plumbing	P	
Electric water heaters	15A Plumbing	P	Power wiring and disconnect by 16A Electrical.
Division 15C Fire Protection			
Fire hoses and fire-hose cabinets	15C Fire Protection	P	Cabinet installation to be coordinated with trade building wall.

P = Provide
F = Furnish
I = Install

Figure 14.17 (continued)

Coordination Between the Trades (continued)

Item	Responsibility		Comments
Division 16A Electrical			
Power wiring for motors for plumbing, mechanical, and fire protection equipment	16A Electrical	P	Motor provided by plumbing, mechanical, or fire protection trade.
Motor controls and starters for motors for plumbing, mechanical, and fire protection equipment	16A Electrical 15A Plumbing 15B HVAC 15C Fire Protection	I F F F	Factory-installed motor starters to be furnished and installed by trade providing motor.
Electric baseboard heat with line voltage thermostats	16A Electrical	P	16A Electrical to provide entire system.
Electric baseboard heat, cabinet heaters, and unit heaters with integral thermostats and/or fans	16A Electrical	P	16A Electrical to provide entire system.
Duct smoke detectors	16A Electrical 15B HVAC	F I	16A Electrical to make all connections to fire alarm system, 15B HVAC to mount detector in ductwork and make connections to HVAC controls.
Electric heater cables for pipe tracing with local thermostat control	16A Electrical	P	Coordinate installation with pipe insulation.
Electric heater cables for pipe tracing with remote or DDC control	16A Electrical	P	Coordinate installation with pipe insulation. 15B HVAC to provide control sensors and connections.
Electric snow melting equipment with local sensor control	16A Electrical	P	Coordinate installation with 3A Concrete.
Electric snow melting equipment with remote or DDC control	16A Electrical	P	Coordinate installation with 3A Concrete 15B HVAC to provide control sensors and connections.
Division 15B HVAC			
Electric duct heaters	15B HVAC	P	16A Electrical to provide power wiring to and including disconnect.
Electric duct heater controls	15B HVAC	P	15B HVAC to provide power wiring from disconnect to unit and all controls.
Through-wall A/C and electric heating units	15B HVAC	P	Sleeve installation by trade building wall. Sleeve furnished by 15B HVAC. Power wiring by 16A Electrical.
Miscellaneous			
Access doors	Trade requiring access Trade constructing wall or ceiling	F I	
Masonry shafts, drywall shafts, tunnels utilized for air ducts			It is mandatory to assure the air tightness of all joints, holes, and other openings to make the air conveyors acceptable for their function.
Thermal and acoustic insulation for mechanical room walls and ceilings	7B Insulation	P	

P = Provide
F = Furnish
I = Install

Figure 14.17 (continued)

Typical Annual O&M Expenses

Item	Description
Service Contracts	
1	HVAC service contracts
2	Control systems
3	Filters
4	Pest control
5	Alarm system testing
6	Radio service
7	Water treatment
8	Artwork rental
9	Sprinkler and standpipe testing
10	Elevator maintenance
11	Escalator maintenance
12	Building cleaning
13	Trash removal
14	Plant maintenance
15	Security system maintenance
16	Window washing
17	Roof maintenance
18	Water coolers
19	Garage maintenance
20	Plaza maintenance/landscaping
21	Telephone
22	Electric maintenance
23	Interior design services
24	Automatic door equipment
25	Towel and uniform cleaning
Energy Costs	
E1	Electric utility
E2	Gas utility
E3	Steam
E4	Oil
Taxes, Fees, and Utilities	
T1	Property tax
T2	Sewer
T3	Water
T4	Fees and licenses
T5	Inspection fees—elevator, building, etc.
Service Contracts	
S1	Property management services
S2	Leasing agent services
S#	Legal services
Insurance	
I1	Property insurance
I2	Workers' Compensation
I3	General Liability

Figure 14.18

Typical Annual O&M Expenses (continued)	
Payroll of Staff	
P1	Building maintenance
P2	Building security
P3	Administration

Figure 14.18 (continued)

Typical Capital Improvement Expenses	
Enclosure	
E1	Roof
E2	Pointing
E3	Glazing/windows
E4	Caulking
E5	Building cleaning
Interiors	
I1	Tenant improvements
I2	Painting
I3	New lobby
I4	ADA entry
I5	Toilets
HVAC	
H1	Boiler
H2	Chiller
H3	Cooling tower
H4	DHW tanks
H5	Control system
H6	Smoke control
Hazardous Materials	
HM1	Asbestos removal
HM2	PCB removal
Electrical	
E1	Fire alarm
E2	Security

Figure 14.19

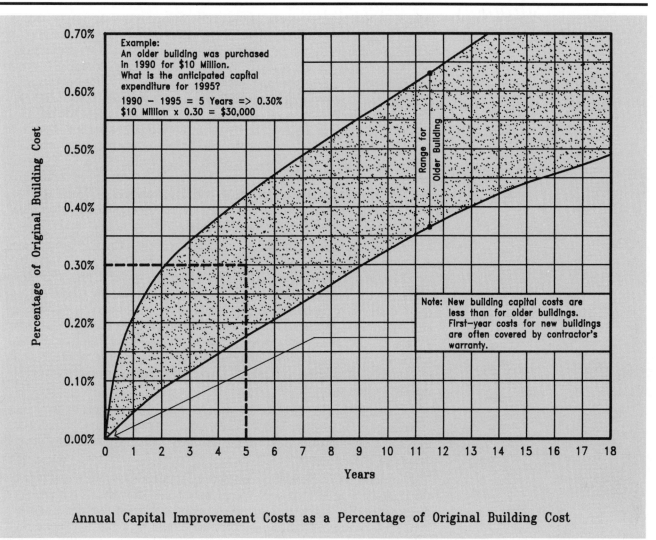

Figure 14.20

Summary

This chapter has reviewed some of the major subsystems in HVAC design, including an analysis of fire and smoke control design, which requires special attention because of the effect on life safety. Incorporation of design considerations for smoke evacuations, which is an important part of fire management, is useful for all building types. A review of the more complicated life safety systems, including those in high-rise buildings, atriums, and covered malls, are analyzed in this chapter in some detail. These building types have many systems that are also tied to the fire alarm system and that are designed to try to improve chances for survival in a fire. The overall results for survival in buildings are generally good, especially when compared to other areas. In a recent survey by NFPA for the years 1986–1990, excluding one- and two-family residences, there are less than 200 deaths per year in building fires; compared to over 500,000 deaths per year resulting from alcohol or drugs; 50,000 per year in automobile accidents; and over 50,000 from gunshots.

As health care buildings continue to specialize, the HVAC requirements will need to be even more carefully considered. The health care industry is rapidly changing, but the need for good design is evermore present. Proper control of the environment for patients who are medically fragile does more than make them comfortable; it promotes their overall well-being and contributes to the success of the medical procedures. Hospitals have a great number of different types of spaces and systems, and this chapter has covered the basic considerations for isolation rooms, waiting areas, laboratories, sterilizers, and overall layouts depending on whether or not the building is sprinklered.

The chapter has also discussed a variety of common specialty areas, such as kitchen and laboratory hood design, and a general review of fire codes, fire and smoke partitioning, and fire dampers, which supplements the code discussions in Chapter 2.

The importance of clear and complete specifications coordinated between the trades has been reviewed. Specifications and drawings are integral documents that make up a significant part of the contract documents. Figure 14.17 represents informaton that should be a part of every design professional's vocabulary; coordination can help solve a number of such problems during design and construction.

Problems

14.1 Using recent publications, locate a building that has an atrium. From the plans,
   a. Calculate the smoke exhaust criteria.
   b. Draw the schematic diagram of a system that will handle smoke exhaust in accordance with the governing model code in your community.
   c. Size the intake air louver(s).

14.2 Write the key points that the governing code in your community requires for high-rise buildings. Include smoke exhaust criteria, elevator and gurney criteria, fire alarm notification, and fire command criteria.

14.3 Sketch a design for a UV light system in ductwork for a waiting room. Show all access panels, testing, and safety monitoring. How would the cleanliness of the light be determined?

14.4 A cooking hood is to be designed for a restaurant in the middle of the dining area, where patrons can see their meals prepared. The cooking surface will be 3' × 8' [1 m × 2.5 m].
   a. Show a plan and elevation of the proposed cooking area.

    b. Size the hood, determine the cfm exhaust, and the size of the grease filter.

14.5 The dining area of the restaurant in problem 4 has a cooling load of 20 tons [70 kW] exclusive of patrons and cooking equipment. It is expected that there may be 75 patrons and 10 staff.
    a. Estimate the total cooling load, including for patrons, the central cooking area, and ventilation.
    b. Does the required ventilation for the patrons exceed the exhaust requirements for the hood?
    c. What energy-saving measures can be utilized in the design of the space?

14.6 Which trade performs the following work?
    a. Internal wiring for elevator controls.
    b. Coring through a foundation wall for a sprinkler pipe.
    c. Scaffolding or ladders for installation of ductwork in a space that is 14 feet to the underside of the ceiling.
    d. Furnishing duct smoke detectors.
    e. Motor starter for a pump.
    f. Installing electric duct heaters.

14.7 Classify the following as capital expenditures or operating and maintenance costs:
    a. New condenser pump
    b. Lubricating fans
    c. New fan belt
    d. New roof
    e. Changing all air filters
    f. Fees for HVAC service contract

14.8 A building was built in 1990 at a cost of 13 million dollars. What is an estimate of probable capital expenditures in the year 2000?

14.9 An atrium for a 10-story hotel is 100′ × 80′ × 110′ high. The atrium is surrounded by hotel rooms that open onto a balcony that overlooks the atrium. Select one of the codes and size the smoke evacuation system and sketch the layout in sections to conform to code provisions.

Chapter Fifteen

ENVIRONMENTAL CONSIDERATIONS

As we move into the twenty-first century, the condition of the environment will be the overriding issue throughout the world. The deterioration of the earth over the last 100 years has affected each one of us. Issues such as the need for clean water, breathable air, food untainted by toxins, and chemicals that do not disturb reproductive genes are well-known even today.

As we describe how to design successful environments inside buildings, we must also review the techniques for protecting the outdoor environment. There are consequences to using chemicals such as CFC's and HCFC's; there are health problems associated with modern building materials and indoor air quality; and certain areas of building design can involve radon or containment of hazardous materials such as oil tanks. In this chapter, some of the related techniques available in the design of building systems are presented.

CFC's and HCFC's: An Introduction

As noted in Chapter 1, refrigeration systems typically utilize refrigerants as fluids that circulate between the evaporator, compressor, condenser, and expansion valve. Refrigerants have a low boiling point at atmospheric pressure – this feature allows them to be effective in providing cooling.

The vast majority of refrigeration machines currently use either CFC's or HCFC's as refrigerants. CFC's are *chlorofluorocarbons*. HCFC's are *hydrochlorofluorocarbons*. Both are commonly used in air conditioning equipment and chillers. Other fluids, such as glycol or brine, are used in industrial applications where cold liquids (liquids that flow at temperatures below the freezing point of water) are moved from place to place to avoid the consequences of freezing. The use of these fluids, however, is confined to absorption chillers and for heat transfer. Ammonia is also used in refrigeration applications.

The three most common CFC refrigerants used today are CFC-11 (R-11), which is used mainly in negative-pressure centrifugal chillers of about 250 – 1,000 ton [900 – 3600 kW]* capacity; and CFC-12 (R-12) and CFC-500 (R-500), which are used primarily in positive-pressure chillers. Negative-pressure chillers operate below atmospheric pressure and leak inward; positive-pressure chillers operate above atmospheric pressure and leak to the atmosphere.

*Throughout the book, SI metric equivalents are provided for most imperial units. The SI units appear in brackets [ ].

The most common HCFC refrigerant is HCFC-22 (R-22), which is used mainly in positive-pressure reciprocating chillers and most small commercial and residential air conditioners.

Ozone Depletion

The ozone layer protects the earth from harmful and carcinogenic ultraviolet radiation. During the second half of this century the ozone layer has rapidly deteriorated, largely because man-made chemicals and emissions from man-made technology have been released to the stratosphere. These emissions include CFC's and HCFC's from HVAC equipment and industrial processes. The emission of CFC's and HCFC's and other chlorine-based chemicals – such as carbon tetrachloride, methyl chloroform, methyl bromide, and halons – is the major contributor to ozone depletion in the stratosphere.

There are natural phenomena that continuously make and deplete ozone. As solar ultraviolet radiation bombards the atmosphere, it produces ozone. A variety of chemical reactions changes the ozone back to molecular oxygen, thus completing the balance of ozone that is produced and destroyed. Man-made chemicals react with the ozone, breaking down the molecule and adding substantially to the ozone destruction rate.

Oxygen (O_2) and ozone (O_3) molecules absorb most of the ultraviolet radiation from the sun. After absorbing UV radiation, some oxygen molecules split into two oxygen (O) atoms, which react with other oxygen (O_2) molecules to form more ozone (O_3). Studies on the effects of ultraviolet radiation have shown that only one chlorine atom is released from a CFC molecule for every 60,000 ozone molecules created. Thus CFC and HCFC molecules absorb only a small proportion of the ultraviolet rays. However, one chlorine atom can cause a chain reaction, which can remove 100,000 ozone molecules. With only 60,000 ozone molecules being created and 100,000 ozone molecules being removed for each chlorine atom released from CFC and HCFC molecules, there is a negative impact of 40,000 ozone molecules for each chlorine molecule in the stratospheric ozone layer.

When they are released in the atmosphere, CFC and HCFC molecules absorb ultraviolet radiation from the sun, which breaks down the molecules and releases chlorine atoms as a by-product. The chlorine atom released from the CFC or HCFC molecule chemically reacts with the ozone layer, breaking down the ozone molecules and forming a gap in the ozone layer. International agreements have called for a discontinuation of these chemicals and have established a production phaseout schedule. These agreements are discussed in the next section.

By 1993, the gap in the ozone layer above the Antarctic measured over 9,000,000 square miles [23 000 000 square kilometers] – approximately three times the size of the continental United States. A method must be found to replenish the ozone. One concept is to produce more ozone and send it up to the stratosphere. This is impractical because the power involved in the natural formation of the stratospheric ozone by the sun's ultraviolet radiation is 24 terawatts. The total power level of all U.S. activities, including oil, gas, coal, nuclear, and hydropower, is only 10 terawatts. Also, it takes 3 electron volts to produce one molecule of ozone; only one chlorine atom and no energy is needed to destroy 100,000 ozone molecules. This is a classic case that shows the consequences of tampering with nature.

CFC's and HCFC's have also been linked to global warming. When CFC or HCFC molecules absorb ultraviolet rays, the molecules break down and release carbon atoms as well as the chlorine atoms already mentioned. The carbon atoms react with oxygen molecules to form carbon dioxide.

Global warming is occurring primarily as a result of the buildup of carbon dioxide in the earth's atmosphere. When the buildup of carbon dioxide occurs, the sun's ultraviolet rays pass into the atmosphere, where they are trapped as a result of the low emissitivity of the carbon dioxide. This causes a "greenhouse effect" and can produce global warming.

International Solutions

The reduction in ozone depletion will involve many industries and all nations. A significant benefit can be realized by reducing the emissions of CFC's and HCFC's. CFC's and HCFC's currently leak into the environment as a result of repair and maintenance on cooling systems that use CFC's and HCFC's as refrigerants, and by other industries that have used these chemicals as cleaning agents, mainly in the cleaning of computer boards.

In keeping with worldwide environmental concerns, most industrialized nations have adopted policies that reflect concern for the environment, as outlined in the Montreal Protocol discussed later in this chapter. In 1990 the *Clean Air Act (CAA)* amendment, a sweeping piece of legislation that calls for a complete phaseout of the production of all such ozone-depleting chemicals, was adopted as legislation in the United States. In the U.S., the Environmental Protection Agency (EPA) is the regulatory and enforcement agency for the phaseout of CFC and HCFC production.

According to the timetable for phasing out the use of such chemicals, production is to be halted on CFC's initially, and the production of HCFC's will be gradually phased out. CFC's will be phased out first because they contain more chlorine than HCFC's and therefore are more hazardous to the ozone layer. Stockpiled CFC's and HCFC's can be used for repairs and installation. New or existing equipment that utilizes CFC's can remain and be maintained utilizing the existing stock of chemicals.

Clean Air Act of 1970 (CAA) (Amendments, Reauthorized in 1990) – Public Law 101-549

The Clean Air Act is divided into 11 titles. Titles IV, V, VI, and VII are directly related to HVAC systems operation and maintenance.

Title IV focuses on a reduction in acid rain. A general reduction in the emission of sulphur dioxide (SO_2) and nitrogen oxides (NO) is mandated. *Acid rain* is formed when sulphur dioxide and nitrogen oxides react with atmospheric chemicals to form acids such as sulfuric and nitric acids. SO_2 and NO are products of combustion from automobiles, incinerators, power plants, and boilers that burn high sulphur-content hydrocarbons.

Title V mandates the permitting and certification of recovery equipment and personnel who work with CFC and HCFC refrigerants and associated equipment, the permitting of recovery and recycling equipment, and the permitting of all major sources of air pollution. Fines are set at a minimum of $10,000 per day per violation.

Title VI pertains to stratospheric ozone protection and schedules the phaseout of CFC and HCFC refrigerants as well as halons, carbon tetrachloride (CCl_4), and methyl chloroform (CH_3CCl_3). Until these chemicals have been fully phased out and substitute materials become available, special taxes and surcharges are being levied; the monies are used to fund research on alternative chemicals and the environment.

Title VII pertains to the enforcement of the CAA and the consequences of nonconformance, which are discussed further below.

EPA Stratospheric Ozone Protection (Summary of Section 608 of the CAA): Under section 608 of the CAA, the EPA has established regulations that:

- Require service practices that maximize recycling of ozone-depleting compounds (both CFC's and HCFC's) during the service and disposal of air-conditioning and refrigeration equipment.
- Set certification requirements for recycling and recovery equipment, technicians, and reclaimers.
- Restrict the sale of refrigerants to certified technicians.
- Require persons servicing or disposing of air-conditioning and refrigeration equipment to certify to the EPA that they have acquired recycling or recovery equipment and are complying with the requirements of the Act.
- Require the repair of substantial leaks in air-conditioning and refrigeration equipment with a charge of greater than 50 pounds [23 kg].
- Establish safe disposal requirements to ensure removal of refrigerants from goods that enter the waste stream with the charge intact (e.g., motor vehicle air conditioners, home refrigerators, room air conditioners).

Prohibition on venting of CFC's and HCFC's: Until July 1, 1992, venting of CFC and HCFC refrigerants to the atmosphere during service or disposal was permitted. Losses of refrigerant to the atmosphere result largely from system discharging, leaks, improper purging, mishandling, and contamination from oil, water acid, or motor burn.

After July 1, 1992, Section 608 of the CAA prohibits individuals from knowingly venting ozone-depleting compounds used as refrigerants into the atmosphere while maintaining, servicing, repairing, or disposing of air-conditioning or refrigeration equipment. There are four exceptions to the venting rule:

- Minimum quantities of refrigerant released in the course of making good faith attempts to recapture, recycle, or recover refrigerant.
- Refrigerants emitted during the course of normal operation of air-conditioning and refrigeration equipment, such as from mechanical purging and leaks. However, the EPA requires the repair of substantial leaks.
- Mixtures of nitrogen and R-22 that are used as holding charges or as leak-test gases (in these cases, the ozone-depleting compound is not used as a refrigerant). However, a technician may not avoid recovering refrigerant by adding nitrogen to a charged system; before nitrogen is added, the system must be evacuated. Otherwise, the CFC or HCFC vented along with the nitrogen is considered a refrigerant.
- Small releases of refrigerant that result from purging hoses or from connecting or disconnecting hoses to charge or service appliances. However, recovery and recycling equipment manufactured after November 15, 1993, must be equipped with low-loss fittings.

Service Practice Requirements:

1. *Evacuation of refrigerant from equipment:* As of July 13, 1993, technicians are required to evacuate air-conditioning and refrigeration equipment to the vacuum levels shown in Figure 15.1. Recovery or recycling

equipment manufactured before November 15, 1993, must be evacuated to the levels listed in column one in Figure 15.1. If the recovery or recycling equipment was manufactured on or after November 15, 1993, the equipment must be evacuated to the levels listed in column two, and the recovery or recycling equipment must be certified and EPA-approved by an approved equipment testing organization.

Technicians repairing small appliances such as household refrigerators, freezers, and water coolers must recover 80-90% of the refrigerant in the system, depending on the status of the compressor.

There are limited exceptions to the evacuation requirements described above. These exceptions apply to 1) repairs to leaky equipment, and 2) repairs that are not major and that are not followed by an evacuation of the equipment to the environment.

2. *Reclamation requirement:* Refrigerant that has been recovered and/or recycled can be returned to the same system or other systems owned by the same person without restriction. However, if a refrigerant changes ownership, that refrigerant must be *reclaimed* – it must be cleaned to the American Refrigerant Institute (ARI) 700 standard of purity and be chemically analyzed to ensure that it meets this standard.

Equipment Certification: As mentioned above, the EPA requires that recovery and recycling equipment manufactured on or after November 15, 1993, be tested by an EPA-approved testing agency. All recovery and recycling equipment must be tested under ARI 740-1993 test protocol. Recovery equipment used with small appliances must be capable of recovering 90% of the refrigerant when the compressor is operating and 80% of the refrigerant when the compressor is not operating.

Required Levels of Evacuation for Refrigeration Appliances		
Type of Appliance	Inches of Mercury [kilopascals] Vacuum Using Equipment Manufactured	
	Before Nov. 15, 1993	After Nov. 15, 1993
HCFC-22 appliance normally containing less than 200 pounds [90 kg] of refrigerant	0	0
HCFC-22 appliance normally containing 200 pounds [90 kg] or more of refrigerant	4 [13.5]	10 [34]
Other high-pressure appliance normally containing less than 200 pounds [90 kg] or more of refrigerant (CFC-12, 500, 502, 114)	4 [13.5]	10 [34]
Other high-pressure appliance normally containing 200 pounds [90 kg] or more of refrigerant (CFC-12, 500, 502, 114)	4 [13.5]	10 [51]
Very high-pressure appliance (CFC-13, HCFC-503)	0	0
Low-pressure appliance (CFC-11, HCFC-123)	25 [85]	25 mm [83] Hg absolute

All numbers shown in brackets are metric equivalents.

Figure 15.1

Equipment manufactured before November 15, 1993, including homemade equipment, will be "grandfathered" if it meets the requirements of column one in Figure 15.1.

Refrigerant Leaks: Owners of equipment with refrigerant charges of 50 pounds [23 kg] or more are required to repair substantial leaks. A maximum leak rate of 35% of charge per year has been established for equipment in industrial processes and 15% of charge per year for cooling chillers and equipment with a charge of 50 pounds or more. In addition, owners of equipment with a charge of 50 pounds or more are required to keep records of the quantity of the refrigerant added to their equipment during servicing and maintenance procedures.

Mandatory Technician Certification: Four types of certification have been developed by the EPA:

Type I:	For servicing small appliances
Type II:	For servicing or disposing of high- or very-high-pressure appliances, except small appliances and Mechanical Vapor Compression (MVAC) appliances
Type III:	For servicing or disposing of low-pressure appliances
Universal:	For servicing all types of equipment

Anyone who removes refrigerant from small appliances and motor vehicle air conditioners for purposes of disposal does not have to be certified.

Technicians are required to pass an EPA-approved certification test given by an EPA-approved certifying organization to become certified under the EPA mandatory certification program. A list of EPA-approved testing organizations is available from the EPA Stratospheric Ozone Protection Hotline.

After November 14, 1994, technicians must be certified in order to work on the above-noted equipment.

Refrigerant Sales Restrictions: The sale of refrigerant in any size container after November 14, 1994, is restricted to technicians certified under the EPA's mandatory certification program. Only those certified to work on refrigerants may purchase them.

Record-keeping Requirements: To track the amounts and types of refrigerants being added to equipment, transferred between owners, and reclaimed by reclaiming agencies, the EPA has established stringent record-keeping requirements.

- Technicians servicing appliances that contain 50 or more pounds [23 kg] of refrigerant charge must provide the owner with an invoice that indicates the amount of refrigerant added to the appliance.
- Owners of equipment that contain 50 or more pounds [23 kg] of refrigerant charge must keep service records documenting the date of service and the quantity of refrigerant added.
- Wholesalers who sell CFC and HCFC refrigerants must retain invoices that indicate the name of the purchaser, date of sale, and the quantity of refrigerant purchased.
- Reclaimers must maintain records of the names and addresses of persons sending them material for reclamation as well as the quantity of material sent for reclamation. Within 30 days of the end of the calendar year, reclaimers must report to the EPA the mass of material sent to them that year for reclamation, the mass of

refrigerant reclaimed that year, and the mass of waste products generated that year.

CFC and HCFC Phaseouts: In September 1987, the United Nations Environmental Program (UNEP) signed the *Montreal Protocol*, which established the production phaseout of CFC and HCFC refrigerants. This phaseout schedule has been accelerated twice: first by the London Amendment in June 1990 and then by the Copenhagen Amendment in November 1992. The 1990 Clean Air Act was amended to adopt and further accelerate the UNEP production phaseout regulations.

As of the date of this publication, the EPA phaseout schedule is summarized as shown in Figure 15.2.

Phaseout of other ozone-depleting substances: CFC and HCFC refrigerants are not the only ozone-depleting substances in use today. Halons, carbon tetrachloride, methyl chloroform, and methyl bromide also have a powerful effect on the ozone layer. These substances, along with CFC's, are designated as Class I substances. Their phaseout schedule according to the CAA is outlined in Figure 15.3.

HCFC's are regarded as Class II substances. Their phaseout schedule is outlined in Figure 15.2.

Recycling and Disposing of CFC and HCFC Refrigerants: Although production of CFC and HCFC refrigerants will be phased out according to the EPA Clean Air Act, disposing of CFC and HCFC refrigerants is not permitted. In fact, the phaseout of CFC's and HCFC's does not restrict the continued use of these materials in existing equipment or their use in new equipment. As long as the refrigerants are recovered, recycled, and/or reclaimed within the limits of the EPA and the CAA, CFC's and HCFC's may be used over and over again.

CFC and HCFC Production Phaseout Schedule

Deadline	CFC Schedule	HCFC Schedule
January 1, 1996	Ban on production and imports of CFC's, including R-11, R-12, and R-500*.	Cap at 3.1% of 1989 CFC production and imports, plus 1989 HCFC production and imports.
January 1, 2004		65% of 1996 levels
January 1, 2010		35% of 1996 levels
January 1, 2013		10% of 1996 levels
January 1, 2020		0.5% of 1996 levels. Ban on HCFC-22 and HCFC-123 production and imports.
January 1, 2030		Total ban on all HCFC production and imports.

* There is a limited extension for medical users and sterilizers utilizing ETO (ethylene oxide) who demonstrate need; however, this extension will expire on or before the end of 1999.

Figure 15.2

Refrigerants may not be disposed of or released into the atmosphere. The refrigerants from equipment that is converted to non-CFC or -HCFC products must be reclaimed and stored for recycling. CFC "banks" are expected to be established. The rapidly increasing cost of CFC and HCFC refrigerants will greatly affect their use.

Owners of air-conditioning and refrigeration equipment using CFC's or HCFC's are encouraged to convert or replace systems now. This may not be economically viable for most owners, but the restrictions and additional housekeeping required by the EPA for equipment using CFC's or HCFC's, the expected life of the equipment, and the rapidly increasing cost of CFC and HCFC refrigerants will influence the choice of converting or replacing equipment.

Consequences of Nonconformance to the Clean Air Act

Environmental: Environmental consequences of nonconformance to the Clean Air Act can result in a continuing depletion of the ozone layer and an increase in global warming, both of which will cause severe environmental and health damage.

Legal: Under the Clean Air Act, penalties for nonconformance are very severe. Civil penalties can range as high as $25,000 per violation per day, up to a maximum of $200,000. Certain violations (e.g., when responsible parties had "prior knowledge" of a condition that posed a potential for significant injury or death) carry criminal penalties that can include fines of up to $250,000 per day ($500,000 for corporations) and, as felonies, imprisonment of up to 5 years.

Alternatives to CFC's and HCFC's

Containment, Replacement, or Conversion

When faced with the need to deal with CFC's or HCFC's, there are basically three choices: containment, replacement, or conversion.

Because the environmental effects of CFC's and HCFC's are now better understood, and as the requirements for phaseouts and regulations

| Class I Substance Production Phaseout Schedule |||||
| Allowable Production of Class I Substances (Percentage of baseline production) |||||
Date: January 1,	Halons	Carbon Tetrachloride	Methyl Chloroform	Methyl Bromide
1994	0%	50%	50%	100%
1995	0%	15%	30%	100%
1996	0%	0%	0%	100%
1997	0%	0%	0%	100%
1998	0%	0%	0%	100%
1999	0%	0%	0%	100%
2000	0%	0%	0%	0%

Figure 15.3

regarding emissions begin to take effect, choices and alternatives for refrigerants will be necessary. There is a group of refrigerants that do not contain chlorine – HFC's *(hydrofluorocarbons)*. They are not an immediate threat to the ozone layer. This class of HFC can be a suitable replacement for CFC's and HCFC's in certain circumstances.

Containment

Containment refers to continuing to use the existing CFC or HCFC refrigerant, but modifying existing equipment so emissions are minimized and refrigerant can be recovered and reused.

Recovering and recycling the existing CFC or HCFC refrigerant should be carried out only with EPA-approved and -certified equipment, as already discussed. The recovery and recycling process should be done in a way that minimizes the loss or accidental venting of the refrigerant to the atmosphere.

Containment is the least expensive option in the phaseout of CFC's and HCFC's. It involves investing in recovery and recycling equipment and leak-monitoring equipment, and providing adequate ventilation to the space, all of which are relatively low-cost. In addition, obtaining recovery, recycling, ventilation, and leak-detection equipment is good engineering and facility practice – this equipment should be installed whether the equipment is maintained, converted, or replaced.

Replacement

Replacement refers to removing the existing equipment and replacing it with new equipment using HFC (which is permitted) or HCFC (which is allowed for a limited time) refrigerants, or even an absorption chiller or alternate cycle unit.

Replacement is potentially the most expensive option from a first-cost standpoint; it may cost as much as four times more than containment. However, replacement is by far the most reliable solution. There are obvious advantages to a new, warranted, reliable, high-efficiency unit with an expected long life span.

The issue of using HCFC's does involve a risk that the timetables for use of HCFC's will not be further shortened. For institutional users, it may mean another replacement after the year 2030, when HCFC consumption is prohibited.

Conversion

Conversion refers to removing the CFC refrigerant and replacing it with an HFC, HCFC (while permitted), or other alternative refrigerant.

Conversion costs may range somewhere between the two options already discussed or in some cases could be even more expensive than replacement. The range in cost for conversion of chillers depends on:

- Type of chiller
- The refrigerant in use
- Type of lubricant used in the chiller
- The alternative refrigerant
- Type of lubricant required for the alternative refrigerant
- The compatibility of the chiller with the alternative refrigerant and lubricant
- The mechanical and physical change required to the chiller, including replacing parts
- The change in chiller capacity

- Costs to access, remove, and replace the existing chiller
- Costs for refrigerant monitors and new ventilation systems
- The toxicity of the alternative refrigerant

Compatibility of the chiller with the alternate refrigerant and lubricant is the governing factor for the extent of work to be carried out on the existing chiller. If the alternative refrigerant and lubricant is compatible with the chiller, then minimal changes are required. Conversion may be possible by replacing only the refrigerant, seals, oil, gaskets, o-rings, seats, and sealants. This would be relatively inexpensive.

On the other hand, if the alternative refrigerant and lubricant are not compatible, the conversion could also require replacing heat exchanger tubes and impellers, changing the compressor speed or even the compressor, providing larger passages for increased flow to achieve compatible efficiency, flushing out and replacing the oil, and adding economizers to the compressors. This would be much more expensive and, depending on other factors such as chiller age, life-cycle cost analysis, phaseout of the replacement refrigerant, and additional monitoring and maintenance, replacement may be a better choice.

Another factor to consider is the change in chiller capacity. Many chillers are oversized at the design stages of the job; in this case a reduction in capacity may not have any adverse effect. However, some chillers are sized with closer tolerances; in this case, a reduction in capacity may not be advisable. A conversion may also reduce the efficiency of the chiller or refrigeration equipment.

When considering conversion of an existing chiller, it is important to contact the chiller manufacturer for advice and information regarding the correct route to take.

Alternatives to Refrigerants

As mentioned earlier in this chapter, the most common refrigerants used in air-conditioning and refrigeration equipment are CFC-11 (R-11), CFC-12 (R-12), CFC-500 (R-500), and HCFC-22 (R-22). The replacements for these refrigerants are shown in Figure 15.4.

At this time, one replacement for CFC-11 is an HCFC refrigerant, HCFC-123; the replacement for CFC-12 and CFC-500 is HFC-134a. HCFC-22 as yet has no replacement. As mentioned earlier, HCFC

CFC and HCFC Alternate Refrigerants		
Existing Refrigerant	Replacement Refrigerant	Replacement Refrigerant Phaseout Date
CFC-11	HCFC-123	2030
CFC-12	HFC-134a	None
CFC-500	HFC-134a	None
HCFC-22	HCFC-22	2020

Figure 15.4

refrigerants will be phased out by 2030; it is expected that a replacement will be produced by that time.

When replacing a refrigerant the following factors must be considered:
- Efficiencies of the replacement refrigerant and payback for conversion cost
- Costs of applying the new refrigerant
- Ozone-depletion potential of the replacement refrigerant
- Global-warming potential of the replacement refrigerant
- Toxicity of the replacement refrigerant

Summary

There is a strong worldwide consensus that maintenance of the environment is both desirable and economical. If every person can be made aware of the world's environmental problems and of the responsibility of us all to make an effort to solve these problems, the burden will be greatly relieved. HVAC systems use energy and chemicals that by their nature add a burden to the overall environment. Some techniques for more environment-conscious designs are discussed in Chapter 4.

Indoor Air Quality: An Introduction

Indoor air quality (IAQ), which refers to the general healthfulness of the air in a building, is a topic that has gained attention in the building industry. Building owners, architects, engineers, building managers, and building occupants now actively seek to ensure that indoor environments are safe and productive. Proper ventilation and air filtration systems have evolved in an attempt to keep buildings habitable and to vent out the fumes that make people sick. There has been an increased awareness that buildings may contain indoor environments that can make people feel unwell – a condition that has come to be known as *sick building syndrome (SBS)*. The primary reason for poor IAQ recently is that modern materials pollute the indoor environment because they are not vented properly.

The effects of poor IAQ are extensive. People who work in buildings with poor IAQ have experienced respiratory illnesses, allergic reactions, headaches, drowsiness, poor health, and, consequently, low production, lost work days, and low morale.

While ventilation standards have increased in response to IAQ problems, it is important to realize that ventilation is not the cause and is often a weak cure. If there were no indoor-contaminating pollutants, there would be no indoor air quality problem. Poor IAQ is caused by any combination of the following:
- Excessive emissions of *Volatile Organic Compounds (VOC)* – chemicals that contain carbon molecules and are volatile enough to evaporate from material surfaces into indoor air at normal temperatures. These originate in large measure from:
  - "wet" building materials, e.g., plywoods and particle boards containing urea-formaldehyde resins, paints, adhesives, and caulking
  - cleaning, waxing, and polishing agents
  - equipment
  - people
  - office products
- Products in the environment, e.g.,
  - tobacco smoke
  - nitrogen dioxide
  - carbon dioxide

- carbon monoxide
- radon
- formaldehyde
- sulphur dioxide
- ozone
- asbestos
- Odors and emissions from occupants
- Lack of adequate fresh air ventilation to the building or space in question
- Products of combustion from heating plants and vehicles that enter the building through the ventilation system or through windows

Poor IAQ is often improved by increasing the fresh air supply and exhaust and by controlling emissions from materials. However, extensive investigations and testing are often required to find the actual source of the poor air quality.

"Acceptable" IAQ can be very subjective. The simple question "Are you comfortable?" can be answered in a variety of ways by different occupants in the same space. ASHRAE Standard 62 determines that acceptable IAQ is air quality that is deemed satisfactory to at least 80% of the occupants at any one time.

Fresh Air Ventilation

Ventilation Standards

When deciding on ventilation rates, the building structure and materials should be analyzed. It is important to provide fresh air to compensate for VOC emissions as well as to provide adequate fresh air for occupants.

Currently the only codes and regulations intended to improve IAQ are those codes that establish minimum values for the volume of fresh air and the air-change rate for a space. In the early 1970s, the required fresh-air rate was considered high – 15 – 20 cfm [7 – 10 L/s] per person – but following the oil crises in the mid-'70s the requirement was reduced considerably to 5 cfm [2.5 L/s] per person for most buildings. As awareness of poor IAQ increased, ASHRAE revised their Standard (62-1989 Ventilation For Acceptable Indoor Air Quality) to increase the recommended fresh-air supply to 15 – 35 cfm [7 – 18 L/s] per person for most occupancies (see Figures 2.4 and 2.5). It is expected that most state and local codes will adopt the ventilation rates ASHRAE recommends.

Adequate fresh-air ventilation to neutralize most indoor contaminants is thought to be in the region of 0.5 to 1.0 air changes per hour (ACH). Most new mechanical systems are designed to provide an average of 0.85 ACH. However, the actual supply of fresh air to the space is often well below these design numbers as a result of inadequate air distribution, inadequate air filtration, poorly operating drain pans and drain lines, contaminated ductwork and duct liners, malfunctioning humidifiers, inappropriate control strategies, and inadequate maintenance. In addition, since many commercial buildings are equipped with VAV systems, the design fresh-air supply is always reduced on part load, which occurs 95% of the time of occupancy.

Purging and Constant Air Change Rates

Purging air from a space to clear it of all stale air and to introduce 100% fresh air is a good method of instantaneously improving IAQ. This method should be employed only in places of relatively short occupancy, such as theaters, opera houses, and lecture halls. The average occupancy of these areas is only one to two hours, and the volume of the rooms is often so large that the buildup of pollutants is quickly diluted before becoming objectionable. Purging should occur just before the space is scheduled to be occupied.

Purging would be the ultimate method of controlling IAQ for all building types and occupancies; however, the cost of the resulting energy bills would seriously influence the overall environmental economies. For most buildings that have occupancy rates of 8–10 hours, purging would be incredibly expensive. In these buildings, a constant air-change rate would be much more acceptable. Codes require constant fresh-air change rates for most types of occupancy; as mentioned earlier, these air-change rates translate to supply-air values ranging from 15–35 cfm [7–18 L/s] per person.

Exhaust

Just as supplying fresh air to a space is necessary to dilute the concentration of pollutants in the air, it is of equal or greater importance to *exhaust* equal amounts of air from the space. The exhaust rids the space of pollutants at a greater rate (it is hoped) than pollutants are emitted into the air.

Exhaust outlets and grilles should be located at points of greatest concentration within a building. For example, all of the air from toilets, bathrooms, kitchens, smoking rooms, laboratories, photocopy rooms, and similar rooms should be exhausted. If exhaust systems are properly designed and installed, many potential air pollutants and contaminants will be removed from the building before they can mix with the common airstream.

Kitchens and Laboratories

Both kitchens and laboratories require special ventilation and exhaust systems. Kitchens have a large buildup of grease-laden smoke and fumes, all of which need immediate removal for fire safety and air pollution reasons. Kitchen hoods should be provided over all cooking equipment; the hoods should exhaust the air directly outdoors through fire-rated ducts. The layout and design of kitchen hoods is discussed in detail in Chapter 14.

Most laboratories have fume hoods for handling toxic and dangerous chemicals and materials. A fume hood will exhaust air directly from the hood to the outdoors. All fumes and products of experiments should be conveyed quickly and directly to outdoors.

Kitchens and laboratories should both be kept under negative pressure to prevent contaminated air from entering surrounding spaces. Keeping a room under negative pressure implies that more air is exhausted than is supplied to the space.

Indoor Contaminants

Although increased fresh air ventilation will dilute the air and consequently improve the quality of the air, it will not stop pollutant emission. Indoor air quality is as much an architectural issue as it is a mechanical or ventilation issue.

Architectural Design and Emissions from Building Materials

Traditional construction materials such as masonry and plaster are relatively benign once installed. Many modern construction materials, however, contain volatile organic compounds. Most of the VOC's emitted from building materials are mucus-membrane irritants; consequently, there is a high rate of mucus-membrane irritation for people in new, remodeled, retrofitted, or refurnished buildings where VOC's are present.

"Natural" Materials

The selection of materials in building construction requires careful consideration. While natural materials and sustainable design are desirable (Chapter 4 discusses some of these principles), there are disadvantages.

Many natural materials are prone to decay. Wood studs rot and wool carpeting attracts moths. To combat this decay, chemicals and preservatives are used, which are often deadlier than some of the artificial substitutes (metal studs, nylon carpeting) that can be used. Some natural materials, such as lead, arsenic, asbestos, and formaldehyde, should not be used in any case. Also, while newer materials may be less polluting, they are often less durable. Therefore, in all likelihood they will be reapplied more often, usually while a building is occupied, thereby creating a higher frequency of emissions.

"Toxin-Free" Products

In the early stages of design, the architect must be aware of the building products, their content, and their VOC emission rates. The products specified should have low pollutant-emission characteristics. As noted earlier, many adhesives, paints, and sealants have very high VOC emissions. Carpeting, too, can have a high emission of compounds such as formaldehyde and fibers. Most manufacturers now produce equipment and materials with lower emission rates; many claim that their products are "low-polluting", "non-toxic", and/or "environmentally-safe."

"Baking" VOC's during Construction

During construction there are some activities that can considerably reduce VOC emissions. It is typically necessary to run the HVAC system (particularly heating) during construction to provide warmth for plasterers and other finish trades. The heat often bakes the VOC's, which has the beneficial effect of driving them out of the building before occupancy. A building may be deliberately "baked" for a week at warm temperatures with a high ventilation rate to help alleviate pollutants.

Architecture

While much of the responsibility for controlling indoor air quality belongs to the HVAC designer, IAQ can be improved by overall good design, including architectural considerations. How a building is laid out and which materials are used by architects and interior designers can have a significant impact on the HVAC design.

Architectural Layout

Good internal building layout contributes to good IAQ. It is important to separate areas for pollutant-generating activities such as food preparation, graphic arts, physical exercise, smoking, and photography from other areas of the building. These areas should be equipped with 100% exhaust and ventilation air systems to convey contaminants directly outdoors. The pollutant-generating areas should be kept under negative pressure with respect to abutting occupied spaces.

It is important to avoid installing "fleecy" products in which microbiological organisms can reside, such as rugs, woven wallcoverings, heavy curtains, and fiberglass duct insulation.

Office layout and structure have a considerable effect on IAQ. Most fabric-covered office partitions absorb VOC's when the building or space is completed; the partitions release a constant rate of VOC's into the environment, hence avoiding the opportunity for many of the VOC's to be removed by the ventilation system at an early stage of occupancy. Office partitions also interfere with good air distribution and circulation, causing air pockets or voids with stale and contaminated air. In addition, open shelves containing piles of paper products are a source of VOC's and microbiological organisms.

Figure 15.5 illustrates some basic building layout principles for good IAQ.

Operable Windows

Operable windows in offices and new buildings were discouraged in the 1960s; mechanical ventilation systems were relied on to control energy consumption. However, studies have suggested that incidences of poor IAQ, building-related illnesses, and sick-building syndrome have been more frequent in mechanically-ventilated buildings than in naturally-ventilated buildings. Operable windows in buildings can increase the IAQ and give occupants more control over their environment.

Site Layout

Site layout is also an important factor in controlling IAQ. Situating loading areas, drop-off areas, parking areas, and other high vehicular traffic areas away from air-intake openings and any building openings can greatly reduce contamination of the indoor air.

Location of a building's services in relation to internal spaces can be critical. For example, a loading dock with diesel-powered trucks should not be adjacent to the staff cafeteria. Operable windows and air intakes should be located as far as possible from heavily vehicled roads.

Mechanical Systems and IAQ

Mechanical ventilation systems can be either an effective method of controlling IAQ *or* a source of poor IAQ. Cleanliness of system components, filtering, air distribution, humidity control, recirculation of air, transfer of air, system layout, and ventilation rates all can have positive or negative effects on IAQ. For example, exhaust outlets, toilet vents, and kitchens should be as remote as practicable from HVAC units and intake louvers.

Cleanliness

The cleanliness of the intake louvers, ductwork, intake and recirculation filters, diffusers, and registers is important to maintain clean circulation of ventilation air. Dust and particulate matter can easily build up in diffusers and registers. A proper housekeeping program can ensure frequent cleaning.

System components such as cooling towers, cooling coils, drain pans, humidifiers, mixing boxes, and manmade fibrous insulations and liners are areas where many microbiological organisms can flourish. Organisms such as Legionella Pneumophillia, the bacteria that causes Legionnaires Disease, may be found in these components. The harmful organisms may be picked up in the airstream and circulated throughout the building. Cooling towers must be cleaned and treated with chemicals to avoid contamination.

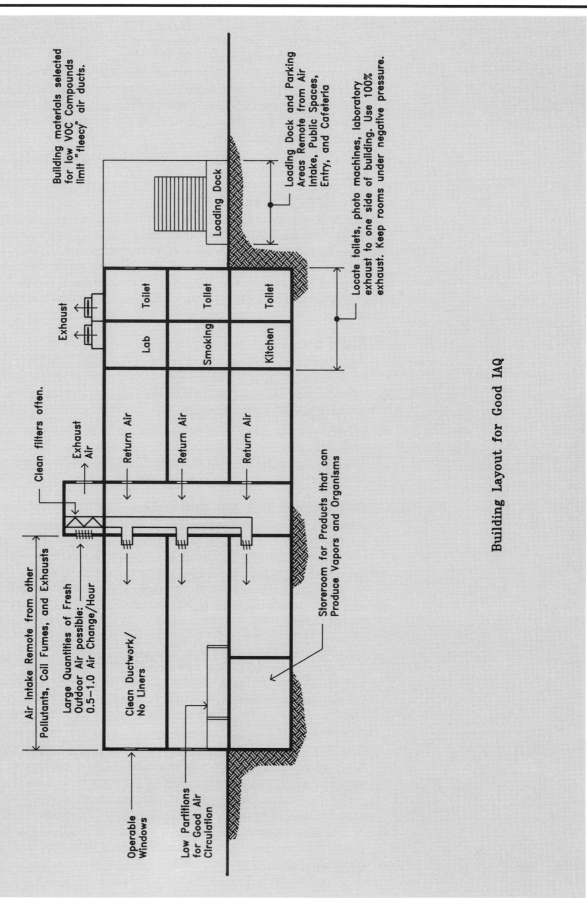

Figure 15.5

Building Layout for Good IAQ

Water introduced into a duct system is a major cause of mold. Leakage of coils, backed-up condensate lines, or steam leaks should all be fixed. Cooling coils and their drain pans should be cleaned regularly; the slope of drain pans should also be checked. Humidifiers should be the dry-steam type, not the water-spray type. Unitary portable humidifiers should be maintained and cleaned regularly.

Mixing boxes should be inspected and cleaned regularly. Fibrous insulations should not be used as duct liners and should be avoided where possible. Chapter 14 reviews the impact and design criteria of these areas for hospitals.

Filtration

System filters should be designed and selected for recirculation air and for outside air. Low-efficiency filters are not recommended. Recirculation filters should be capable of removing all VOC's from the airstream. Outdoor-air filters should sufficiently remove from the air contaminants that are common in the outdoor environment of that area; for example, city projects require filters adequate for vehicular fumes. Specialized or HEPA filters may be required for industrial applications or for special applications in hospitals.

Air Distribution and Exhaust

Proper air distribution is critical to overall air quality. Not only should the proper quantity of air be delivered to a space, but the distribution of the air must match the space layout and the requirements of the occupants.

Moveable office partitions can cause problems; they can be too close to the ceiling and block the distribution of air to occupants. In addition, moveable partitions that worked well in an initial layout are likely to be less successful in a new layout.

Air movement, transfer of air from space to space, local exhaust, and dedicated supplies can prevent bad air from circulating through the wrong spaces. Kitchens and laboratories should be equipped with hoods and kept under negative pressure; graphic arts rooms and smoking lounges should have dedicated exhaust fans and should also be kept under negative pressure. Offices and other concentrated areas should be kept under positive pressure in relation to surrounding areas. Air should not be recirculated from any areas such as kitchens, bathrooms, cafeterias, laboratories, or print rooms, where there is a buildup of pollutants.

Basements often have a buildup of humidity where microbiological organisms and mold can thrive. As these contaminants develop, they eventually propagate throughout a building. Basements should be dehumidified and ventilated.

Regulatory Controls for IAQ

Currently there are few regulatory guidelines or limits for most air pollutants found indoors. The number of chemicals used in building construction is vast, and individual sensitivity to any one chemical varies widely. These factors have made it difficult to establish general guidelines for building chemicals and IAQ standards.

Noise is also classified as a pollutant in several jurisdictions. Sensitive areas, such as hospital patient wards, should not be located next to an airport, quarry, or industrial processing plant. Noise and vibration control for mechanical equipment is generally controlled with isolation systems (see Chapter 13, Accessories).

IAQ and Energy Conservation

One consequence of increased ventilation standards is the increased energy cost for additional fresh air. Studies have indicated that while there is indeed an increase, the total increase in energy consumption for most buildings is under 10%. Schools, hotels, and retail buildings will experience the highest increase in costs because of the increased ventilation required for the high number of occupants and the higher quantities of ventilation air required per occupant. However, for most buildings some form of inexpensive energy recovery could be employed. See Chapter 4 for an outline of energy recovery devices.

As ventilation standards are implemented, the volume of fresh air and the energy consumed to condition that fresh air will be evaluated and new technologies will surely be developed to improve the IAQ of buildings.

IAQ and HVAC Controls

HVAC controls have an effect on the IAQ in some large buildings. There are instances in which one thermostat or fan switch controls the temperature and exhaust systems for an entire 10,000-square-foot [930-square-meter] area. This results in hot and cold spots in a room; also, some areas that require large quantities of exhaust may not have any.

A classic problem with a poorly-designed control system occurs when all occupants have access to the controller or thermostat. If each occupant can adjust the thermostat, there may be constantly fluctuating conditions within a space and very uncomfortable conditions; this can also damage equipment.

A good control system will manage many zones. For larger areas, the system will average space temperatures and adjust systems and air flows accordingly for overall comfort. For small spaces, individual room control is preferred and should be provided where possible. Thermostats and controllers will be accessible only to authorized personnel. For more information regarding the layout and design of control systems see Chapter 12, "Controls."

Oil Storage and Containment Piping

Oil storage and containment is a critical environmental issue because of the potential for soil and ground-water contamination. Several states have very strict codes and regulations regarding the installation and construction of new oil storage tanks and the testing of existing tanks. Underground oil piping is also under strict regulation.

Oil storage tanks are generally classified as either above-grade or underground (buried). Some of the regulations and requirements to be expected when planning the installation of a new above-grade or underground storage tank and associated piping are outlined below.

Typical Design and Code Requirements for Oil Tanks

Before proceeding with any other part of a project, the local fire department and other governing bodies should be contacted, as they usually control the storage of fuel-burning material. They typically review the location, flammability, and quantity of stored materials and issue permits for their use.

General Requirements for All Storage Tanks

- Permits are required for tanks in excess of 10,000 gallons [40 000 liters] capacity.
- A dike may be required to surround the above-grade tank. The capacity of the dike should be at least 125% of the tank capacity.
- Manholes should be provided for access to the tank.

- All tanks should be at least 10 feet [3.05 m] away from all buildings.
- Maximum diameters and lengths of tanks for different tank capacities, whether horizontal or vertical, should be within code limits. For example, a 2,500-gallon [9500-liter] vertical tank should be no greater than 72" [1.8 m] in diameter and no greater than 12' [3.65 m] high.
- Exceptions for tank sizes are permitted, when approved by local governing bodies, for tanks of very large capacities (25,000+ gallons) [95 000+ liters].
- All tanks should have filler and vent pipes.
- Contractors installing tanks and associated piping should be fully certified.
- All tanks should be tested and inspected regularly.

Special Requirements for Underground Storage Tanks

- All tanks should be designed and constructed to minimize the risk of corrosion and leakage.
- All tanks should be double-walled fiberglass or steel or a combination thereof.
- Splash plates should be installed at the bottom of the tank at each opening.
- Steel tanks should have cathodic protection and electrical isolation.
- Tanks should have a spill-containment manhole and an overfill-protection device.
- Some local regulations may require leak-detection equipment and cathodic-protection monitoring equipment on a new underground tank.

Special Requirements for Underground Piping

- All piping should have secondary containment, such as impervious liners or double-walled piping.
- All new piping should be constructed of noncorrodible material.
- Piping should be installed in a trench on a 6" [150 mm] bed of well-compacted, noncorrosive material, such as clean-washed sand or gravel.
- Pipes at the tank should be installed so as to prevent siphoning.
- Where practical, piping should slope back toward the tank.

Figure 15.6 shows a detail of an above-grade fuel oil tank located in a bunker. Even though the tank is actually underground, it is housed in a concrete room, so it is considered above-grade. Tanks in basements are also usually considered above-grade because they can easily be visually inspected.

Figure 15.7 shows a detail of a buried (underground) fuel oil tank. It requires a double wall, monitoring equipment, and tie-downs to prevent heaving when the tank is empty and the ground is wet. The detail also shows a heating element that would be used for low-grade oil, which needs to be heated when cold before it can flow through the piping to the burners.

Radon Gas

Radon is an odorless, colorless radioactive gas with a short half-life (less than 4 days). It seeps into spaces, primarily basements, from the earth below. Radon attaches itself to airborne particles, which can be inhaled.

Once inhaled, the particles may cling to the mucus lining of the throat area where they emit damaging radiation as they decay, which has been linked to lung cancer.

There are several accepted means of drawing radon gas out of basements. First, a pipe can be placed from below the basement slab up to the roof. A fan installed in the pipe runs continuously, drawing the gas through the gravel under the slab and out of the roof vent. This system can also be installed without the fan, relying on a stack effect to draw the gas out.

Another method of removing radon is to place an air-to-air heat exchanger in the basement, which will draw in fresh air and exhaust the radon-laden air.

The basement could be positively pressurized with a fan. Negative pressurization is not acceptable, as it can draw more radon up through the slab and potentially distribute it throughout the building.

The systems of venting radon gas are shown in Figure 15.8.

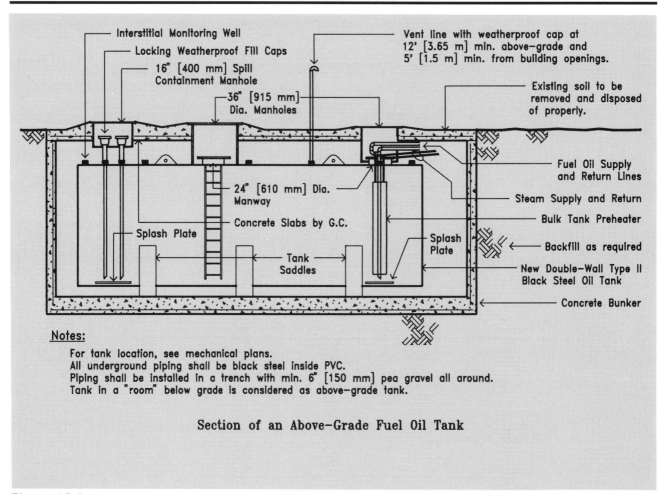

Figure 15.6

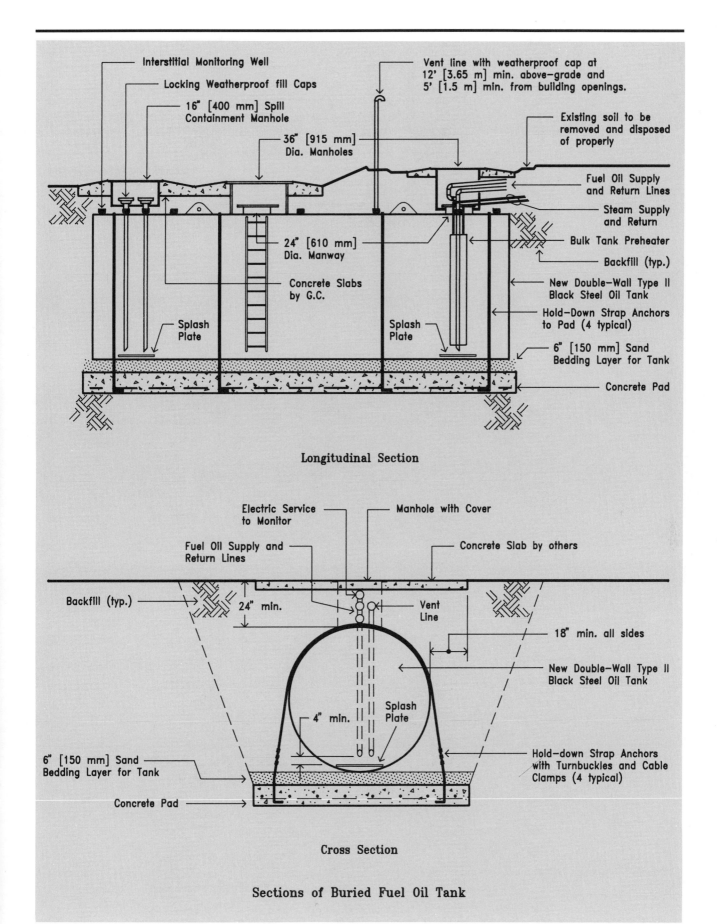

Figure 15.7

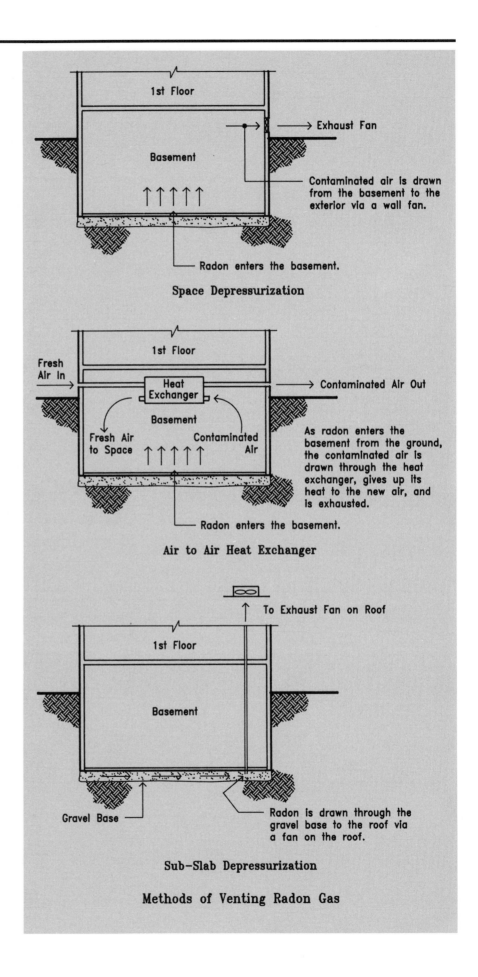

Figure 15.8

Summary

Environmental considerations for mechanical systems is a relatively new area. Until recently the effects on the atmosphere and on the overall environment of the earth were ignored when the indoor comforts of buildings were designed and built. This is no longer true. Each time a gallon of oil is burned, a kilowatt of electricity is used, or a pound of refrigerant is injected into a chiller – there are regulations to follow and real long-term effects and costs for the inevitable environmental cleanup that will result from this energy use.

By reducing the use of CFC's and HCFC's we are helping to prolong our survival on earth. We need to also watch the consequences of our solutions. Poor indoor air quality has resulted from materials that last. Some of the materials we have created as a substitute for natural materials cause poor indoor air quality by releasing toxins. We are partners with the earth that houses us. By combining the technical controls necessary as outlined in this chapter with an awareness of sustainable design as outlined in Chapter 4, we can produce better buildings that meet our needs.

Problems

15.1 True or false?
   a. Ozone is hazardous to the earth's stratosphere.
   b. CFC's are more hazardous to the stratosphere than HCFC's.
   c. The full life of radon is eight days.
   d. Splash plates in oil tanks prolong the life of the tank.
   e. A good ventilation system will solve poor indoor air quality.

15.2 Lay out a two-story office bulding and ventilation system for a building to include the following areas: Ten offices (100 square feet [9 square meters] each), open plan office area (1,000 square feet [90 square meters]), conference room (500 square feet [45 square meters]), men's and women's toilet rooms (200 square feet [18 square meters] each), storage and loading area (2,000 square feet [180 square meters]), trash collection area, and parking lot for 20 cars.

15.3 Outline the procedures necessary for a facility to conform to the Clean Air Act 1990 Amendments.

15.4 Sketch a 1,000-gallon oil-storage tank for an installation in the basement of a house.

15.5 Determine the net effect to the ozone layer when 1,000 molecules of CFC's are released to the stratosphere. Determine the amount of carbon dioxide formed.

15.6 Name three types of radon mitigation, and for the systems shown in Figure 15.8, list the advantages and disadvantages, and estimate relative costs.

Part Three

SAMPLE ESTIMATES

Part Three contains two sample projects: a residential building requiring heating only and a commercial building requiring some heating and year-round cooling. The projects illustrate HVAC system design and selection. These last two chapters of the book illustrate the design selection process from generation to distribution to termination, and produce complete projects from start to finish.

In Chapter 16, the residential building's heating system is configured in SI metric units. The commercial building design in Chapter 17 is configured in Imperial units. In both examples, extensive references to charts and tables are made to facilitate use of the book and applicable reference data.

Chapter Sixteen

MULTI-FAMILY HOUSING MODEL

In this chapter, an HVAC system is designed utilizing metric [SI] units for a low-rise multi-family building, using the principles discussed in the previous chapters. The sample building is a four-story, multi-family apartment building with eight units on each upper floor and six units on the first floor, in Boston, Massachusetts. It has a brick facade with wood casement windows. The windows have both operable and fixed glass sections. The plans for the building are illustrated in Figures 16.1 (and 16.1a) through 16.3.

Each of the steps for designing an HVAC system is discussed in detail in the sections that follow. Included are references to design tables and equipment data sheets; examples of calculations for heating and cooling loads, boiler output, and tank and pipe sizing; and a system summary sheet to guide the reader through the design process.

The four basic steps to take in order to establish the type of HVAC system for a building are listed below. The chapters in which each process is fully discussed are also listed.

1. Make the initial selections (Chapters 5 and 6).
2. Determine the design criteria (Chapters 3 and 4).
3. Compute the loads (Chapter 3).
4. Select the equipment (Chapters 6 through 13).

Initial Selection

As described in Chapters 5 and 6, the initial selections to make are: (1) the type of system, (2) the type of fuel, (3) the type and make-up of the distribution system, and (4) the type of generation system.

Type of System

The types of heating and cooling systems for buildings are listed in Figure 5.2; a forced hot-water system with separate window or through-the-wall air conditioners is recommended for multi-family residences. Combined systems utilizing water source heat pumps are also common. In this example, it is assumed that the building does not require cooling as part of the base system, and that the electrical designer and architect will provide for future through-the-wall air conditioners supplied by the tenants. Therefore, only heating by hot water will be provided. Thus the type of system chosen is a separate hot water heating system.

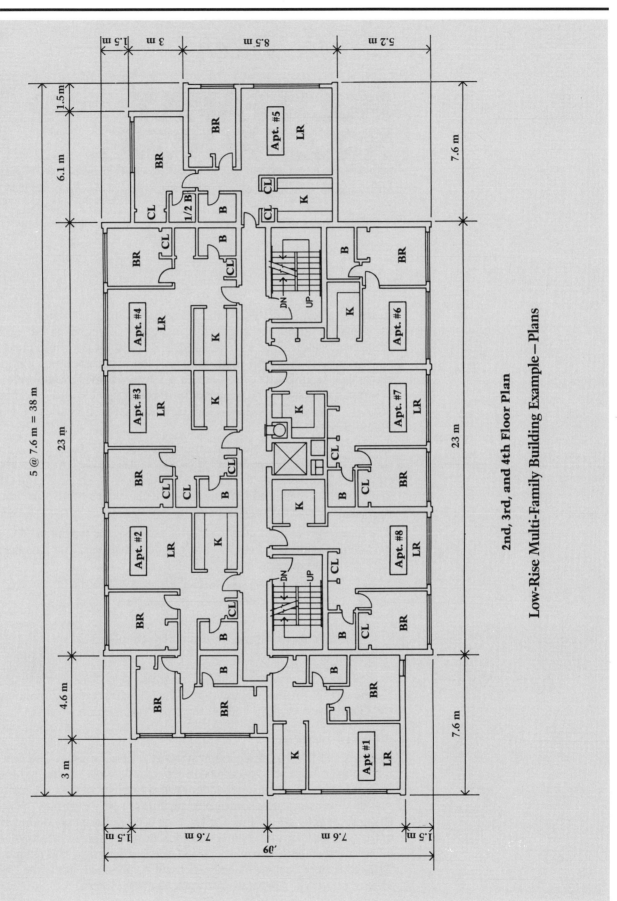

Figure 16.1

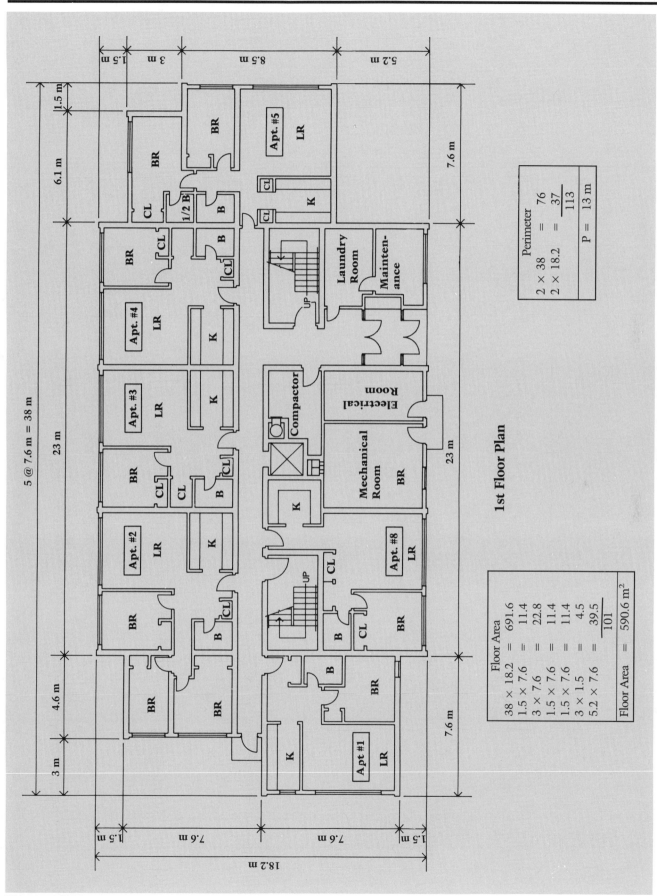

Figure 16.1a

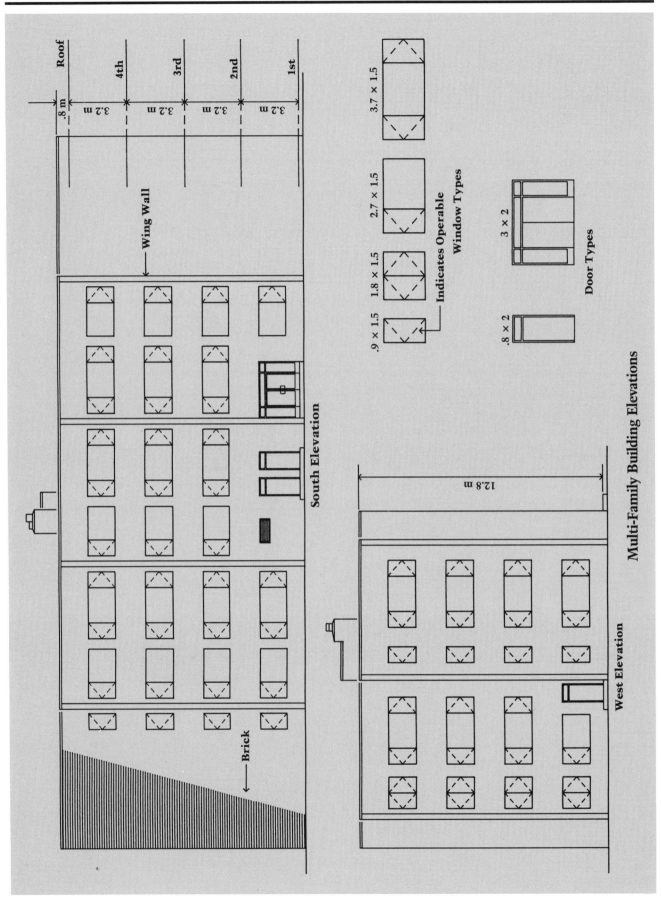

Figure 16.2

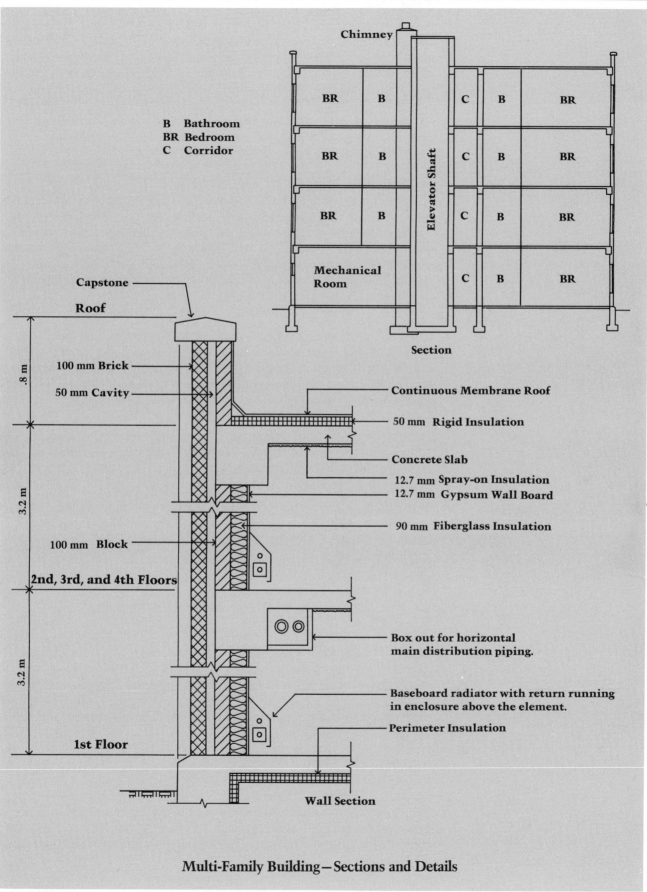

Figure 16.3

Fuel

A new system should have high fuel efficiencies. In Chapter 5, Figure 5.3 lists basic properties and costs of fuel and indicates that #2 oil is the most economical. Therefore, #2 oil is selected. (Fuel efficiencies vary; a different fuel could be a better choice for a particular geographic location.)

Distribution System

In Figure 5.4, a hot water distribution system is recommended for continuously used buildings. Therefore, the distribution system selected for this building is a hot water distribution system.

Generation System

Hot water is generated by a steam boiler, a high-temperature water boiler, or a hot water boiler (see Figure 1.6). In this example, direct steam from a utility company is not available. Any system other than a hot water boiler also requires a heat exchanger, which adds to the overall cost of the system because of the relatively equal cost of steam and hot water boilers. Therefore, a hot water boiler is selected, as it is the most economical (see Figure 16.4).

Design Criteria

Once the initial selections have been made – a separate hot water system, #2 fuel oil, hot water distribution, and a hot water boiler – the design criteria must be determined. Design criteria are the indoor and outdoor conditions of temperature, humidity, and ventilation around which the building will be designed. The design criteria for this project are listed in Figure 16.5.

Design criteria are used to compute the heating and cooling loads. In all cases, check such criteria against local codes, manufacturers' literature, and overall architectural requirements.

Computation of Loads

The next step is to compute the total heating load (H_t). The total heating load for a building is the sum of all conduction and convection heat losses (see Figure 3.14).

$H_t = H_c + H_e + H_v + H_s$
$H_c = UA\,\Delta T$
$H_e = FP\,\Delta T$
$H_v = 1.27\,L/s\,\Delta T$ (Use 1.27 for heating, 1.23 for cooling)
$H_s = KA$

Conduction Heat Losses

To compute the conduction heat losses (H_c, H_e, and H_s) through walls, floors, doors, windows, and slabs, first establish the U values and related coefficients. (Refer to Chapter 3, "Heating and Cooling Loads," for more information on the computation of U values.) The U values for this model are listed in Figure 16.6.

Convection Heat Losses

Convection heat losses (H_v) are caused by infiltration and kitchen and bathroom ventilation. Generally, the amount of convection loss resulting from infiltration is larger than the loss resulting from ventilation, since exhaust fans run only intermittently. Therefore, infiltration is used to compute heat loss and the crack coefficients must be determined (see Figure 16.6).

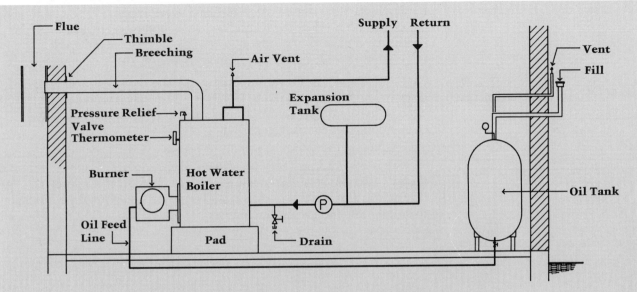

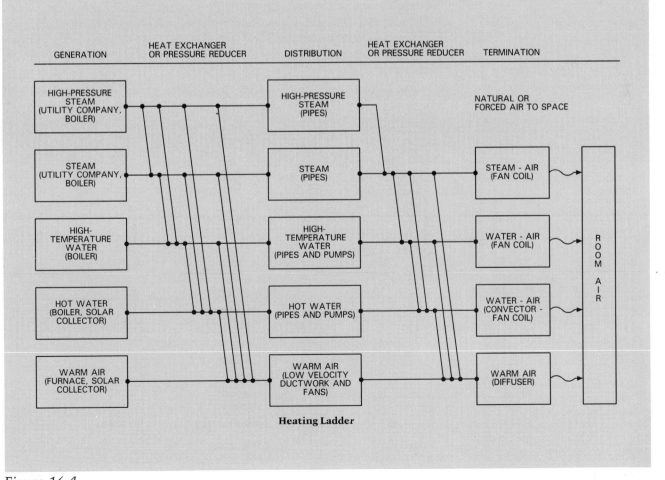

Figure 16.4

The areas of the roof, walls, windows, doors, and floors must be calculated to obtain the linear value of the cracks and perimeter. These quantities are then multiplied by the respective values of temperature or constant to obtain the building's convection heat losses. The area and crack calculations for the building are shown in Figure 16.7.

The heat losses are now computed using the design temperatures determined in Figure 16.5, the U values and coefficients from Figure 16.6, and the areas and perimeters from Figure 16.7. The total heat loss calculation is shown in Figure 16.8.

The convection heat loss is from infiltration and has been computed for all windows simultaneously. As discussed in Chapter 3, only a portion of the air infiltrates simultaneously (see Figure 3.13). It is assumed that 50 percent [31 kW] of the convection load occurs at any given time. Therefore, the total heat load for the building is

$$H_t = H_c + H_e + H_s + H_v \text{ (kW)}$$
$$= 49.6 + 3 + 3.8 + [31/2] = 71.9 \text{ kW}$$
$$= 71.9 \text{ kW (load/m}^2 = 71\,900/4\,(594.6) = 30 \text{ W/m}^2$$

The average load of 30 W/m² is a very low heat load compared to many existing buildings, because current strict energy codes require much better building insulation than they have in the past. Older buildings with large areas of glass, single-pane glass, and little roof insulation often have heat loads three times as large.

Equipment Selection

The final step in the design process is the selection of equipment, including generation, distribution, and termination equipment. The proposed layout is first drawn as a system diagram. The apartments of the building are characterized by a living room and bedroom adjacent to each other, with a considerable amount of glass in both rooms. Because glass is the source of greatest heat loss, it makes sense to place hot water radiators under the windows.

It also is possible to use baseboard radiators with the supply and return piping fed from the same end by running the return piping back under the

Design Criteria					
Item	**Reference Section**	**Value**	**Comments**	**Recommendation**
Indoor Design Temp.	Fig. 3.8		Local code 19° C	19° C
Outdoor Design Temp.	Fig. 3.9		Local code −12.8° C	−12.8° C
ΔT Adjoining Spaces	Fig. 3.10		Lobby	Use fan coil unit
Ventilation Standards	Fig. 1.22	4% floor area	Windows meet criteria	
		Exhaust toilets and kitchens	Use as required	
U Valves (Numbers shown are recommended values)	Fig. 2.6	W/m² K		
	Walls	.45	Verify	Compute
	Roof	.34	Actual	Actual
	Windows	3	Actual	Actual

Figure 16.5

U Values and Constants

Reference	Material	$U = \dfrac{1}{R_t}$	Remarks
Walls			
Fig. 3.7	100 mm brick	.28	Face brick
Fig. 3.5	50 mm cavity	.16	Allow 2.54 – 10.2 cm
Fig. 3.7	150 mm block	.29	Sand and gravel
Fig. 3.7	90 mm insulation	2.29	Blanket type
Fig. 3.7	12.7 mm gypsum	.079	Plaster
Fig. 3.5	Outside film	.03	
Fig. 3.5	Inside film	.12	
		$R_t = 3.25$	
		$U = 1/R_t = .31$ (.45 req'd.)	
		U walls = .31 W/m² K	
Roof			
Fig. 3.7	Roof membrane	.026	Assume roll roof
Fig. 3.7	100 mm rigid insulation	1.94	Polyisocyanurate
Fig. 3.7	153 mm concrete	.71	Sand and gravel
Fig. 3.7	12.7 mm spray-on insulation	3.3	Cement fiber-wood
Fig. 3.5	Outside film	.03	
Fig. 3.5	Inside film	.11 (horiz. surface)	
		$R_t = 6.1$	
		$U = .16$ (.34 req'd.)	
		U roof = .16 W/m² K	
Windows			
Fig. 3.6	Double insulated glass	U = 3.14 (3 req'd.)	
		U windows = 1.7 W/m² K	
Door			
Fig. 3.6	Wood 25 mm	U = 2 W/m² K	
	Front door with storm door	U = 1.7 W/m² K	
Edge Coefficient (F)			
Fig. 3.11			F = .85 W/m²
			100 mm
			150 mm
Slab Constant (k)			
Fig. 3.10	Slab on grade		k = 6.3 W/m²
Crack Coefficient Windows			
Fig. 3.12	Residential casement/ 24 km/hr.	Window coefficient = 1.35 L/s/m .8 mm crack	
Doors			
Fig. 3.12	Weather-stop doors/24 km/hr.		Door coefficient = 1.4 L/s/m

Figure 16.6

enclosure and above the element. The layout of the radiation with a typical detail is illustrated in Figure 16.9.

Distribution

The object in laying out a distribution system is to supply from the generating system to each terminal unit the minimum length of piping (or ductwork) in order to save on both the installation cost of the pipe and the operating cost of pumping through extra pipe.

As a general rule, the least amount of piping is needed when one horizontal main, located on the mechanical floor, distributes from the main vertically to each terminal unit. Rising vertically approximately 3 meters to each radiator uses considerably less piping than running horizontally on each floor, where the average distance between radiators is approximately 4.5 meters. When radiators are adjacent to each other (e.g., less than 3 meters), they can be supplied from a common riser. Systems to consider are illustrated in Figure 16.10.

Areas and Crack Calculations

Facade	(1) Gross Wall Area (m²)	(2) Window Area (m²)	(3) Door Area (m²)	(4) Miscellaneous	Net Area (1)-(2)-(3)-(4)	Crack Length (m)
South	$12.8 \times 38 =$ 486.4 m²	$4 \times .9 \times 1.5 +$ $11 \times 2.7 \times 1.5 +$ $10 \times 3.7 \times 1.5$ 105.5 m²	$2 \times .8 \times 2 =$ 3.2 m²	Front door $3 \times 2 =$ 6 m²	371.7 m²	Windows $35 \times (2 \times .9 + 2 \times 1.5) =$ 168 m Doors $4 \times (2 \times .8 + 2 \times 2) =$ 22.4 m
North	$12.8 \times 38 =$ 486.4 m²	$16 \times 2.7 \times 1.5 +$ $12 \times 3.7 \times 1.5$ 131.4 m²	—	—	355 m²	Windows $40 \times (2 \times .9 + 2 \times 1.5) =$ 192 m Doors —
East	$12.8 \times 18.2 =$ 233 m²	$4 \times 3.7 \times 1.5 +$ $4 \times 2.7 \times 1.5$ 38 m²	—	—	195 m²	Windows $12 \times (2 \times .9 + 2 \times 1.5) =$ 57.6 m Doors —
West	$12.8 \times 18.2 =$ 233 m²	$9 \times 2.7 \times 1.5 +$ $7 \times 3.7 \times 1.5$ 75.3 m²	$.8 \times 2 =$ 1.6 m²	—	156.1 m²	Windows $23 \times (2 \times .9 + 2 \times 1.5) =$ 110.4 m Doors $(2 \times .8) + (2 \times 2) = 5.6$ m
		350.1 m² Windows	4.8 m² Doors	6 m² Front door	1077.8 m² Walls	528 m Windows 28 m Doors

Transom over doors assumed same as wall.
Intake louver at boiler room ignored.
Elevator penthouse ignored (not a heated space).

Figure 16.7

Two of the systems shown in Figure 16.10 combine direct and reverse piping systems (see Chapter 10, "Distribution and Driving Systems," for an explanation of direct and reverse return systems). In this example, the vertical risers are piped as direct return and the horizontal main is piped as a reverse return system. While a complete reverse return system is theoretically superior for overall distribution, a stack of two to six radiators can be effectively attached in a direct return stack and properly balanced with valves and still maintain similar flow characteristics of the reverse return system, but without the additional third riser. Figure 16.10 shows that the two complete reverse return systems have more piping than the modified systems, and that the combined piping system has the least.

The above analysis results in the following design objectives:

1. Place radiators under windows.
2. Distribute vertically.
3. Combine reverse return on the horizontal main with direct return on the risers.
4. Pipe radiators with supply and return fed at the same end (where possible).
5. Connect radiators to the same riser at living rooms and bedrooms.

Heat Loss Calculations

Conduction $H_c = U A \Delta T$	U (W/m²·K) (Fig. 16.6)	A (m²) (Fig. 16.7)	ΔT (°C) (Fig. 16.5)		Heat Loss (Watts)
Walls	.31	1077.8	19−(−13)		10 700
Windows	3.14	350.1	19−(−13)		35 200
Doors	2	4.8	19−(−13)		310
Front Door	1.7	6	19−(−13)		330
Roof	.16	594.6	19−(−13)		3050
					49 590
$H_e = FP$	F	P	ΔT °C		Heat Loss (Watts)
Perimeter	.85 W/m²·K/m	112.4 m (See plans for calculations)	−		3057 W
$H_s = kA$	k	A	−		Heat Loss (Watts)
Slab on Grade	6.3 W/m²	594.6 m²	−		3746 W
			Total Conduction Loss = 56 393 W rounded to 57 kW		
Convection $H_v = 1.27$ L/s/m ΔT	Crack Length (m)	L/s/m	T		Heat Loss (Watts)
Windows	528	× 1.35	× 32	× 1.27	= 28 968
Doors	28	× 1.40	× 32	× 1.27	= 1593
			Total Convection Loss = 30 561 W rounded to 31 kW		

All numbers shown in brackets are metric equivalents.

Figure 16.8

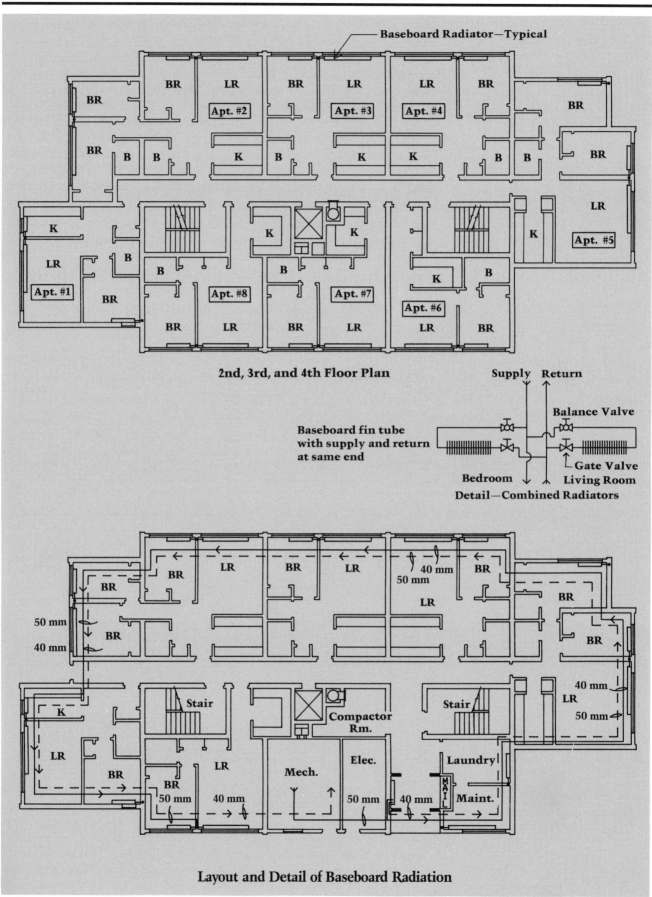

Figure 16.9

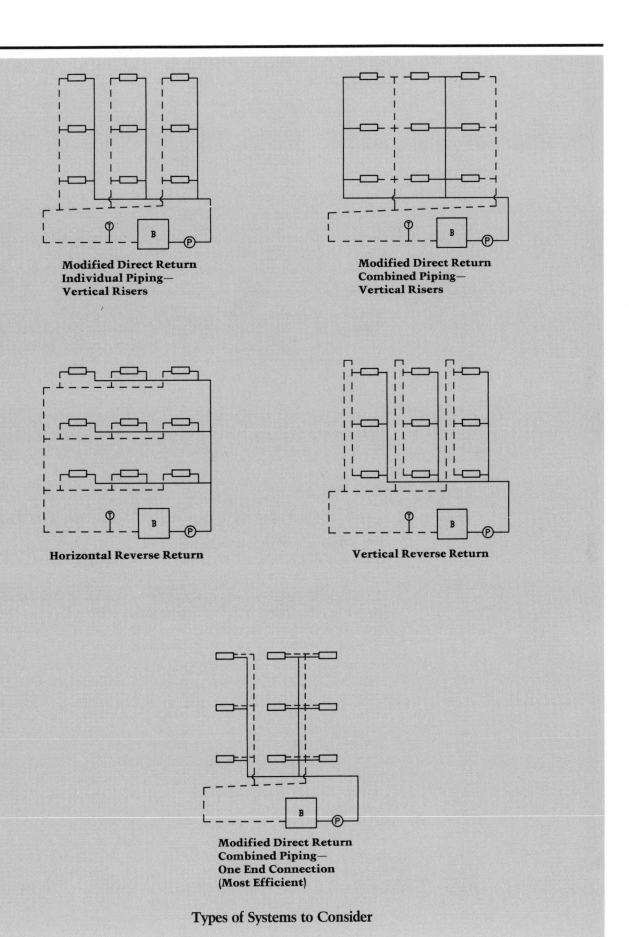

Figure 16.10

The riser diagram shown in Figure 16.11 incorporates most of these objectives and will be used to complete the design and selection of the heating system.

Generation Equipment

The following generation equipment is selected for this model (see Figure 16.4):

- Hot-water boiler (oil-fired cast iron)
- Oil tank and accessories
- Expansion tank
- Circulating pump(s)

Each piece of generation equipment is discussed below and is summarized in Figure 16.15 later in this chapter. A Means line number is provided in order to determine costs in the current edition of *Means Mechanical Cost Data*.

Hot-Water Boiler – Oil (71.9 kW Load)

The capacity of the hot-water boiler should be sized approximately 25 percent larger than the design load to allow for a reserve on excessively cold days and to account for a lowering of the boiler efficiency over its useful life (see Figures 6.1 and 6.3). The design output for the boiler is shown below:

Net building heat loss	71.9	kW
Allowance for pipe loss (5%)	3.6	(insulated pipe)
Pickup allowance (10%)	7.2	(for pickup)
Total load	82.7	
Allow 25% oversizing	20.8	
Total design boiler output	103.5	round to 110 kW

For oil-fired boilers with capacities in the 110 kW output range, a cast-iron sectional boiler is selected. Cast iron is the customary boiler used in buildings of this type because of its life cycle, cost, and longevity.

The generation equipment data sheet on cast-iron sectional boilers can be used to make the final selection (see Figure 6.11a). An oil-fired cast-iron boiler comes in capacities ranging from 30 to 2050 kW. (See line numbers 155-120-1000/3000 from *Means Mechanical Cost Data*.)

Because this type of building may require a back-up system, current practice often uses two boilers, one sized for one-third and the second for two-thirds of the total load, controlled to sequence as necessary to meet demand. If the boiler should fail for any reason during the heating season, great discomfort to the inhabitants, as well as considerable damage to the property from freezing, could result. Furthermore, the efficiencies noted for the equipment are based on continuous full load. On mild days, a single boiler would waste considerable heat warming itself and the chimney on each start-up. Using two pieces of equipment is effective because each boiler, when running, will be close to peak efficiency. Further, in the event of a boiler failure, the second boiler can be relied on for at least some heat. Thus, for this building, two boilers are chosen, one close to one-third and the other close to two-thirds of the required output load. By referring to *Means Mechanical Cost Data* and manufacturers' literature, two boilers, one boiler at 36 kW and the other at 74 kW, are selected. The total selected output now equals 110 kW.

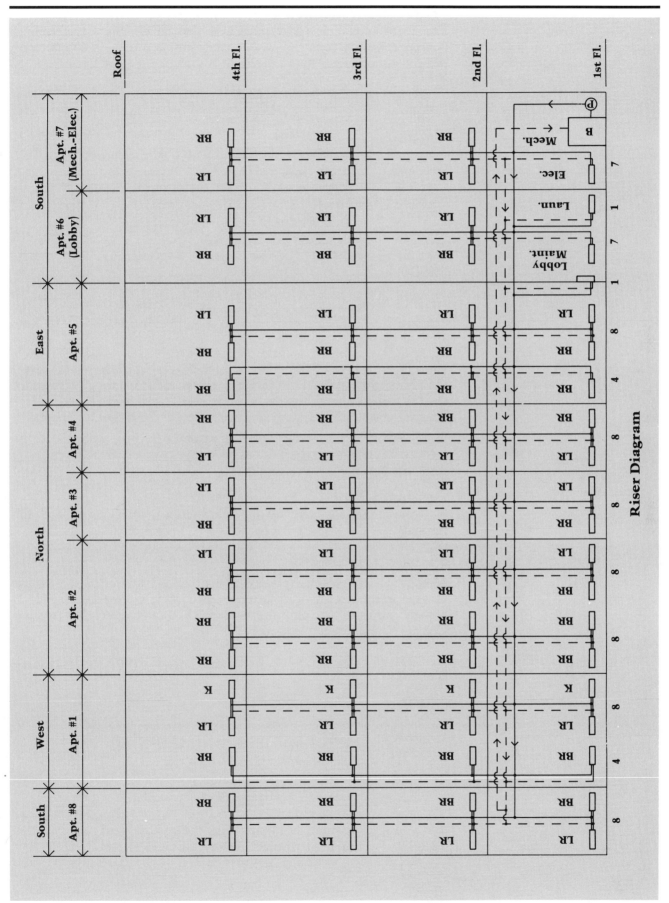

Figure 16.11

Oil Tank and Accessories

The size of an oil tank is determined from the total amount of fuel required during the heating season. The following formula is used for computing fuel consumption (see also Chapter 6, "Generation Equipment"):

$$F = \frac{24 \times DD \times H_t \times C}{E \, \Delta T \times V}$$

where

DD = degree days (3,114 DD Boston [SI]) (Figure 3.9)
H_t = design heat loss 71.9 kW
e = boiler efficiency – assume 85% (0.85) (Figure 6.1)
ΔT = 19 – (–13) = 32 °C (Figure 16.5)
V = heating value of fuel 11.2 kW/L (Figure 5.3)
C = coefficient (see Figure 6.3)

For a ΔT of 9 and H_G/H_N = 370/258.6 = 1.43, C = 1.42
Thus,

$$F = \frac{24 \times 3114 \times 71.9 \times 1.42}{.85 \times 32 \times 11.2} = 25\,047 \text{ L/yr.} = 6{,}617 \text{ gals.}$$
approx. 7,000 gals.

From Figure 6.11a, a 7,000-gallon steel tank is selected after comparing the advantages and disadvantages of fiberglass and steel tanks. (Accessories such as cathodic protection anchors, excavation, pad, pumps, and fuel lines are noted on the system summary sheet at the end of the chapter.)

In this example, because the tank is sized for the full year heating load, the fuel can be purchased off-season in bulk delivery at a savings of ten to thirty percent below peak season prices.

Expansion Tank

Expansion tanks are sized after the total volume of water necessary to fill the system is determined. For convenience, it is assumed that this was done after the piping was sized, and the total volume of water was determined to be 1893 liters.

Although the size of the expansion tank depends primarily on the volume of water in the system, other factors, such as system temperature and pressure, also affect the tank size, as shown in the equation below (see also Chapter 6, "Generation Equipment").

$$VT = \frac{(0.00052\,T - 0.014)\,V_s}{\dfrac{P_a}{P_f} - \dfrac{P_a}{P_o}}$$

where

T = 86.7 °C (assumed average water temperature)
V_s = liters of water in system (1893 assumed)
P_a = atmospheric pressure 101.3 kPa
P_o = maximum tank operating pressure 207 kPa + P_a – 10%
  i.e., 207 + 101.3 – 31 = 277.3 kPa
P_f = initial tank pressure = P_a + P_{pump} + P_{height} + $P_{residual}$
  = assume the pump is after the expansion tank and provides no head. Also assume that the tank is 1.8 m above the floor of a 12.8 m return riser and 10.3 kPa residual pressure exists.

$$= 101.3 + 0 + ((12.8 - 1.8) \times 10) + 10.3 = 221.6 \text{ kPa}$$

$$V_t = \frac{((0.00052 \times 86.7 - 0.014)\,1893}{\dfrac{101.3}{221.6} - \dfrac{101.3}{277.3}} = 619 \text{ L}$$

Use a 950-liter expansion tank to allow for a 50 percent air cushion. Note that if the tank were placed at the top of the building, the 11 m height would be eliminated from P_f, resulting in a smaller tank.

Circulating Pumps

Although most pumps are suitable for this application of hot water heating, a hot water circulation pump with high-temperature seals is selected, because it is recommended for pumps with a temperature over 66 °C in order to prevent leaks. A frame-mounted type is chosen because the motor, which is separate, can be easily replaced. Multiple stages are not necessary for heating, since the boiler controls can be used to vary output temperature.

The pump capacity (in L/s) and the head (in meters) must be specified. The L/s is determined from the heat load:

$$1 \text{ L/s} = 46.5 \text{ kW at 11 °C temperature drop}$$

$$\frac{71.9 \text{ kW}}{46.5 \text{ kW/L/s}} = 1.55 \text{ L/s}$$

Based on this equation, a load of 71.8 kW, 1.65 L/s is required.

The pump head is determined by examining the friction loss around the longest loop, as shown below.

Straight run of pipe
Basement loop (Figure 16.9) 2 × perimeter = 2 × 113 m	=	226 m
Boiler room piping, at 10% of basement	=	22.6
Vertical piping, one loop (Figure 16.11) 2 × 12.8 m	=	25.6
One radiator 4 elbow equiv. × 1.6*	=	6.4
One control valve at radiator 0.7 elbow equiv. × 1.12*	=	1.12
One balancing valve at radiator 17 elbow equiv. × 27.2*	=	27.2

Total straight run of pipe	308.92
Add 50% for fittings	154.46
Total equivalent length	463.38 m

The total equivalent length is rounded to 470 meters.

*See Figure 10.13 – use 1 m/s on 1" pipe for elbow, copper pipe.

The 50-percent allowance for fittings is a convenient short cut, but should be verified against a thorough takeoff of all elbows, tees, strainers, valves, meters, and other components in the system that may cause friction losses.

Using 470 meters as the total equivalent straight run of pipe, the pump head required to drive water around the loop can be determined by selecting the maximum loss per meter of pipe that will be permitted. A design head loss of 249 Pa of head per meter is selected. This maintains a maximum flow of 1.2 m/s in the pipes, which is considered acceptable. Therefore, the total head for the pump is:

$$H_p = 249 \times 470 = 117{,}030 \text{ Pa}$$

A pump with 1.65 L/s (26 gpm) and 117 Kilopascals (37.5 feet) of head is required; therefore, a 60-16 type pump is selected (see Figure 8.3).

Accessories

In the mechanical room, additional fittings, thermometers, pressure gauges, and specialties are required (see Figure 16.12).

Distribution Equipment

The major component of the distribution system is the piping. The pipe sizes are determined by the flow (gpm) in each pipe. The flow for each line is determined by the heat load being supplied. Use Figure 16.11, the distribution system layout, to size the piping after the heat load for each room has been determined.

Room-by-Room Load Breakdown

A detailed room-by-room analysis is usually performed on a computer. However, because this building is fairly uniform, the loads are easily determined and then adjusted to the special conditions of the first floor and top floor. The room-by-room load breakdown is outlined in Figure 16.13.

As expected, the second and third floor have identical heat losses. The first-floor heat load is higher because of door, slab, and perimeter losses. The fourth-floor heat load is also higher because of roof losses. For convenience, the floor loads are divided by the perimeter to obtain the loads per foot that the active length of baseboard radiators will be designed to carry.

Now the loads must be determined for each room containing a radiator. The riser for apartment #8 (see Figure 16.11) will be used to illustrate the room-by-room loads, piping loads, and pipe sizes (see Figure 16.14).

Because the building loads are relatively low, a 20 mm diameter pipe will be sufficient for each of the risers. This size is selected because it exceeds the design requirement of 15 mm and it matches the diameter of the typical radiator connections. Copper tubing is selected for its ease of installation and adaptability to the copper element in the radiation.

The main horizontal loop in the first-floor ceiling must also be sized. Steel pipe is used for the main and for all piping in the boiler room. The total load of 71.9 kW requires 1.65 L/s, which at 249 Pa/m of friction loss results in a 50 mm diameter pipe for the horizontal supply and return. This could be reduced to 40 mm approximately halfway around the loop, where the required flow drops below .95 L/s. Note that as the supply pipe size decreases, the return pipe size increases (see Figure 16.9).

The results of the sizing of the distribution system are outlined below.

20 mm pipe	11 risers × 2 pipes each riser × 12.8 m	=	924 l.f.
20 mm pipe	80 radiators × 3 m average each	=	800 l.f.
40 mm pipe (steel)	horizontal main × 2 pipes × 56.4 m	=	370 l.f.
50 mm pipe (steel)	horizontal main × 2 pipes × 56.4 m	=	370 l.f.

Terminal Units

The terminal units for this model are supplied with forced hot water. Fin tube baseboard radiation is selected because it is economical, will fit the room dimensions, and can be supplied and returned from the same end. Cast-iron baseboard could have been selected at a slight cost premium. Fan coil units are used for public spaces (e.g., lobbies).

By referring to Figure 11.1, the length of the radiator elements can be determined. For the unit selected at 88 °C average operating temperature (82 – 93 °C), this type of radiator yields 841 W/m (with no enclosure).

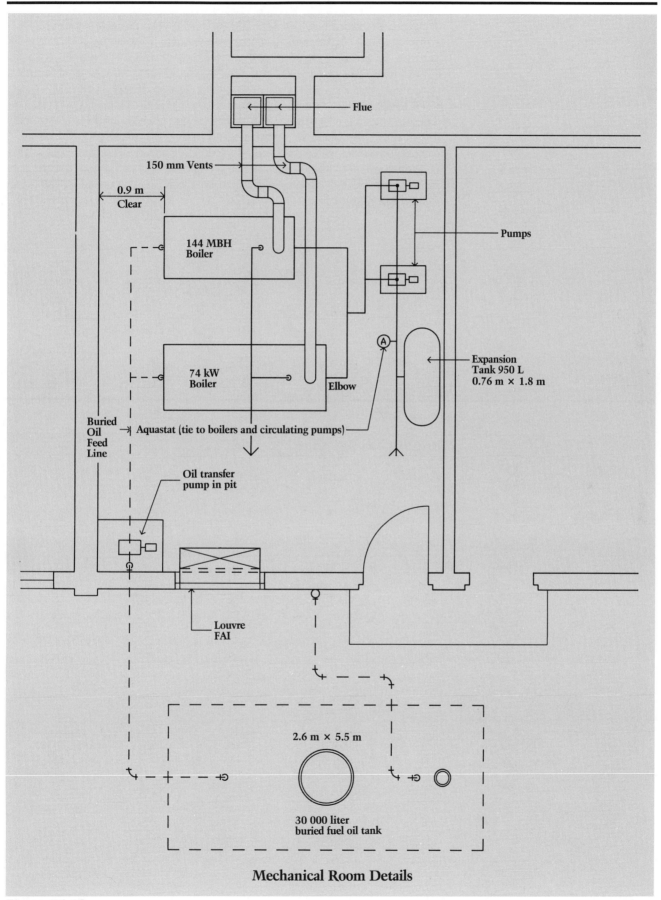

Mechanical Room Details

Figure 16.12

Figure 16.14 shows the selection of fin tube baseboard radiators based on design loads and an output of 750–875 W/m. All radiators selected are 1 meter of active fin length except the fourth floor living room, which requires 1.25 meter radiators. Of the 80 terminal units, 8 will be 1.25 meters long. Fan coil units for the lobby are selected in a similar manner.

Controls

The controls for a heating system are straightforward. In the mechanical room, the controls turn the pumps and boilers on or off and, if desired, may regulate the water temperature according to the outdoor temperature. In each apartment there may also be room or unit thermostats to regulate room temperature. Each of the controls selected is discussed in the following sections.

Heating Load Breakdown by Floor						
Conduction $H_c = UA \Delta T$	Total Load (Watts)	1st Floor	2nd Floor	3rd Floor	4th Floor	Remarks
Walls	10 700	2675	2675	2675	2675	Approx. equal
Windows	35 200	8800	8800	8800	8800	Approx. equal
Doors	310	310			3050	1st floor only
Front Door	330	330				1st floor only
Roof	3050					4th floor only
Total	49 590	12 115	11 475	11 475	14 525	100 000 W (rounded)
$H_c = FP$ $H_c = kA$	Total Load (Watts)	1st Floor	2nd Floor	3rd Floor	4th Floor	Remarks
Perimeter	3057	3057	—	—	—	1st floor only
Slab on grade	3746	3746				1st floor only
Total	6803	6803				6800 W (rounded)
Convection (50% at one time)	Total Load (Watts)	1st Floor	2nd Floor	3rd Floor	4th Floor	Remarks
Windows	28 968 x .5	3621	3621	3621	3621	Approx. equal
Doors	1593 x .5	797				1st floor only
Total	15 281	4418	3621	3621	3621	15 300 W
Floor-by-Floor Summary		23 336	15 096	15 096	18 146	72 000 W (rounded)
Loss per Linear Foot of Perimeter (P = 113 m)		207	134	134	161	

Figure 16.13

System On/Off

The heating system operates during the colder weather (heating season) and is off during the rest of the year. A master control switch is used to deactivate both boilers during the off-season.

Outdoor Reset

During the heating season, the boiler water temperature is set at 200°F for an outdoor design temperature of −13 °C. On warmer days, the boiler water temperature can be automatically reset to a lower temperature, down to 60 °C, for example, when the outdoor temperature is 16 °C.

Loads and Pipe Sizes for Risers

Riser Diagram Apt. #8[a]	Perimeter Load (W/m)[b]	Room Load (Watts) BR	Room Load (Watts) LR	Flow Req'd. (1 L/s = 46.5 kW)	Riser Size Based on Copper Type L 249 Pa/ Meter Loss[c]	Baseboard Length Based on 841 W/m
4th FL: BR 0.6 kW, LR 1 kW (3 m, 4.5 m); 0.03 L/s; 20 mm	207	207 × 3 = 621 rounded to 0.6 kW	207 × 4.5 = 932 rounded to 1 kW	BR = 0.01 L/s LR = 0.02 L/s ――― 0.03 L/s Riser below = 0.03 L/s	0.03 L/s 15 mm	BR 0.74 use 1 m LR 1.1 use 1.25 m
3rd FL: 0.4 kW, 0.6 kW; 0.05 L/s; 20 mm	134	134 × 3 = 402 rounded to 0.4 kW	134 × 4.5 = 603 rounded to 0.6 kW	BR = 0.01 L/s LR = 0.01 L/s ――― 0.02 L/s 0.02 + 0.03 Riser below = 0.05 L/s	0.05 20 mm	BR 0.48 use 1 m LR 0.72 use 1 m
2nd FL: 0.4 kW, 0.6 kW; 0.07 L/s; 20 mm	134	134 × 3 = 402 rounded to 0.4 kW	134 × 4.5 = 603 rounded to 0.6 kW	BR = 0.01 L/s LR = 0.01 L/s ――― 0.02 L/s 0.02 + 0.05 Riser below = 0.07 L/s	0.07 20 mm	BR 0.48 use 1 m LR 0.72 use 1 m
1st FL: 0.5 kW, 0.8 kW; 0.03 L/s; 20 mm; Main Supply/Return	161	161 × 3 = 483 rounded to 0.5 kW	161 × 4.5 = 725 rounded to 0.8 kW	BR = 0.01 L/s LR = 0.02 L/s ――― 0.03 L/s Riser below = 0.03 L/s	103 15 mm	BR 0.57 use 1 m LR 0.86 use 1 m

The flow rates in each length of pipe are indicated on the diagram. Since the standard radiator uses 20 mm diameter tube, all sizes are increased to 20 mm minimum for compatibility and to eliminate the cost of furnishing and installing reducers.

[a] Reference figure 16.11
[b] Reference 16.13
[c] Reference 10.11

Figure 16.14

Room or Unit Thermostats

Depending on the requirements of the building as established by the client, there may or may not be thermostats to control heat in each apartment. Individual nonelectric thermostatically-controlled valves are usually provided at each fin tube radiator. These operate by sensing the room air and, depending on the type of valve, use a refrigerant gas to operate the position of the valve. By using the self-contained thermostatic valves on each radiator, not only is each room individually controlled, but elaborate control wiring and extensive use of motorized valves is eliminated.

Boiler Sequencing

Two boilers have been selected to complement each other and to stage the heating sequence. A simple temperature control first activates the small boiler. When the small boiler can no longer handle the load, the large boiler cycles on and the small boiler shuts down. In extreme conditions, when the large boiler alone cannot handle the load, both boilers are called online. With these controls in place, the heating system can be expected to function with good overall efficiency.

Controllers

Whenever the water temperature falls below the preset design temperature, the boilers are activated by an aquastat mounted on the boiler (see Figure 16.12). When the boiler water temperature is at the design temperature, a signal is generated to activate the circulating pumps. The controller for the oil transfer pump is integral with the transfer pump unit.

Night Set-Back/Time Clock

Because this is a residential building occupied 24 hours a day, 7 days a week, it is not possible to predict when the entire system could be set back for heating. Some building codes require that public hallways be ventilated, in which case supply or exhaust fans would be controlled to shut down at night. However, for this building, it is assumed that they are not required.

Accessories

The heating system for this building has the accessory devices and equipment described below.

- Breeching
- Motor starters
- Vibration isolators
- Hangers and supports
- Insulation

Breeching

The breeching from the boilers to the vertical flue should not be less than the diameter of the boiler breeching connection. It is assumed that the breeching size for the boilers selected is 150 mm. The boilers are individually connected to the flue and each run uses three elbows. The 150 mm diameter can be determined using the following formula:

$$D^2 = \frac{HMT_f}{26 P_{(atm)} V}$$

H = 110 kW (Boiler sizes selected)
M = 1.9
T_f = 525 °C
$P_{(atm)}$ = 101.3 kPa
V = 6.1 m/s (assumed – conservative)

Therefore,

$$D^2 = \frac{110 \times 1.9 \times 525}{1.02 \times 101.3 \times 6.1} = 171.4$$

$$D = 13 \text{ cm diameter}$$

Figure 16.12 shows approximately 8.2 m of 15 mm flue piping in addition to six elbows.

Motor Starters

In this system, the electrically driven pieces of equipment that require motor starters are the two burners, the two circulating pumps, and the oil transfer pump. Each of these starters is automatic (magnetic) and activated by a controller. The starters are in NEMA I (normal indoor) enclosures.

Vibration Isolators

The circulating pumps should be mounted so as to minimize transmission of vibrations. The pumps selected run at 1,750 rpm and weigh approximately 300 pounds, or 34.5 kg per spring.

$$1{,}750 \text{ rpm} = \frac{1{,}750}{60} \text{ or } 29 \text{ Hz}$$

The isolator should have a frequency of one-third to one-fifth 29 Hz, or 8 Hz. From this figure, the recommended spring constant (k) is determined using the following formula:

$$k = \frac{(\text{nat. freq. of isolator (Hz)})^2 \times (\text{equipment (kg.)})}{25}$$

$$k = \frac{8^2 \times 75}{9.8} = 88, \text{ round to } 90 \text{ kg/cm maximum}$$

The piping from the pump should be isolated by flexible connectors and the entire assembly placed on a pad similar to that shown in Figure 13.3.

Hangers and Supports

Pipe hangers and supports are selected for the diameter of the pipe and noise control. In the boiler room and on the horizontal mains (steel pipe), the pipes are supported on clevis-type hangers. Note that the cost of hanger assemblies 3 meters on center are included in the cost of piping. Riser clamps are also required.

$$22 \text{ vertical pipes} \times 3 \text{ supports per riser} = 66 \text{ clamps}$$

Insulation

All piping is insulated with fiberglass insulation and an ASJ (all service jacket) cover. The 20 mm, 40 mm, and 50 mm lengths of pipe are entered on the system summary sheet.

Summary

All components of the system – generation equipment, distribution equipment, terminal units, controls, and accessories – have now been determined. The basic system is shown in Figures 16.4 and 16.11, and the layout of the mechanical room is shown in Figure 16.12. Figure 16.15 outlines each major component of the system, including a Means line number to determine costs in the current annual edition of *Means Mechanical Cost Data*.

System Summary Sheet

Item	Size	Quantity	Type	Accessories	R.S. Means Line No.	Remarks
Generation						
Oil-Fired Boiler	36 kW	1	Cast iron	Insulated jacket	155-120	
Oil-Fired Boiler	74 kW	1	Cast iron	Insulated jacket	155-120	
Oil Tank	30 000 L	1	Steel	Underground ST1-P3	155-671	Excavate and pour pad
Oil Pump and Piping	10 m	1	Copper	Remote gauge	155-250	
Expansion Tank	950 L	1	Steel	ASME	155-671	
Circulating Pumps	1.55 L/sec	2	Frame mounted	117 030 Pa	152-410	High temperature seals
Distribution						
Copper Tubing	20 mm	530 m	Type L	Supports and anchors	151-401	
Steel Pipe	40 mm	112.8 m	Schedule 40	Supports and anchors	151-701	Dielectric fittings
Steel Pipe	50 mm	112.8 m	Schedule 40	Supports and anchors	151-701	
Valves and Fittings	Various	112.8 m	Various			
Terminal Units						
Fin Tube Baseboard	20 mm	106 m	180 mm high	8 – 1.25 m lengths, 72 – 1 m lengths	155-630	
Lobby, Fan Coil Unit		142 L/sec	Wall hung	Heating only	157-150	
Controls						
On/off Switch	Unit	1	Winter "on"	Limit controls	157-420	Tie to boiler and pumps
Outdoor Reset	Unit	1	Automatic	Adjust boiler temperature	157-420	Adjust for electronic
Thermostatic Valves	Unit	81	Individual	Self-contained	156-240	Non-electric
Controllers	Unit		Aquastat	Set at 81° C	157-420	Tie to boilers and pumps
Accessories						
Breeching	150 mm	8.2 m	Single wall	6 elbows	155-680	
Motor Starters	Unit	5	NEMA O, 3.75 W		163-130	
Vibration Isolation	Unit	8	k – 90 kg/cm		157-485	4 each pump
Flexible Hose	Unit	4	50 mm pipe/pump		156-235	2 each pump
Riser Clamps	Unit	66	20 mm		151-901	
Pipe Insulation	20 mm	526 mm	Fiberglass		155-651	With ASJ Cover
Pipe Insulation	40 mm	113 m	Fiberglass		155-651	With ASJ Cover
Pipe Insulation	50 mm	113 m	Fiberglass		155-651	With ASJ Cover

Figure 16.15

Chapter Seventeen

COMMERCIAL BUILDING MODEL

In this chapter, an HVAC system is designed, using Imperial units only, for a low-rise commercial office building. The principles discussed in the previous chapters are the basis for the design, and references to tables and charts are provided to make the discussion easier to follow. The sample building, a three-story office building with a lower-level garage, is located in Albuquerque, New Mexico, and thus requires cooling. The building is an exposed concrete frame with fixed windows, hung ceilings, and a penthouse, which is provided to house the HVAC equipment. Each floor has an open plan. The entire building will be occupied by one tenant. The plans for the building are illustrated in Figures 17.1 through 17.3.

The four basic steps to take in order to establish the appropriate type of HVAC system for a building are listed below. The chapters in which each process is discussed are also listed.

1. Make the initial selections (Chapters 5 and 6).
2. Determine the design criteria (Chapters 3 and 4).
3. Compute the loads (Chapter 3).
4. Select the equipment (Chapters 6 through 13).

Initial Selection

As described in Chapters 5 and 6, the initial selections to make are: (1) the type of system, (2) the type of fuel, (3) the type and make-up of the distribution system, and (4) the type of generation system.

Type of System

The types of heating and cooling systems for buildings are listed in Figure 5.2 in Chapter 5; for a multi-story business occupied by one tenant, one central system is recommended. Because the building has fixed glass, it is necessary to supply fresh outdoor air for ventilation. A hung ceiling plenum is provided for air ducts. An air distribution system will heat, cool, and ventilate, which are required, as well as filter and humidify/dehumidify, which are desirable. Thus an air-conditioning system is provided.

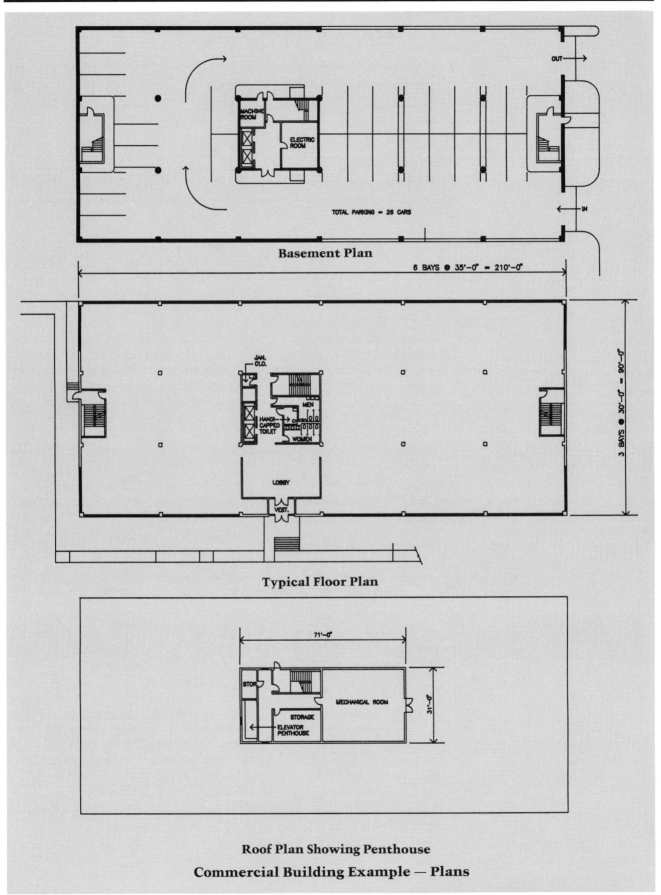

Figure 17.1

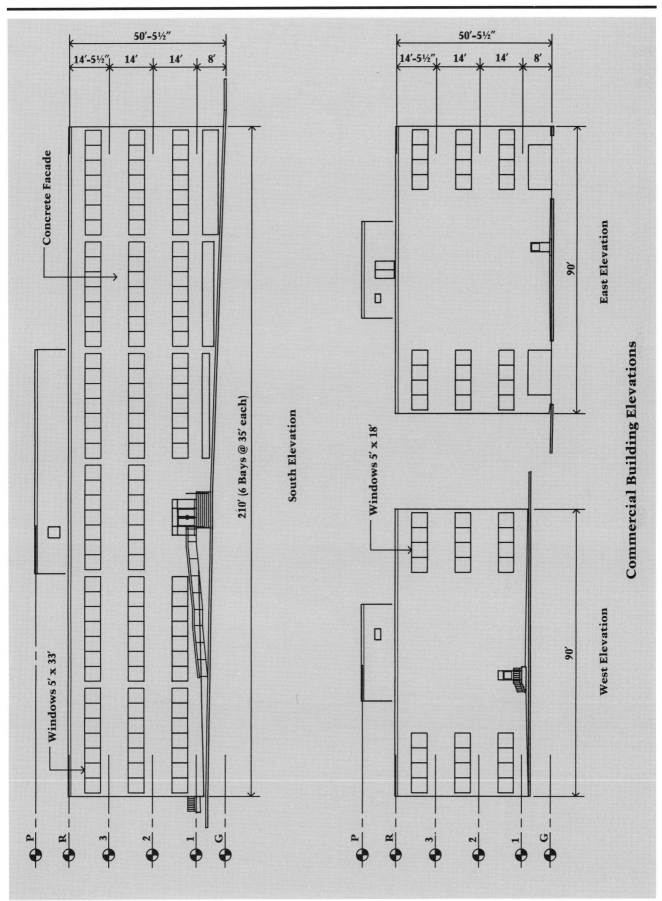

Figure 17.2

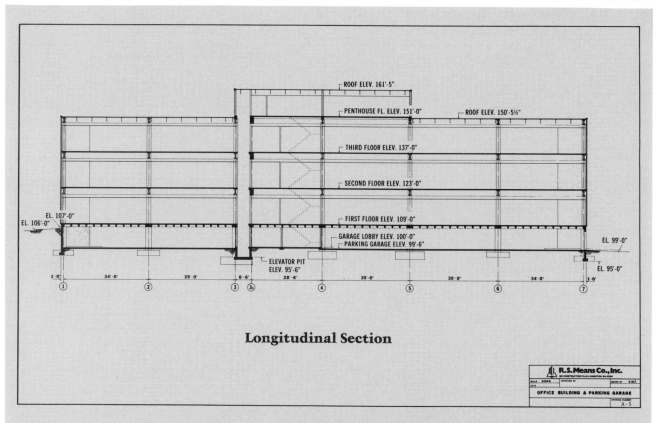
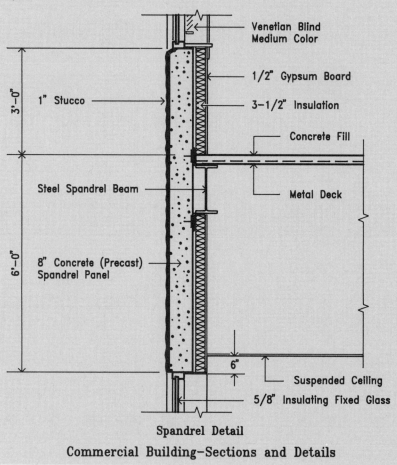

Figure 17.3

Fuel

Figure 5.3 lists the basic properties and cost of fuels. Even though electricity is listed as the most expensive fuel, it is the likely choice for cooling for this building, because electricity must be purchased in sizable quantities for the light and power of the building and is an essential fuel for cooling. In winter, the actual heat load of the office building may in fact be low, as people, lights, and equipment add warmth to the space, thereby reducing the heat load. However, it is also assumed that gas is available; and since gas is less expensive than electricity, it will be used for heating.

Distribution System

An air distribution system is selected for this building because it can heat and cool, as well as provide the required ventilation. Moving the heating and cooling medium from the mechanical room to the terminal units can be done with ductwork as an all-air system or by piping from the mechanical room to air handling units. A piping distribution system from the mechanical room to the terminal units is chosen for the following reasons:

- To save floor space that would be consumed by larger vertical air duct risers
- To allow for future flexibility, because additional piping can be run above the ceiling to new or relocated units with minimal disturbance to the floor layout
- To independently operate each floor during off-hours with minimum fan power

Therefore, the distribution system consists of both hot- and chilled-water piping to distribute hot/chilled water from the mechanical room to air handling units on each floor, and ductwork from each unit to distribute the conditioned air to the ceiling diffusers.

Generation System

For this model, the generation equipment selected is a gas-fired hot water boiler; it is the most direct choice that does not involve heat exchangers (see Figure 1.23 in Chapter 1). Chilled water will be made by a chiller, requiring a piece of equipment to reject heat, such as a cooling tower or air-cooled condensing unit.

Figure 17.4 shows the path selected for this system on the air conditioning ladder as well as the schematic diagram for the mechanical system.

Design Criteria

With the overall basic parameters now established, the design criteria must be determined. The design criteria for this sample project are listed in Figure 17.5. These criteria are used in the computation of the heating and cooling loads. Local codes, manufacturer's literature, and architectural requirements must also be checked to verify building design criteria.

Computation of Loads

The next step is to compute the total heating and cooling loads for the building. (For more information, refer to Chapter 3, "Heating and Cooling Loads.")

Heating Loads

The total heating load, H_t, for a building is the sum of all conduction and convection heat losses (see Figure 3.14).

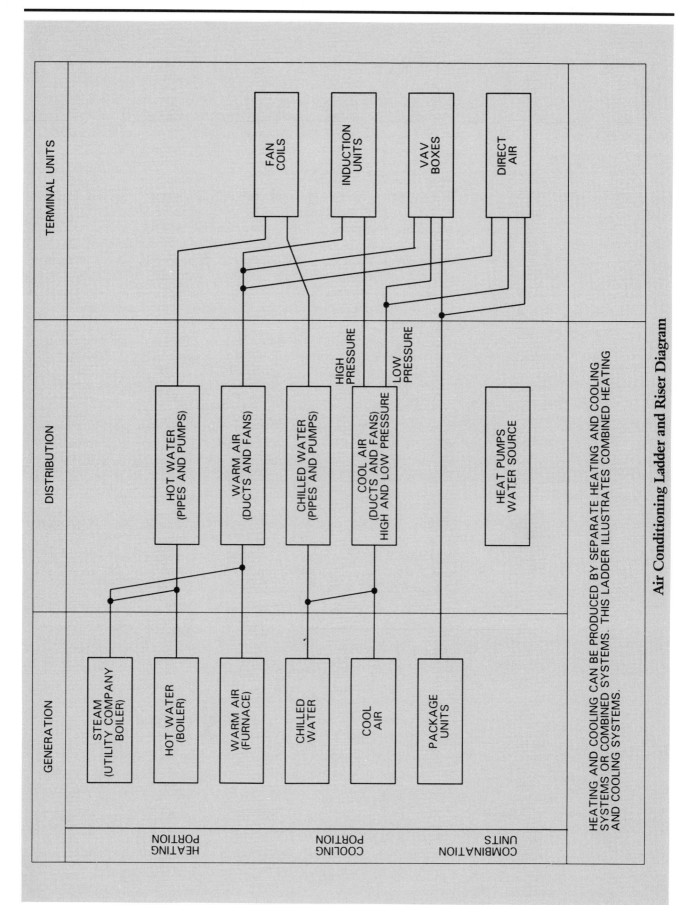

Figure 17.4

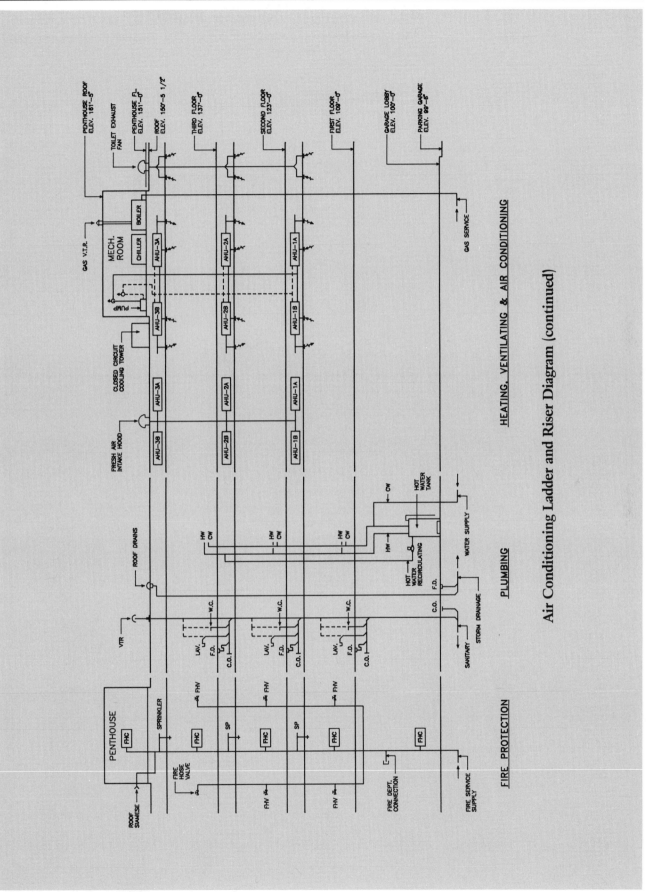

Figure 17.4 (continued)

Conduction Heat Losses

The conduction heat losses through all surfaces – walls, floors, doors, windows, and slabs – are determined by first calculating the area of the surfaces and then multiplying by the temperature differences and U values (see Figure 17.6).

Convection Heat Losses

Because the windows are fixed and the building has a forced air system, the infiltration losses are zero, except at the front door. The ventilation losses, however, are substantial, because the building has an occupancy of 524 persons, with 15 cfm of fresh outdoor air recommended (see Figure 17.5). The crack lengths (in linear feet) for this project are shown in Figure 17.7. The total heat loss is outlined in Figure 17.8.

$$H_t = H_c + H_e + H_v + H_s = 620 + 0 + 452 + 0 = 1{,}072 \text{ MBH}$$

\multicolumn{5}{c}{**Design Criteria**}				
Item	**Reference**	**Value**	**Comments**	**Recommendations**
Indoor Design Temp.	Fig. 3.8			
Business	Chap. 2			
Summer		74–78°F/30–55% RH	78°F	78°F/40% RH
Winter		70–74°F	68°F typical code	68°F
Garage		N/A		Not heated or cooled
Outdoor Design Temperature	Fig. 3.9			
Summer (2-1/2%)		94°F DB/61°F WB		94°F DB/61°F WB
Winter (97-1/2%)		16°F		16°F
ΔT Adjoining Spaces	Fig. 3.10			
Stairs			Assume 10°F outdoor	10°F
Garage				98°F Summer
				16°F Winter
Ventilation Standards	Fig. 1.22 Fig. 2.4 Fig. 2.5	15–25 cfm/person	Toilet exhaust serves to relieve pressure	15 cfm/person
U Values	Chap. 2			
Walls		0.08 Btu/hr./s.f./°F		0.08 Btu/hr./s.f./°F
Roof		0.06 Btu/hr./s.f./°F		0.06 Btu/hr./s.f./°F
Windows		0.53 Btu/hr./s.f./°F		0.53 Btu/hr./s.f./°F
Population	Chap. 3	100 s.f./person Floors 3 × 90′ × 210′ = 56,700 s.f. −Stairs 3 × 2 × 10′ × 20′ = −1,200 −Core 3 × 30′ × 35′ = −3,150 _____ 52,350 s.f.		524 People

Figure 17.5

U Values and Coefficients

Reference	Material	$U = \dfrac{1}{R_t}$	Remarks
Walls	1″ stucco	0.2	
Fig. 3.7	8″ concrete	0.88	Sand and gravel
Fig. 3.7	3-5/8″ insulation	13.00	Blanket type
Fig. 3.7	1/2″ gypsum	0.32	(plaster)
Fig. 3.5	Outside air film	0.17	
Fig. 3.5	Inside air film	0.68	
		$R_t = 15.25$	
		$U = 0.065$ (0.08 req'd) OK	
		U walls = 0.065 Btu./hr./s.f./°F	
Stair Wall	Double partition (Fig 2.6)	U stair wall = 0.34 Btu./hr./s.f./°F	Interior wall — 3/8″ gypsum board
Roof			
Fig. 3.7	Roof membrane	0.15	Assume roll roof
Fig. 3.7	2″ rigid insulation	14.40	Polyisocyanurate 2″
Fig. 3.7	2″ concrete	0.22	Sand and gravel
Fig. 3.7	1/2″ spray-on insulation	1.65	Cement fiber-wood
Fig. 3.5	Outside film	0.17	
Fig. 3.5	Inside film	0.61 (horiz. surface)	
		$R_t = 17.10$	
		$U = 0.059$ (0.06 req'd) OK	
		U roof = 0.059 Btu./hr./s.f./°F	
Floor (First)			
Fig. 3.7	3″ concrete and 3″ average for joists	0.67	Equiv. to waffle slab
Fig. 3.7	Carpet and fibrous pad	2.08	Add insulation recommended
Fig. 3.5	Inside film	0.61 (horiz. surface)	
Fig. 3.5	Outside film	0.17	Conserves little wind
		$R_t = 3.53$	
		U floor = 0.28 Btu/hr./s.f./°F	
Windows (Fig. 3.6)	Double insulated glass	U windows = 0.55 Btu/hr./s.f./°F	
Doors (Fig. 3.6)	Glass door — single	$U = 1.1$ Btu/hr./s.f/°F	
Edge Coeff. (F)	Not applicable — building above grade		F = N/A
Slab Const. (k)	Not applicable		k = N/A
Crack Coeff.	Windows — fixed — no leakage		Window coeff. = 0
Fig. 3.12	Doors — weatherstop		Door coeff. = 0.90
Fig. 3.12	15 mph (winter)		Door coeff. = (0.45 + 0.60)/2
	7-1/2 mph (summer)		= 0/53 cfm/ft.

All numbers shown in brackets are metric equivalents.

Figure 17.6

Cooling Loads

The cooling loads are now determined for the building and are summarized in Figure 17.9.

Several times of day and several days of the year could be selected to establish the design temperature. The ASHRAE method used to determine cooling factors utilizes average coefficients based on the latitude. Other methods actually take into account time of day and day of year. For a more exact analysis, several times and days would be considered to determine the overall worst condition. The ability to analyze several dozen trials is an obvious benefit of using a computer. Because the cooling load temperature difference (CLTD), solar cooling load factor (SCL), shading coefficient (SC), and cooling load factors (CLF) depend on orientation, the heat gain calculations can be done for each orientation separately and then added together to obtain the total cooling load. This example utilizes the average coefficients developed by ASHRAE to simplify the calculations required.

$$H_t = H_c + H_v + H_s + H_i + H_p$$
$$H_c = \text{conduction heat gain} = UA \,(CLTD) \,(\text{roof, walls, glass, and floors}$$

Areas and Crack Calculations

Facade	(1) Gross Wall Area (s.f.)	(2) Window Area (s.f.)	(3) Door Area (s.f.)	(4) Misc. Area (s.f.)	Net Area (1)-(2)-(3)-(4)	Crack Length (l.f.)
North	50.46' x 210' 10,596 s.f.	18 x 5' x 33' 2,970 s.f.	—	—	7,626 s.f.	Windows - 0 Doors - 0
South	50.46' x 210' 10,596 s.f.	17 x 5' x 33' 10' x 10' – 42 2,863 s.f.	2 x 3' x 7' 42 s.f.	—	7,691 s.f.	Windows - 0 Doors 2 x 6 + 3 x 7 33 l.f.
East	50.46' x 90' 4,541 s.f.	6 x 5' x 18' 540 s.f.	—	—	4,001 s.f.	Windows - 0 Doors - 0
West	50.46' x 90' 4,541 s.f.	6 x 5' x 18' 540 s.f.	—	—	4,001 s.f.	—
Total		Windows 6,913 s.f.	Doors 42 s.f.		23,319 s.f.	Doors 33 l.f.

Penthouse and stair doors ignored.
Roof Area = 90' x 210' = 18,900 s.f.
1st Floor Area = 90' x 210' = 18,900 s.f.
Stair Towers – Wall Area = 2 (towers) x 50.46' (high) x [10' + 20' + 10'] (interior wall) = 4,036 s.f.

Figure 17.7

H_v = convection heat gain = 1.1 cfm ΔT + 4,840 cfm ΔW (ventilation and infiltration)
H_s = solar heat gain = A × (SCL)(SC)
H_i = internal heat gain = (sensible load) × (CLF) + (latent load)
H_p = people heat gain = N_o × P_s × (CLF) + N_o × P_i

Conduction Heat Gain

The conduction heat gain (U × A × CLTD), or the heat transmitted to the building through the enclosure walls, is computed for the roof, walls, glass, doors, and floor. The U and A values have already been determined in Figures 17.6 and 17.7.

Heat Loads

Conduction
$H_c = UA\Delta T$

	U Btu/hr./s.f./°F (Fig. 17.6)	A (s.f.) (Fig. 17.7)	ΔT (°F) (68 – 16) = 52°F (Fig. 17.5)	Heat Loss (Btu/hr.)
Walls (net)	0.06	23,319	52	72,755
Windows	0.55	6,913	52	197,712
Doors	1.1	42	52	2,402
Roof	0.059	18,900	52	57,985
First Floor	0.28	18,900	52	275,184
Stairs	0.34	4,036	10	13,722

H_c = 619,760 Btu/hr.

H_e = FP = 0 (F = 0, no perimeter losses— building above grade over garage)

H_s = kA = 0 (k = 0, no slab on grade in heated space)

H_e = 0 Btu/hr.
H_s = 0 Btu/hr.

Total Conduction = 619,760 Btu/hr.
H_c = 0 Btu/hr.
= 620 MBH

Convection
H_v = 1.1 cfm ΔT

	Crack Length (Fig 17.7)	cfm/ft. (Fig 17.5)	ΔT (Fig. 17.5)	Constant	Heat Loss Btu/hr.
Windows		0			0
Doors	33' ×	0.90 ×	52 ×	1.1 =	1,699 Btu/hr.
Ventilation (Fig. 17.5)	524 people ×	15/person ×	52 ×	1.1 =	449,592

Total Convection = 451,291 Btu/hr.
= 452 MBH

Note that all of the 452 MBH from ventilation infiltrates simultaneously; no reduction is taken, as was done for the windows in the multi-family housing model in Chapter 16.

Figure 17.8

Figure 3.17 in Chapter 3 is used to determine the cooling load temperature difference. The roof has a suspended ceiling with a metal deck and concrete slab approximately 2" thick inside the insulation; roof No. 13, with a total R value of 17, is selected. The cooling load temperature difference can be corrected as noted in the table.

Cooling Loads

Conduction Heat Gain
$H_c = U \times A \times (CLTD)$ [1]

	U (Btu/hr./s.f./°F) (Fig. 17.6)	A (s.f.) (Fig. 17.7)	CLTD (°F) (Fig. 3.17)	$CLTD_{corr}$ (°F)	H_t	Remarks Solar Time 1,500 hrs.
Roof	0.059	18,900	38	44	49,064	Suspended ceiling Roof #9, Type B
Walls						
North	0.065	7,626	13	22	10,905	K = 1.0, T_r = 78, T_o = 94
South	0.065	7,691	17	26	12,988	
East	0.065	4,001	31	40	10,103	
West	0.065	4,001	15	24	6,242	
Floor	0.28	18,900		22	116,424	Use north CLTD T_o for under floor
Glass						
Doors	1.1	42		16	739	
Windows	0.55	6,913		16	60,834	94−78 = 16
Partitions	0.34	4,036		10	13,722	Fig. 3.6—double partition-stair

Total = 281,331 Btu/hr. H_c = 282 MBH

Convection Heat Gain
$H_v = 1.1$ cfm $\Delta T + 4,840$ cfm ΔW

	Sensible (Btu./hr.)	+	Latent (Btu/hr.)	
Windows	0	+	0	= 0
Doors	[1.1 × (0.53 × 33) × 20] = 385 +		[4,840 × (0.53 × 33) × (−0.004)]	= −339
Ventilation	[1.1 × (524 × 15) × 20] = 172,920 +		[4,840 × (524 × 15) × (−0.004)]	= −152,170
	H_v (sensible) = 173,305		H_v (latent) =	−152,509
	= 174 MBH			−153
				−77

Computation of Moisture Content (ΔW)

Outdoor (94°F DB/61°F WB) =	0.004 lb./lb. air [2]
Indoor (78°F DB/40°F WB) =	0.0082 lb./lb. air [2]
	ΔW = −0.0042 lb. moisture/lb. dry air

[1] The CLTD numbers used are taken from charts for 35° lat.
[2] Refer to Figures 3.18 and 17.5.

Figure 17.9

$$\text{CLTD}_{corr} = \text{CLTD} + (78 - T_r) + (T_m - 85)$$

CLTD = 38°F (This value is taken from 35° latitude tables, which are not included in this book because of the voluminous data therein.)

T_r = indoor room design temperature = 78°F (Figure 17.5)

T_m = maximum outdoor design temperature = 94°F (Figure 17.5)

Cooling Loads (continued)

Solar Heat Gain
$H_s = A (SCL)^2 (SC)$

	A (s.f.) (Fig. 17.7)	Zone Type (Fig. 3.19)	SC (Fig. 3.20)	SCL (Fig. 3.19)	H_s (Btu/hr.)
Windows (July – 36° lat.)					
North	2,970	A	0.56	36	59,875
South	2,863	A	0.56	49	78,561
East	540	A	0.56	39	11,794
West	540	A	0.56	159	48,082
Doors (south)	42		1.0	45	1,890

Total H_S = 200,202 = 201 MBH

Internal Heat Gain
H_i + (Sensible Load) × CLF + (Latent Load)

		H_i (Sensible) (Btu/hr.)	CLF (Fig. 3.21)	H_i (Latent) (Btu/hr.)
Lights	2.25 watts/s.f./× 3.41 Btu/watt × 3 floors (18,900 s.f./floor)	435,030	1.0	0
Computers	262 × 450	117,900	1.0	0
Office Machines	3 × 10,000	30,000	1.0	0
Coffee	3 × 3,580	10,740	1.0	3 × 1,540 = 4,620
Power	2.0 watts/s.f. × 3.41 Btu/watt × 3 floors (18,900 s.f./floor)	386,694		

H_i (Sensible) = 980,364 = 981 MBH H_i (Latent) = 4,620 = 5 MBH

People Heat Gain +
$H_p = N_o P_s (CLF) + N_o P_l$

Activity	N_o	P_s (Fig. 3.23)	CLF	H_p (Sensible)	P_l (Fig. 2.23)	H_p (Latent)
Office work	524	245	1.0	128,380	105	55,020

H_p(Sensible) = 129 MBH
H_p(Latent) = 55 MBH

[1] The SCL numbers used are taken from charts for 35° lat., available from ASHRAE on computer disc. The charts printed in Figure 3.19 are for 40° lat. only.

Figure 17.9 (continued)

Therefore,

$$\text{CLTD}_{corr} = 38 + (78 - 78) + (94 - 85)$$
$$\text{CLTD}_{corr} = 47$$

For the heat gained by the walls, the cooling load temperature difference depends on the type of wall, time of day, and orientation. As noted in Figure 17.3, the principal wall material is 8" HW concrete with code C8; the secondary material is stucco. The R value of the wall is 15.25. According to Figure 3.17, a type II wall is selected. The time selected is 1,500 hours; the cooling load temperature difference is listed for each orientation in Figure 17.9. The correction factor for the walls is similar to that for the roof. The correction factor is listed at the bottom of the CLTD tables for walls and computed below for each orientation:

$$\text{CLTD}_{corr} = \text{CLTD} + (78 - T_r) + (T_m - 85)$$

CLTD = north 13°F, south 17°F, east 31°F, west 15°F (Figure 3.17)

T_r = 78°F (Figure 17.5)

T_m = 94°F (Figure 17.5)

Thus, for the north wall

$$\text{CLTD}_{corr} = 13 + (78 - 78) + (94 - 85) = 22°F$$

The CLTD_{corr} for south, east, and west walls are computed similarly and are listed in Figure 17.9.

The floor over the garage is a special case. The cooling load temperature difference for the north wall is used to compute the heat gain for the underside of the floor, because it acts as a shaded surface in much the same way as the north wall does. Windows are thin and transparent compared to building walls, because they do not store heat during the day regardless of their orientation. Using the detailed (ASHRAE) method, the temperature may be varied during the day. However, many designers prefer to take the full load through the glass. This method is used here. Thus the conduction through all windows is determined simultaneously, and the temperature difference equals

$T_o - T_r$

$94 - 78 = 16°F$

Note: At 1,500 hours, this method produces results nearly equal to the results of the ASHRAE method. At other times of the day, it produces a slightly larger (conservative) load when compared to the ASHRAE method.

Convection Heat Gain

The quantities for infiltration (in cfm) in the summer are lower than they are in the winter, because the design velocity of the air in the summer is 7.5 mph; in the winter it is 15 mph. In this example, only the doors infiltrate air and the majority of the convection load comes from the ventilation air requirement of 15 cfm per person. The convection sensible heat load (1.1 cfm ΔT) is computed separately from the convection latent heat load (4,840 cfm ΔW), and they are added together. From Figure 3.18, the moisture content of the design room condition (78°F/40% relative humidity) is 0.0082 pounds of moisture per pound of dry air. This is larger than the moisture content of the outdoor air (94°F/61°F wet bulb), which is .004 pounds of moisture per pound of dry air. Therefore, the drier outdoor air will minimize some of the latent heat load, which is indicated by the negative values for ΔW in Figure 17.9. Normally, outdoor air adds dramatically to the latent and overall load, but Albuquerque is unusual in this regard.

Humidification would generally be indicated; however, in the summer, dry air is preferable, so the unusual choice of dehumidification is used to offset internal latent-heat gains. The toilet exhaust will remove some moisture – dry outdoor air will be introduced to replace it.

Solar Heat Gain

Solar heat gain – the heat gain resulting from radiant solar energy through glass – varies with orientation. Therefore, the loads for each of the four sides (and skylights, if any) are computed individually. The solar cooling load factor (SCL) is obtained from Figure 3.19. The zone type is also chosen from Figure 3.19, assuming the floor covering is carpet and the partitions are gypsum walls. The zone type for SCL factors is A. For each orientation in July, the selected month of the design temperature, 35° latitude is chosen for Albuquerque. (Note: The table in Figure 3.19 represents SCL factors at 40° latitude. Actual factors used in the example are from 35° latitude tables.)

The shading coefficient is obtained from Figure 3.20. For single-pane glass, the transmissibility is 86 percent. One pane of glass with medium-color venetian blinds has a shading coefficient of 0.65. Double-insulated glass, in this example, reduces the shading coefficient by 86 percent, or:

$$SC = 0.65 \times 0.86 = 0.56$$

For the door, the shading coefficient equals 1.0. Although the shading coefficient could be reduced further because the windows are recessed and a portion of the head and one jamb are in shadow, this would account for only a small overall savings and therefore is ignored for this model.

The solar cooling load factor for the doors is assumed to be 45 because the glazing is recessed.

Internal Heat Gain

The main internal heat gains in offices are from lights, computers, office equipment, and small appliances. The lighting load should be obtained from the architect, who often consults a lighting designer or the electrical engineer. A typical layout for offices uses $2' \times 4'$ fluorescent fixtures arranged in a suspended ceiling on an $8' \times 12'$ spacing, or 96 square feet for each fixture. If each fixture has four 40-watt fluorescent lamps, the energy density would be $(4 \times 40 \text{ watts})/96$ square feet or 1.67 watts per square foot. Increasing this by 15 percent for the ballast energy yields 1.92 watts per square foot for general lighting. This should be adjusted for task lights and special lighting to 2.0 to 2.25 watts per square foot (2.015 is used in the example). Lighting is one of the most significant heat gain loads to be considered when calculating the overall cooling load.

Computers and other office machines add heat to a building. For this example, personal computers are assumed to be dispersed throughout the building for office tasks. The building does not have a specific computer room. Each personal computer produces 450 Btu/hr. sensible heat. One personal computer for every 2 occupants, or 262 computers, are assumed to be operating at any one time.

An allowance of 10,000 Btu/hr. on each floor will be made for copy machines, typewriters, and other office machines. It is also assumed that each floor will have an electric coffee warmer. From Figure 3.22, the sensible load for this appliance is 3,580 Btu/hr. and the latent load is 1,540 Btu/hr.

Power allowances must also be made for receptacles and machine power, such as fans and portions of pump energy that will wind up in the occupied spaces and add to the heat load. To account for this, 1-1/2 watts per square foot is assumed for general receptacles and 1/2 watt per square foot for machine loads, or a total of two watts per square foot for power.

People Heat Gain

For office work, the sensible heat gain is equal to 245 Btu/hr. per person and the latent heat gain is equal to 105 Btu/hr. per person. The cooling load factor equals 1.0, a figure that is slightly conservative and should be used when the HVAC system does not operate 24 hours per day.

Thus, the total heat gain for the building is:

Gain		Sensible Load (MBH)	Latent Load (MBH)
Conduction	H_c	282	
Convection	H_v	82	-77
Solar	H_s	201	
Internal	H_i	981	5
People	H_p	129	55
	H_t	1,767 MBH	-17 MBH

$H_t = 1{,}767$ MBH $= 147$ tons (round to 150 tons)

Note: The negative latent load should not be added to the sensible load in order to get correct unit selections.

Equipment Selection

The basic system shown in Figure 17.4 indicates a mechanical room on the roof with a boiler, chiller, and air-cooled condenser. The overall simplicity of the open layout lends itself to minimizing the number of pieces of equipment. All ductwork and air handling units can be easily accommodated above the suspended ceiling. The ductwork above the ceiling provides for heating or cooling and allows for a degree of flexibility in future layouts. By placing the air handling units near the central core, the piping is minimized and accessible, and fresh outdoor air can be easily supplied to each unit. The supply-air ducts distribute air out to the perimeter as well as to some interior zones, and the return-air ducts are located at the center of the interior zones. In this way, supply air is delivered to the perimeter, where most of the heat loss and gain originate, and supplies cooling to the general office area. The central return air removes warmed air for cooling. Toilet exhaust fans are sized for 80 percent of the fresh air supply requirements to keep the building slightly pressurized. A layout of the first floor is shown in Figure 17.10. A layout for the piping is shown in Figure 17.11.

Because the piping to a cooling tower or air-cooled condenser is located outside the building, consideration must be given to freeze protection. The choices are to either drain the condenser water in the winter (which is not a possibility if water operation is necessary) or use a glycol/water solution in the exposed piping network. This office building requires cooling in cold weather, so glycol is used in the outside loop. This is somewhat conservative, but it will avoid a severe repair cost should a power failure in winter cause the pipes to freeze. The building size is considered small to intermediate, and would probably not have a full-time attendant on duty for that reason; the extra glycol protection is good overall insurance. The use of glycol means that an air-cooled condenser (closed circuit) must be used to cool the condenser glycol/water solution from the chiller. The glycol/water solution will be in a closed loop. If plain water were used, a cooling tower would be selected, which causes the condenser water to spill

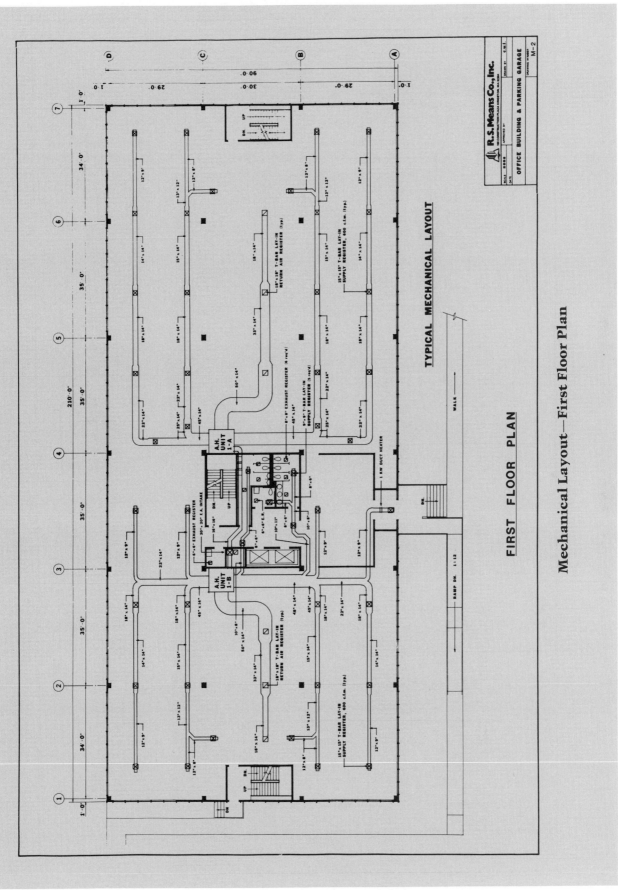

Figure 17.10 Mechanical Layout—First Floor Plan

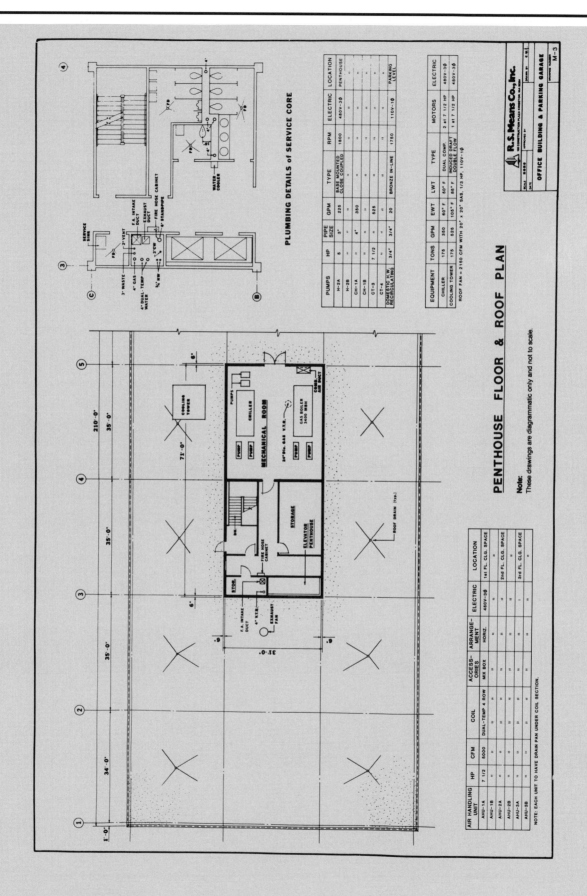

Figure 17.10 (continued)

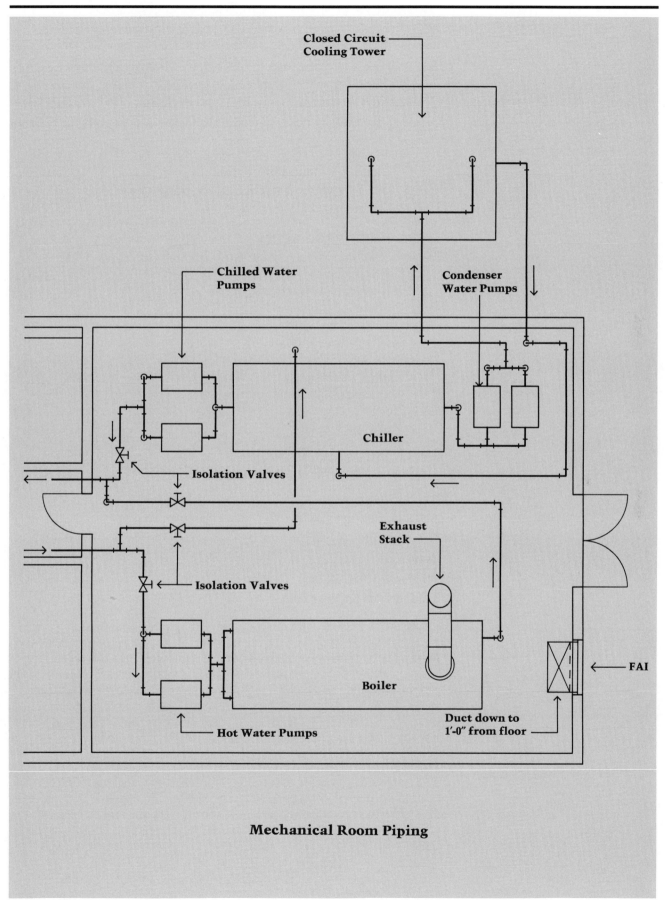

Figure 17.11

over the tower fill in an open loop. The objection to using a cooling tower with glycol is that it mandates the use of the closed loop, such as an air-cooled condenser. The selection is made later in this chapter.

The previous analysis results in the following design objectives:
1. Provide supply ducts in ceiling near perimeter and exterior zones.
2. Provide return air at center of interior zones.
3. Distribute hydronic piping from mechanical room to air handling units.
4. Provide fresh air for ventilation from roof to air handling units – exhaust air from toilet rooms.
5. Use a glycol/water solution in the condenser loop.
6. Use an air-cooled condenser.

Generation Equipment

The following generation equipment is selected (see Figure 17.4):
- Hot water boiler – gas-fired
- Hot-water expansion tank
- Chiller
- Chilled-water expansion tank
- Air-cooled condenser
- Hot-water circulating pumps
- Chilled water pumps
- Condenser glycol/water pumps
- Fresh air supply fan
- Toilet exhaust fan

Hot-Water Boiler – Gas-Fired

The boiler output size is determined as follows:

Net building heat loss	1,072 MBH	(H_t)
Allowance for pipe losses (5%)	54	(insulated pipe)
Pickup allowance (5%)	56	(small startup short piping)
Total load	1,182	
Allow 25% oversizing	296	
Total design boiler output	**1,478 MBH**	

Using the boiler selection chart (Figure 6.1), cast-iron boilers are chosen for this example because they are available in the range needed. Gas-fired boilers are commonly available with a low-fire and high-fire operation as an option, and this feature is selected. This option gives a boiler the same general efficiency as two boilers, but at a substantial cost savings. Since the building has substantial heat gains from lights, people, and computers, the risk of a heat loss problem from boiler failure is minimized. A gas-fired hot water boiler with an output of 1,500 MBH is selected.

Hot-Water Expansion Tank

The hot-water system is assumed to contain 1,000 gallons. This should be double-checked during a final review of all piping totals. The appropriate tank in terms of volume is obtained from the expansion tank data sheet. The following equation is used to calculate the required volume of the tank:

$$V_t = \frac{(0.00041t - 0.0466) \times V_s}{P_a/P_f - P_a/P_o} \text{ (for water between 160°F and 280°F)}$$

V_t = tank volume
t = 190°F average hot water temperature
V_s = 1,000 gallons of water in system
P_a = 14.7 atmospheric pressure in psia
P_f = P_a + P_{height} + $P_{residual}$ initial tank pressure. With the tank at top of building, P_{height} = 0.
14.7 + 0 + 1.5 = 16.2 psia
P_o = maximum tank pressure – 30 psi (relief valve) + 14.7 (P_a) – 10% = 39 psig

$$V_t = \frac{(0.00041 \times 190 - 0.0466) \times 1,000}{14.7/16.2 - 14.7/39} = \frac{31.3}{0.91 - 0.38} = 59.1 \text{ gal.}$$

An 80-gallon ASME-approved steel tank is selected.

Chilled-Water Expansion Tank

The size of the chilled-water expansion tank is determined similar to a hot water expansion tank, but using the adjusted formula. The volume of water is assumed to be the same (1,000 gallons) and the average system temperature, t, is assumed to be 50°F.

$$V_t = \frac{(0.006)V_s}{14.7/16.2 - 14.7/39} = \frac{6}{0.53} = 11.3 \text{ gal.}$$

A 30-gallon ASME steel expansion tank (the smallest practical size) is selected.

Chiller

The chiller is sized with a 15 percent reserve capacity.

Heat gain	150 tons –	1,800 MBH
15% reserve	23 tons –	270 MBH
Chiller size	173 tons –	2,070 MBH

Therefore, use a 175 ton chiller (2,100 MBH).

The chiller produces water at 45 – 50°F, which is sent to the coils in the air handling units. The warmed return water (52 – 60°F) is sent back to the chiller for cooling. The chiller cools the water using a refrigeration cycle, which takes the heat from the water in the air handling unit loop and transfers the heat to the water, which gives it to the air-cooled condenser, which dumps the heat to the outside. The condenser water is sent to the air-cooled condenser to be cooled at approximately 88 – 90°F, and returns to the chiller at approximately 80 – 85°F. As previously discussed, a glycol/water mixture will be used instead of water to cool the system to prevent freezing.

Using the selection criteria for chillers (from Figure 1.14), either a reciprocating or centrifugal type could be selected. The small size indicates that an absorption cycle chiller is too impractical to use and the required steam is not available. Therefore, a reciprocating water-cooled chiller at 175 tons with dual compressors and direct drive is chosen. The chiller will be glycol-cooled with adjustments made for glycol in sizing the pumps. The dual compressors allow for some load modulation, and the direct drive is available from the manufacturer. The manufacturer's literature notes that the chiller handles 600 gpm of water, requires 28 feet of head to drive the water through the chiller at design conditions, and the head loss for the condenser water through the chiller is 18 feet.

Air-Cooled Condenser

The air-cooled condenser is sized to match the requirements of the chiller, which, in this case, is 175 tons. The condenser water from the chiller could be cooled by a cooling tower or by an air-cooled condenser. In a cooling tower, the water spills over a series of baffles (fill), and the droplets are cooled by circulating air, which also causes some evaporation, adding to the cooling effect. A cooling tower is an open loop, because the water leaves the piping to spill over the fill (see Figure 10.1). A cooling tower is not effective in this case, however, because the glycol does not operate well in an open cooling tower and requires a closed loop to contain the glycol/water mixture. Therefore, a 175-ton closed-loop air-cooled condenser is selected to cool the glycol/water mixture as it passes through coils. Manufacturer's literature indicates that approximately 800 gpm of water against a 57-foot head is the necessary flow through the condenser.

Hot-Water Circulating Pump(s)

Two hot water circulating pumps will be provided, one as a standby. The hot-water pumps should have high-temperature seals. Using Figure 8.2, a frame-mounted pump is selected. To specify the pump, the capacity in gpm and the head in feet must be determined.

Capacity: Using the relationship that 1 gpm = 10 MBH at a 20°F temperature drop, the capacity to move the design heat load of 1,072 MBH is 107.2 gpm, rounded to 108 gpm.

Head: The head is calculated by determining the friction loss around the longest loops:

Straight run of pipe

Mechanical room supply and return 2 × 30	=	60.0 l.f.
Vertical supply and return to AHU (156' − 114')	=	42.0
Horizontal supply and return at AHU 2 × 6'	=	12.0
Straight run	=	114.0 l.f.

Fittings

1 elbow equivalent = 2.7 feet (see Figure 10.12) – 3 fps on 1" pipe

Boiler Room

1 boiler (Fig. 10.12) 3 × 1	3.0 l.f.
10 elbows 10 × 2.7' × 1	27.0
5 gate valves 5 × 2.7' × 0.7	9.5
2 globe valves 2 × 2.7' × 17	91.8

AHU Fittings

4 elbows 4 × 2.7' × 1	10.8
2 gate valves 2 × 2.7' × 0.7	3.8
2 globe valves 2 × 2.7' × 17	91.8
	237.7

Add 50% for misc. fittings, strainers		118.9
Straight run of pipe		114.0
Total equivalent pipe length	=	470.6 l.f.

The 50 percent allowance for fittings is a short cut. A detailed take-off of the final loop would be required in an actual job to double-check this figure. The air handling unit coil also has a head loss. For heating, it will be taken at 3 feet for the anticipated flow – check with specific manufacturers for other specific recommendations.

The head loss on 471 linear feet of pipe at the general design loss of 2.5'/100' equals:

$$Hp = 471' \times 2.5/100 + 3'(AHU) = 14.8, \text{ or } 15 \text{ feet of head}$$

This is very conservative, since the chilled water demand is much higher and, if the piping that is used for heating is also sized for the chilled water capacity, the heating loop will be oversized, resulting in lower head for the hot water pump. A pump with 108 gpm and 15 feet of head is therefore required. From Figure 8.3, a 60–14 or 60–19 type pump is selected.

Chilled-Water Pump(s)

The chilled-water system piping loop is identical to the hot water loop, except that the chilled water will flow through the chiller instead of the boiler. This produces a different friction loss through the system, which circulates at approximately 50°F instead of 200°F.

Capacity: The required flow through the system is determined from the heat gain plus the reserve, which totals 175 tons (2,100 MBH). In cooling, the supply and return water temperatures typically differ by approximately 7°F, instead of the 20°F drop associated with heating. Thus, the standard flow equation, 1 gpm = 10 MBH at 20°F drop, must be modified by the ratio of 20 to 7 to correct for this difference. The seven-degree differential and required flow must be verified with the particular chiller selected.

$$\text{gpm req'd} = \frac{2,100 \text{ MBH}}{10 \text{ MBH/gal.}} \times \frac{20}{7} = 600 \text{ gpm}$$

The 600 gpm agrees with the chiller data previously stated.

Head: The head from the heating loop can be adjusted to account for the minor differences of the chiller and the temperature. From the chiller selection, the chiller has a head loss of 28 feet, the boiler 0.08 feet (3 × 2.5/100). The head loss for the air handling unit coil in cooling is assumed to be nine feet for cooling, which is greater than the three feet needed for the heating coils.

Since the friction loss flow charts, Figure 10.11, are based on 60°F water, no temperature correction is necessary for the ± 50°F chilled water. Therefore, the friction head required for the chilled water piping is:

Head loss piping loop (same as hot water pumps)	=	15.0 feet
Correction for chiller (+ 28' – 0.08')	=	28.0
Correction for AHU (+ 9' – 3 cts)	=	6.0
Design chilled water pump	=	49.0 feet
Friction head required	=	50.0 feet

The chilled-water pump is designed for 600 gpm flow and 50 feet of head. Beyond the range of those shown in Figure 8.3, manufacturers' literature should be consulted. The requirements of the pumps are listed on the system summary sheet at the end of this chapter.

Condenser Glycol/Water Pumps

The condenser glycol/water pumps circulate the glycol/water mixture between the chiller and the air condenser. The capacity and head are now determined.

Capacity: In this example, the condenser water enters at 88°F and leaves at 83°F – a 5°F drop in temperature, which is about the average temperature drop. In some cases, particularly Albuquerque, evaporative coolers perform well because of the relatively dry air. However, an air-cooled condenser has been selected for this example, so the 5°F temperature drop is used.

$$\text{gpm req'd} = \frac{2{,}100 \text{ MBH}}{10 \text{ MBH/gal.}} \times \frac{20}{5} = 840 \text{ gpm}$$

Because glycol is used, the flow must also be corrected to account for the specific heat of the mixture, using Figure 10.12. For a 100°F fluid temperature, the flow increases by a factor of 1.16. Hence, the capacity of the pumps required is: 840 gpm × 1.16 = 975 gpm.

Head: The manufacturer's literature for this particular air-cooled condenser indicates a head loss of 57 feet, and the manufacturer's literature for the chiller indicates a head loss of 18 feet. In addition to these losses, the following piping losses must be accounted for:

Straight run of pipe			
Supply and return chiller to condenser		=	80.0 l.f.
Fittings (1 elbow equivalent = 2.7 feet)			
6 gate valves	6 × 2.7' × 0.7		11.3
2 globe valves	2 × 2.7' × 17		91.8
10 elbows	10 × 2.7' × 1		27.0
	Equivalent pipe length	=	210.0
	Add 50% for strainers	=	105.0
	Use		315.0 l.f.
Pump head (315 l.f. × 2.5'/100')			7.9 ft.
	+ chiller		57.0
	+ air-cooled condenser		18.0
	Head for water system	=	**82.9 ft.**

This figure must be adjusted for glycol using the formula provided in Figure 10.12. For 100°F water, the pressure drop correction factor is 1.49. Thus, the design head loss is:

82.9 × 1.49 = 123.5 ft.

Glycol may dissolve certain types of gaskets; therefore, glycol-resistant seals are specified. A 975 gpm, 124-foot head pump with glycol-resistant seals is entered on the system summary sheet.

Fresh-Air Supply Fan

The purpose of the fresh-air supply fan is to supply ventilation air to the building. The design requirement is 15 cfm of air per person for the design capacity of 524 persons, which has already been accounted for as part of the convection loss for heating and cooling (Figures 17.8 and 17.9). The selection guide for fan types (Figure 8.7) indicates that a centrifugal fan with forward curved or radial blades would be appropriate for this low-volume application. The fan requires curbs and flashing accessories for roof mounting. In order to avoid heating coil freeze-up in winter, an electric preheat coil set at 55°F minimum must be provided. The capacity and head of the supply fan are now determined.

Capacity: 524 persons × 15 cfm each = 7,860 cfm, or 7,900 cfm.

Head

The fan will direct air vertically down to the six air handling units. The length of the ductwork is:

Straight run of duct
   Roof to first floor (151 − 120)   = 31 l.f.
Fittings
   roughly 7,800 cfm @1,000 fpm = 7.8 s.f.
   duct = 36″ × 24″
   3 elbows (smooth elbow (Figure 10.22))
   L/D = 9 D = 2′ (24″) L = 9 × 2′   = 18
                                                                       $\overline{49}$
                          Add 50% fittings   = 25
                          Total duct length   = $\overline{74}$ l.f. (say 75′)

Minimum fan pressure head:

   75 l.f. × 0.1″/100′ = 0.075″ (pressure including the duct heater)

This is low when compared to the capability of such fans. Therefore, a 7,800 cfm at 0.5″ pressure supply fan is selected.

Toilet Exhaust

The exhaust requirements are now sized for 80 percent of the supply air:

   0.8 × 7,800 = 6,240 cfm

The pressure requirements are similar to the supply air and, therefore, a roof exhaust fan is selected. Louvers should be provided in the doors or transoms to the toilet rooms to permit the air to move from the space to the exhaust system.

Distribution Equipment

There are two major components that make up a distribution system — the piping and the ductwork. For both, the cooling loads govern the design of the distribution system.

The cooling load is broken down by floor using the data from Figure 17.9, which shows the computation of the cooling load for this commercial building model.

Generally, the loads on the three floors are equal except for adjustments for the roof, first-floor slab, and door loads. The load breakdown for the building is shown in Figure 17.12.

Using these loads, the loads for each air handling unit are established. (The loads and pipe sizes are detailed in Figure 17.13.)

     AHU 1A and 1B = 804 MBH/2 = 402 MBH
     AHU 2A and 2B = 609 MBH/2 = 305 MBH
     AHU 3A and 3B = 664 MBH/2 = 332 MBH

The flow to each air handling unit is based on a 7°F temperature drop. Therefore:

$$\text{AHU 1A and 1B} = \frac{402 \text{ MBH}}{10 \text{ MBH/gal.}} \times \frac{20}{7} = 115 \text{ gpm}$$

$$\text{AHU 2A and 2B} = \frac{305 \text{ MBH}}{10 \text{ MBH/gal.}} \times \frac{20}{7} = 87 \text{ gpm}$$

$$\text{AHU 3A and 3B} = \frac{332 \text{ MBH}}{10 \text{ MBH/gal.}} \times \frac{20}{7} = 95 \text{ gpm}$$

The total flow to all air handling units matches the 600 gpm supplied by the chiller. The pipe sizes are indicated in Figure 17.13. Steel pipe is used throughout.

Ductwork

The ductwork consists of the following (see Figure 17.10):
- Supply air from each air handling unit
- Return air to each air handling unit
- Fresh air supply from the roof
- Toilet exhaust

The constant pressure drop method is used to size the ductwork, which, for architectural consideration, has been limited to 14 inches deep in this building. Because all six air handling units are relatively close in size (305 to 402 MBH), the ductwork is designed for only one air handling unit, 1A, to illustrate the procedure. Figure 17.14 illustrates the schematic layout of the ductwork and sizing calculations.

The supply registers are laid out approximately 30 feet on center around the building perimeter, and the return air is taken from the center of each area. This draws the warmer inner core air to the air handling unit for

Load Breakdown by Floor					
Adjustments					
Roof Loads (third floor)					
Conduction					48 MBH
First Floor Slab and Door Loads					
Conduction					116.5 MBH
Doors					.7 MBH
Convection—Doors					.4 MBH
Solar					
Doors					1.9 MBH
					167.5 MBH
	MBH		**MBH**		**MBH**
Roof	48	Slab & Door	167.5		
Third Floor	529.0	First Floor	529.0	Second Floor	529.0
Roof	577.0	First Floor	696.5	Second Floor	529.0
Increase by 15%	+86.5		+104.8		+79.4
Roof/Third Floor	604				
First Floor	804				
Second Floor	607				
Total	2,077 MBH				

All numbers shown in brackets are metric equivalents.

Figure 17.12

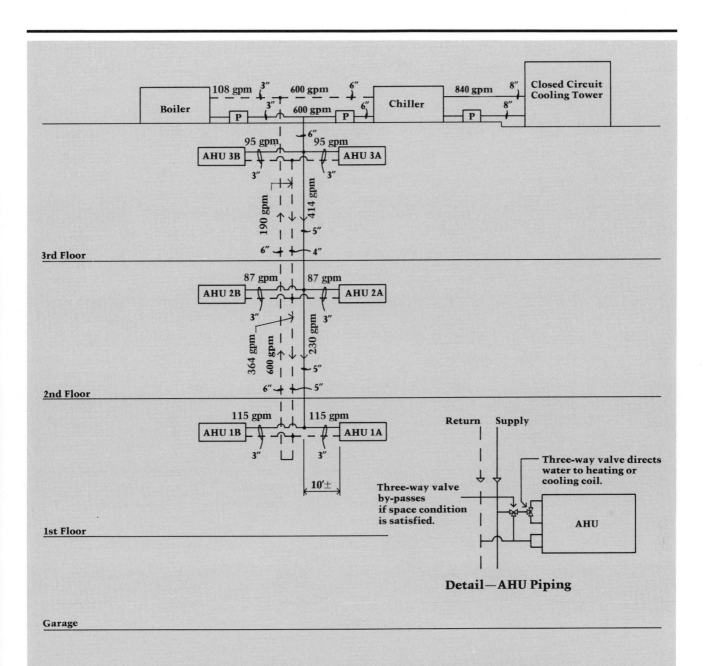

Loads and Pipe Sizes for Distribution Systems

Figure 17.13

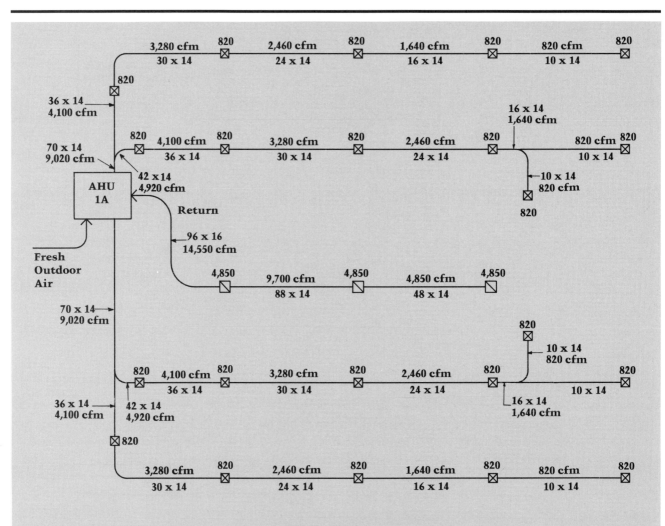

Duct Sizing (cfm)	Loss per 100' (Fig. 10.22)	Round Duct (Fig. 10.20)	Rect. Equiv. (Fig. 10.21)	Length (l.F.)	Perimeter (in.)	14 Ga. Alum. (lbs./l.f.)	Weight of Ductwork per Unit (lbs.)
Supply Air							
820	.1	13	10 x 14	2 x 30 + 2 x 25 + 2 x 10 = 130	48	4	520
1,640	.1	16.5	16 x 14	2 x 30 + 2 x 5 = 70	60	5	350
2,460	.1	19.5	24 x 14	4 x 30 = 120	76	7	840
3,280	.1	22	30 x 14	2 x 30 + 2 x 35 = 130	88	7.5	975
4,100	.1	24	36 x 14	2 x 30 + 2 x 25 = 110	100	8	880
4,920	.1	25.5	42 x 14	2 x 15 = 30	112	9	270
9,020	.1	32	70 x 14	8 + 30 = 38	168	13	494
Return Air							
4,850	.07	27	48 x 14	30	124	9.5	285
9,700	.07	35	88 x 14	30	204	15	450
14,550	.07	40	96 x 16	30	224	16	480

5,544
10% Fittings 554
6098 lbs.
Rounded to 6100

Figure 17.14

cooling directly, distributes the cool air to the perimeter, and draws it across the floor to the inner core.

Fresh air from the roof is injected into the air handling unit along with the return air.

Air handling unit 1A has a heat load of 402 MBH. The amount of air to be distributed is determined using the following equation:

H_v = 1.1 cfm ΔT
H_v = sensible heat gain
cfm = air quantity
ΔT = $T_{room} - T_{supply}$

The sensible load for air handling unit 1A equals the total load (402,000 Btu/hr.) minus the latent load. The latent load for the building equals 55,000 Btu/hr. for people, plus 5,000 Btu/hr. for equipment (see Figure 17.9). Air handling unit 1A will take 1/6 of the 60,000, or 10,000 Btu/hour. Therefore,

H_s = 402,000 − 10,000 = 392,000 Btu/hr.

Because of the relatively dry air in Albuquerque, the latent load is small. Generally, approximately 25 percent of the cooling load is latent heat. Thus,

392,000 = 1.1 cfm (78 − 58)

$$\text{cfm} = \frac{392,000}{1.1 \times 18} = 17,818 \text{ cfm, rounded to } 18,000 \text{ cfm}$$

Note: The ΔT of 20°F is on the high end of the recommended range of 15 − 20°F, and is used both to reduce the size of ducts and to decrease costs. However, poor air distribution design will result in cool air pockets and uncomfortable zones in the cooled space.

From the layout, there are 22 supply air diffusers, each handling approximately equal areas and loads. Thus, each diffuser must supply 18,000/22, or 820 cfm each. The cfm for each run of duct is shown in Figure 17.14.

The recommended friction loss is 0.1"/100' for supply ducts and 0.07"/100' for return ducts (see Figure 10.22). For the duct carrying 820 cfm, Figure 10.20 indicates a 13" duct at 0.1"/ 100'. This is shown in Figure 17.14, along with other circular duct sizes. After each round duct is determined, Figure 10.21 is used to convert it to an equivalent rectangular duct. The 13" round duct is equivalent to a 10" × 14" rectangular duct.

The return duct system is designed to accommodate 80 percent of the supply air, or

18,000 × 0.80 = 14,400 cfm

Each of the three return-air registers takes 14,400/3 = 4,800. Return-air ducts are sized for 0.07"/100'.

The ductwork for air handling unit 1A weighs approximately 6,100 pounds, which is typical for all six units. Thus, the total ductwork equals:

6 × 6,100 = 36,600 lbs.

The fresh-air and toilet-exhaust ductwork is sized in a similar manner. The total fresh air requirement is 7,900 cfm, 2,633 cfm to each floor. The exhaust air equals 6,320 cfm, 2,107 cfm from each floor. They are sized as follows:

Location		cfm	Loss per 100'	Round Duct	Length Use	(l.f.)
Fresh air	1-2	2,633	0.1	20"	20"	14'
	2-3	5,266	0.1	27"	28"	14'
	3-roof	7,900	0.1	30"	30"	17'
Exhaust air	1-2	2,107	0.07	20"	20"	14'
	2-3	4,214	0.07	26"	26"	14'
	3-roof	6,320	0.07	30"	30"	17'

Spiral preformed aluminum is selected for all ductwork.

Terminal Units

The terminal units for this building consist of the following:
- Air handling units (6)
- Supply air diffusers (11 per air handling unit)
- Return air grilles (3 per air handling unit)
- Toilet exhaust grilles (12)

Sizing for each of these components is described in the following sections.

Air Handling Units

The six air handling units are sized for 30 tons each to make them uniform and, therefore, easier to maintain. Each has a heating and a cooling coil, two-speed fan, and two connections, one for the mixture of fresh and return air, and one for supply air. The low fan speed is used for heating.

Supply-Air Diffusers

Each of the 11 supply-air diffusers has been designed for 820 cfm. The actual diffuser characteristics vary with each manufacturer. The design considerations for this system are to keep the exit velocity low for noise control (under 800 fpm), to select the diffuser for the proper throw (3/4 × 30/2 or ± 12'), and to keep the pressure losses through the diffuser between 0.05" and 0.15", if possible. A review of manufacturers' catalogs for available sizes of diffusers indicates that 21" × 21" four-way diffusers with volume control meet these criteria. For the 6 air handling units, 66 diffusers are required.

Return-air Grilles

Eighteen return-air grilles are required, three per air handling unit. Each grille is designed for 4,850 cfm. The return air is assumed to travel at 1,200 fpm (see Figure 10.18). This results in a free area of:

$Q = VA$

$50 = 1,200 \times A$

$A = 4.04$ s.f.

The free area must be increased by fifty percent to account for the blade area of the grilles:

$A = 4.04 \times 1.5 = 6.06$ s.f.

A 30" x 30" grille meets the area requirement. Note that a velocity of 1,200 fpm is high; lower values are often used.

Toilet-Exhaust Grilles

The total toilet room exhaust of 6,320 cfm will be removed by the exhaust grilles with capacities of 506 cfm each. The free area required is:

$$Q = VA$$
$$526 = 1{,}200 \times A$$
$$A = 0.44 \text{ s.f.}$$

The free area must be increased by fifty percent to account for the blade area of the grilles:

$$A = 0.44 \times 1.5 = 0.66 \text{ s.f.}$$

A 10″ × 10″ grille is adequate. However, use 12″ × 12″ to reduce velocity noise.

Controls

The control system for this model is more sophisticated than the controls for the apartment model (Chapter 16), because both the heating and cooling systems must be controlled; the building has occupied and unoccupied hours; there is opportunity for "free cooling"; and both piping and ductwork must be controlled.

Time Clock

A 24-hour, 7-day time clock is provided to switch equipment from occupied to unoccupied settings. During occupied hours, the equipment operates at an indoor temperature of 68°F in the winter and 78°F in the summer. During unoccupied hours (6 p.m. to 6 a.m. week nights and all weekend), the fresh-air supply fan, the toilet-exhaust fan, and all cooling equipment are turned off. In the winter, the night/weekend setback temperature is 55°F. The time clock has a 16-hour spring-wound reserve so that it does not need to be reset in the event of a power failure. As an alternate, a thermostat with built-in 7-day control can be used.

Summer/Winter/Off Switch

The piping to the air handling units is used for both heating and cooling. The boiler and chiller cannot run simultaneously in this layout, which is designed to save energy. The summer/winter/off switch activates the boiler and its pumps in the winter and simultaneously deactivates the chiller, air-cooled condenser, and pumps. The reverse happens in the summer. The summer/winter/off switch also sets the proper temperature, and sets the fan speed of the air handling units to low in winter.

Outdoor Reset

The outdoor reset control measures the outdoor temperature and modulates the boiler water temperature accordingly to achieve maximum efficiency.

Automatic Control Valves

The hot- and chilled-water coil to each air handling unit has an automatic three-way valve. On demand for heating or cooling, the valve diverts water into the respective coil; otherwise, it bypasses the unit (see Figure 17.13). Twelve valves, two per unit, are required.

Return Air Sensors

The room temperature is measured by sensors in the return air ductwork. Each air handling unit has one temperature sensor tied to the control system to activate the three-way valves.

Controllers

Aquastats are used for the boiler, chiller, and air-cooled condenser to modulate the internal controls. It is assumed that these come with the equipment selected.

Accessories

The HVAC system for this building has accessory devices and equipment as described below. (See Chapter 13 for more information on accessories.)

- Breeching
- Motor starters
- Vibration isolators
- Hangers and supports
- Insulation

Breeching

The breeching from the boiler to the chimney should equal at least the diameter of the breeching outlet. It is assumed that the breeching size for the boiler is ten inches. The boiler is connected to the flue and vented directly through the roof. The ten-inch diameter is verified using the following formula:

$$D^2 = \frac{H \times M \times T_f}{26 \, P_{(atm)} \, V}$$

H = 1,500 MBH (boiler size selected)
M = 1.1
T_f = 800°F
$P_{(atm)}$ = 29.92 inches H_g
V = 20 feet/second (assumed – conservative)

therefore,

$$D^2 = \frac{1,500 \times 1.1 \times 800}{26 \times 29.92 \times 20} = 84.7$$

D = 9.5" diameter

Use 10" diameter as the next manufactured size larger than 8" diameter. Figure 17.4 shows approximately 10 linear feet of 10" flue.

Motor Starters

In this system, the equipment items that require motor starters are the boiler, supply fan, toilet fan, chiller, air-cooled condenser, the six circulating pumps, and the air handling units. Each of these starters are automatic (magnetic) and activated by a controller. The starters are in NEMA I (normal indoor) enclosures.

Vibration Isolators

The circulating pumps are mounted so as to minimize transmission of vibrations. The pumps selected run at 1,750 rpm and weigh approximately 500 pounds, or 125 pounds per spring.

$$1,750 \text{ rpm} = \frac{1,750}{60} \text{ or } 29 \text{ Hz}$$

The isolators should have a frequency of one-third to one-fifth 29 Hz, or 8 Hz. From this, the recommended spring constant, k, can be found using the following formula:

$$k = \frac{(\text{nat. freq. of isolator (Hz)})^2 \times (\text{equipment (lbs.)})}{9.8}$$

$$k = \frac{8^2 \times 125}{9.8} = 816, \text{ rounded to } 1,000 \text{ lbs./in.}$$

The piping connections at the pump are isolated by flexible connectors and the entire assembly placed on a pad (see Figure 13.5). The chiller and air-cooled condenser also require vibration isolation. Details vary with the specific equipment selected.

Hangers and Supports

Hangers and supports are selected for the ducts and pipes and for noise control. In the mechanical room and on all horizontal mains, the supports used are clevis-type hangers, with rods into concrete inserts. Note: The cost of hangers 10 feet on center is included in the cost of piping. Riser supports are also required.

Insulation

All piping is insulated with fiberglass insulation and an ASJ jacket. This is entered on the system summary sheet (Figure 17.15). Ductwork is sound-insulated within ten feet of each air handling unit.

Summary

All components of the system – generation equipment, distribution equipment, terminal units, controls, and accessories – have now been determined. The basic system is illustrated in Figures 17.5 and 17.10. Figure 17.15 lists the major components, sizes, quantities, models, and accessories, as well as line numbers from *Means Mechanical Cost Data* to determine the cost of the overall system.

System Summary Sheet

Item	Size	Quantity	Type/Model	Remarks	R.S. Means Line No.
Generation					
Boiler—gas	1,500 MBH	1	Gas, hot water	Two-stage	155-115
Expansion tank	80 gal.	1	Steel, ASME	Galvanized Hot water	151-671
Expansion tank	30 gal.	1	Steel, ASME	Galvanized Chilled water	155-671
Chiller	175 tons	1	Glycol-cooled	Dual compressors Dir. drive/reciprocating	157-190
Air-cooled condenser	175 tons	1	Air-cooled	Closed circuit cooling tower	157-225
Pumps—hot water	108 gpm	2	Frame-mounted	15 ft. head High-temp. seals	152-410
Pumps—chilled water	600 gpm	2	Frame-mounted	50 ft. head Casing drain	152-410
Pumps—condenser	975 gpm	2	Frame-mounted	124 ft. head Glycol-resistant seals	152-410
Fresh air fan	7,900 cfm	1	Centrifugal	Roof curb Elec. preheat coil	157-290
Toilet exhaust fan	6,320 cfm	1	Roof exhaust	Roof curb Self-flashing curb	157-290
Distribution					
Steel pipe	3"	200 l.f.	Sched. 40	Supports and anchors	151-701
	4"	12 l.f.	Sched. 40	" " "	151-701
	5"	40 l.f.	Sched. 40	" " "	151-701
	6"	70 l.f.	Sched. 40	" " "	151-701
	8"	40 l.f.	Sched. 40	" " "	151-701
Ductwork	Varies	36K lbs.	14 ga. alum.	Aluminum	157-250
	16"	28 l.f.	Spiral 18 ga.	Aluminum	157-250
	20"	28 l.f.	Spiral 18 ga.	Aluminum	157-250
	24"	34 l.f.	Spiral 18 ga.	Aluminum	157-250
Valves and fittings	Varies				
AHU	30 ton	6	Ceiling-mounted	Hot water coil	157-150
Diffusers	21" x 21"	66	4-way	Volume damper	157-450
Grilles, office	30" x 30"	18	Grille	Aluminum	157-460
Grilles, toilet	12" x 12"	6	Grille	Aluminum	157-460

Figure 17.15

| \multicolumn{5}{c}{**System Summary Sheet (continued)**} |
Item	Size	Quantity	Remarks	R.S. Means Line No.
Controls				
Time clock	6 contacts	1	24 hr.-7 day Spring-wound	
Switch	Main	1	Summer/Winter Manual	
Outdoor reset	Unit	1	Automatic adj. temp.	157-420
3-way motor-operated valves	3"	12	Automatic	157-420
Thermostat	Unit	6	Return air duct Tie to control valves	157-425
Accessories				
Breeching	10"	10 l.f.	Double wall	155-680
Motor starters	Unit	17 l.f.	NEMA 1	163-130
Vibration isolation	Unit	24 l.f.	k = 1,000 lbs./in., 4 ea. pump	157-485
Flexible hose conn.	Unit	12 l.f.	Two 3", two 6", two 8"	156-235
Riser clamps	Unit	—	Varies, alternate floors	151-901
Pipe insulation	Varies	—	Fiberglass w/ASJ	155-651
Duct liner	1/2"	—	Liner 2 lb. density	155-651

Figure 17.15 (continued)

APPENDIX, GLOSSARY AND INDEX

APPENDIX

Table of Contents

Appendix A: HVAC Symbols	566
Appendix B: Abbreviations	569
Appendix C: SI Conversion Tables	572
Appendix D: Solutions to Problems	581

Appendix A

HVAC

Valves, Fittings & Specialties

Symbol	Symbol
Gate	Pipe Pitch Up or Down (Up/Dn)
Globe	Expansion Joint
Check	Expansion Loop
Butterfly	Flexible Connection
Solenoid	Thermostat (T)
Lock Shield	Thermostatic Trap
2-Way Automatic Control	Float and Thermostatic Trap (F&T)
3-Way Automatic Control	Thermometer
Gas Cock	Pressure Gauge
Plug Cock	Flow Switch (FS)
Flanged Joint	Pressure Switch (P)
Union	Pressure Reducing Valve
Cap	Humidistat (H)
Strainer	Aquastat (A)
Concentric Reducer	Air Vent
Eccentric Reducer	Meter (M)
Pipe Guide	Elbow
Pipe Anchor	Tee
Elbow Looking Up	
Elbow Looking Down	
Flow Direction	

Appendix A

Plumbing			HVAC			HVAC (Cont.)
	Floor Drain	◻	—— FOG ——	Fuel Oil Gauge Line		
	Indirect Waste	—— W ——	—o— PD —o—	Pump Discharge		
	Storm Drain	—— SD ——	— — — — —	Low Pressure Condensate Return		
	Combination Waste & Vent	—— CWV ——				
	Acid Waste	—— AW ——	—— LPS ——	Low Pressure Steam		
	Acid Vent	— — AV — —	—— MPS ——	Medium Pressure Steam		
	Cold Water	—— CW ——	—— HPS ——	High Pressure Steam		
	Hot Water	—— HW ——				
	Drinking Water Supply	—— DWS ——	—— BD ——	Boiler Blow-Down		
	Drinking Water Return	—— DWR ——	—— F ——	Fire Protection Water Supply		Fire Protection
	Gas-Low Pressure	—— G ——	—— WSP ——	Wet Standpipe		
	Gas-Medium Pressure	—— MG ——	—— DSP ——	Dry Standpipe		
	Compressed Air	—— A ——	—— CSP ——	Combination Standpipe		
	Vacuum	—— V ——	—— SP ——	Automatic Fire Sprinkler		
	Vacuum Cleaning	—— VC ——	—o————o—	Upright Fire Sprinkler Heads		
	Oxygen	—— O ——				
	Liquid Oxygen	—— LOX ——	—●————●—	Pendent Fire Sprinkler Heads		
	Liquid Petroleum Gas	—— LPG ——	⌀	Fire Hydrant		
HVAC	Hot Water Heating Supply	—— HWS ——	⊢	Wall Fire Dept. Connection		
	Hot Water Heating Return	—— HWR ——				
	Chilled Water Supply	—— CHWS ——	⊷	Sidewalk Fire Dept. Connection		
	Chilled Water Return	—— CHWR ——				
	Drain Line	—— D ——	FHR o—|||||||	Fire Hose Rack		
	City Water	—— CW ——				
	Fuel Oil Supply	—— FOS ——	▨ FHC	Surface Mounted Fire Hose Cabinet		
	Fuel Oil Return	—— FOR ——				
	Fuel Oil Vent	—— FOV ——	▬▨▬ FHC	Recessed Fire Hose Cabinet		

567

Appendix A

HVAC Ductwork Symbols

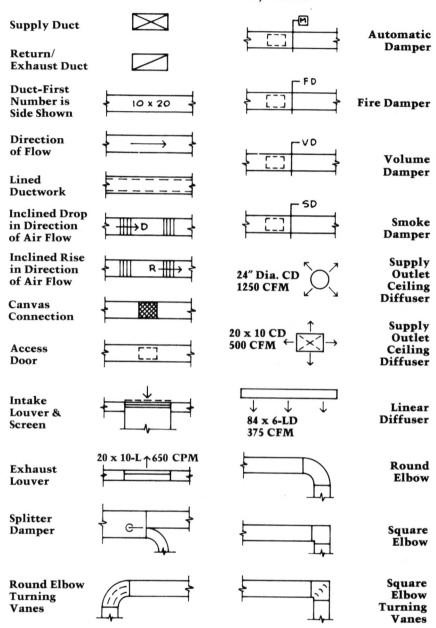

Double Duct Air System

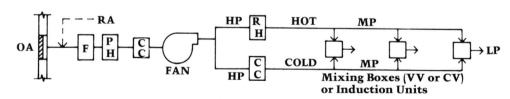

OA = Outside Air
RA = Return Air
F = Filter
PH = Preheat Coil

CC = Cooling Coil
RH = Reheat Coil
HP = High Pressure Duct
MP = Medium Pressure Duct

LP = Low Pressure Duct
VV = Variable Volume
CV = Constant Volume

Appendix B

A	Area Square Feet; Ampere	Calc	Calculated	D.H.	Double Hung
ABS	Acrylonitrile Butadiene Styrene; Asbestos Bonded Steel	Cap.	Capacity	DHW	Domestic Hot Water
A.C.	Alternating Current; Air Conditioning; Asbestos Cement	Carp.	Carpenter	Diag.	Diagonal
		C.B.	Circuit Breaker	Diam.	Diameter
		C.C.F.	Hundred Cubic Feet	Distrib.	Distribution
		cd	Candela	Dk.	Deck
A.C.I.	American Concrete Institute	cd/sf	Candela per Square Foot	D.L.	Dead Load; Diesel
Addit.	Additional	CD	Grade of Plywood Face & Back	Do.	Ditto
Adj.	Adjustable	CDX	Plywood, grade C&D, ext. glue	Dp.	Depth
af	Audio-frenquency	Cefi.	Cement Finisher	D.P.S.T.	Double Pole, Single Throw
A.G.A.	American Gas Association	Cem.	Cement	Dr.	Driver
Agg.	Aggregate	CF	Hundred Feet	Drink.	Drinking
A.H.	Ampere Hours	C.F.	Cubic Feet	D.S.	Double Strength
A hr	Ampere-hour	CFM	Cubic Feet per Minute	D.S.A.	Double Strength A Grade
A.I.A.	American Institute of Architects	c.g.	Center of Gravity	D.S.B.	Double Strength B Grade
AIC	Ampere Interrupting Capacity	CHW	Commercial Hot Water	Dty.	Duty
Allow.	Allowance	C.I.	Cast Iron	DWV	Drain Waste Vent
alt.	Altitude	C.I.P.	Cast in Place	DX	Deluxe White, Direct Expansion
Alum.	Aluminum	Circ.	Circuit	dyn	Dyne
a.m.	ante meridiem	C.L.	Carload Lot	e	Eccentricity
Amp.	Ampere	Clab.	Common Laborer	E	Equipment Only; East
Approx.	Approximate	C.L.F.	Hundred Linear Feet	Ea.	Each
Apt.	Apartment	CLF	Current Limiting Fuse	Econ.	Economy
Asb.	Asbestos	CLP	Cross Linked Polyethylene	EDP	Electronic Data Processing
A.S.B.C.	American Standard Bld. Code	cm	Centimeter	E.D.R.	Equiv. Direct Radiation
Asbe.	Asbestos Worker	CMP	Corr. Metal Pipe	Eq.	Equation
A.S.H.R.A.E.	American Society of Heating, Refrig. & AC Engineers	C.M.U.	Concrete Masonry Unit	Elec.	Electrician; Electrical
		Col.	Column	Elev.	Elevator; Elevating
A.S.M.E.	American Society of Mechanical Engineers	CO_2	Carbon Dioxide	EMT	Electrical Metallic Conduit; Thin Wall Conduit
		Comb.	Combination		
A.S.T.M.	American Society for Testing and Materials	Compr.	Compressor	Eng.	Engine
		Conc.	Concrete	EPDM	Ethylene Propylene Diene Monomer
Attchmt.	Attachment	Cont.	Continuous; Continued		
Avg.	Average			Eqhv.	Equip. Oper., heavy
Bbl.	Barrel	Corr.	Corrugated	Eqlt.	Equip. Oper., light
B.&B.	Grade B and Better; Balled & Burlapped	Cos	Cosine	Eqmd.	Equip. Oper., medium
		Cot	Cotangent	Eqmm.	Equip. Oper., Master Mechanic
B.&S.	Bell and Spigot	Cov.	Cover	Eqol.	Equip. Oper., oilers
B.&W.	Black and White	CPA	Control Point Adjustment	Equip.	Equipment
b.c.c.	Body-centered Cubic	Cplg.	Coupling	ERW	Electric Resistance Welded
B.F.	Board Feet	C.P.M.	Critical Path Method	Est.	Estimated
Bg. Cem.	Bag of Cement	CPVC	Chlorinated Polyvinyl Chloride	esu	Electrostatic Units
BHP	Brake Horse Power	C. Pr.	Hundred Pair	E.W.	Each Way
B.I.	Black Iron	CRC	Cold Rolled Channel	EWT	Entering Water Temperature
Bit.; Bitum.	Bituminous	Creos.	Creosote	Excav.	Excavation
		Crpt.	Carpet & Linoleum Layer	Exp.	Expansion
Bk.	Backed	CRT	Cathode-ray Tube	Ext.	Exterior
Bkrs.	Breakers	CS	Carbon Steel	Extru.	Extrusion
Bldg.	Building	Csc	Cosecant	f.	Fiber stress
Blk.	Block	C.S.F.	Hundred Square Feet	F	Fahrenheit; Female; Fill
Bm.	Beam	C.S.I.	Construction Specification Institute	Fab.	Fabricated
Boil.	Boilermaker			FBGS	Fiberglass
B.P.M.	Blows per Minute	C.T.	Current Transformer	F.C.	Footcandles
BR	Bedroom	CTS	Copper Tube Size	f.c.c.	Face-centered Cubic
Brg.	Bearing	Cu	Cubic	f'c.	Compressive Stress in Concrete; Extreme Compressive Stress
Brhe.	Bricklayer Helper	Cu. Ft.	Cubic Foot		
Bric.	Bricklayer	cw	Continuous Wave	F.E.	Front End
Brk.	Brick	C.W.	Cool White	FEP	Fluorinated Ethylene Propylene (Teflon)
Brng.	Bearing	Cwt.	100 Pounds		
Brs.	Brass	C.W.X.	Cool White Deluxe	F.G.	Flat Grain
Brz.	Bronze	C.Y.	Cubic Yard (27 cubic feet)	F.H.A.	Federal Housing Administration
Bsn.	Basin	C.Y./Hr.	Cubic Yard per Hour	Fig.	Figure
Btr.	Better	Cyl.	Cylinder	Fin.	Finished
BTU	British Thermal Unit	d	Penny (nail size)	Fixt.	Fixture
BTUH	BTU per Hour	D	Deep; Depth; Discharge	Fl. Oz.	Fluid Ounces
BX	Interlocked Armored Cable	Dis.; Disch.	Discharge	Flr.	Floor
c	Conductivity			F.M.	Frequency Modulation; Factory Mutual
C	Hundred; Centigrade	Db.	Decibel		
		Dbl.	Double	Fmg.	Framing
C/C	Center to Center	DC	Direct Current	Fndtn.	Foundation
Cab.	Cabinet	Demob.	Demobilization	Fori.	Foreman, inside
Cair.	Air Tool Laborer	d.f.u.	Drainage Fixture Units	Foro.	Foreman, outside

569

Appendix B

Fount.	Fountain	I.P.S.	Iron Pipe Size	M.C.M.	Thousand Circular Mils
FPM	Feet per Minute	I.P.T.	Iron Pipe Threaded	M.C.P.	Motor Circuit Protector
FPT	Female Pipe Thread	J	Joule	MD	Medium Duty
Fr.	Frame	J.I.C.	Joint Industrial Council	M.D.O.	Medium Density Overlaid
F.R.	Fire Rating	K.	Thousand; Thousand Pounds	Med.	Medium
FRK	Foil Reinforced Kraft	K.D.A.T.	Kiln Dried After Treatment	MF	Thousand Feet
FRP	Fiberglass Reinforced Plastic	kg	Kilogram	M.F.B.M.	Thousand Feet Board Measure
FS	Forged Steel	kG	Kilogauss	Mfg.	Manufacturing
FSC	Cast Body; Cast Switch Box	kgf	Kilogram force	Mfrs.	Manufacturers
Ft.	Foot; Feet	kHz	Kilohertz	mg	Milligram
Ftng.	Fitting	Kip.	1000 Pounds	MGD	Million Gallons per Day
Ftg.	Footing	KJ	Kiljoule	MGPH	Thousand Gallons per Hour
Ft. Lb.	Foot Pound	K.L.	Effective Length Factor	MH	Manhole; Metal Halide; Man Hour
Furn.	Furniture	Km	Kilometer		
FVNR	Full Voltage Non Reversing	K.L.F.	Kips per Linear Foot	MHz	Megahertz
FXM	Female by Male	K.S.F.	Kips per Square Foot	Mi.	Mile
Fy.	Minimum Yield Stress of Steel	K.S.I.	Kips per Square Inch	MI	Malleable Iron; Mineral Insulated
g	Gram	K.V.	Kilo Volt	mm	Millimeter
G	Gauss	K.V.A.	Kilo Volt Ampere	Mill.	Millwright
Ga.	Gauge	K.V.A.R.	Kilovar (Reactance)	Min.	Minimum
Gal.	Gallon	KW	Kilo Watt	Misc.	Miscellaneous
Gal./Min.	Gallon Per Minute	KWh	Kilowatt-hour	ml	Milliliter
Galv.	Galvanized	L	Labor Only; Length; Long	M.L.F.	Thousand Linear Feet
Gen.	General	Lab.	Labor	Mo.	Month
Glaz.	Glazier	lat	Latitude	Mobil.	Mobilization
GPD	Gallons per Day	Lath.	Lather	Mog.	Mogul Base
GPH	Gallons per Hour	Lav.	Lavatory	MPH	Miles per Hour
GPM	Gallons per Minute	lb.; #	Pound	MPT	Male Pipe Thread
GR	Grade	L.B.	Load Bearing; L Conduit Body	MRT	Mile Round Trip
Gran.	Granular	L. & E.	Labor & Equipment	ms	millisecond
Grnd.	Ground	lb./hr.	Pounds per Hour	M.S.F.	Thousand Square Feet
H	High; High Strength Bar Joist; Henry	lb./L.F.	Pounds per Linear Foot	Mstz.	Mosaic & Terrazzo Worker
		lbf/sq in.	Pound-force per Square Inch	M.S.Y.	Thousand Square Yards
H.C.	High Capacity	L.C.L.	Less than Carload Lot	Mtd.	Mounted
H.D.	Heavy Duty; High Density	Ld.	Load	Mthe.	Mosaic & Terrazzo Helper
H.D.O.	High Density Overlaid	L.F.	Linear Foot	Mtng.	Mounting
Hdr.	Header	Lg.	Long; Length; Large	Mult.	Multi; Multiply
Hdwe.	Hardware	L. & H.	Light and Heat	MVAR	Million Volt Amp Reactance
Help.	Helper average	L.H.	Long Span High Strength Bar Joist	MV	Megavolt
HEPA	High Efficiency Particulate Air Filter			MW	Megawatt
		L.J.	Long Span Standard Strength Bar Joist	MXM	Male by Male
Hg	Mercury			MYD	Thousand yards
H.O.	High Output	L.L.	Live Load	N	Natural; North
Horiz.	Horizontal	L.L.D.	Lamp Lumen Depreciation	nA	nanoampere
H.P.	Horsepower; High Pressure	lm	Lumen	NA	Not Available; Not Applicable
H.P.F.	High Power Factor	lm/sf	Lumen per Square Foot	N.B.C.	National Building Code
Hr.	Hour	lm/W	Lumen Per Watt	NC	Normally Closed
Hrs./Day	Hours Per Day	L.O.A.	Length Over All	N.E.M.A.	National Electrical Manufacturers Association
HSC	High Short Circuit	log	Logarithm		
Ht.	Height	L.P.	Liquefied Petroleum; Low Pressure	NEHB	Bolted Circuit Breaker to 600V.
Htg.	Heating			N.L.B.	Non-Load-Bearing
Htrs.	Heaters	L.P.F.	Low Power Factor	nm	nanometer
HVAC	Heating, Ventilating & Air Conditioning	Lt.	Light	No.	Number
		Lt. Ga.	Light Gauge	NO	Normally Open
Hvy.	Heavy	L.T.L.	Less than Truckload Lot	N.O.C.	Not Otherwise Classified
HW	Hot Water	Lt. Wt.	Lightweight	Nose.	Nosing
Hyd.; Hydr.	Hydraulic	L.V.	Low Voltage	N.P.T.	National Pipe Thread
		M	Thousand; Material; Male; Light Wall Copper	NQOB	Bolted Circuit Breaker to 240V.
Hz.	Hertz (cycles)			N.R.C.	Noise Reduction Coefficient
I.	Moment of Inertia	m/hr	Manhour	N.R.S.	Non Rising Stem
I.C.	Interrupting Capacity	mA	Milliampere	ns	nanosecond
ID	Inside Diameter	Mach.	Machine	nW	nanowatt
I.D.	Inside Dimension; Identification	Mag. Str.	Magnetic Starter	OB	Opposing Blade
		Maint.	Maintenance	OC	On Center
I.F.	Inside Frosted	Marb.	Marble Setter	OD	Outside Diameter
I.M.C.	Intermediate Metal Conduit	Mat.	Material	O.D.	Outside Dimension
In.	Inch	Mat'l.	Material	ODS	Overhead Distribution System
Incan.	Incandescent	Max.	Maximum	O & P	Overhead and Profit
Incl.	Included; Including	MBF	Thousand Board Feet	Oper.	Operator
Int.	Interior	MBH	Thousand BTU's per hr.	Opng.	Opening
Inst.	Installation	M.C.F.	Thousand Cubic Feet	Orna.	Ornamental
Insul.	Insulation	M.C.F.M.	Thousand Cubic Feet per Minute	O.S.&Y.	Outside Screw and Yoke
I.P.	Iron Pipe			Ovhd	Overhead

570

Appendix B

Oz.	Ounce	S.	Suction; Single Entrance; South	T.S.	Trigger Start
P.	Pole; Applied Load; Projection	Scaf.	Scaffold	Tr.	Trade
p.	Page	Sch.;		Transf.	Transformer
Pape.	Paperhanger	Sched.	Schedule	Trhv.	Truck Driver, Heavy
PAR	Weatherproof Reflector	S.C.R.	Modular Brick	Trlr.	Trailer
Pc.	Piece	S.D.R.	Standard Dimension Ratio	Trlt.	Truck Driver, Light
P.C.	Portland Cement; Power Connector	S.E.	Surfaced Edge	TV	Television
P.C.F.	Pounds per Cubic Foot	S.E.R.; S.E.U.	Service Entrance Cable	T.W.	Thermoplastic Water Resistant Wire
P.E.	Professional Engineer; Porcelain Enamel; Polyethylene; Plain End	S.F.	Square Foot	UCI	Uniform Construction Index
		S.F.C.A.	Square Foot Contact Area	UF	Underground Feeder
Perf.	Perforated	S.F.G.	Square Foot of Ground	U.H.F.	Ultra High Frequency
Ph.	Phase	S.F. Hor.	Square Foot Horizontal	U.L.	Underwriters Laboratory
P.I.	Pressure Injected	S.F.R.	Square Feet of Radiation	Unfin.	Unfinished
Pile.	Pile Driver	S.F.Shlf.	Square Foot of Shelf	URD	Underground Residential Distribution
Pkg.	Package	S4S	Surface 4 Sides	V	Volt
Pl.	Plate	Shee.	Sheet Metal Worker	VA	Volt/amp
Plah.	Plasterer Helper	Sin.	Sine	V.A.T.	Vinyl Asbestos Tile
Plas.	Plasterer	Skwk.	Skilled Worker	VAV	Variable Air Volume
Pluh.	Plumbers Helper	SL	Saran Lined	Vent.	Ventilating
Plum.	Plumber	S.L.	Slimline	Vert.	Vertical
Ply.	Plywood	Sldr.	Solder	V.G.	Vertical Grain
p.m.	Post Meridiem	S.N.	Solid Neutral	V.H.F.	Very High Frequency
Pord.	Painter, Ordinary	S.P.	Static Pressure; Single Pole; Self Propelled	VHO	Very High Output
pp	Pages			Vib.	Vibrating
PP; PPL	Polypropylene	Spri.	Sprinkler Installer	V.L.F.	Vertical Linear Foot
P.P.M.	Parts per Million	Sq.	Square; 100 square feet	Vol.	Volume
Pr.	Pair	S.P.D.T.	Single Pole, Double Throw	W	Wire; Watt; Wide; West
Prefab.	Prefabricated	S.P.S.T.	Single Pole, Single Throw	w/	With
Prefin.	Prefinished	SPT	Standard Pipe Thread	W.C.	Water Column; Water Closet
Prop.	Propelled	Sq. Hd.	Square Head	W.F.	Wide Flange
PSF; psf	Pounds per Square Foot	S.S.	Single Strength; Stainless Steel	W.G.	Water Gauge
PSI; psi	Pounds per Square Inch	S.S.B.	Single Strength B Grade	Wldg.	Welding
PSIG	Pounds per Square Inch Gauge	Sswk.	Structural Steel Worker	Wrck.	Wrecker
PSP	Plastic Sewer Pipe	Sswl.	Structural Steel Welder	W.S.P.	Water, Steam, Petroleum
Pspr.	Painter, Spray	St.; Stl.	Steel	WT, Wt.	Weight
Psst.	Painter, Structural Steel	S.T.C.	Sound Transmission Coefficient	WWF	Welded Wire Fabric
P.T.	Potential Transformer	Std.	Standard	XFMR	Transformer
P. & T.	Pressure & Temperature	STP	Standard Temp. & Pressure	XHD	Extra Heavy Duty
Ptd.	Painted	Stpi.	Steamfitter, Pipefitter	Y	Wye
Ptns.	Partitions	Str.	Strength; Starter; Straight	yd	Yard
Pu	Ultimate Load	Strd.	Stranded	yr	Year
PVC	Polyvinyl Chloride	Struct.	Structural	Δ	Delta
Pvmt.	Pavement	Sty.	Story	%	Percent
Pwr.	Power	Subj.	Subject	~	Approximately
Q	Quantity Heat Flow	Subs.	Subcontractors	∅	Phase
Quan.; Qty.	Quantity	Surf.	Surface	@	At
Q.C.	Quick Coupling	Sw.	Switch	#	Pound; Number
r	Radius of Gyration	Swbd.	Switchboard	<	Less Than
R	Resistance	S.Y.	Square Yard	>	Greater Than
R.C.P.	Reinforced Concrete Pipe	Syn.	Synthetic		
Rect.	Rectangle	Sys.	System		
Reg.	Regular	t.	Thickness		
Reinf.	Reinforced	T	Temperature; Ton		
Req'd.	Required	Tan	Tangent		
Resi	Residential	T.C.	Terra Cotta		
Rgh.	Rough	T.D.	Temperature Difference		
R.H.W.	Rubber, Heat & Water Resistant Residential Hot Water	TFE	Tetrafluoroethylene (Teflon)		
		T. & G.	Tongue & Groove; Tar & Gravel		
rms	Root Mean Square	Th.; Thk.	Thick		
Rnd.	Round	Thn.	Thin		
Rodm.	Rodman	Thrded	Threaded		
Rofc.	Roofer, Composition	Tilf.	Tile Layer Floor		
Rofp.	Roofer, Precast	Tilh.	Tile Layer Helper		
Rohe.	Roofer Helpers (Composition)	THW	Insulated Strand Wire		
Rots.	Roofer, Tile & Slate	THWN; THHN	Nylon Jacketed Wire		
R.O.W.	Right of Way				
RPM	Revolutions per Minute	T.L.	Truckload		
R.R.	Direct Burial Feeder Conduit	Tot.	Total		
R.S.	Rapid Start				
RT	Round Trip				

Appendix C

Standard and Metric Linear and Area Conversion Tables

Linear Conversions

Inches	Feet	Yards	Rods	Miles	Centimeters	Meters	Kilometers
1	0.083	0.028	0.005	—	2.540	0.0254	—
12	1	0.333	0.061	0.0002	30.480	0.305	0.0003
36	3	1	0.182	0.0006	91.440	0.914	0.0009
0.3937	0.033	0.011	—	—	1	0.01	—
39.37	3.281	1.094	0.199	0.0006	100	1	0.001
					Furlongs		
198	16.5	5.5	1	0.003	0.025	5.029	0.005
	5280	1760	320	1	8	1 609 347	1.609
	660	220	40	0.125	1	201.168	0.201
	3280.83	1093.61	198.838	0.621	4.971	1000	1

Area Conversions

Square Inches	Square Feet	Square Yards	Acres	Square Centimeters	Square Meters	Hectares	Square Kilometers
1	0.007	—	—	6.452	0.0006	—	—
144	1	0.111	0.00002	929.034	0.093	—	—
1296	9	1	0.0002	8361.31	0.836	—	—
0.155	0.001	—	—	1	0.0001	—	—
1549.997	10.764	1.196	0.0002	10 000	1	0.0001	—
				Square Miles			
	43 560	4840	1	0.002	4046.87	0.405	0.004
	27 878 400	3 097 600	640	1	2 589 998	258.999	2.590
	107 638.7	11 959.9	2.471	0.004	10 000	1	0.01
	10 763 867	1 195 985	247.104	0.386	1 000 000	100	1

Basic Metric Units and Prefixes

Basic Metric Units		Prefixes for Metric Units			
Quantity	Unit	Multiple and Submultiple		Prefix	Symbol
length	meter (m)	1 000 000 000 000 =	10^{12}	tera	T
mass	kilogram (kg)	1 000 000 000 =	10^{9}	giga	G
time	second(s)	1 000 000 =	10^{6}	mega	M
electric current	ampere (A)	1000 =	10^{3}	kilo	k
temperature (thermodynamic)	kelvin (K)	100 =	10^{2}	hecto	h
amount of substance	mole (mol)	10 =	10	deka	da
luminous intensity	candela (cd)	0.1 =	10^{-1}	deci	d
		0.01 =	10^{-2}	centi	c
		0.001 =	10^{-3}	milli	m
		0.000 001 =	10^{-6}	micro	μ
		0.000 000 001 =	10^{-9}	nano	n
		0.000 000 000 001 =	10^{-12}	pico	p
		0.000 000 000 000 001 =	10^{-15}	femto	f
		0.000 000 000 000 000 001 =	10^{-18}	atto	a

Appendix C

Metric Conversion Table

	Inches to Centimeters	Feet to Meters	Pounds to Kilograms		Inches to Centimeters	Feet to Meters	Pounds to Kilograms
1	2.54	0.304 8	0.453 6	51	129.54	15.544 8	23.133 6
2	5.08	0.609 6	0.907 2	52	132.08	15.849 6	23.587 2
3	7.62	0.914 4	1.360 8	53	134.62	16.154 4	24.040 8
4	10.16	1.219 2	1.814 4	54	137.16	16.459 2	24.494 4
5	12.7	1.524	2.268	55	139.7	16.764	24.948
6	15.24	1.828 8	2.721 6	56	142.24	17.068 8	25.401 6
7	17.78	2.133 6	3.175 2	57	144.78	17.373 6	25.855 2
8	20.32	2.438 4	3.628 8	58	147.32	17.678 4	26.308 8
9	22.86	2.743 2	4.082 4	59	149.86	17.983 2	26.762 4
10	25.4	3.048	4.536	60	152.4	18.288	27.216
11	27.94	3.352 8	4.989 6	61	154.94	18.592 8	27.669 6
12	30.48	3.657 6	5.443 2	62	157.48	18.897 6	28.123 2
13	33.02	3.962 4	5.896 8	63	160.02	19.202 4	28.576 8
14	35.56	4.267 2	6.350 4	64	162.56	19.507 2	29.030 4
15	38.1	4.572	6.804	65	165.1	19.812	29.488
16	40.64	4.876 8	7.257 6	66	167.64	20.116 8	29.937 6
17	43.18	5.181 6	7.711 2	67	170.18	20.421 6	30.391 2
18	45.72	5.486 4	8.164 8	68	172.72	20.726 4	30.844 8
19	48.26	5.791 2	8.618 4	69	175.26	21.031 2	31.298 4
20	50.8	6.096	9.072	70	177.8	21.336	31.752
21	53.34	6.400 8	9.525 6	71	180.34	21.640 8	32.205 6
22	55.88	6.705 6	9.979 2	72	182.88	21.945 6	32.659 2
23	58.42	7.010 4	10.432 8	73	185.42	22.250 4	33.112 8
24	60.96	7.315 2	10.886 4	74	187.96	22.555 2	33.566 4
25	63.5	7.62	11.34	75	190.5	22.86	34.02
26	66.04	7.924 8	11.793 6	76	193.04	23.164 8	34.473 6
27	68.58	8.229 6	12.247 2	77	195.58	23.469 6	34.927 2
28	71.12	8.534 4	12.700 8	78	198.12	23.774 4	35.380 8
29	73.66	8.839 2	13.154 4	79	200.66	24.079 2	35.834 4
30	76.2	9.144	13.608	80	203.2	24.384	36.288
31	78.74	9.448 8	14.061 6	81	205.74	24.688 8	36.741 6
32	81.28	9.753 6	14.515 2	82	208.28	24.993 6	37.195 2
33	83.82	10.058 4	14.968 0	83	210.82	25.298 4	37.648 8
34	86.36	10.363 2	15.422 4	84	213.36	25.603 2	38.102 4
35	88.9	10.668	15.876	85	215.9	25.908	38.556
36	91.44	10.972 8	16.329 6	86	218.44	26.212 8	39.009 6
37	93.98	11.277 6	16.783 2	87	220.98	26.517 6	39.463 2
38	96.52	11.582 4	17.236 8	88	223.52	26.822 4	39.916 8
39	99.06	11.887 2	17.690 4	89	226.06	27.127 2	40.370 4
40	101.6	12.192	18.144	90	228.6	27.432	40.824
41	104.14	12.496 8	18.597 6	91	231.14	27.736 8	41.277 6
42	106.68	12.801 6	19.051 2	92	233.68	28.041 6	41.731 2
43	109.22	13.106 4	19.504 8	93	236.22	28.346 4	42.184 8
44	111.76	13.411 2	19.958 4	94	238.76	28.651 2	42.638 4
45	114.3	13.716	20.412	95	241.3	28.956	43.092
46	116.84	14.020 8	20.865 6	96	243.84	29.260 8	43.545 6
47	119.38	14.325 6	21.319 2	97	246.38	29.565 6	43.999 2
48	121.92	14.630 4	21.772 8	98	248.92	29.870 4	44.452 8
49	124.46	14.935 2	22.226 4	99	252.46	30.175 2	44.906 4
50	127	15.24	22.68	100	254	30.48	45.36

Conversion Formulas: Inches x 2.54 = Centimeters
                           Feet x 0.3048 = Meters
                           Pounds x 0.4536 = Kilograms

Appendix C

To convert from	to	Multiply by
abampere	ampere (A)	1.000 000*E+01
abcoulomb	coulomb (C)	1.000 000*E+01
abfarad	farad (F)	1.000 000*E+09
abhenry	henry (H)	1.000 000*E−09
abmho	siemens (S)	1.000 000*E+09
abohm	ohm (Ω)	1.000 000*E−09
abvolt	volt (V)	1.000 000*E−08
acre foot[13]	cubic metre (m^3)	1.233 489 E+03
acre[13]	square metre (m^2)	4.046 873 E+03
ampere hour	coulomb (C)	3.600 000*E+03
angstrom	metre (m)	1.000 000*E−10
are	square metre (m^2)	1.000 000*E+02
astronomical unit	metre (m)	1.495 979 E+11
atmosphere, standard	pascal (Pa)	1.013 250*E+05
atmosphere, technical (= 1 kgf/cm^2)	pascal (Pa)	9.806 650*E+04
bar	pascal (Pa)	1.000 000*E+05
barn	square metre (m^2)	1.000 000*E−28
barrel (for petroleum, 42 gal)	cubic metre (m^3)	1.589 873 E−01
board foot	cubic metre (m^3)	2.359 737 E−03
British thermal unit (International Table)[14]	joule (J)	1.055 056 E+03
British thermal unit (mean)	joule (J)	1.055 87 E+03
British thermal unit (thermochemical)	joule (J)	1.054 350 E+03
British thermal unit (39°F)	joule (J)	1.059 67 E+03
British thermal unit (59°F)	joule (J)	1.054 80 E+03
British thermal unit (60°F)	joule (J)	1.054 68 E+03
Btu (International Table)·ft/(h·ft^2·°F) (thermal conductivity)	watt per metre kelvin [W/(m·K)]	1.730 735 E+00
Btu (thermochemical)·ft/(h·ft^2·°F) (thermal conductivity)	watt per metre kelvin [W/(m·K)]	1.729 577 E+00
Btu (International Table)·in/(h·ft^2·°F) (thermal conductivity)	watt per metre kelvin [W/(m·K)]	1.442 279 E−01
Btu (thermochemical)·in/(h·ft^2·°F) (thermal conductivity)	watt per metre kelvin [W/(m·K)]	1.441 314 E−01
Btu (International Table)·in/s·ft^2·°F) (thermal conductivity)	watt per metre kelvin [W/(m·K)]	5.192 204 E+02
Btu (thermochemical)·in/(s·ft^2·°F) (thermal conductivity)	watt per metre kelvin [W/(m·K)]	5.188 732 E+02
Btu (International Table)/h	watt (W)	2.930 711 E−01
Btu (International Table)/s	watt (W)	1.055 056 E+03
Btu (thermochemical)/h	watt (W)	2.928 751 E−01
Btu (thermochemical)/min	watt (W)	1.757 250 E+01
Btu (thermochemical)/s	watt (W)	1.054 350 E+03
Btu (International Table)/ft^2	joule per square metre (J/m^2)	1.135 653 E+04
Btu (thermochemical)/ft^2	joule per square metre (J/m^2)	1.134 893 E+04

---

[13] The U.S. Metric Law of 1866 gave the relationship, 1 metre equals 39.37 inches. Since 1893 the U.S. yard has been derived from the metre. In 1959 a refinement was made in the definition of the yard to bring the U.S. yard and the yard used in other countries into agreement. The U.S. yard was changed from 3600/3937 m to 0.9144 m exactly. The new length is shorter by exactly two parts in a million.

At the same time it was decided that any data in feet derived from and published as a result of geodetic surveys within the U.S. would remain with the old standard (1 ft = 1200/3937 m) until further decision. This foot is named the U.S. survey foot.

All conversion factors for units of land measure in these tables referenced to this footnote are based on the U.S. survey foot and the following relationships: 1 fathom = 6 feet; 1 rod (pole or perch) = 16½ feet; 1 chain = 66 feet; 1 mile (U.S. statute) = 5280 feet.

[14] The Fifth International Conference on the Properties of Steam in 1956 defined the calorie (International Table) as 4.1868 J. Therefore, the exact conversion for Btu (International Table) is 1.055 055 852 62 E+03 J.

Copyright ASTM. Reprinted with permission.

Appendix C

To convert from	to	Multiply by
Btu (International Table)/(ft$^2 \cdot$s)	watt per square metre (W/m^2)	1.135 653 E+04
Btu (International Table)/(ft$^2 \cdot$h)	watt per square metre (W/m^2)	3.154 591 E+00
Btu (thermochemical)/(ft$^2 \cdot$h)	watt per square metre (W/m^2)	3.152 481 E+00
Btu (thermochemical)/(ft$^2 \cdot$min)	watt per square metre (W/m^2)	1.891 489 E+02
Btu (thermochemical)/(ft$^2 \cdot$s)	watt per square metre (W/m^2)	1.134 893 E+04
Btu (thermochemical)/(in$^2 \cdot$s)	watt per square metre (W/m^2)	1.634 246 E+06
Btu (International Table)/(h$\cdot$ft$^2 \cdot$°F) (thermal conductance)[15]	watt per square metre kelvin [W/(m$^2 \cdot$K)]	5.678 263 E+00
Btu (thermochemical)/(h$\cdot$ft$^2 \cdot$°F) (thermal conductance)[15]	watt per square metre kelvin [W/(m$^2 \cdot$K)]	5.674 466 E+00
Btu (International Table)/(s$\cdot$ft$^2 \cdot$°F)	watt per square metre kelvin [W/(m$^2 \cdot$K)]	2.044 175 E+04
Btu (thermochemical)/(s$\cdot$ft$^2 \cdot$°F)	watt per square metre kelvin [W/(m$^2 \cdot$K)]	2.042 808 E+04
Btu (International Table)/lb	joule per kilogram (J/kg)	2.326 000*E+03
Btu (thermochemical)/lb	joule per kilogram (J/kg)	2.324 444 E+03
Btu (International Table)/(lb$\cdot$°F) (heat capacity)	joule per kilogram kelvin [J/(kg$\cdot$K)]	4.186 800*E+03
Btu (thermochemical)/(lb$\cdot$°F) (heat capacity)	joule per kilogram kelvin [J/(kg$\cdot$K)]	4.184 000*E+03
Btu (International Table)/ft^3	joule per cubic metre (J/m^3)	3.725 895 E+04
Btu (thermochemical)/ft^3	joule per cubic metre (J/m^3)	3.723 402 E+04
bushel (U.S.)	cubic metre (m^3)	3.523 907 E-02
calorie (International Table)[14]	joule (J)	4.186 800*E+00
calorie (mean)	joule (J)	4.190 02 E+00
calorie (thermochemical)	joule (J)	4.184 000*E+00
calorie (15°C)	joule (J)	4.185 80 E+00
calorie (20°C)	joule (J)	4.181 90 E+00
calorie (kilogram, International Table)	joule (J)	4.186 800*E+03
calorie (kilogram, mean)	joule (J)	4.190 02 E+03
calorie (kilogram, thermochemical)	joule (J)	4.184 000*E+03
cal (thermochemical)/cm^2	joule per square metre (J/m^2)	4.184 000*E+04
cal (International Table)/g	joule per kilogram (J/kg)	4.186 800*E+03
cal (thermochemical)/g	joule per kilogram (J/kg)	4.184 000*E+03
cal (International Table)/(g$\cdot$°C)	joule per kilogram kelvin [J/(kg$\cdot$K)]	4.186 800*E+03
cal (thermochemical)/(g$\cdot$°C)	joule per kilogram kelvin [J/(kg$\cdot$K)]	4.184 000*E+03
cal (thermochemical)/min	watt (W)	6.973 333 E-02
cal (thermochemical)/s	watt (W)	4.184 000*E+00
cal (thermochemical)/(cm$^2 \cdot$s)	watt per square metre (W/m^2)	4.184 000*E+04
cal (thermochemical)/(cm$^2 \cdot$min)	watt per square metre (W/m^2)	6.973 333 E+02
cal (thermochemical)/(cm$^2 \cdot$s)	watt per square metre (W/m^2)	4.184 000*E+04
cal (thermochemical)/(cm$\cdot$s$\cdot$°C)	watt per metre kelvin [W/(m$\cdot$K)]	4.184 000*E+02
cd/in^2	candela per square metre (cd/m^2)	1.550 003 E+03
carat (metric)	kilogram (kg)	2.000 000*E-04
centimetre of mercury (0°C)	pascal (Pa)	1.333 22 E+03
centimetre of water (4°C)	pascal (Pa)	9.806 38 E+01
centipoise (dynamic viscosity)	pascal second (Pa$\cdot$s)	1.000 000*E-03
centistokes (kinematic viscosity)	square metre per second (m^2/s)	1.000 000*E-06
chain[13]	metre (m)	2.011 684 E+01
circular mil	square metre (m^2)	5.067 075 E-10
clo	kelvin square metre per watt (K$\cdot$m^2/W)	1.55 E-01
cup	cubic metre (m^3)	2.365 882 E-04
curie	becquerel (Bq)	3.700 000*E+10
darcy[16]	square metre (m^2)	9.869 233 E-13
day	second (s)	8.640 000*E+04
day (sidereal)	second (s)	8.616 409 E+04
degree (angle)	radian (rad)	1.745 329 E-02
degree Celsius	kelvin (K)	$T_K = t_{°C} + 273.15$
degree centigrade	[see 3.4.2]	
degree Fahrenheit	degree Celsius (°C)	$t_{°C} = (t_{°F} - 32)/1.8$

[15] In ISO 31 this quantity is called *coefficient of heat transfer*.
[16] The darcy is a unit for measuring permeability of porous solids.

Copyright ASTM. Reprinted with permission.

Appendix C

To convert from	to	Multiply by
degree Fahrenheit	kelvin (K)	$T_K = (t_F + 459.67)/1.8$
degree Rankine	kelvin (K)	$T_K = T_R/1.8$
°F·h·ft²/Btu (International Table) (thermal resistance)[17]	kelvin square metre per watt (K·m²/W)	1.761 102 E−01
°F·h·ft²/Btu (thermochemical) (thermal resistance)[17]	kelvin square metre per watt (K·m²/W)	1.762 280 E−01
°F·h·ft²/[Btu (International Table)·in] (thermal resistivity)	kelvin metre per watt (K·m/W)	6.933 471 E+00
°F·h·ft²/[Btu (thermochemical)·in] (thermal resistivity)	kelvin metre per watt (K·m/W)	6.938 113 E+00
denier	kilogram per metre (kg/m)	1.111 111 E−07
dyne	newton (N)	1.000 000*E−05
dyne·cm	newton metre (N·m)	1.000 000*E−07
dyne/cm²	pascal (Pa)	1.000 000*E−01
electronvolt	joule (J)	1.602 19 E−19
EMU of capacitance	farad (F)	1.000 000*E+09
EMU of current	ampere (A)	1.000 000*E+01
EMU of electric potential	volt (V)	1.000 000*E−08
EMU of inductance	henry (H)	1.000 000*E−09
EMU of resistance	ohm (Ω)	1.000 000*E−09
ESU of capacitance	farad (F)	1.112 650 E−12
ESU of current	ampere (A)	3.335 6 E−10
ESU of electric potential	volt (V)	2.997 9 E+02
ESU of inductance	henry (H)	8.987 554 E+11
ESU of resistance	ohm (Ω)	8.987 554 E+11
erg	joule (J)	1.000 000*E−07
erg/(cm²·s)	watt per square metre (W/m²)	1.000 000*E−03
erg/s	watt (W)	1.000 000*E−07
faraday (based on carbon-12)	coulomb (C)	9.648 70 E+04
faraday (chemical)	coulomb (C)	9.649 57 E+04
faraday (physical)	coulomb (C)	9.652 19 E+04
fathom[13]	metre (m)	1.828 804 E+00
fermi (femtometre)	metre (m)	1.000 000*E−15
fluid ounce (U.S.)	cubic metre (m³)	2.957 353 E−05
foot	metre (m)	3.048 000*E−01
foot (U.S. survey)[13]	metre (m)	3.048 006 E−01
foot of water (39.2°F)	pascal (Pa)	2.988 98 E+03
ft²	square metre (m²)	9.290 304*E−02
ft²/h (thermal diffusivity)	square metre per second (m²/s)	2.580 640*E−05
ft²/s	square metre per second (m²/s)	9.290 304*E−02
ft³ (volume; section modulus)	cubic metre (m³)	2.831 685 E−02
ft³/min	cubic metre per second (m³/s)	4.719 474 E−04
ft³/s	cubic metre per second (m³/s)	2.831 685 E−02
ft⁴ (second moment of area)[18]	metre to the fourth power (m⁴)	8.630 975 E−03
ft/h	metre per second (m/s)	8.466 667 E−05
ft/min	metre per second (m/s)	5.080 000*E−03
ft/s	metre per second (m/s)	3.048 000*E−01
ft/s²	metre per second squared (m/s²)	3.048 000*E−01
footcandle	lux (lx)	1.076 391 E+01
footlambert	candela per square metre (cd/m²)	3.426 259 E+00
ft·lbf	joule (J)	1.355 818 E+00
ft·lbf/h	watt (W)	3.766 161 E−04
ft·lbf/min	watt (W)	2.259 697 E−02
ft·lbf/s	watt (W)	1.355 818 E+00
ft-poundal	joule (J)	4.214 011 E−02
g, standard free fall	metre per second squared (m/s²)	9.806 650*E+00

[17] In ISO 31 this quantity is called *thermal insulance* and the quantity *thermal resistance* has the unit K/W.
[18] This is sometimes called the moment of section or area moment of inertia of a plane section about a specified axis.

Copyright ASTM. Reprinted with permission.

Appendix C

To convert from	to	Multiply by
gal	metre per second squared (m/s²)	1.000 000*E−02
gallon (Canadian liquid)	cubic metre (m³)	4.546 090 E−03
gallon (U.K. liquid)	cubic metre (m³)	4.546 092 E−03
gallon (U.S. dry)	cubic metre (m³)	4.404 884 E−03
gallon (U.S. liquid)	cubic metre (m³)	3.785 412 E−03
gallon (U.S. liquid) per day	cubic metre per second (m³/s)	4.381 264 E−08
gallon (U.S. liquid) per minute	cubic metre per second (m³/s)	6.309 020 E−05
gallon (U.S. liquid) per hp·h (SFC, specific fuel consumption)	cubic metre per joule (m³/J)	1.410 089 E−09
gamma	tesla (T)	1.000 000*E−09
gauss	tesla (T)	1.000 000*E−04
gilbert	ampere (A)	7.957 747 E−01
gill (U.K.)	cubic metre (m³)	1.420 653 E−04
gill (U.S.)	cubic metre (m³)	1.182 941 E−04
grade	degree (angular)	9.000 000*E−01
grade	radian (rad)	1.570 796 E−02
grain	kilogram (kg)	6.479 891*E−05
grain/gal (U.S. liquid)	kilogram per cubic metre (kg/m³)	1.711 806 E−02
gram	kilogram (kg)	1.000 000*E−03
g/cm³	kilogram per cubic metre (kg/m³)	1.000 000*E+03
gf/cm²	pascal (Pa)	9.806 650*E+01
hectare	square metre (m²)	1.000 000*E+04
horsepower (550 ft·lbf/s)	watt (W)	7.456 999 E+02
horsepower (boiler)	watt (W)	9.809 50 E+03
horsepower (electric)	watt (W)	7.460 000*E+02
horsepower (metric)	watt (W)	7.354 99 E+02
horsepower (water)	watt (W)	7.460 43 E+02
horsepower (U.K.)	watt (W)	7.457 0 E+02
hour	second(s)	3.600 000*E+03
hour (sidereal)	second (s)	3.590 170 E+03
hundredweight (long)	kilogram (kg)	5.080 235 E+01
hundredweight (short)	kilogram (kg)	4.535 924 E+01
inch	metre (m)	2.540 000*E−02
inch of mercury (32°F)	pascal (Pa)	3.386 38 E+03
inch of mercury (60°F)	pascal (Pa)	3.376 85 E+03
inch of water (39.2°F)	pascal (Pa)	2.490 82 E+02
inch of water (60°F)	pascal (Pa)	2.488 4 E+02
in²	square metre (m²)	6.451 600*E−04
in³ (volume)[19]	cubic metre (m³)	1.638 706 E−05
in³ (section modulus)[19]	metre cubed (m³)	1.638 706 E−05
in³/min	cubic metre per second (m³/s)	2.731 177 E−07
in⁴ (second moment of area)[18]	metre to the fourth power (m⁴)	4.162 314 E−07
in/s	metre per second (m/s)	2.540 000*E−02
in/s²	metre per second squared (m/s²)	2.540 000*E−02
kayser	1 per metre (1/m)	1.000 000*E+02
kelvin	degree Celsius (°C)	$t_{°C} = T_K - 273.15$
kilocalorie (International Table)	joule (J)	4.186 800*E+03
kilocalorie (mean)	joule (J)	4.190 02 E+03
kilocalorie (thermochemical)	joule (J)	4.184 000*E+03
kilocalorie (thermochemical)/min	watt (W)	6.973 333 E+01
kilocalorie (thermochemical)/s	watt (W)	4.184 000*E+03
kilogram-force (kgf)	newton (N)	9.806 650*E+00
kgf·m	newton metre (N·m)	9.806 650*E+00
kgf·s²/m (mass)	kilogram (kg)	9.806 650*E+00
kgf/cm²	pascal (Pa)	9.806 650*E+04
kgf/m²	pascal (Pa)	9.806 650*E+00
kgf/mm²	pascal (Pa)	9.806 650*E+06
km/h	metre per second (m/s)	2.777 778 E−01

[19] The exact conversion factor is 1.638 706 4*E−05.

Copyright ASTM. Reprinted with permission.

Appendix C

To convert from	to	Multiply by
kilopond (1 kp = 1 kgf)	newton (N)	9.806 650*E+00
kW·h	joule (J)	3.600 000*E+06
kip (1000 lbf)	newton (N)	4.448 222 E+03
kip/in^2 (ksi)	pascal (Pa)	6.894 757 E+06
knot (international)	metre per second (m/s)	5.144 444 E−01
lambert	candela per square metre (cd/m^2)	1/π *E+04
lambert	candela per square metre (cd/m^2)	3.183 099 E+03
langley	joule per square metre (J/m^2)	4.184 000*E+04
light year	metre (m)	9.460 55 E+15
litre[20]	cubic metre (m^3)	1.000 000*E−03
lm/ft^2	lumen per square metre (lm/m^2)	1.076 391 E+01
maxwell	weber (Wb)	1.000 000*E−08
mho	siemens (S)	1.000 000*E+00
microinch	metre (m)	2.540 000*E−08
micron (deprecated term, use micrometre)	metre (m)	1.000 000*E−06
mil	metre (m)	2.540 000*E−05
mile (international)	metre (m)	1.609 344*E+03
mile (U.S. statute)[13]	metre (m)	1.609 347 E+03
mile (international nautical)	metre (m)	1.852 000*E+03
mile (U.S. nautical)	metre (m)	1.852 000*E+03
mi^2 (international)	square metre (m^2)	2.589 988 E+06
mi^2 (U. S. statute)[13]	square metre (m^2)	2.589 998 E+06
mi/h (international)	metre per second (m/s)	4.470 400*E−01
mi/h (international)	kilometre per hour (km/h)	1.609 344*E+00
mi/min (international)	metre per second (m/s)	2.682 240*E+01
mi/s (international)	metre per second (m/s)	1.609 344*E+03
millibar	pascal (Pa)	1.000 000*E+02
millimetre of mercury (0°C)	pascal (Pa)	1.333 22 E+02
minute (angle)	radian (rad)	2.908 882 E−04
minute	second (s)	6.000 000*E+01
minute (sidereal)	second (s)	5.983 617 E+01
oersted	ampere per metre (A/m)	7.957 747 E+01
ohm centimetre	ohm meter (Ω·m)	1.000 000*E−02
ohm circular-mil per foot	ohm metre (Ω·m)	1.662 426 E−09
ounce (avoirdupois)	kilogram (kg)	2.834 952 E−02
ounce (troy or apothecary)	kilogram (kg)	3.110 348 E−02
ounce (U.K. fluid)	cubic metre (m^3)	2.841 306 E−05
ounce (U.S. fluid)	cubic metre (m^3)	2.957 353 E−05
ounce-force	newton (N)	2.780 139 E−01
ozf·in	newton metre (N·m)	7.061 552 E−03
oz (avoirdupois)/gal (U.K. liquid)	kilogram per cubic metre (kg/m^3)	6.236 023 E+00
oz (avoirdupois)/gal (U.S. liquid)	kilogram per cubic metre (kg/m^3)	7.489 152 E+00
oz (avoirdupois)/in^3	kilogram per cubic metre (kg/m^3)	1.729 994 E+03
oz (avoirdupois)/ft^2	kilogram per square metre (kg/m^2)	3.051 517 E−01
oz (avoirdupois)/yd^2	kilogram per square metre (kg/m^2)	3.390 575 E−02
parsec	metre (m)	3.085 678 E+16
peck (U.S.)	cubic metre (m^3)	8.809 768 E−03
pennyweight	kilogram (kg)	1.555 174 E−03
perm (0°C)	kilogram per pascal second square metre [kg/(Pa·s·m^2)]	5.721 35 E−11
perm (23°C)	kilogram per pascal second square metre [kg/(Pa·s·m^2)]	5.745 25 E−11
perm·in (0°C)	kilogram per pascal second metre [kg/(Pa·s·m)]	1.453 22 E−12
perm·in (23°C)	kilogram per pascal second metre [kg/(Pa·s·m)]	1.459 29 E−12
phot	lumen per square metre (lm/m^2)	1.000 000*E+04

[20] In 1964 the General Conference on Weights and Measures reestablished the name litre as a special name for the cubic decimetre. Between 1901 and 1964 the litre was slightly larger (1.000 28 dm^3); in the use of high-accuracy volume data of that time interval, this fact must be kept in mind.

Copyright ASTM. Reprinted with permission.

Appendix C

To convert from	to	Multiply by
pica (printer's)	metre (m)	4.217 518 E−03
pint (U.S. dry)	cubic metre (m^3)	5.506 105 E−04
pint (U.S. liquid)	cubic metre (m^3)	4.731 765 E−04
point (printer's)	metre (m)	3.514 598*E−04
poise (absolute viscosity)	pascal second (Pa·s)	1.000 000*E−01
pound (lb avoirdupois)[21]	kilogram (kg)	4.535 924 E−01
pound (troy or apothecary)	kilogram (kg)	3.732 417 E−01
lb·ft^2 (moment of inertia)	kilogram square metre (kg·m^2)	4.214 011 E−02
lb·in^2 (moment of inertia)	kilogram square metre (kg·m^2)	2.926 397 E−04
lb/ft·h	pascal second (Pa·s)	4.133 789 E−04
lb/ft·s	pascal second (Pa·s)	1.488 164 E+00
lb/ft^2	kilogram per square metre (kg/m^2)	4.882 428 E+00
lb/ft^3	kilogram per cubic metre (kg/m^3)	1.601 846 E+01
lb/gal (U.K. liquid)	kilogram per cubic metre (kg/m^3)	9.977 637 E+01
lb/gal (U.S. liquid)	kilogram per cubic metre (kg/m^3)	1.198 264 E+02
lb/h	kilogram per second (kg/s)	1.259 979 E−04
lb/hp·h (SFC, specific fuel consumption)	kilogram per joule (kg/J)	1.689 659 E−07
lb/in^3	kilogram per cubic metre (kg/m^3)	2.767 990 E+04
lb/min	kilogram per second (kg/s)	7.559 873 E−03
lb/s	kilogram per second (kg/s)	4.535 924 E−01
lb/yd^3	kilogram per cubic metre (kg/m^3)	5.932 764 E−01
poundal	newton (N)	1.382 550 E−01
poundal/ft^2	pascal (Pa)	1.488 164 E+00
poundal·s/ft^2	pascal second (Pa·s)	1.488 164 E+00
pound-force (lbf)[22]	newton (N)	4.448 222 E+00
lbf·ft	newton metre (N·m)	1.355 818 E+00
lbf·ft/in	newton metre per metre (N·m/m)	5.337 866 E+01
lbf·in	newton metre (N·m)	1.129 848 E−01
lbf·in/in	newton metre per metre (N·m/m)	4.448 222 E+00
lbf·s/ft^2	pascal second (Pa·s)	4.788 026 E+01
lbf·s/in^2	pascal second (Pa·s)	6.894 757 E+03
lbf/ft	newton per metre (N/m)	1.459 390 E+01
lbf/ft^2	pascal (Pa)	4.788 026 E+01
lbf/in	newton per metre (N/m)	1.751 268 E+02
lbf/in^2 (psi)	pascal (Pa)	6.894 757 E+03
lbf/lb (thrust/weight [mass] ratio)	newton per kilogram (N/kg)	9.806 650 E+00
quart (U.S. dry)	cubic metre (m^3)	1.101 221 E−03
quart (U.S. liquid)	cubic metre (m^3)	9.463 529 E−04
rad (absorbed dose)	gray (Gy)	1.000 000*E−02
rem (dose equivalent)	sievert (Sv)	1.000 000*E−02
rhe	1 per pascal second [1/(Pa·s)]	1.000 000*E+01
rod[13]	metre (m)	5.029 210 E+00
roentgen	coulomb per kilogram (C/kg)	2.58 000*E−04
rpm (r/min)	radian per second (rad/s)	1.047 198 E−01
second (angle)	radian (rad)	4.848 137 E−06
second (sidereal)	second (s)	9.972 696 E−01
shake	second (s)	1.000 000*E−08
slug	kilogram (kg)	1.459 390 E+01
slug/ft·s	pascal second (Pa·s)	4.788 026 E+01
slug/ft^3	kilogram per cubic metre (kg/m^3)	5.153 788 E+02
statampere	ampere (A)	3.335 640 E−10
statcoulomb	coulomb (C)	3.335 640 E−10
statfarad	farad (F)	1.112 650 E−12
stathenry	henry (H)	8.987 554 E+11
statmho	siemens (S)	1.112 650 E−12
statohm	ohm (Ω)	8.987 554 E+11
statvolt	volt (V)	2.997 925 E+02

[21] The exact conversion factor is 4.535 923 7*E−01.
[22] The exact conversion factor is 4.448 221 615 260 5*E+00.

Copyright ASTM. Reprinted with permission.

Appendix C

To convert from	to	Multiply by
stere	cubic metre (m³)	1.000 000*E+00
stilb	candela per square metre (cd/m²)	1.000 000*E+04
stokes (kinematic viscosity)	square metre per second (m²/s)	1.000 000*E−04
tablespoon	cubic metre (m³)	1.478 676 E−05
teaspoon	cubic metre (m³)	4.928 922 E−06
tex	kilogram per metre (kg/m)	1.000 000*E−06
therm (European Community)[23]	joule (J)	1.055 06 E+08
therm (U.S.)[23]	joule (J)	1.054 804*E+08
ton (assay)	kilogram (kg)	2.916 667 E−02
ton (long, 2240 lb)	kilogram (kg)	1.016 047 E+03
ton (metric)	kilogram (kg)	1.000 000*E+03
ton (nuclear equivalent of TNT)	joule (J)	4.184 E+09[24]
ton of refrigeration (= 12 000 Btu/h)	watt (W)	3.517 E+03
ton (register)	cubic metre (m³)	2.831 685 E+00
ton (short, 2000 lb)	kilogram (kg)	9.071 847 E+02
ton (long)/yd³	kilogram per cubic metre (kg/m³)	1.328 939 E+03
ton (short)/yd³	kilogram per cubic metre (kg/m³)	1.186 553 E+03
ton (short)/h	kilogram per second (kg/s)	2.519 958 E−01
ton-force (2000 lbf)	newton (N)	8.896 443 E+03
tonne	kilogram (kg)	1.000 000*E+03
torr (mmHg, 0°C)	pascal (Pa)	1.333 22 E+02
unit pole	weber (Wb)	1.256 637 E−07
W·h	joule (J)	3.600 000*E+03
W·s	joule (J)	1.000 000*E+00
W/cm²	watt per square metre (W/m²)	1.000 000*E+04
W/in²	watt per square metre (W/m²)	1.550 003 E+03
yard	metre (m)	9.144 000*E−01
yd²	square metre (m²)	8.361 274 E−01
yd³	cubic metre (m³)	7.645 549 E−01
yd³/min	cubic metre per second (m³/s)	1.274 258 E−02
year (365 days)	second (s)	3.153 600*E+07
year (sidereal)	second (s)	3.155 815 E+07
year (tropical)	second (s)	3.155 693 E+07

[23] The therm (European Community) is legally defined in the Council of the European Communities Directive 80/181/EC of December 20, 1979. The therm (U.S.) is legally defined in the *Federal Register*, Vol 33, No. 146, p. 10756, of July 27, 1968. Although the European therm, which is based on the International Table Btu, is frequently used by engineers in the U.S., the therm (U.S.) is the legal unit used by the U.S. natural gas industry.

[24] Defined (not measured) value.

Copyright ASTM. Reprinted with permission.

Appendix D

Solutions to Problems

Following are answers for most of the problems presented throughout the book. Answers that require an essay-type answer or a drawing are not provided here.

Chapter 1
1.1 a. Basement mechanical room
     b. Basement next to boilers in mechanical room
     c. Roof
     d. In existing closet or area served
     e. Roof
1.2 P = 283 psi
1.3 Answers vary.
1.4 Answers vary.
1.5 Answers vary.
1.6 Answers vary.
1.7 Answers vary.

Chapter 2
2.1 Answers vary.
2.2 20 cfm per person, all conditions
2.3 Answers vary.
2.4 Space conforms if 400 s.f. operable.

Chapter 3
3.1 60,000 Btu/hr.
3.2 22.6 MBH
3.3 2.6 tons
3.4 1.18, 1.05, and 0.50 Btu/hr./s.f./°F
3.5 0.0085 lbs. water per lb. of dry air
3.6 4.9 Btu/lb. dry air; 44%
3.7 Answers vary.
3.8 a. H_c = 29,097 Btu/hr.
     b. zero
     c. zero
     d. 29,097 Btu/hr.
     a. H_c = 3,816 Btu/hr.
     b. zero c. 122,292 Btu/hr.
     d. zero
     e. 126,108 Btu/hr.
3.9 a. 182 people; 3,640 cfm of fresh outdoor air
     b. 112,112 Btu/hr.
     c. 196,618 Btu/hr.
3.10 All except the fourth statement are true.
3.11 Yes; no
3.12 3"; 6"; double-glazed
3.13 2875 Kcal food fuel
3.14 0.0018 lbs. moisture per lb. dry air
3.15 a. 50 people; 715 people
     b. 1,000 cfm; 48.9 MBH
     c. 75,000 cu. ft. of air; 75 minutes of occupancy without additional fresh air
     d. Installing exhaust and supply fans with multiple speed settings tied to a manual override
3.16 2,800 Btu/min.

Chapter 4
Answers for all problems vary.

Chapter 5
Answers for all problems vary.

Chapter 6
6.1 Answers vary.
6.2 1,500 gallons/year
6.3 a. 500,000 gallons/year
     b. 40,000 gallons/week
     c. 123 cubic yards

Chapter 7
Answers for all problems vary.

Chapter 8
8.1 a. The inlet and discharge sides of the pump
     b. 1-1/4" pump and 4-11/16" impeller
     c. 23-1/2 gpm and 21 feet of head
8.2 a. 83 gpm
     b. 2.89 H.P.
     c. Answers vary.
8.3 a. In-line hot water circulating pump
     b. Base-mounted, multi-stage centrifugal pump
     c. Pump specifically designed for steam condensate
8.4 a. Propeller type
     b. Centrifugal fans with spark arrestors
     c. Vaneaxial fan
     d. Airfoil blades

Chapter 9
9.1 a. 8" diameter, 6' long
9.2 a. Float and thermostatic trap
     b. Inverted bucket
     c. Float and thermostatic trap
9.3 Answers vary.
9.4 1" angle trap

Chapter 10
10.1 a. H_n = 275 MBH
     b. Answers vary.

Appendix D

      c. Answers vary.
      d. 22 gpm and 10 feet of head
10.2 10" diameter pipe and 3-1/2" riser; 1/16" pressure drop per 100 feet
10.3 a. 11.4 gpm
      b. 44.25 feet of head
10.4 9,600 feet of pipe; 16,400 feet of pipe
10.5 250 gpm with 75 feet of head
10.6 285 gpm and 88.5 feet of head
10.7 1,200 gpm capacity with 86.25 feet of head, assuming 20 feet for the chiller
10.8 a. Answers vary.
      b. Answers vary.
      c. Answers vary.
      d. 4,400 cfm and 3/4" pressure
10.9 15" × 48" by the velocity method; 15" × 60" duct for the constant pressure drop method
10.10 a. 16.5 tons
       b. Answers vary.
       c. Fan uses 12,000 cfm at 3/4" pressure

Chapter 11

11.1 a. 36 feet of single-row baseboard
      b. 26 feet of double-row baseboard
      c. A 56" × 25" × 9" fan coil delivers 40 MBH heating.
11.2 1,500 cfm in the space; six 9" × 9" 4-way diffusers
11.3 Answers vary.
11.4 6-2/3 tons

Chapter 12

12.1 a. sensor
      b. controller
      c. actuator
      d. sensor
      e. controller
      f. controller
      g. actuator
      h. sensor
12.2 Answers vary.
12.3 Answers vary.
12.4 Answers vary.
12.5 Answers vary.

Chapter 13

13.1 All of the statements are true except *c* and *J*.
13.2 Use 12".
13.3 a. A clevis-type hanger with no special vibration device
      b. Steel beams and pads
      c. Flexible pipe isolators
      d. Rubber-mount pad and sound isolate enclosure

Chapter 14

14.1 Answers vary.
14.2 Answers vary.
14.3 Answers vary.
14.4 4' × 9'; 5,400 cfm; grease filters 2,700 sq. in.
14.5 a. 27.6 tons
      b. The hood exhausts more than the patron ventilation requirement.
      c. Answers vary.
14.6 a. Elevator
      b. Sprinkler fitter
      c. HVAC
      d. HVAC
      e. HVAC
      f. HVAC
14.7 a. Capital
      b. O&M
      c. O&M
      d. Capital
      e. O&M
      f. O&M
14.8 $58,500 in the year 2000
14.9 Answers vary.

Chapter 15

15.1 *a*, *b*, and *d* are true; *c* and *e* are false.
15.2 Answers vary.
15.3 Answers vary.
15.4 Answers vary.
15.5 40 million ozone molecules will be removed.
15.6 Answers vary.

GLOSSARY

Glossary

Ambulatory Care
Health care facility for very short-term (less than 24-hour), noncritical care.

ANSI
American National Standards Institute.

API
American Petroleum Institute.

ASHRAE
American Society of Heating, Refrigerating, and Air Conditioning Engineers.

ASME
American Society of Mechanical Engineers.

ASTM
American Society for Testing Materials.

Atrium
An occupied space that includes a floor opening or several floor openings that connects two or more stories.

Backing Ring
A metal ring used during the welding process to prevent melted metal from entering a pipe when making a butt-weld joint, and to provide a uniform gap for welding. Often referred to as a "chill ring."

BE
Beveled end. (The end of a pipe or fitting prepared for welding.)

Bell and Spigot Joint
The most commonly used joint in cast iron soil pipe. Each length is made with an enlarged bell at one end into which the spigot end of another piece is inserted. The joint is then made tight by lead and oakum or a rubber ring caulked into the bell around the spigot.

Black Steel Pipe
Steel pipe that has not been galvanized.

Blank Flange
A flange in which the bolt holes are not drilled.

Blind Flange
A flange used to close off the end of a pipe.

Branch
The outlet or inlet of a fitting that is not in line with the run, and takes off at an angle to the run (e.g., tees, wyes, crosses, laterals, etc.)

British Thermal Unit (Btu)
The heat required to heat 1 pound of water 1°F.

Building Sewer
The pipe running from the outside wall of the building drain to the public sewer.

Building Storm Drain
A drain for carrying rain, surface water, condensate, etc., to the building drain or sewer.

Bull Head Tee
A tee with the branch larger than the run.

Buoyancy
The tendency of high-temperature smoke to rise above the denser ambient air.

Butt Weld Joint
A welded pipe joint made with the ends of the two pipes butting each other.

Butt Weld Pipe
Pipe welded along the seam and not lapped.

CAA
Clean Air Act. Legislation calling for a complete phaseout of the production of all ozone-depleting chemicals.

Carbon Steel Pipe
Steel pipe that owes its properties chiefly to the carbon it contains.

CFC's
Chlorofluorocarbons.

Cogeneration
Providing two forms of energy from one piece of equipment.

Companion Flange
A pipe flange that connects with another flange or with a flanged valve or fitting. This flange differs from a flange that is an integral part of a fitting or valve.

Compartmentation
Containment of smoke by smoke barriers within compartments throughout a building.

Cooling Tower
A device for cooling water by evaporation. A natural draft cooling tower is one where the air flow through the tower results from its natural chimney effect. A mechanical draft tower employs fans to force or induce a draft.

Conservation
The protection or preservation of a valuable resource.

Couplings
Fittings for joining two pieces of pipe.

Covered Mall
A roofed-over common pedestrian area serving more than one tenant.

Drainage System
The piping system in a building, up to the point of discharge to the sewer system.

ECM
Energy conservation measure. ECM procedures involve a capital investment and have a simple payback of more than one or two years.

Elbow
A fitting that makes an angle in a pipe run. The angle is 90 degrees unless another angle is specified.

Elevator Recall
The automatic delivery of all elevators to the main lobby in an alarm.

Energy Audit
A comprehensive energy conservation study of a building or facility.

EPA
Environmental Protection Agency.

Equivalent Length
The resistance of a duct or pipe elbow, valve, damper, orifice, bend, fitting, or other obstruction to flow, expressed in the number of feet of straight

Glossary

duct or pipe of the same diameter that would have the same resistance.

ERW
Electric resistance weld. (A method of welding pipe in the manufacturing process.)

Expansion joint
A joint whose primary purpose is to absorb the longitudinal expansion and contraction in the line resulting from temperature changes.

Expansion Loop
A large-radius loop in a pipeline that absorbs the longitudinal expansion and contraction in the line resulting from temperature changes.

Fire Safety Plans
Instructions pertaining to fire emergencies for building staff members or tenants.

Flange
A ring-shaped plate at right angles to the end of the pipe. The flange has holes for bolts to allow fastening of the pipe to a similar flange.

Flat Face
Pipe flanges that have the entire face of the flange faced straight across and that use a full face gasket. These are commonly employed for pressures less than 125 pounds.

FSES
Fire Safety Evaluation System.

Galvanizing
Coating iron or steel surfaces with a protective layer of zinc.

Gate Valve
A valve utilizing a gate, usually wedge-shaped, which allows fluid flow when the gate is lifted from the seat. Gate valves have less resistance to flow than globe valves, and should always be used fully open or fully closed.

Globe Valve
A valve with a rounded body utilizing a manually raised or lowered disc which, when closed, seats so as to prevent fluid flow. Globe valves are ideal for throttling in a semi-closed position.

Header
A large pipe or drum into which each of a group of boilers, chillers, or pumps are connected. (See Manifold.)

HCFC's
Hydrochlorofluorocarbons.

HFC's
Hydrofluorocarbons.

Horizontal Branch
In plumbing, the horizontal line from the fixture drain to the waste stack.

Hospital
A health care facility where a patient may stay for more than 24 hours or be treated while incapacitated.

Hot Water Heating System
A system in which hot water is the heating medium. Flow is either gravity or forced circulation.

IAQ
Indoor air quality. General healthfulness of the air in a building.

ID
Inside diameter.

IDHA
International District Heating Association.

Insulation
Thermal insulation is a material used for covering pipes, ducts, vessels, etc., to reduce heat loss or gain.

Lapped Joint
A lapped joint, like a van stone joint, is a type of pipe joint made using loose flanges on lengths of pipe. The ends of this pipe are lapped over to give a bearing surface for a gasket or metal-to-metal joint.

Lap Weld Pipe
Pipe made by welding along a scarfed longitudinal seam in which one part is overlapped by another.

LCL
Less carload lot. (Pipe is usually ordered from the mill by the carload, or LCL.)

Lead joint
A joint made by pouring molten lead into the space between a bell and spigot, and making the joint tight by caulking.

Malleable Iron
Cast iron that has been heat-treated to reduce its brittleness.

Manifold
A fitting with several branch outlets.

Mill Length
Also known as "random length". The usual run-of-the-mill pipe is 16 to 20 feet in length. Line pipe for power plant or oil field use is often made in double random lengths of 30 to 35 feet.

Nipple
A piece of pipe less than 12 inches long and threaded on both ends. Pipe over 12 inches long is regarded as a cut measure.

Nominal
Name given to standard pipe size designations through 12 inches nominal OD. For example, 2" nominal is 2-3/8" OD.

OD
Outside diameter.

O & M
Operation and maintenance. O & M procedures do not involve any initial capital cost, but do require good "housekeeping", staff cooperation, awareness, and dedication.

O.S. & Y.
Outside screw and yoke. A valve configuration where the valve stem, having exposed external threads supported by a yoke, indicates the open or closed position of the valve.

Passive Design
Design that minimizes energy consumed by burning fuel or using power; a form of sustainable design.

Glossary

P.E.
Plain end. (Used to describe the ends of the pipe that are shipped from the mill with unfinished ends. These ends may eventually be threaded, beveled, or grooved in the field.)

Performance Specifications
Describe the *result* to be achieved by the products, rather than the products themselves.

Pipe
A hollow cylinder or tube for conveyance of a fluid.

Plug Valve
A valve containing a tapered plug through which a hole is drilled so that fluid can flow through when the holes line up with the inlet and the outlet. However, when the plug is rotated 90 degrees, the flow is stopped.

Plumbing Fixtures
Devices that receive water, liquid, or water-borne wastes, and discharge the wastes into a drainage system.

Plumbing System
Arrangements of pipes, fixtures, fittings, valves, and traps in a building, which supply water and remove liquid-borne wastes, including storm water.

Potable Water
Water suitable for human consumption.

Power Factor
The ratio of true power (watts) to the apparent power (volt-amperes); occurs when the current in a circuit leads or lags the voltage as a result of inductance or capacitance.

Proprietary Specifications
Indicate the exact product or equipment – manufacturer, options, and accessories by make and model.

PSI
Pounds per square inch.

Radon
An odorless, colorless radioactive gas with a short half-life (less than four days).

Reducer
A pipe coupling with a larger size at one end than the other. The larger size is designated first. Reducers are threaded, flanged, welded, etc. Reducing couplings are available in either eccentric or concentric configurations.

Riser
A vertical pipe extending one or more floors.

Roof Drain
A fitting that collects water on the roof surface and discharges it into the leader.

SBS
Sick building syndrome. Condition in which buildings' indoor environments make people feel unwell.

Schedule Number
Schedule numbers are American Standards Association designations for classifying the strength of pipe. Schedule 40 is the most common form of steel pipe used in the mechanical trades.

Screwed Joint
A pipe joint consisting of threaded male and female parts.

Seamless Pipe
Pipe or tube formed by piercing a billet of steel and then rolling, rather than having welded seams.

Service Pipe
A pipe connecting water or gas mains into a building from the street.

Slip-on Flange
A flange slipped onto the end of a pipe and then welded in place.

Socket Weld
A pipe joint made by use of a socket weld fitting that has a female end or socket for insertion of the pipe to be welded.

Stack Effect
Natural movement of air resulting from temperature differentials between interior and exterior environments.

Stainless Steel
An alloy steel having unusual corrosion-resisting properties, usually imparted by nickel and chromium.

Storm Sewer
A sewer carrying surface or storm water from roofs or exterior surfaces of a building.

Street Elbow
An elbow with a male thread on one end and a female thread on the other end.

Sustainable Design
Design that minimizes the impact of the use of materials and energy in the production of a building.

Swing Joint
An arrangement of screwed fittings to allow for movement in a pipe line.

Swivel Joint
A special pipe fitting designed to be pressure-tight under continuous or intermittent movement of the equipment to which it is connected.

TBE
Thread both ends. Term used when specifying or ordering cut measures of pipe.

T & C
Threaded and coupled; an ordering designation for threaded pipe.

Tee
A pipe fitting that has a side port at right angles to the run.

TOE
Thread one end. Term used when specifying or ordering cut measures of pipe.

Union
A fitting used to join pipes. It commonly consists of three pieces. Unions are extensively used, because they allow dismantling and reassembling of piping assemblies with ease and without distorting the assembly.

Van Stone
A type of joint made by using loose flanges on lengths of pipe

whose ends are lapped over to give a bearing surface for the flange.

Vents
Vents are used to permit air to escape from hydronic systems, condensate receivers, fuel oil storage tanks, as a breather line for gas regulators, etc.

Vent Stack
A vertical vent pipe that provides air circulation to and from the drainage system.

Vent System
Piping that provides a flow of air to or from a drainage system to protect trap seals from siphonage or back pressure.

VOC
Volatile organic compounds. Chemicals that contain carbon molecules and are volatile enough to evaporate from material surfaces into indoor air at normal temperatures.

Welding Fittings
Wrought steel elbows, tees, reducers, saddles, and the like, beveled for butt welding to pipe. Forged fittings with hubs or with ends counter-bored for fillet welding to pipe are used for small pipe sizes and high pressures.

Welding Neck Flange
A flange with a long neck beveled for butt welding to pipe.

Wye
A pipe fitting with a side outlet that is any angle other than 90 degrees to the main run or axis.

INDEX

Index

A

Absolute pressure, 12, 291
Absorption
  of odors, 407
Absorption cycle
  equipment, 25, 248
  solar, 24-25
Absorption cycle systems, 20, 23-26
Absorption-type chillers, 229-230
Accessories, 397-422
  commercial, 558-559
  data sheets, 418-422
  residential, 518, 520, 524-525
Actuators, 376
Air
  heat capacity, 14
Air cleaning, 3, 313, 404-409
Air conditioning
  basic systems, 3-32
  defined, 29
Air conditioning ladder, 29
  illust., 32
Air-cooled condensers, 230-231, 548
Air cycle, 18
Air distribution, 313-339
  advantages, 313
  applications, 313
  control, 313
  disadvantages, 313
  high velocity, 315
  layout, 315
  outlets
    number and location, 317
  static pressure, 332-338
  variable air volume (VAV), 315
Air flow
  friction loss, 266
Air handling
  central station, 233-234
Air handling units, 254
Air systems, 201-202, 359
  single and dual ducts, 359
  terminal units, 351-371
  types, 351
Air vents, 352, 355
  hydronic systems, 311
Air washing, 407

Algorithms, 379
Aluminum, 397
American National Standards Institute (ANSI), 45
American Society of Heating Refrigeration, and Air Conditioning Engineers (ASHRAE), 33, 34, 44-45, 46, 407, 488
American Society of Mechanical Engineers (ASME), 8, 34, 45
American Society for Testing and Materials (ASTM), 34, 45
Annual Fuel Utilization Efficiency (AFUE), 221-222
Appliances
  heat gain, 85
Approach temperature, 279
Architects
  role of in HVAC, 37-38
Atriums, 435
Axial flow fans, 268

B

Background noise
  reducing, 402-404
Bituminous coatings
  tanks, 225
Bleed valves
  hydronic systems, 311
"Boiler and Pressure Vessel Code," 45
Boilers, 3, 4, 6-7, 219-222, 247, 291
  accessibility, 222
  accessories, 223
  cast iron, 219, 222
  cost of installation, 222
  cost of operation, 222
  efficiency, 220, 221-222, 223
  electric, 219
  fuel, 166, 219
    modular, 166
    dual, 166
  high-pressure, 14, 220
  hot water
    gas-fired, 546
    oil-fired, 516
  low-pressure, 13-14
  hot water, 220
  steam, 220

  materials, 219, 220
  output size, 222, 223
  pulse condensing, 221
  rating, 221
  safety, 223
  selection conditions, 222
  sequencing, 524
  sizing, 222-223
  steel, 219
  wall-hung, 221
  water tube, 219
Bottled gas
  see propane
Boxes
  variable air volume, 237, 359
Breeching, 525-526, 558
Brick, 397
British thermal unit (Btu)
  defined, 6
Bronze
  in pumps, 261
Building automation system (BAS), 389
Building codes
  fire protection, 35-37, 39
  model, 33
  typical requirements, 33, 38-43
Building life, 193
Building Official's and Code Administrator's (BOCA) Code, 33, 34
Burners
  types, 225
Burner safety, 225
Burner shut down, 225
Butane, 205

C

Calcium silicate
  rigid, 414
Carbon monoxide, 205, 407
Cast iron
  in pumps, 261
  radiators, 351
Cathodic protection tanks, 225
Cavitation, 262
Cellular glass, 414
Centrifugal compressor, 229
Centrifugal fans, 268-269

590

Index

Centrifugal (turbo) compressors, 23
Charcoal
 activated, 407
Chilled water systems, 304
Chillers, 229-230, 247
Chimneys, 397-399
 codes, 397
 materials, 397
Chlorofluorocarbons (CFC's), 477-487, 499
 alternatives to
  containment, 485
  replacement, 485
  conversion, 485-486
 and ozone depletion, 478-479
 reduction of, 479-484
Circulating pump
 chilled water, 549
 hot water, 548
Clay tile, 397
Clean Air Act, 34, 479-484, 499
 consequences of nonconformance, 484
Closed loop systems
 defined, 259
Coal, 206
Codes,
 complications, 35
 general requirements, 38-43
 HVAC system energy requirements, 42-44
 life safety requirements, 39
 national standards, 34
 see also building codes
Coefficient of performance (COP), 16-17, 20, 42
 defined, 16-17
Cogeneration, 157-159
Compartmentation, 423, 429
Compression cycle, 18-23
 energy use, 18-20
 operating cost, 18-20
Compression cycle equipment, 18, 22, 247
Compressors, 22, 23
 costs, 18
 types, 22-23
Computers, 143, 375, 520
Computer rooms, 203

Concrete, 397
Concrete anchors
 for underground tanks, 225-226
Condensate flow, 293
Condensate pumps, 8
Condensers, 26-28, 230-231, 247
 air-cooled, 26, 27
 evaporative, 26, 27
 types, 26
 water-conserving, 231
 water-cooled, 26, 27
 water recycling systems, 230-231
Conduction, 54
Conduction heat gain, 86-89
 commercial, 537-540
Conduction heat loss, 54, 55-74
 calculation, 55
Conservation, 149-150
 defined, 149
 see also energy conservation
Constant volume air systems, 359
Containment areas, 225
Contaminants, indoor
 see indoor air quality
Controllers, 524, 376
 see also motor starters
Controls, 42-43, 166-168, 359, 373-396
 and indoor air quality, 493
 equipment data sheets, 393-395
 individual, 237
 internal vs. external, 376-377
 selection
  commercial, 557
  residential, 524
 types, 157-159, 373-376
Control loop, 376
Control sequences, 382
Convection heat gain, 98
 commercial, 540-541
Convection heat loss, 74-84
Convectors, 3, 351
Cooling, 14-28
 design standards, 42
 method, 14
 process, 15
 units, 15

 year-round, 212
Cooling equipment
 fuel, 209
Cooling ladder, 14, 209
 illust., 15, 211
Cooling load, 85-107
 calculation, 86
Cooling load factors, 116-117, 536
Cooling loads
 commercial, 536-542
Cooling load temperature difference (CLTD), 536, 539, 540
 calculation, 89
Cooling systems, 3
 selection, 193-216
 types, 18-26
 see also heating and cooling systems; refrigeration; specific systems
Cooling tower, 26, 27, 231-233
 cost, 232-233
 indoor installation, 232
  location, 232
 materials, 231
 recent developments, 233
 types, 231
 see also generation equipment data sheets
Cooling tower fill, 231
Copper
 water pipes, 304
 in water tube boilers, 219
Covered malls, 435
CPVC
 water pipes, 304
Cut-off
 low water, 223

D

Deadband programs, 387
Demand-limiting programs, 387-388
Design temperatures, 44, 60, 74
Diameter
 equivalent round, 322, 323
Diffusers, 3, 317, 362
 sizing, 362-363
Direct digital controls, 375-376

591

Index

Direct expansion coils (DX)
  see heat exchangers, fin and tube
Dirt pockets, 299
Distribution
  defined, 287
Distribution equipment, 4, 287-288
  data sheets, 345-347
  selection
    commercial, 551-556
    residential, 512-516, 520
  types, 287
Distribution loops
  open and closed, 287
Distribution systems, 287-350
  selection, 206
    commercial, 551
    residential, 508
Downfed
  defined, 290
Drag
  see friction
Drain cocks, 299
Driving equipment
  see distribution systems; fans; pumps
Duct
  equivalent straight run, 331-332
Duct heaters, 228
Ducts, 3, 28, 552-556
  cooling, 28
  cutting and coring for, 466
  heating, 28
Duct sizing, 317-339
  balanced pressure drop method, 317, 332
  constant pressure drop method, 317, 322-332
  static regain method, 317, 332-338
  velocity method, 317, 318-322
Duty-cycling programs, 387

E

Economizer system, 235
Edge factor, 60
Elastometric
  flexible, 414
Electrical equipment
  enclosures, 402

Electric controls, 373
Electricity, 203
  advantages, 203
  efficiency, 203
  installation and maintenance cost, 203
Electric motors, 401
  overloading, 401
Electric system terminal units, 363-366
Electronic controls, 225, 373
Electronic ignition, 225
Energy audit, 172-190, 194
  defined, 172-173
Energy conservation, 42, 43, 237
  and indoor air quality, 494
  codes, 42
  techniques, 155-172
Energy conservation measures (ECM)
  in energy audit, 173, 181, 184-188
  rebate programs, 190
Energy consumption, 149, 178, 179
  monitoring, 382
Energy efficient ratio (EER), 42
  defined, 18-20
Energy management systems, 382-388
Engineers, 38
Enthalpy programs, 389
Equipment selection
  commercial, 542-559
  residential, 510-525
Equivalent direct radiation (EDR), 220, 288
Evaporative condensers, 230
Exhaust
  in commercial kitchens, 454-458
Expansion joints, 297-298
Expansion loops, 297-298
Expansion tanks, 226-227, 357-359, 518-519
  chilled water, 547
  hot water, 546-547

F

Facilities, 168, 436-458
  ambulatory care, 439
  colleges, 168
  commercial kitchens, 454-458
  energy recovery systems in, 168-172
  hospitals, 168, 439-455
Fan coils, 351
Fan curves, 269
Fan laws, 270
Fan pitch
  variable, 268
Fans, 248, 266-272, 288, 315
  arrangement, 271-272
  data sheets, 274-275
  fresh-air supply, 550
  location, 266-268
  overload, 401
  oversizing, 266-268
  performance, 269
  size, 266-268
  types, 268-269
Fan speeds, 268-269
Feasibility, of a system, 194
Fiberglass, 414
Filters, 409
Filtration, 313, 404-409
  in hospitals, 445
Fin tube baseboards, 351
  see also heat exchangers, fin and tube
Fire and smoke control, 423-431
  compartmentation, 429
    single injection system, 426
    multiple injection system, 426
  dampers, 431
  design considerations, 426-429
  in atriums and malls, 435
  in high-rise buildings, 431-433
  stair pressurization, 426
Fire ratings, 36
Fire safety, 35-37, 39, 247
  standards, 45-46
Fire Safety Evaluation System (FSES), 35-37
Fire separation, 39
Flexible connections, 404
Float and thermostatic (combination) traps, 352

Index

Flow, 262
Flue
  defined, 397
  linings, 399
  sizing, 398-399
  temperatures, 397
  ventilation, 399
Flue gas, 397
  pressures, 397-398
  temperatures, 397
  velocity, 398-399
Forced draft cooling towers, 231
Free cooling, 234-235
Free-cooling options, 3
Friction (drag), 259
Fuel
  selection, 203-206
    commercial, 531
    residential, 508
  selection criteria, 203
  sulphur content, 399
  types, 203
Fuel consumption
  calculation, 223-225
Fuel tanks, 225-227
Furnaces, 227, 247
  warm air, 227, 248

G

Gas
  types, 205
Gauge pressure, 12, 291
Generation equipment, 4, 237-245, 247
  data sheets, 239-245
  primary, 247
  secondary, 247
  selection, 207
    commercial, 546-550
    residential, 516-520
Generation equipment assemblies, 247-257
Generation system selection
  commercial, 531
  residential, 508
Glycol/water mixtures, 306
Gravity steam heat, 291
  venting, 291
Grilles, 317, 362

Grounding, 401

H

Hartford Loop, 297
Head, 262
  total, 306
  units, 259
Heat
  units, 6
Heat capacity
  air, 14, 313-315
  water, 304
Heat of condensation
  see latent heat
Heat distribution, 3, 4
Heat exchangers, 7-8, 227, 248, 277-285, 288
  data sheets, 285
  fin and tube, 279-281
  fluids, 277-279
  location, 277
  performance, 279
  plate coil, 233
  selection criteria, 277-279
  shell and tube, 277-279
  sizing, 279
  temperature
    approach, 279
    primary, 279
    secondary, 279
  where required, 277
Heat extractors, 221
Heat flow
  see heating ladder
Heat gain, 14, 86, 98, 110, 117
  commercial, 541-546
  latent, 107
  types, 117
Heating
  basic systems, 6-14
  most common systems, 8-14
  system selection, 193-216
  units, 53
Heating and cooling
  combined, 233-234, 254
  packaged units, 236
  rooftop units, 236
  split systems, 236
  through-the-wall units, 236
  simultaneous, 209-212, 237
Heating ladder, 7, 209, 210

illust., 9
Heating load, 53-85, 223
  commercial, 533-536
  defined, 53
  residential, 508-512
  room-by-room breakdown, 520-522
Heating systems
  see specific systems
Heat loss, 53-85
  conduction, 508
  convection, 508
Heat pumps, 237
Heat sources
  ambient, 14
Heat transfer
  phase change, 18
Heat of vaporization, 9
  water, 24
Hospitals, 168, 436-454
  airborne particle control, 441-445
  HVAC design of, 445-457
    laboratories, 448-450
    negative pressure isolation rooms, 446-447
    positive pressure immune suppressed rooms, 447-448
    psychiatric units, 448
    special procedure rooms, 450
    waiting areas, 448
  HVAC layout and equipment for, 450-458
  infection control, 440-441
  ultraviolet (UV) light radiation in, 445
Hot water heating
  see hydronic heating systems
Hot water supply
  heat loss, 304
Hot water systems, 12-14, 247
  forced, 12-14
  terminal units, 351-354
  types, 351
  see also specific components
Humidity control, 3, 14, 29, 53, 233-234, 247, 313
HVAC
  common systems, 198-202
  defined, 3
  life safety code requirements, 39
  residential model, 503-528

Index

system selection, 195-198
  criteria, 193-195
Hydrochlorofluorocarbons (HCFC's), 477-487, 499
  alternatives to
    containment, 485
    conversion, 485-486
    replacement, 485
  and ozone depletion, 478-479
  reduction of, 479-484
Hydrofluorocarbons (HFC's), 484-485
  as substitutions for CFC's, 484
Hydrogen, 205
Hydronic heating systems, 3, 219-227
  special considerations, 311
  types, 219

I

Ice storage, 162-165
Ignition
  intermittent, 221
  loss, 225
  types, 225
Indigenous building types, 150-154
  pueblo, 151
  peristyle, 151
  takhtabush, 151
  mashrabiya, 151
  tent, 151
  grass hut, 153
  igloo, 153
  wigwam, 153
  tepee, 154
Indoor air quality (IAQ), 487-488
  and architecture, 490-491
  and energy conservation, 494
  and HVAC controls, 494
  and mechanical systems, 491-493
  and sick building syndrome (SBS), 487
  and ventilation, 488-490
  and use of "natural" materials, 490
  in kitchens, 489
  in laboratories, 489
  indoor contaminants, 489-490
  regulatory controls for, 493
Induced draft cooling towers, 231

Infiltration, 79
Insulation, 42, 414-415, 525
  blanket, 414
  fan coil casing, 414
  flexible, 414
  high temperature, 414-415
  materials, 414
  piping supports, 415
  rigid, 414
  thermal, 413
Internal heat gain, 117
  commercial, 541

K

Kitchens, commercial, 454-458
  exhaust hoods in, 454-458
    air volumes, 457
    fire systems, 457
    grease removal, 457
    indoor air quality in, 487

L

Laboratories
  indoor air quality in, 487
Ladder diagrams, 379
Latent heat, 86, 107
Latent heat loads, 107-119
L/D ratio, 332
Leak detection
  tanks, 225
Licenses
  for designers and contractors, 37
Lighting, 44
  energy efficient, 155-156
Lithium bromide, anhydrous, 24
Louvers, 360
Low pressure steam, 291, 293

M

Mains, 290, 291, 295, 298
  pitch, 295
Makeup air, 228-229
MasterFormat divisions
  see specifications
Materials
  "natural", 490
  standards, 45
MBH, 6

see also British thermal unit
Mechanical draft cooling towers, 231
Metering
  individual, 237
Methane, 205
Metric, SI, 6
  common units, 6
  conversion tables, Appendix D
Mixing boxes, 315
Monoflow loops, 299, 301
Motor starters, 401-402, 525, 558
  automatic, 401
  magnetic, 401-402
  manual, 401
  types, 401
Mountings (vibration isolators), 402-404

N

National Electric Code, 46
National Fire Protection Association (NFPA), 45
National Fuel Gas Code, 46
Natural draft cooling towers, 231
Natural gas, 205
Net positive suction head (NPSH), 262
Night-purge programs, 389
Noise control, 3, 247, 248, 261, 268-269, 295, 402-404

O

Odor control, 404-409
Odor masking, 407
Oil
  grades, 203
  storage, 205
Oil storage
  regulation, 205
Oil tank
  containment, 205
  leak detection systems, 205
  materials, 205
Oneline schematic diagrams, 379
Open systems, 259-260

Index

Operation and management (O&M) measures, 181-188, 190
Optimum start/stop programs, 388-389
Organizations, professional, 44-46
  see also specific organizations
Outdoor air, 234
Outdoor reset, 523, 557
Output
  gross, 220
Overall heat transfer coefficient, 55
Ozone layer, 478
  depletion of, 478-479
    solutions, 479-484
  see also CFC's, HCFC's

P

Passive design
  defined, 150
  in solar energy, 160
Payback time, 193-194
People heat gain, 117, 542
Perimeter heat loss, 60-74
Permits, building, 37
Phenolic foam, 414
Pickup load, 220, 223, 229
Pilot ignition, 225
Pipe
  cutting and coring for, 466
  total equivalent run, 306
Pipe length
  calculations, 293
Pipe materials, 304
  steam heat, 293
Pipe sizing
  steam heat, 295-297
Piping
  expansion, 411-412
  pitch, 311
Piping hangers
  see piping supports
Piping supports, 409-414, 525, 559
  anchoring, 409-413
  location, 409
  materials, 409

Piping tax, 220, 223
Pneumatic controls, 374
Pollution control, 407-409
Polyurethane
  closed cell, 414
Polyurethane foam, 414
Positive displacement compressors
  see reciprocating
Pressure
  units, 259
Pressure drop, 293
Problems, chapter, 29-30, 46, 144-146, 191, 216, 239, 254, 272, 284, 339-347, 367, 396, 416, 474-475, 499
Propane, 205
Pump bodies
  materials, 261
Pump curves, 261-262
Pump laws, 262, 266
Pumping,
  primary, 265-266
  secondary, 265-266
Pumps, 248, 259-266, 288, 289
  arrangement, 262-265
  auxiliary, 262
  centrifugal, 261
  circulating, 519
  data sheets, 273
  efficiency, 262
  glycol/water, 549-550
  oversizing, 262
  size, 262
  staging, 265
  types, 259, 261
  see also heat pumps, specific types
PVC
  water pipes, 304, 305, 306

R

Radiation heat gain, 110
Radiation heat loss, 55, 84
Radiation shielding, 453
Radiators, 3, 295, 351
Radon gas, 495-498
  removing, 496

Reciprocating compressor chiller, 229
Reciprocating compressors, 22
Refrigerants
  alternatives to, 486-487
  boiling points, 18
Refrigeration, 18-26
  cycle selection, 19
Refrigeration equipment, 247
Registers, 317, 359
Reheat coils, 317
Renovation projects
  distribution system, 206
  system selection, 194-195
  see also fuel conversion
Replacement cost, 194-195
Replacement equipment
  compatibility with existing equipment, 195
Return air grilles, 556
Return air sensors, 557
Risers, 290, 291, 293, 299, 352, 355
Rock wool, 414
Roll hangers, 413
Runouts, 290, 295

S

Scrubbers, 407
Self-contained controls, 373
Sensible energy, 141
Sensible heat, 53, 107
Sensors, 376
Series loops, 301
Shading coefficient, 110, 536
Sheet Metal and Air Conditioning Contractors National Association (SMACNA), 34, 45
Short circuits, 401
Sick building syndrome, 487
  see also indoor air quality
Siphon, 297
Slab heat loss, 74
Smoke shafts, 429
Solar collectors
  flat plate, 162

Index

concentrating, 162
Solar cycle, 18
Solar energy, 159-162, 206
   active, 160-162
   for heating, 159
   passive, 160
Solar heat gain factor, 110, 541
Space coordination, 464-467
Spark ignition, 225
Specifications, 458-467
   coordination between trades, 460-467
   MasterFormat divisions, 458
   organization, 460
   performance, 459
   proprietary, 459
   scope of work, 461
   substitutions, 459-460
Split systems, 233, 247-248
Spray pond cooling towers, 231
Spray pond, 26
Square foot of radiation, 288
Stack effect, 425
Stack temperature monitor, 225
Steam, 203
   purchased, 7
   supplied, 248, 291
   waste, 248
Steam generation, 248
Steam heat, 8-12, 247
   advantages, 12, 222
   applications, 290
     high pressure, 291
     low pressure, 291
   boilers
     venting, 291
   codes, 8
   distribution, 288-289
   high pressure, 291
   mains
     venting, 291
   operation, 293-295
   pipe capacity, 295
   pipe insulation, 298
   pipe sizing, 293-297
   pipe support, 298
   piping layout, 290
   pressure, 291, 293
   pressure reduction, 291
   quantity of heat supplied, 13
   return systems, 352

risers
   venting, 291
safety features, 297-299
sludge buildup, 298-299
terminal units, 351-352
   venting, 291
   see also radiators
types, 290-291
venting, 291
see also hydronic heating systems; specific components
Steam jet cycle, 18
Steel, 399
   galvanized, 399
   stainless, 399
   water pipes, 304
   in water tube boilers, 219, 222
Step firing, 222
Suction
   see net positive suction head
Sulfur dioxide, 399
Sulfuric acid, 399
Supply air, 227
Supply air diffusers, 556
Surface area, 55, 60
Sustainable design, 150
   defined, 150
   see also passive design
Swing joints, 297-298, 311
Switches
   see motor starters
System selection
   commercial, 527-531
   residential, 503-508

T

Tanks, 225-227
   oil, 518-519
   requirements for, 494-495
Temperature drop
   standards, 302-303
Terminal units, 3, 4, 351-369
   advantages, 351
   applications, 351
   data sheets, 367
   disadvantages, 351
   selection
     commercial, 556-557
     residential, 520
   size, 352

Thermoelectric cycle, 18
Thermostatic trap, 352
Thermostats, 203, 522
Three-pipe systems
   see water distribution, reverse return
Throw
   defined, 317
Time clock, 524, 557
Ton (cooling unit)
   defined, 14, 229
Total resistance, 56
Traps, 291, 299, 352

U

Ultraviolet (UV) light radiation
   in hospitals, 445
Underwriters Laboratories (UL), 34, 45
Unoccupied-period programs, 388
Upfed, 290
Urethane
   rigid, 414

V

Vacuum pumps, 8
Vacuum steam heat, 290-291
   venting, 291
Valves, 352, 352-358
   automatic, 352, 557
   pressure relief, 223
Vapor barrier, 414
Vapor steam heat, 290
   venting, 291
Variable frequency drive motors, 156-157
Variable volume air systems, 359
VAV
   see boxes, variable air volume
Ventilation, 79, 315, 399, 407
   basic systems, 3, 28-29
   effect on heating and cooling, 29
   effect on indoor air quality, 487
   methods, 29
   reasons, 28

Index

smoke, 429-430
standards, 407, 491
system types, 3
Vibration, 400
Vibration isolation, 248, 261, 402-404, 525
Vibration isolators, 402, 558
Viscosity
water, 304, 306
Volatile organic compounds (VOC), 487-488

W

Warm air heat, 14
Warrantees
equipment, 464
Waste gas
see flue gas
Water
as heating fluid, 53
makeup, 220
Water chiller, 229-230
sizing, 229
types, 229
Water conservation, 230-231
Water-cooled chillers, 229-230
Water distribution, 299-313
control, 301, 302
direct return, 302
expansion, 311
flow temperature, 306
head loss, 306
heat loss calculation, 302-303
pipe length, 306-311
piping, 304-306
pressure, 313
pump size, 306-311
reliability, 302
reverse return, 301-302
safety features, 313
sizing, 304
types, 299
velocity, 304
viscosity, 306
Water hammer, 298
Water loss
cooling tower, 231-232
Wiring diagrams, 379

Z

Zoning, 377